Springer Series in

MATERIALS SCIENCE 165

Springer Series in
# MATERIALS SCIENCE



The Springer Series in Materials Science covers the complete spectrum of materials physics, including fundamental principles, physical properties, materials theory and design. Recognizing the increasing importance of materials science in future device technologies, the book titles in this series reflect the state-of-the-art in understanding and controlling the structure and properties of all important classes of materials.

Please view available titles in *Springer Series in Materials Science* on series homepage http://www.springer.com/series/856

Steven W. Cranford • Markus J. Buehler

# Biomateriomics

With 159 Figures

Springer

Steven W. Cranford
Civil and Environmental Engineering
Massachusetts Institute of Technology
Cambridge, MA, USA

Markus J. Buehler
Civil and Environmental Engineering
Massachusetts Institute of Technology
Cambridge, MA, USA



ISSN 0933-033X Springer Series in Materials Science
ISBN 978-94-007-1610-0 ISBN 978-94-007-1611-7 (eBook)
DOI 10.1007/978-94-007-1611-7
Springer Dordrecht Heidelberg New York London

Library of Congress Control Number: 2012939523



Printed on acid-free paper

Springer is part of Springer Science+Business Media (www.springer.com)

*For Deborah, LOML—S.W.C.*

*To my family—M.J.B.*

# Preface

The holistic study of biological material systems has emerged as an exciting area of research. While such systems are commonly complex, we frequently encounter similar components—universal building blocks and hierarchical structural motifs—which result in a diverse set of functionalities. Similar to the way music or language arises from a limited set of musical notes and words, the relationships between form and function can be exploited in a meaningful way by recognizing the similarities between Beethoven and bone, or Shakespeare and silk. Through the investigation of material properties, examining fundamental links between processes, structures, and properties at multiple scales and their interactions, materiomics explains system functionality from the level of building blocks. *Bio*materiomics specifically focuses the analysis of the role of materials in the context of biological processes, the transfer of biological material principles towards biomimetic and bioinspired applications, and the study of interfaces between living and non-living systems. Inevitably, materiomics also holds great promise for nanoscience and nanotechnology, where material concepts from biology might enable the bottom-up development of new structures and materials or devices.

The challenges of biological materials are vast, but the convergence of biology, mathematics and engineering as well as computational and experimental techniques have resulted in the toolset necessary to describe complex material systems, from nano to macro. Applying biomateriomics can unlock Nature's secret to high performance materials such as spider silk, bone, or nacre, and elucidate the progression and diagnosis or the treatment of diseases. Similarly, it contributes to develop a *de novo* understanding of biological material processes and to the potential of exploiting novel concepts in innovation, material synthesis and design. With this impetus, the field of biomateriomics attempts to reconcile all aspects of a biological material system—from universal motifs of nano-scale building blocks to macro-scale functional properties—with a focus on studying the mechanisms of deformation and failure by utilizing a multi-scale materials science approach.

This book encompasses the current work reflective of many review articles and journal papers under a common banner, and makes this exciting field of research accessible to the broader engineering and science community. It should provide a valuable reference for engineers, materials scientists, and researchers in both academia and industry and will hopefully ignite extended discourse and inquiry. Indeed, many technical details are omitted *in lieu* of presenting key concepts and simple ideas. Many of the examples are adapted from studies carried out by the authors of this book, and some of the discussion should therefore not be considered as a comprehensive review with respect to the wider range of available results. Rather, they represent a set of specific illustrative examples of materiomics, including theoretical aspects, associated principles, and applications. The primary text provides an overview of the field of materiomics, including earlier work and future opportunities and intellectual challenges for research, and is organized into three main parts:

Part I: A Materiomics Perspective provides an introduction to biomateriomics. This is especially important given that the entire field is being developed and potential applications explored. The outside resources and investigations we henceforth refer were never intended to encompass materiomics *per se*—but yet contribute to its foundation and future progress. Admittedly, we are standing on the shoulders of others and declaring their work to be in a newfangled (and as yet unproven) field. Therein lays the stimulus for such a paradigm: only by the convergence of disparate fields can materiomics find its worth—from the astute combination of advancements in chemistry, biology, physics, materials science and engineering (further discussed in Chap. 1: Introduction). Such a combination is clearly beyond the capabilities of any individual (including these humble authors) but clearly achievable by the scientific community. The chapters constituting Part I present our interpretation of a *materiomic perspective*. The fundamental goals need only to be defined—our intent is to shed light on those goals.

For these reasons, we base this book's content around our own experience—specifically, the mechanical characterization of biological materials founded at the molecular level. We shall see that this is just one aspect of a complex *materiome*, and far from a complete picture desired (and implied) by the "omics" suffix (there is a more detailed discussion of this in subsequent chapters). Nevertheless, a focus on atomistic and molecular mechanics has various advantages:

1. It is based on fundamental principles of physics and chemistry, which are ultimately defined by quantum mechanics, providing a common starting point regardless of the specific material system(s) considered.
2. It is representative of some of the most relevant and critical topics and, most importantly, challenges in the field of biomaterials.
3. It allows us to present some case studies, which, although based on a particular scale, can easily be used as frameworks for other problems.
4. It enables other researchers to contribute to the field of materiomics, in addition to the molecular perspective emphasized here.

If our objective was to encompass all disciplines, bridge all fields, and tie together all scales of biological materials from the atomistic sequence of amino acids to a functional biological tissue or organ—we would never come to completion. Instead, we hope that through a focus on simple examples, the potential of a more holistic perspective of biological materials—discovering the relations between structure and function across multiple scales—will be apparent. As such, Part I presents the emerging field with the associated scope, thematic paradigms, and an outline of essential concepts (Chap. 2: The Materiome), as well as an in depth discussion of biological materials as the motivation for the development of a materiomics framework (Chap. 3: The Challenges of Biological Materials), and the unifying categorization and abstraction necessary for modeling and understanding such complex materiomic systems (Chap. 4: Universality-Diversity Paradigm: Music, Materiomics, and Category Theory).

Part II: Methods and Tools discusses the ever-expanding toolset required for materiomic investigations. A selection of the most promising strategies to investigate materiomics and analyze the properties and behavior of complex materials are reviewed, with examples, case studies, and theoretical background when appropriate.

In order to realize the promising opportunities that arise from an improved understanding of complex biological materials several critical challenges must be overcome. Up until now, theories fully describing hierarchical biological materials are still lacking. Only recently has the understanding about how specific features at distinct scales interact, and for example, participate in mechanical deformation, begun to emerge for complex biological systems. In recent years, the development of new quantitative experimental, analytical, and computational methods have lead to advances in understanding of some details of complex biological and synthetic systems. Theoretical, numerical, and experimental methods now enable the investigation of nanoscale mechanics of materials using quantitative analysis techniques—an area referred to as "nanomechanics". For example, development and application of nanoindentation, atomic force microscopy, and other tools enables scientist to probe the origins of mechanical properties, with forces in the range of piconewtons, and at scales approaching that of individual atoms (Ångstroms) and molecules (nanometers). At the same time, computational methods, computational power, and theoretical approaches have led to significant advances in addressing nanomechanics from a first principles perspective. This combination of experiment, theory, and computation has proven to be very fruitful, and could lead to major advances in materials theories and engineering.

The most recent innovations have occurred in the field of nanotechnology and nanoscience, where cross-disciplinary interactions with the biological sciences present an enormous opportunity for innovative basic research and also technological advancement. Such advances could enable us to provide engineered materials and structures with properties that resemble those of biological systems, in particular the ability to self-assemble, to self-repair, to adapt and evolve, and to provide multiple functions that can be controlled through external cues. However, despite significant advancements in the study of biological materials in the past decade, the fundamental physics of many phenomena in biology continue to pose substantial challenges with respect to model building, experimental studies, and simulation. As materiomics is founded by a combination of multidisciplinary theories and multiscale techniques, approaches that integrate *experiment* and predictive *simulation* are essential to this new paradigm of materials research.

The behavior of biological materials, in particular their mechanical properties, are intimately linked to the atomic microstructure of the material. Different mechanisms operate at larger length scales, where the interaction of extracellular materials with cells and of cells with one another, different tissue types and the influence of tissue remodeling become more evident. The dominance of specific mechanisms is controlled by geometrical parameters, the chemical nature of the molecular interactions, as well as the structural arrangement of the protein elementary building blocks, across many hierarchical scales, from nano to macro. Thus, materiomic investigative approaches must also consider multi-scale schemes, both experimentally and computationally, to link hierarchical effects and mechanisms.

Much of the functionality that biological materials provide occurs through mechanical contact and behavior. Therefore, to completely understand the structure-property-functionality relationships of biological materials it is necessary to quantify the mechanical behavior and influences on biological and *de novo* materials. Thus, Part II includes the means of mechanical investigation, including experimental methods (Chap. 5: Experimental Approaches), computational methods (Chap. 6: Computational Approaches and Simulation), and the interpretation of results (Chap. 7: Mechanical Characterization in Molecular Simulation). Although descriptions of techniques are to be presented, with relevant case studies and applications, specific technical details (*i.e.*, application of molecular dynamics) are only outlined, with commentary of strengths and weaknesses of various approaches, and their applicability at different scales. When appropriate, suggestions will be made for more detailed texts and references in the field. In other words, the objective of the text is not to provide an in-depth handbook for analytical procedures, but rather to discuss the various means of biomateriomic investigation. As anticipated, biomateriomics requires an extensive "toolbox".

Part III: Applied Materiomics illustrates how we can immediately benefit from biomateriomic approaches. Application of materiomic principles and approaches has already been undertaken on a variety of biological systems throughout different fields of research. The combination of high-level structural control of matter as achieved in nanoscience and nanotechnology, multiscale analytical techniques, and integration of living and non-living components into systems and interfaces will lead to the development of new technologies that utilize the advantages of both micro and nanotechnology with the principles of biology. With an inevitable merger of material and structure, with increasing complexity, materials start to resemble dynamic systems or machines, so that the borderlines between conventional concepts such as "machine" and "material" also start to disappear. Such approaches have been used systematically by Nature for millions of years. However, their systematic exploitation for technological applications has so far been severely hindered due to lack of understanding of how to link the atomistic scale with material structure and device properties and function. Like all endeavors, we only get better with practice!

Part III discusses practical applications of materiomic techniques and approaches with three main focuses. Fundamentally, materiomics provides an integrated and holistic approach, advantageous in the investigation of complex biological material system phenomenon and system characterization (Chap. 8: Unlocking Nature: Case Studies). Moreover, materiomics can facilitate the development of novel diagnostic tools for disease and afflictions with mechanistic symptoms, predicting what components and functionalities "fail" under minute changes in material and structural conditions (Chap. 9: Pathological Materiomics). Finally, biomateriomics has a role in the design of *de novo* materials, or the synthesis and manipulation of biological materials, materiomic engineering, and nanomedical devices (Chap. 10: Synthesis and Design). Using natural processes as a guide, substantial advances have already been achieved at the interface of nanomaterials and biology.

Irrespective of the challenges still present in a thorough investigation and complete characterization of the materiome as discussed by prior chapters, current experimental and practical approaches exist that allow the immediate application of materiomics to real problems. This branch of materiomics, termed *applied materiomics*, is still in its infancy, yet has already demonstrated potential as a valuable basis for material design. A materiomic approach is likely to become an integral part of nanomaterials manufacture—where molecular assembly is control to attain macroscale behavior—requiring a deep understanding of individual molecular building blocks, their potential structures, assembly properties, dynamic behavior, and multiscale propagations. We hence focus discussion on broad areas of application that are becoming increasingly widespread (throughout different disciplines) and can be encompassed by the common field of applied materiomics. The applications, undoubtedly, are as variegated as Nature. The text is closed with an outlook to future opportunities in Chap. 11: The Future of Biomateriomics.

The discussions presented in this book are intended to be both a review of current materiomics research as well as a pedagogical discourse. While we embrace the term to encompass our own work, we believe the worth of materiomics will naturally emerge from the shared contributions of many scientists and research groups. It is not a term to lay claim, but a label to encompass a new perspective of chemistry, biology, and materials science. Indeed, any "closed-form" interpretation of materiomics will limit both the growth and potential of materiomics research. As biomateriomics is a relatively new field, it behooves us to include discussion to help define and explicate both the intent and scope with analogous examples, illustrating the integrative nature, universality, and benefits and impact of a materiomics approach. The perspectives and overviews presented throughout this book are intended to provide a broad overview. Further details can be found in the papers cited and recommended readings.

Most importantly, completing this book would not have been possible without the help and support of numerous people and institutions. The authors are indebted to all who have contributed to this book in some way. In particular, sincere gratitude goes to the many students and researchers who have collaborated with the authors within the Laboratory of Atomistic and Molecular Mechanics (LAMM) at MIT, whose enthusiasm and excitement regarding materiomics are unmatched. We are also thankful for many discussions with colleagues and friends that contributed to the development of this book. We gratefully acknowledge the support from the National Science Foundation, Army Research Office, Office for Naval Research,

Defense Advanced Research Projects Agency (DARPA), Air Force Office for Scientific Research, and the Massachusetts Institute of Technology.

Cambridge, MA, USA

Steven W. Cranford
Markus J. Buehler

# Contents

# Part I
# A Materiomics Perspective

# Chapter 1
# Introduction

**Abstract** Biomateriomics refers to the holistic study of biological material systems. We can predict the performance of engineered materials in engineered systems, but there is an inherent disconnect when investigating Nature's materials, with little understanding of how functionality arises from both the material and complex structure with properties and interactions across scales. New developments enable a new perspective through the convergence of many scientific disciplines, and advancements in nanotechnology empower us to investigate material systems from the "bottom-up". If we hope to learn from Nature, we need a new holistic perspective: an "omic" approach. We begin with a definition and introduction of biomateriomics, presenting the emerging field with the associated scope, and thematic paradigms, to the tools required for investigations, to ongoing and future applications.

*All sciences are connected; they lend each other material aid as parts of one great whole, each doing its own work, not for itself alone, but for the other parts; as the eye guides the body and the foot sustains it and leads it from place to place.*

*Roger Bacon, Opus Tertium (1266–1268)*

## 1.1 Introduction

The introductory quotation from the thirteenth century is ideally suited for an introduction to *biomateriomics* for two reasons. Primarily, it encapsulates the intrinsic cooperativity of modern science. Previously disparate research fields now commonly borrow concepts, ideas and approaches from each other and collaborations are deemed essential for technological innovation and to tackle the greatest challenges. Biological systems, for example, are no longer limited to biologists and chemists—engineers design biomimetic devices, while materials scientists collaborate with medical researchers to develop bio-compatible implants. In the world of academia, interdisciplinary efforts are no longer a rarity, but are the current *status quo*.

Secondly, the words of Roger Bacon—whether referring to sciences in general or the human body—encompasses a popular idiom that most of us take for granted:

S.W. Cranford, M.J. Buehler, *Biomateriomics*, Springer Series in Materials Science 165, DOI 10.1007/978-94-007-1611-7_1, © Springer Science+Business Media Dordrecht 2012

*the whole is greater than the sum of its parts*. It is well agreed that we (as human beings) are more than just a combination of limbs, a cardiovascular system, and a functioning brain (among other things, of course). Even a cursory glance can deduce the difference between a dog and cat, which can be described by similar "components". Yet, such a high-level perspective is commonly lacking in more technical problems. While we can easily determine the "whole" for some systems, for others we are limited to a view of the "parts"—we are missing the proverbial *forest* to study the *trees*.[1] Such intellectual barriers are present in a wide variety of scientific challenges—the search for a grand unified theory in physics for example—in which each part of the problem is beyond the capabilities of a single researcher or field. The focus of this text is on yet another fundamental problem—the complexity of biological materials.

## 1.2 The Unpredictable Nature of Materials

Biological materials—neither steel nor concrete—are the most abundantly used materials on earth, yet we know (relatively) little about how they function. They are the main constituents in plant and animal bodies and have a diversity of functions. While biologists and materials scientists alike are impressed by the mechanical properties of silk [1–7] or the toughness of bone [8–11], there are difficulties in replicating the successes of Nature in a synthetic manner. While we can approximate such materials, they are often not as elegant as there natural counterparts. The key difference lies in the long-term "product development" stage of Nature. Whereas we attempt to design a material to suit a particular application (*i.e.*, choosing a material such as silicon to make computer chips due to its semiconducting properties, for example), Nature has implemented the simultaneous development of material and function (more commonly known as *evolution*). A complex biological material like bone was not "selected" to be a supporting structure for our bodies—it has specific material properties and characteristics to serve its own (intended and evolved) function. Unlike engineering materials, the distinction between material *properties* and material *function* is lost. The subtle difference between material function and application is further discussed in Chap. 2: The Materiome.

Another major difference between materials from Nature and engineering is in the way they are made. While an engineer selects a material to fabricate a part according to an exact design, Nature goes the opposite direction and grows both the material and the whole organism (a plant or an animal) using the principles of (biologically controlled) self-assembly—this is more commonly referred to as *growth*. Moreover, biological structures are even able to remodel and adapt to changing environmental conditions during their whole lifetime. This control over the structure

[1]We note that the reductionist approach of science (studying the *trees*) has continuing success in the explanation of fundamental phenomena in physics, chemistry, and biology, and the current discussion is not intended to be a criticism, but rather a complementary perspective.

at multiple scales is certainly the key to the successful use of (relatively) soft protein materials as robust structural components.

The consequence, of course, is that we must consider intended functionality in the investigation and design of biological systems and novel biomaterials. While easy to say, this task is complicated by the complex, hierarchical nature of such materials [12]. Functionality is ultimately rooted at the molecular scale [13, 14]. Through recent advancements in single-molecular assays, analytical chemistry, and computational approaches, we have made great strides in determining what a biological system is composed from the molecular level. Individual molecules and amino acids can be deduced *via* nuclear magnetic resonance (NMR) spectroscopy and segments of DNA and other protein structures can be sequenced. It is relatively easy to compile such information. We are collecting a vast amount of data on such materials—but how can we combine what we know (*i.e.*, what we measure) with what we think we know (*i.e.*, prediction of function)?

With all these advancements, we are unable to predict the behavior of a particular molecular sequence. As a result, we cannot engineer synthetic proteins designed for a specific function or application (such as attacking cancer cells or tissue regeneration). We have copious amounts data, but are unable (at this point) to use it. Unlike traditional structural engineering systems—we can predict the behavior of a building by the analysis of steel trusses, for example—there is an inherent disconnect in our ability to predict functional and mechanical behavior for biological systems (see Fig. 1.1). This has been exemplified by the difficulties in predicting structures from single protein folding—that is, the prediction of secondary, tertiary, and quaternary structure from a primary protein sequence (further discussed in Chap. 3: The Challenges of Biological Materials). Unlike in engineered structures, at the molecular level, the difference between material properties and structural function is not clear.

## 1.3 Differences Between Material and Structure

What is the fundamental difference between material and structure? This question can be alternatively posed, from a structural engineering perspective, *"What constitutes a structure?"* Popular answers will undoubtedly encompass bridges and buildings—the Golden Gate Bridge or the Empire State Building are undeniably structures in the traditional sense of the word. At such scales, it is also very easy to label what the "materials" are—a bridge may be build of concrete and steel, for example. Things get a little fuzzier as we reduce the size—where does the structure turn into the material? If we consider the glass sponge *Euplectella*, a deep-sea, sediment-dwelling sponge from the Western Pacific [15–17], we see a sophisticated hierarchial structure that performs a multitude of functions, yet is predominantly composed of the same constituent material—*silica*—which is intrinsically brittle. It has been shown that spicules in siliceous sponges exhibit exceptional flexibility and toughness compared with brittle synthetic glass rods of similar length scales [18, 19]—but why? Function is derived from the structure, but at what scale can we separate the material from the structure?

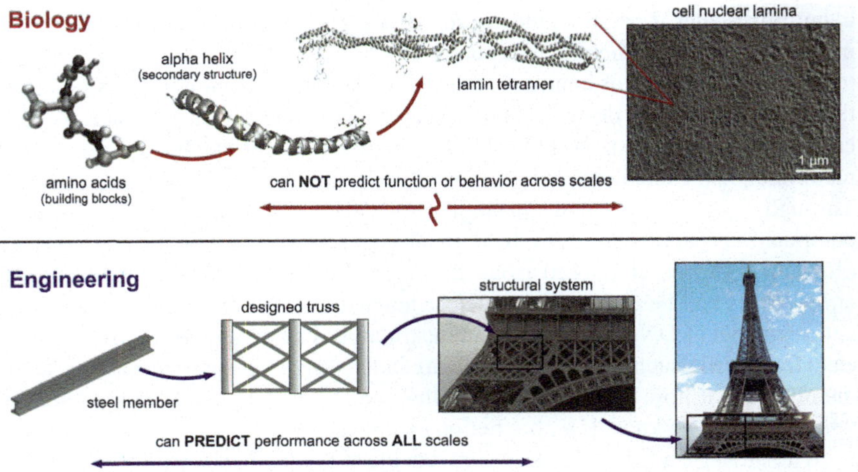


**Fig. 1.1** Example of the inherent disconnect between biological systems and traditional structures. In biology, if we consider a single scale, one phase, with perfect knowledge of composition and sequence, in controlled conditions, we can make predictions. Here, with knowledge of the constitutive amino acids, we can predict an alpha-helical structure for short polypeptide sequences, and its corresponding properties (such as strength). We cannot (yet), however, accurately predict large scale behavior of larger protein assemblies, such as protein networks, let alone the structural role such materials play in a cellular structure (*e.g.*, the nuclear envelope) or in the context of other biological properties. We utterly fail in real-world applications—the exact opposite of the goals of engineering! For engineering, we can design the components of a structure with reliable and repeatable accuracy—the performance of a fabricated steel member can be utilized in the design of a truss, which is subsequently implemented in a structural system

The skeletal system of *Euplectella sp.* (as shown in Fig. 1.2) shows an intricate, cylindrical cage-like structure with lateral (so-called, oscular) openings. At the macroscale, the cylindrical structure is reinforced by external ridges that extend perpendicular to the surface of the cylinder and spiral the cage. The surface of the cylinder consists of a regular square lattice composed of a series of cemented vertical and horizontal struts, each consisting of bundled spicules aligned parallel to one another, with diagonal elements positioned in every second square cell. Cross-sectional analysis of these beams at the micrometer scale reveal that they are composed of collections of silica spicules embedded in a layered silica matrix. The constituent spicules have a concentric lamellar structure with the layer thickness decreasing from the center to the periphery. These layers are arranged in a cylindrical fashion around a central proteinaceous filament and are separated from one another by organic interlayers. At the nanoscale the fundamental construction unit consists of consolidated hydrated silica nanoparticles (50 to 200 nm in diameter). The assembly of a macroscopic, mechanically resistant cylindrical glass cage is possible in a modular, bottom-up fashion comprising at least *seven* hierarchical levels, all contributing to mechanical performance.

Clearly, we need to begin at some fundamental level. The question is at what scale? For the *Euplectella* sponge, we may want to focus on the constitutive ele-

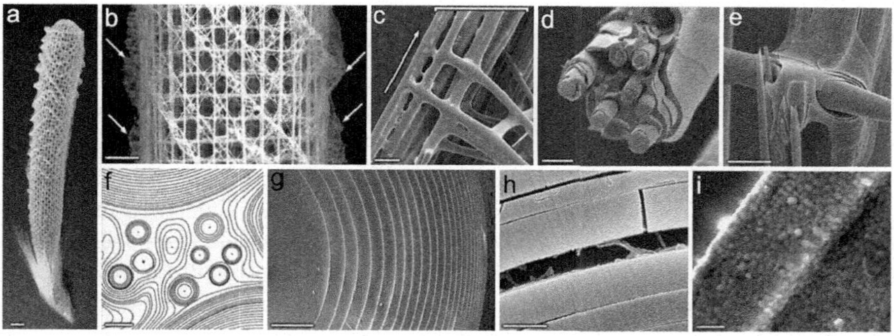


**Fig. 1.2** Structural analysis of the mineralized skeletal system of *Euplectella*: (**a**) Scale: 1 cm; photograph of the entire skeleton, showing cylindrical glass cage; (**b**) Scale: 5 mm; fragment of the cage structure, showing the square grid lattice of vertical and horizontal struts with diagonal elements; (**c**) Scale: 100 μm; scanning electron micrograph (SEM) showing that each strut (enclosed by a *bracket*) is composed of bundled multiple spicules (the *arrow* indicates the long axis of the skeletal lattice); (**d**) Scale: 20 μm; SEM of a fractured single beam revealing its ceramic fiber-composite structure; (**e**) Scale: 25 μm; SEM of the junction area showing that the lattice is cemented with laminated silica layers; (**f**) Scale: 10 μm; contrast-enhanced SEM image of a cross-section through one of the spicular struts revealing that they are composed of a wide range of different-sized spicules surrounded by a laminated silica matrix; (**g**) Scale: 5 μm; SEM of a cross-section through a typical spicule in a strut showing its characteristic laminated architecture; (**h**) Scale: 1 μm; SEM of a fractured spicule, revealing an organic interlayer; (**i**) Scale: 500 nm; bleaching of biosilica surface reveals its consolidated nanoparticulate nature. Reprinted with permission from American Association for the Advancement of Science, *Science*, [15] © 2005

ments at the atomistic scale: silica (also known as chemical compound silicon dioxide, or $SiO_2$). One could subsequently ask, what makes silica a good choice for the sea sponge? Silica, or silicon dioxide, is a material that has been known since antiquity, most commonly found in Nature as *sand* or *quartz*—hardly robust structural materials. Additionally, silica is manufactured in several forms, used in the production of glass, and even optical fibers. We have exploited silica for many uses—but none have the hierarchical structure and intricacy of the skeletal system of a simple sea sponge. It would be trivial to state that the "fundamental" building block for the *Euplectella* sponge is $SiO_2$, yet, undoubtedly, it is the constituent material. If we simply begin at the component elements, we may overlook the necessary hierarchical structure necessary for the sponge to achieve such remarkable properties.

In the vast majority of silica-based materials, the silicon atoms are in a tetrahedral crystal configuration, with four oxygen (O) atoms surrounding a central silicon (Si) atom (the most common example is seen in the quartz crystalline form of silica). Thus, the first level of hierarchy can be said to be this crystalline structure (depicted in Fig. 1.3). Silica has a number of distinct crystalline forms in addition to amorphous forms, but, more importantly, any deviations from these common structures constitute structural differences in the resulting material. Crystalline minerals formed in the physiological environment often show exceptional physical properties (*e.g.*, strength, hardness, fracture toughness) and tend to form hierarchical structures that exhibit microstructural order over a range of scales. Such biominerals are crystallized from an environment that is undersaturated with respect to silicon, and under conditions of neutral pH and low temperature. Simply put, the sea sponge exploits hierarchical arrangements to overcome the brittleness of its constituent material, glass, and does so under accessible conditions that require very low energies. Can we employ the same principles for other materials?

Indeed, it is known that the first level is biologically produced glass composed of consolidated silica nanospheres formed around a protein filament. The resultant structure might be regarded as a textbook example in mechanical engineering, because the seven hierarchical levels in the sponge skeleton represent major fundamental construction strategies such as laminated structures, fiber-reinforced composites, bundled beams, and diagonally reinforced square-grid cells, to name a few. Apparently, the sea sponge is well versed in structural engineering practices and methods!

Again, if we consider the fundamental building blocks—consolidated hydrated silica nanoparticles—we see where "bioglass" fabricated by the sea sponge diverges from other silica-based materials: by implementing collections of nanoparticles rather than continuous crystals, for example, the intrinsically low strength of the glass is balanced at the next structural level. The structure is as important as the material. The structural complexity of the glass skeleton in the sponge *Euplectella sp.* is an example of Nature's ability to improve inherently poor building materials. Moreover, such "bioglass" is not unique to sea sponges, and is also found in diatoms—unicellular algae—able to construct nanoporous silica with 3D precision of tens of nanometers, in a hierarchical manner, and with multifunctional properties [20]. Again, it has been shown that the mechanical properties of such materials can be changed my manipulating the nanostructure [21]. Synthetic mesoporous silica (depicted in Fig. 1.3) is currently being exploited for applications in medicine, biosensors, and imaging [22, 23].

Understanding what a material is composed of and how a material behaves has always been of great importance to enable and advance technologies [24, 25]. As such materials have played a major role in enabling civilization eras, from the stone age to the nano age, and are as such a cornerstone of all engineering disciplines. In the early days materials were obtained and tailored for our purposes from chopping up rocks or using natural resources such as rubber. For example, concrete is a compound material made from sand, gravel and cement. The cement is a mixture of various minerals which when mixed with water, hydrate and rapidly become hard binding the sand and gravel into a solid mass. The Romans found that by mixing a sand-like material (which they obtained from Pozzuoli) with their normal lime-based concretes they obtained a far stronger material. The pink sand turned out to be fine volcanic ash and they had inadvertently produced the first '*pozzolanic*' cement. In the 2,000 or so years since they employed this naturally occurring form of cement to build a vast system of concrete aqueducts and other large edifices, concrete is presently the most widely used construction material in the world, found in large scale structures such as bridges and skyscrapers. Cement is so widely used as a building material that, even in the face of technological advances in materials, it will not be replaced anytime soon. Surely, the chemical details and material properties of such a widely utilized material is well-known from the molecules up?

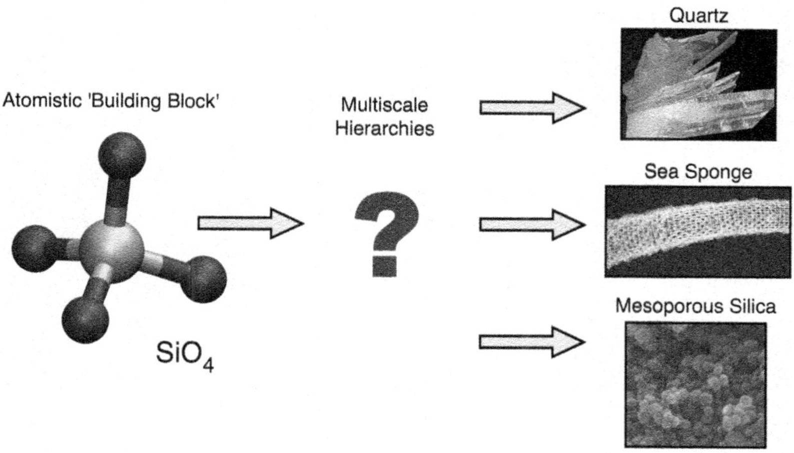


**Fig. 1.3** What defines a material? If we consider the most fundamental atomistic building block—here tetrahedral silica ($SiO_4$)—we cannot predict the properties of the macrostructure. Depending on how silica is arranged in multiscale hierarchies, the resulting material at the macroscale shows extreme variation—from crystalline quartz, to the skeleton of a sea sponge, to synthetic microporous silica. While such structures can be analyzed to determine the structural hierarchies, the functionality of such hierarchies is difficult to predict and engineer. *Inset* SEM image of sea sponge printed with permission from The National Academy of Sciences [17] © 2004

Oddly enough, the three-dimensional crystalline structure of cement hydrate (*i.e.*, calcium silicate hydrate, or C–S–H)—the paste that forms and quickly hardens when cement powder is mixed with water—has eluded scientific attempts at decoding, despite the fact that concrete is the most prevalent man-made material on earth and the focus of a multibillion-dollar industry. The lack of a fundamental multiscale understanding does *not* preclude successful use of concrete as a building material—indeed, it is because of the improvements of concrete design, knowledge of its chemical reactivity, and keen development of additives that we have achieved high strength and corrosion resistant varieties of concrete, unmatched by anything the Romans may have stumbled upon. Such refinements, however, can only optimize a cement/aggregate/water system to a point. If we wish to develop new, stronger, and "greener" concretes, we must have complete knowledge across scales, from "nano" to "macro" (see Fig. 1.4).

Only recently has the three-dimensional structure of the basic unit of cement hydrate been decoded, resulting in a first step toward a consistent model of the molecular structure of cement hydrate [26]. Scientists have long believed that at the atomic level, cement hydrate closely resembles the rare mineral tobermorite, which has an ordered geometry consisting of layers of infinitely long chains of silica tetrahedra interspersed with neat layers of calcium oxide. But it was determined the hydrates in cement aren't really crystalline. They are a hybrid that shares some characteristics with crystalline structures and some with the amorphous structure of frozen liquids, such as glass or ice. Concrete is more disordered and porous (like the silica skeleton of the sea sponge), than ordered and crystalline (like quartz).

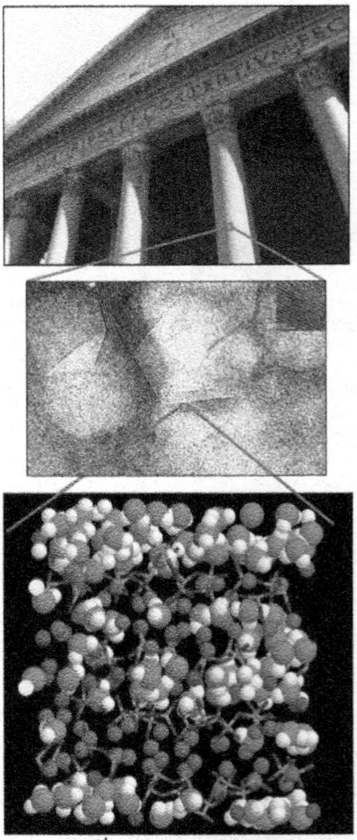


**Fig. 1.4** Concrete, a construction material used for over 2,000 years, and yet only know being fully understood from the atomistic level. *Top*: Photograph of the Pantheon (Rome, Italy, 2008), constructed 126 A.D., an example of Roman concrete construction. *Centre*: TEM image of clusters of C–S–H (courtesy of A. Baronnet, CINaM, CNRS and Marseille Universite, France), the *inset* (*upper-right*) is a TEM image of tobermorite. *Bottom*: Atomistic representation of concrete: the molecular model of C–S–H. The gray and white molecules are oxygen and hydrogen atoms of water, respectively; the individual spheres are inter- and intra-layer calcium ions, respectively; connected sticks are silicon and oxygen atoms in silica tetrahedra. Figure adapted from [26]

This delicate balance between order and disorder within a structure is a concept that resembles many natural biological materials. But why is such disorder beneficial?

It is in this disorder—where breaks in the silica tetrahedra create small voids in the corresponding layers of calcium oxide—that water molecules attach, giving cement its robust quality. These material "flaws" in the otherwise regular geometric structure provide some give to the building material at the atomic scale that transfers up to the macro scale. When under stress, the cement hydrate has the flexibility to stretch or compress just a little, rather than snapping. Whereas water weakens a material like tobermorite, it strengthens the cement hydrate. The *disorder* or *complexity* of its chemistry creates a heterogenic, robust structure. The cement hydrates have a level of hierarchy that helps optimize water content and mechanical performance—analogous to Nature's hierarchical sea sponge. Serendipity was apparently on the Romans' side 2,000 years ago when concrete was discovered!

If we are only now beginning to understand the fundamental behavior and multiscale consequences of a material we have been using for thousands of years, how can we be expected to understand, design, or engineering complex biological materials? Materials that are not cast in place like concrete, but materials that grow and adapt to their environment? Clearly, a new approach is not only warranted, but

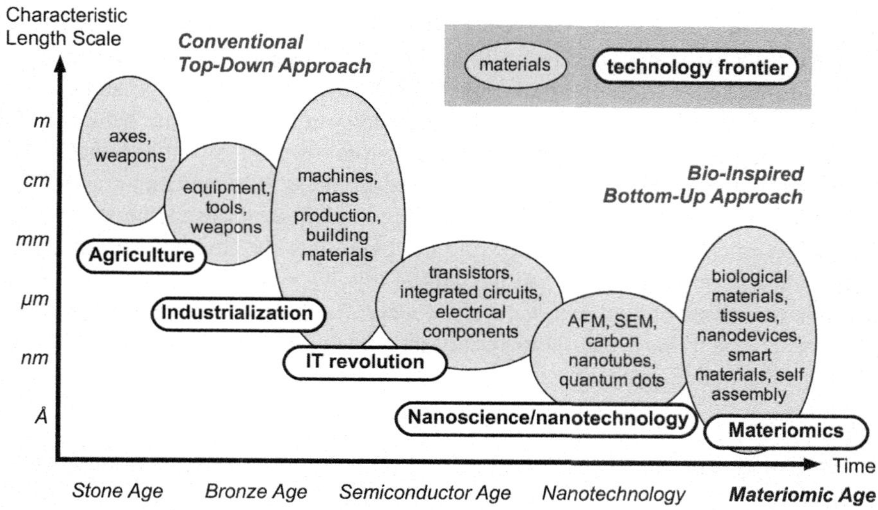


**Fig. 1.5** Characteristic material scales from the Stone Age to nanotechnology and biotechnology. The plot illustrates the trend to create smaller dimensions of materials and structures as the technological frontier progresses. Currently we stand at a crossroads where nanotechnology and biology merge to provide a new bottom-up approach in the development of materials and technologies (Based on graph shown in [27])

necessary to address the challenges we face to support technological advancement and consequent economic growth.

## 1.4 Starting at the Bottom

The quantitative study of biological protein materials is a critical step towards the development of new technological frontiers through smarter use of (limited) resources. Aside from the Romans use of concrete, classes of materials have been used classify stages of civilizations, ranging from stone age more than 300,000 years ago, to the bronze age, and possibly the silicon age in the late twentieth and early twenty-first century. Figure 1.5 schematically displays the various stages of civilization together with an analysis of the characteristic material scales that were used in each period. The plot illustrates the trend to ever smaller material scales as humankind progressed through the ages, and the analysis may suggest that today we may stand at another cross-road in the advancement of technology. This next frontier involves the rigorous understanding of the properties (*e.g.*, mechanical, physical and chemical properties) and mechanisms (*e.g.*, chemomechanical conformation changes, enzymatic processes, mechanotransduction) of biological matter, which may enable us eventually to integrate concepts from living systems into materials and machine design, seamlessly. Solving these challenging problems may transcend the gap that currently exists between engineering and physical sciences and the life sciences.

We have now entered the era of nanoscience and nanotechnology where materials are made with atomistic precision—enabling advances in the design and synthesis of molecular building blocks that we can (theoretically) design and exploit. This bottom-up approach—designing a material/system through the behavior and combination of each constituent element and atom—was envisioned in the 1960's by Richard Feynman, the popular physicist and pioneer of nanotechnology. Feynman hypothesized the direct manipulation of individual atoms as the most powerful form of synthetic chemistry—unlocking the blueprints for atomistic construction. The challenge posed by Feynman is simple [28]: *What would happen if we could arrange the atoms one by one the way we want them*? From a biological perspective, this is exactly how natural materials are formed—the piece-wise combination of molecular building blocks.

**"Feynman paradigm"**: Nanotechnological, bottom-up approach to material design, *via* the direct manipulation of individual atoms and molecules, and precise engineering of functional systems at the molecular scale. In its original sense, Feynman referred to the projected ability to construct items from the bottom-up, whereas the ultimate goal is to control macroscale structure and function from design at the atomistic scale.

The realization of the "Feynman paradigm" (see Fig. 1.6) has opened numerous new opportunities for research, products and development [29]. But its impact for real products and technologies hinges upon a major challenge, the linking of the scales, and to make nanoscale mechanisms visible at larger scales. Indeed, taking a closer look at the vastness of scales in our environment we realize that there are huge opportunities in designing structures and thus functions at multiple length-scales. Developed nanoscale components, once attained, must demonstrate the reproducibility needed to build functional materials and systems, and do so at a size and complexity difficult to achieve by traditional top-down approaches.

We recognize that the scales are separated, and that the scales can be connected by networks in the process of design. This design challenge has been solved by Nature and biology, where scale separation and connection are used effectively to create function from nano to macro through complex functional relations that link seemingly disparate concepts such as individual atoms or amino acids to strength to robustness. This is exemplified in the design of DNA, protein, tissue to organisms, and many others. This paradigm of using hierarchical structures can be used in engineering, to eventually eliminate the border between living and non-living systems. The applications are endless, and include self-healing cement, changeable airplane wings, and others.

But before the realization of what is possible, we first need a complete understanding of what is palpable. One possible approach to improve our understanding of what we can engineer is to turn towards Nature for inspiration. The development of new materials and the discovery of the complexity of existing materials are not

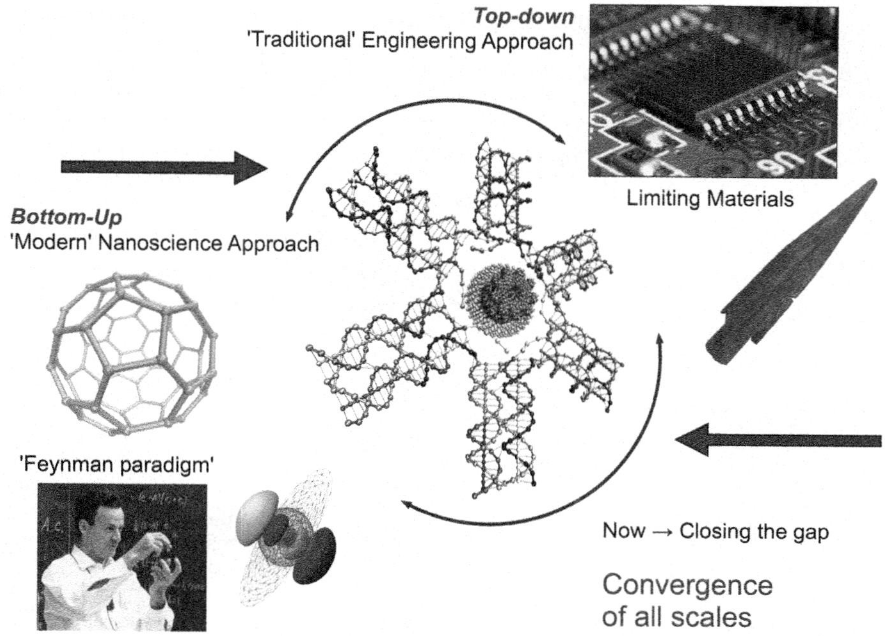


**Fig. 1.6** Juxtaposition of the "Feynman paradigm" with "traditional" engineering approaches. Current top-down methodologies have advanced from simple manipulation of available materials (*e.g.*, crude stone age weaponry) to sophisticated exploitation of material properties (*e.g.*, semiconductors in integrated circuits). Nanotechnology has also developed the ability to investigate and manipulate materials on the atomistic and molecular scale from a bottom-up perspective. Currently, we are at the convergence of both bottom-up and top-down routes, closing the gap between material, structure, and function. As Feynman suggested, precise engineering and control at the nanoscale may dictate the future of material design, but we must also fully understand how nanoscale properties are expressed at the macroscale

mutually exclusive endeavors. Even if a complete and thorough understanding of complex phenomena is not attained, we can still learn lessons and insight from Nature providing guidance for new discoveries and distinct means by which heightened functionality is created in spite of limited resources.

## 1.5 Lessons from Nature: Biological Materials and Biomimetics

Nature exhibits the design guidelines for multi-scale adaption of structure and functionality. An organism evolves to survive because it uses the minimum amount of material to make its structures (be it internal to the organism, such as bone or tissue, or an external structure, such as a spider's web) and also because it can optimize its use of the available environmental sources. Nature thus provides an array of building materials and aptly chooses suitable means for a multitude of natural functions

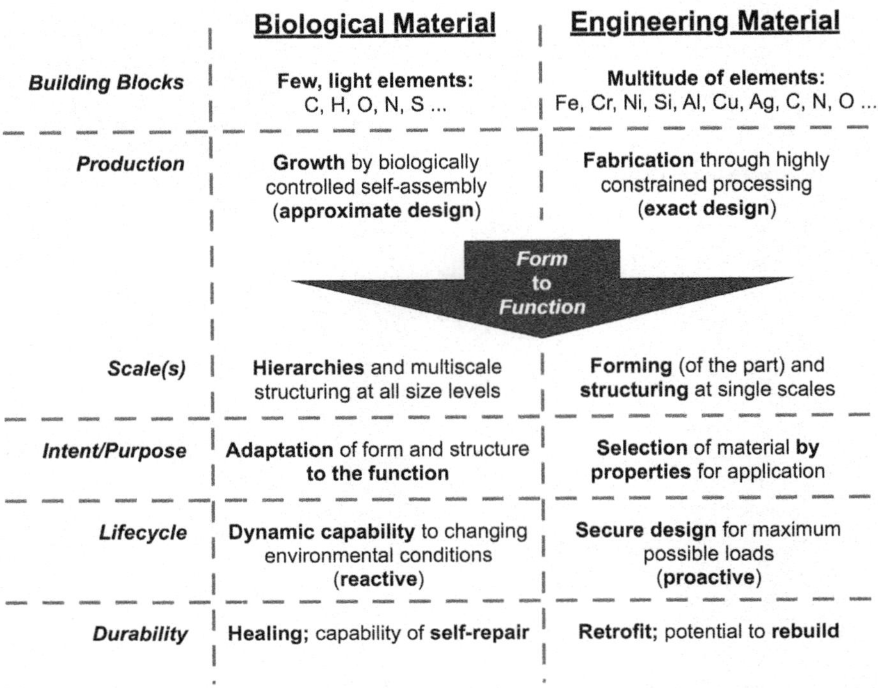


**Fig. 1.7** Biological and engineering materials are governed by a very different choice of base elements (natural materials consist of relatively light elements few whereas engineering materials are characterized by many more elements) and by a different mode of material production (biological growth versus controlled fabrication). From these basic forms, there arise different strategies for materials choice and development (under the *arrow*) of function. Biological materials are inherently multiscale, whereas the performance of engineering materials are typically limited to a single scale. Biological materials have been adapted for a specific biological role/function, whereas an engineering material is typically selected based on desirable properties. Finally, biological materials are dynamic systems, capability of both self-adaptation and healing, whereas engineered systems are typically limited to the design specifications. Additional requirements and incurred damage necessitate reinforcement or retrofitting of a material or structural system. Extended and adapted from [25]

[15, 30–33]. The elasticity of blood vessels, the toughness of bone [8, 9] or the protection of nacre [11, 34–36] illustrate the *apropos* of Nature's material selection. Moreover, Nature has developed such materials with a comparatively poor set of base materials. Why can't we simply copy Nature's systems and substitute materials to maximize performance? If bone is made of proteins and minerals (*i.e.*, collagen and hydroxyapatite), can we simple replace with synthetic materials? Perhaps nylon (a robust synthetic polyamide) and titanium (a metal with high strength-to-weight ratio)? Unfortunately, we cannot. The design strategies of biological materials are neither immediately applicable to, nor compatible with the design of new engineering materials, since there are some remarkable differences between the strategies common in engineering and those used by Nature (see Fig. 1.7).

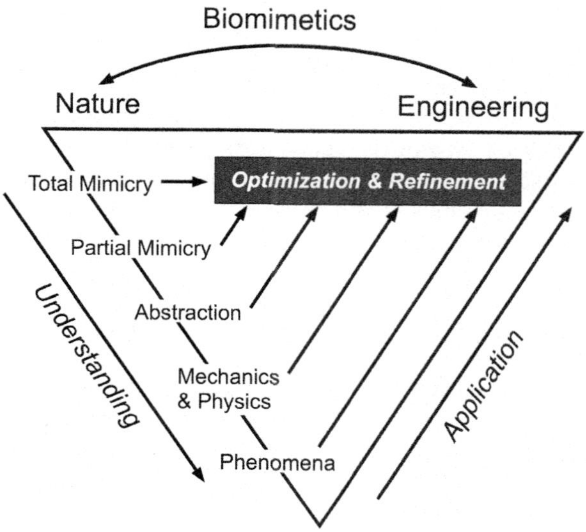


**Fig. 1.8** A biomimetic "map" to illustrate the idea that the more abstract a concept is, the more adaptable it is within another discipline. Adapted from J.F.V. Vincent, "Stealing ideas from nature" in *Deployable Structures* [37]

A holistic knowledge of biological materials offers a unique opportunity to understand how complex materials science, engineering, and chemical principles arise routinely in Nature. Nature has been the motivation factor in a number of texts and studies, and, concurrently, provides the inspiration and stimulation to scientists and engineers for new material concepts, design strategies, and structural optimization. This field defines biomimetics—using ideas from nature to further technology—or, more colloquially, "*. . . the technological outcome of the act of borrowing or stealing ideas from Nature*" [37].

Biomimetics, however, is extending beyond the simple "stealing" of ideas, and evolving to a more didactic role—*i.e.* learning ideas from Nature. The difference lies not just in the abstraction of useful ideas (the invention of Velcro by the observation of sticking plant burrs is a popular example) but also in the detailed and mechanistic understanding of the processes involved. The transfer of ideas from biology is not limited to the ultimate form and function of a biological system—we are not interested in spider silk so we can swing from skyscrapers like Spiderman. Instead, we should look to Nature and biological systems (nay, models), to serve a technical application of practical purpose. The more this application deviates from the biological system, the more basic the analysis has to be in order to generate useful (practical) knowledge and understanding (see Fig. 1.8).

The general concept, as discussed by J.F.V. Vincent ("Stealing ideas from nature" in *Deployable Structures* [37]), is that the further down one can move from the natural origin, the more general and therefore more powerful the concept will be. The goal is the shift from total mimicry (stealing) to an understanding of the process at its basic level (abstraction), defining that process from an analytical perspective (mechanics and physics) and then exploiting the physical phenomena to our own ends. Throughout this text, investigations and studies discussed can be assigned to such categorizations, from the behavior of spider silk in Chap. 8: Unlocking Nature:

Case Studies (mechanics and physics) to the comparison of biological materials and music in Chap. 4: Universality-Diversity Paradigm: Music, Materiomics, and Category Theory (abstraction). We note that optimization and refinement are key components to allow robust designs and applications, avoiding the idiosyncrasies of idealized experimental conditions not present in Nature. A thorough understanding of the phenomena may result in applications that are completely deviant from the original biological system. We will no longer be an apprentice to Nature, but our own journeyman to the trade.

## 1.6 Advancement and Convergence

The systematic investigation of biological materials has advanced considerably in recent years, along with the advancement in the tools required for analysis. Indeed, there are constant updates and refinements of techniques providing new, more accurate means to measure, interpret, quantify, and model the relationships between chemistry, structures, design and function. Advances in information technology, imaging, nanotechnology and related fields coupled with advances in computing, modeling and simulation have transformed investigative approaches to material systems. Only recent developments in imaging methods over the past decades have revealed that biology creates intricate hierarchical structures, where initiated at nanoscales, result in macro or physiological multifunctional materials to provide structural support, force generation, catalytic properties or energy conversion [13, 31, 33, 38, 39]. This is exemplified in a wide range of biological materials such as hair, skin, bone, spider silk or cells, which play important roles in providing key functions to biological systems [12].

The motivation has come from a vast assortment of disciplines: medicine (mechanical and physiological properties of soft and hard tissues, including skin, tendons, bone, etc., for prosthetic devices, replacement materials, and tissue engineering applications); biology (material aspects of adaptation, evolution, functionality, etc.); materials science (thermal and electrical properties of novel hierarchical nanosystems, functionality performance of microscale devices, etc.) to name a few. The complexity of biological materials elicits contributions from a multitude of fields, and is an exciting field for biologists, chemists, and engineers alike. Further advancement is hindered, however, by such a "divide and conquer" approach, and dictates the convergence of scientific disciplines.

The historical progression of science has been typified by the continuous evolution of disciplines and subdisciplines. Fundamental advancements—such as the development quantum theory or the discovery of DNA—have paved the way for research directions that were previously unthinkable. Such paradigm shifts frequently defined a new subdiscipline in their fields, and progressed independently alongside their brethren. Quantum mechanics did not negate the need for Newtonian mechanics, for example, it merely gave another (more precise) piece of the puzzle of physics.

Traditionally, the science and engineering of synthetic material systems have been separated into classes of structures, length scales, and functionality that are used to differentiate disciplines. Nature, however, does not conform to disciplinary boundaries, and effortlessly balances chemistry, materials, structure, and function across a full range of length scales in order to react to a variety of environmental requirements and optimal functionality. Likewise, biomateriomics is inherently multi-disciplinary, borrowing from fundamental physics and chemistry at the atomistic scale, integrating biological mechanisms (such as protein interactions and cellular processes) at the molecular and cellular level, traversing hierarchical scales, and linking a material's components, structure, and mechanical properties with its natural requirements and functionalities. Biomateriomics involves the rigorous understanding of the properties and mechanisms of biological matter. While this is an easy statement, integration of multiple disciplines is difficult in practice. Due to the rich history and unique perspectives of the fields that contribute to advances in fields contribution to materiomics, questions of interest, approaches, tools and even vocabulary particular to each community (the "disciplinary lexicon", so-to-speak) can impede communication and progress in this inherently interdisciplinary venture.

Recent advances in science and technology have seen the creation of a multitude of biologically "themed" interdisciplinary research areas including bioinfomatics, nanobiology, biomaterials, and systems biology. Such fields share a comparable, underlying research paradigm, recently defined as *convergence*.[2] Convergence extends beyond simple collaboration between disciplines to evoke true disciplinary integration—transcending fields through the integration of approaches typically viewed as distinct and potentially contradictory. The impact that convergence is already having in a broad array of fields. Just as advances in information technology, materials, imaging, nanotechnology and related fields—coupled with advances in computing, modeling and simulation—have transformed the physical sciences, so are they are beginning to transform life science. The result is critical new biology-related fields, such as bioengineering, computational biology, synthetic biology and tissue engineering. At the same time, biological models (understanding complex, self-arranged systems) are already transforming engineering and the physical sciences, making possible advances in biofuels, food supply, viral self-assembly and much more.

> **Convergence** [40]: A broad rethinking of how all scientific research is conducted, involving the integration of disciplinary approaches that were originally viewed as separate and distinct. Through the merging of technologies, processes and devices into a unified whole, new pathways and opportunities for scientific and technological advancements are created inaccessible to any single discipline or knowledge bases. The goal of convergence is not a particular scientific advance, but on a new integrated approach for achieving advances.

[2]The third revolution: the convergence of the life sciences, physical sciences, and engineering, http://web.mit.edu/dc/Policy/MIT%20White%20Paper%20on%20Convergence.pdf.

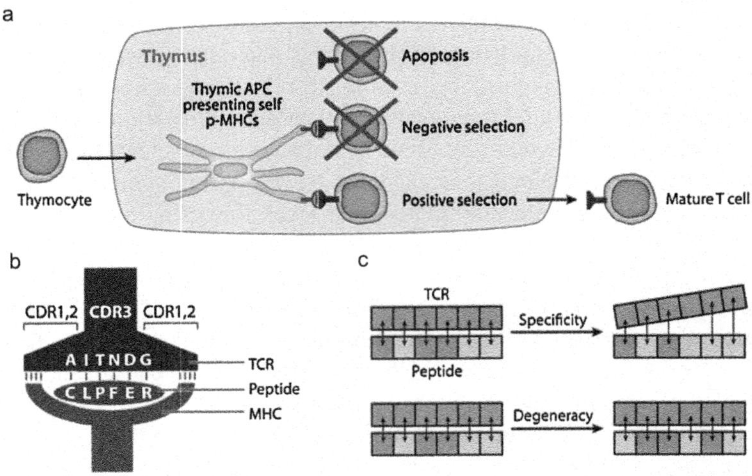


**Fig. 1.9** Statistical mechanical concepts in immunology—the convergence of biology, modeling, and mechanics. Schematic depiction of thymic selection of T cells, and its consequences for the antigen-recognition properties of the T cell repertoire [41]. (**a**) Immature T cells (thymocytes) develop in the thymus which migrate through the thymus and interact with self peptide major histocompatibility complexes (self-pMHCs) presented on the surface of thymic antigen presenting cells (APCs). A T cell's receptor (TCR) must bind to at least one of these self-pMHCs weakly to exit the thymus and become a part of the individual's T cell repertoire (positive selection). Binding with an affinity that exceeds a sharply defined threshold results in negative selection (apoptosis). (**b**) Schematic representation of the interface between TCR and pMHCs. The region of the TCR contacting the peptide and the peptide itself are modeled by string of amino acids to enable the calculation of binding free energy. (**c**) A mechanism for the puzzle of how TCR recognition of pathogen-derived peptides is both specific and degenerate emerges from statistical mechanical theory and is illustrated in the schematic. Peptide amino acids of different shades are not identical. Sufficiently strong interactions required for recognition are mediated by several moderate interactions. Adapted from Chakraborty *et al.* [41], used with permission, copyright © 2010 Annual Review of Physical Chemistry

This trend of integration has already been realized in various studies in which engineering theories have been "borrowed" to model biological systems. As an example, we consider the implementation of computational biology to explore complex biological systems such as, for example, our immune response to HIV. Recently, Chakraborty and colleagues combined computer science, physics, and engineering with molecular and genetic biology to elucidate mechanistic principles underlying the adaptive immune response to pathogens [41, 42]. Such investigations are possible by linking statistical mechanics together with an engineering approach to the chemical kinetics, genetic, biochemical, and imaging experiments, as well as clinical data (see Fig. 1.9).

Modern research has already shed the labels in which science was pigeonholed, with traditional disciplines such as physics and chemistry being replaced by modern interdisciplinary fields of biomechanics or nanotechnology, for example. In his "Lectures on Physics" (1964), Richard Feynman quips:

> …we humans cut Nature up in different ways, and we have different courses in different departments, such compartmentalization is really artificial…

Since the physical universe (living and non-living) is subject to the same fundamental physical laws, it follows that this set of laws serves as common ground for the transfer of information between the disciplines. This is one of the primary motivations for the concentration on atomistic and molecular behavior throughout this text. The merger of science and engineering (*i.e.*, discovery and invention), and the convergence of many fields can pave the way for tomorrow's supermaterials. Ultimately, we may unlock new theories and scientific approaches that allow advanced functional, adaptable, and robust materials that exceed those known in the natural world. In the interim, we look to Nature for proven systems and inspiration.

## 1.7 A New Perspective: Materiomics

Traditional engineering methods encompassing such fields as structural analysis and continuum theory (*e.g.*, continuum mechanics) lack the framework required for the complexities introduced by multi-scale interactions, their discrete hierarchical composition, and structure-property dependencies at all scales, as found in many natural materials. Mechanical notions of stress and strain, fracture and plasticity, toughness and robustness, are inherently linked to material behavior at the macro-scale. Similarly, chemical and biological techniques (including such fields as microbiology, proteomics, and condensed matter physics) shed vast insights on nano-scale phenomena, such as the chemical composition of materials or the interactions of residues, but lack an explicit association with mechanical properties. Significant advances have been made in many disciplines and research areas, ranging throughout a variety of scales, from atomistic, molecular to continuum. In tandem, experimental studies have attained molecular precision, lending insights to molecular defects and mechanisms. As a result of these advances in disparate fields of science, in the spirit of convergence, a fully integrated and holistic paradigm is necessary to elucidate nature's design principles and facilitate the design of materials with exceptional material properties.

The study of material properties of biological protein materials has witnessed an exciting development over the past several years, partly due to the emergence of physical science based approaches in the biological sciences. Specifically, there has been significant effort directed towards the explanation and control of observed macroscopic mechanical and optical behavior of complex polymer composites [43, 44] while concurrently the structure of many protein-based (polymeric) materials are being discovered, motivating the design of novel "*synthetic biological*" materials [45]. The rapid expansion in the scope of materials science and engineering has led to incorporation of such fields as experimental and computational biology, biomedical engineering, and genetics in the context of natural and synthetic materials. Recent progress provides insight into biological mechanisms and enables us a peek into how biology works at the ultimate, molecular scale, and how this relates to macroscopic phenomena such as cell mechanics, tissue behavior or functions provided by entire organisms. This has resulted in the cross-disciplinary investigation of protein materials and structures, diseases as well as the development of novel treatment and diagnostics methods [32, 33, 46–50]. As discussed in the previous section, there is undoubtedly an accelerated progression and convergence of biology, chemistry, materials science, and engineering, each contributing different aspects of the complexity of Nature's design. Is it possible for each field to share a common "mission statement"? The merger of such perspectives is mutually beneficial: materials scientists have extensive experience in treating structures, processes and properties of materials systematically and with rigorous mathematical methods, whereas biologists have gained a detailed understanding of biological systems and structures and related functions by utilizing both physiological models and powerful statistical correlations between, for example, genetics, physiology and pathology.

In addition, there is accelerating interest in the discovery and understanding of Nature's structural design rules, in particular for nanoscopic hierarchical molecular structures and to make them available to engineers in order invent and design tomorrow's supermaterials (*e.g.*, mechanomutable materials, advanced composites, low density low energy structural materials, etc.), seamlessly blending synthetic materials with biological systems (*e.g.*, tissue and biomedical engineering), and using basic biological systems as templates for design (*e.g.*, biomimetic and bio-inspired materials). There is also a surprising relationship between these material design issues and the understanding (or rather lack thereof) of genetic diseases and disorders, where structural changes are due to mutations on the molecular level which lead to changed chemical and mechanical properties, which in turn lead to a malfunction of the protein network under mechanical load.

This type of effort, the linking of mechanisms across multiple scales by using a materials science approach to provide structure-process-property links characterizes the emerging field of materiomics [51]. The term materiomics has been proposed with various definitions in the past (see Akita *et al.*, [52], Buehler *et al.*, [51, 53–55], and the work of de Boer and van Blitterswijk *et al.* [56–59]), where all of the definitions deal broadly with the complexities of synthesis and function of materials and structures. For example, Akita *et al.* propose materiomics as the systematic study of material composition and structure to determine material properties of metal/metal oxide catalysts [52], whereas de Boer and van Blitterswijk *et al.* proposes materiomics as dealing with the complexities of tissue engineering [56, 57] and high-throughput screening of potential bio-compatible material platforms [59]. It is evident that both definitions encompass the intricacies of complex materials, yet limit scope to specific material systems and applications. We believe, that although slightly different, the definition of materiomics proposed in earlier works fall under a much broader perspective, encompassing even quantum scales [60, 61], or natural porous granular materials such as shale and concrete [26, 62, 63], as examples, and thus incorporates a vast array of potential future applications in science and engineering.

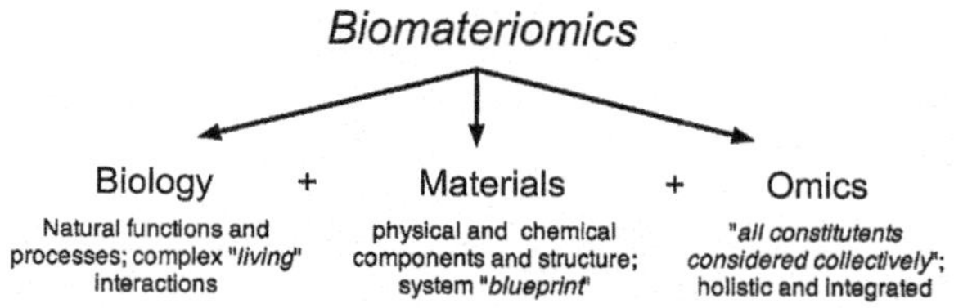


**Fig. 1.10** The etymology of biomateriomics can be easily understood as the integration of biology, materials, and the suffix "omics". Complementary to the field of biomaterials, biomateriomics represents a holistic approach to the investigation of biological material systems, through the integration of natural functions and processes (biological, or "living" interactions) with traditional materials science perspectives (physical properties, chemical components, hierarchical structures, mechanical behavior, etc.)

**Materiomics**: the holistic study of material systems. Materiomics examines links between physicochemical material properties and material characteristics and function. While typically applied to complex biological systems and biomaterials, materiomics is equally applicable to non-biological systems. Materiomics considers a complete material system and the affect on the macroscopic function and failure in a mechanical context, linking processes, structure and properties at multiple scales, through a materials science perspective, integrating experimental, theoretical, and computational methods. A portmanteau of "*material*" and the suffix "*omics*" which refers to "*all constituents considered collectively*".

Materiomics can refer to the study of a broad range of materials, which includes metals, ceramics and polymers in addition to biological materials and tissues (and their interactions with synthetic materials), as well as the study of purely biological structures, such as bone, skin, cells, silk, or ectopic materials (such as amyloids). In a material context, materiomics is not limited to protein based materials, but rather includes the whole range of materials in a biological system (*e.g.*, mineralized protein scaffolds, metal-protein material systems, etc.), encompassing the analysis of the role of materials in the context of biological processes, the transfer of biological material principles in biomimetic and bioinspired applications, and the study of interfaces between living and non-living systems.

Thus far, we have only mentioned the term "materiomics", yet the title of this book is, conspicuously, *bio*materiomics. Biomateriomics can be viewed as a simple juxtaposition of biology, materials, and the Greek suffix *-omics* (see Fig. 1.10). A simple combination of biology and materials naturally yields biomaterials—a relatively well-defined classification in materials science that encompasses Natural materials and synthetic materials that interact with biological systems (biomedical devices, for example). Biological processes introduce complex living interactions to material systems, and thereby require certain tact to analyze mechanistic behavior and material properties. Materials science has been making tremendous advances in this area in recent years. Biomateriomics, however, is not intended to be a trivial rebranding or subfield of biomaterials (although there is indeed significant overlap). The critical difference—as may have been anticipated—is the addition of that pesky suffix: *-omics*.

**Biomateriomics**: A *materiomic* approach to biological systems and materials, including biomimetic and bio-inspired materials. Biomateriomics integrates the structure and material properties of *biological* materials with the inherent functionality and environmental adaptation.

The addition of *-omics* adds a subtle layer of complexity to the material systems we are interested in. The term *omics* generally refers to "*all constituents considered collectively*". For instance, *genomics* and the study of the human genome refers to all of our genes, not just a single subset.

The combination of materials and omics—*materiomics*—thus refers to a similar perspective of a material system: one in which all properties, functions, and interactions are considered collectively, so-to-speak. Biomateriomics applies this perspective to natural and other biomaterials. Such materials—as opposed to more traditional (and benign) engineering materials—are characterized by hierarchical structures, cross-scale interactions, multi-functionality, dynamic feedback, adaptation, and are notoriously difficult to understand, let alone design and exploit (*i.e.*, engineer). While a materials science approach can probe the physical and chemical components and structure of the system—a systematic "blueprint"—the whole is indeed greater than the sum of its parts. Such features (nay, challenges) necessitate a new, holistic and integrated perspective—an *omic* perspective—*biomateriomics*.

## References

1. N. Du, X.Y. Liu, J. Narayanan, L.A. Li, M.L.M. Lim, D.Q. Li, Design of superior spider silk: from nanostructure to mechanical properties. Biophys. J. **91**(12), 4528–4535 (2006)
2. J.M. Gosline, P.A. Guerette, C.S. Ortlepp, K.N. Savage, The mechanical design of spider silks: from fibroin sequence to mechanical function. J. Exp. Biol. **202**(23), 3295–3303 (1999)
3. S. Keten, M.J. Buehler, Atomistic model of the spider silk nanostructure. Appl. Phys. Lett. **96**(15), 153701 (2010)
4. S. Keten, M.J. Buehler, Nanostructure and molecular mechanics of spider dragline silk protein assemblies. J. R. Soc. Interface **7**(53), 1709–1721 (2010)
5. F.G. Omenetto, D.L. Kaplan, New opportunities for an ancient material. Science **329**(5991), 528–531 (2010)
6. F. Vollrath, Spider webs and silks. Sci. Am. **266**(3), 70–76 (1992)
7. F. Vollrath, Spider silk: evolution and 400 million years of spinning, waiting, snagging, and mating. Nature **466**(7304), 319 (2010)
8. P. Fratzel, *Collagen: Structure and Mechanics* (Springer, New York, 2008)
9. S. Weiner, H.D. Wagner, The material bone: structure mechanical function relations. Annu. Rev. Mater. Sci. **28**, 271–298 (1998)
10. M.J. Buehler, Nature designs tough collagen: explaining the nanostructure of collagen fibrils. Proc. Natl. Acad. Sci. USA **103**, 12285–12290 (2006)

11. H.D. Espinosa, J.E. Rim, F. Barthelat, M.J. Buehler, Merger of structure and material in nacre and bone—perspectives on de novo biomimetic materials. Prog. Mater. Sci. **54**(8), 1059–1100 (2009)
12. P. Fratzl, R. Weinkamer, Nature's hierarchical materials. Prog. Mater. Sci. **52**, 1263–1344 (2007)
13. M.J. Buehler, T. Ackbarow, Nanomechanical strength mechanisms of hierarchical biological materials and tissues. Comput. Methods Biomech. Biomed. Eng. **11**(6), 595–607 (2008)
14. M.J. Buehler, S. Keten, Failure of molecules, bones, and the earth itself. Rev. Mod. Phys. **82**(2), 1459–1487 (2010)
15. J. Aizenberg, J.C. Weaver, M.S. Thanawala, V.C. Sundar, D.E. Morse, P. Fratzl, Skeleton of *Euplectella sp.*: structural hierarchy from the nanoscale to the macroscale. Science **309**(5732), 275–278 (2005)
16. A. Woesz, J.C. Weaver, M. Kazanci, Y. Dauphin, J. Aizenberg, D.E. Morse, P. Fratzl, Micromechanical properties of biological silica in skeletons of deep-sea sponges. J. Mater. Res. **21**, 2068–2078 (2006)
17. J. Aizenberg, V.C. Sundar, A.D. Yablon, J.C. Weaver, G. Chen, Biological glass fibers: correlation between optical and structural properties. Proc. Natl. Acad. Sci. USA **101**(10), 3358–3363 (2004)
18. C. Levi, J.L. Barton, C. Guillemet, E. Bras, P. Lehuede, A remarkably strong natural glassy rod: the anchoring spicule of the monorhaphis sponge. J. Mater. Sci. Lett. **8**(3), 337–339 (1989)
19. M. Sarikaya, H. Fong, N. Sunderland, B.D. Flinn, G. Mayer, A. Mescher, E. Gaino, Biomimetic model of a sponge-spicular optical fiber—mechanical properties and structure. J. Mater. Res. **16**(5), 1420–1428 (2001)
20. D. Losic, J.G. Mitchell, N.H. Voelcker, Diatomaceous lessons in nanotechnology and advanced materials. Adv. Mater. **21**, 2947–2958 (2009)
21. A. Garcia, M.J. Buehler, Bioinspired silicon nanoporous bulk material provides great toughness at great deformability. Comput. Mater. Sci. **48**(2), 303–309 (2010)
22. N.G. Pinto, A. Katiyar, S. Yadav, P.G. Smirniotis, Synthesis of ordered large pore sba-15 spherical particles for adsorption of biomolecules. J. Chromatogr. A **1122**(1–2), 13–20 (2006)
23. V.S.Y. Lin, B.G. Trewyn, J.A. Nieweg, Y. Zhao, Biocompatible mesoporous silica nanoparticles with different morphologies for animal cell membrane, penetration. Chem. Eng. J. **137**(1), 23–29 (2008)
24. M.F. Ashby, *Materials Selection in Mechanical Design*, 4th edn. (Butterworth–Heinemann, Oxford, 2011)
25. P. Fratzl, Biomimetic materials research: what can we really learn from natures structural materials. J. R. Soc. Interface **4**, 637–642 (2007)
26. R. Pellenq, A. Kushima, R. Shahsavari, K. Van Vliet, M.J. Buehler, S. Yip, F.-J. Ulm, A realistic molecular model of cement hydrates. Proc. Natl. Acad. Sci. USA **106**(38), 16102–16107 (2009)
27. M.J. Buehler, *Atomistic Modeling of Materials Failure* (Springer, Berlin, 2008)
28. R.P. Feynman, There's plenty of room at the bottom. Caltech Eng. Sci. **23**, 22–36 (1960)
29. K.E. Drexler, *Nanosystems: Molecular Machinery, Manufacturing, and Computation* (Wiley, New York, 1992)
30. H. Gao, B. Ji, I.L. Jger, E. Arzt, P. Fratzl, Materials become insensitive to flaws at nanoscale: lessons from nature. Proc. Natl. Acad. Sci. USA **100**(10), 5597–5600 (2003)
31. U.G.K. Wegst, M.F. Ashby, The mechanical efficiency of natural materials. Philos. Mag. **84**(21), 2167–2181 (2004)
32. M.A. Meyers, P.Y. Chen, A.Y.M. Lin, Y. Seki, Biological materials: structure and mechanical properties. Prog. Mater. Sci. **53**(1), 1–206 (2008)
33. M.J. Buehler, Y.C. Yung, Deformation and failure of protein materials in physiologically extreme conditions and disease. Nat. Mater. **8**(3), 175–188 (2009)
34. S. Kamat, X. Su, R. Ballarini, A.H. Heuer, Structural basis for the fracture toughness of the shell of the conch strombus gigas. Nature **405**(6790), 1036–1040 (2000)

35. Z. Tang, N.A. Kotov, S. Magono, B. Ozturk, Nanostructured artificial nacre. Nat. Mater. **2**, 413–418 (2003)
36. E. Munch, M.E. Launey, D.H. Alsem, E. Saiz, A.P. Tomsia, R.O. Ritchie, Tough, bio-inspired hybrid materials. Science **322**, 1516–1520 (2008)
37. J.F.V. Vincent, Stealing ideas from nature, in *Deployable Structures*, ed. by S. Pellegrino (Springer, Vienna, 2001)
38. R. Lakes, Materials with structural hierarchy. Nature **361**, 511–515 (1993)
39. M.J. Buehler, Y.C. Yung, How protein materials balance strength, robustness and adaptability. HFSP J. **4**(1), 26–40 (2010)
40. P.A. Sharp, C.L. Cooney, M.A. Kastner, J. Lees, R. Sasisekharan, M.B. Yaffe, S.N. Bhatia, T.E. Jacks, D.A. Lauffenburger, R. Langer, P.T. Hammond, M. Sur, The third revolution: the convergence of the life sciences, physical sciences, and engineering. Massachusetts Institute of Technology (2011)
41. A.K. Chakraborty, A. Kosmrlj, Statistical mechanical concepts in immunology. Annu. Rev. Phys. Chem. **61**, 283–303 (2010)
42. A. Andrej Kosmrlj, E.L. Read, Y. Qi, T.M. Allen, M. Altfeld, S.G. Deeks, F. Pereyra, M. Carrington, B.D. Walker, A.K. Chakraborty, Effects of thymic selection of the T-cell repertoire on HLA class I-associated control of HIV infection. Nature **465**(7296), 350–354 (2010)
43. C.K. Ober, S.Z.D. Cheng, P.T. Hammond, M. Muthukumar, E. Reichmanis, K.L. Wooley, T.P. Lodge, Research in macromolecular science: challenges and opportunities for the next decade. Macromolecules **42**(2), 465–471 (2009)
44. M.A.C. Stuart, W.T.S. Huck, J. Genzer, M. Muller, C. Ober, M. Stamm, G.B. Sukhorukov, I. Szleifer, V.V. Tsukruk, M. Urban, F. Winnik, S. Zauscher, I. Luzinov, S. Minko, Emerging applications of stimuli-responsive polymer materials. Nat. Mater. **9**(2), 101–113 (2010)
45. M.J. Doktycz, M.L. Simpson, Nano-enabled synthetic biology. Mol. Syst. Biol. 3(125) (2007)
46. G. Bao, S. Suresh, Cell and molecular mechanics of biological materials. Nat. Mater. **2**(11), 715–725 (2003)
47. K.N. Dahl, P. Scaffidi, M.F. Islam, A.G. Yodh, K.L. Wilson, T. Misteli, Distinct structural and mechanical properties of the nuclear lamina in Hutchinson–Gilford progeria syndrome. Proc. Natl. Acad. Sci. USA **103**(27), 10271–10276 (2006)
48. S.E. Cross, Y.-S. Jin, J. Rao, J.K. Gimzewski, Nanomechanical analysis of cells from cancer patients. Nat. Nanotechnol. **2**, 780–783 (2007)
49. N. Kepper, D. Foethke, R. Stehr, G. Wedemann, K. Rippe, Nucleosome geometry and internucleosomal interactions control the chromatin fiber conformation. Biophys. J. **95**, 3692–3705 (2008)
50. M. Dao, C.T. Lim, S. Suresh, Mechanics of the human red blood cell deformed by optical tweezers. J. Mech. Phys. Solids **53**(2), 493–494 (2005)
51. M.J. Buehler, S. Keten, T. Ackbarow, Theoretical and computational hierarchical nanomechanics of protein materials: deformation and fracture. Prog. Mater. Sci. **53**, 1101–1241 (2008)
52. T. Akita, A. Ueda, Y. Yamada, S. Ichikawa, K. Tanaka, M. Kohyama, T. Kobayashi, Analytical tem observations of combinatorial catalyst libraries for hydrogen production—as a part of "materiomics". Mater. Res. Soc. Proc. **804**, 211–216 (2004)
53. M.J. Buehler, S. Keten, Elasticity, strength and resilience: a comparative study on mechanical signatures of alpha-helix, beta-sheet and tropocollagen domains. Nano Res. **1**(1), 63–71 (2008)
54. M.J. Buehler, Computational and theoretical materiomics: properties of biological and de novo bioinspired materials. J. Comput. Theor. Nanosci. **7**(7), 1203–1209 (2010)
55. S. Cranford, M.J. Buehler, Materiomics: biological protein materials, from nano to macro. Nanotechnology Sci. Appl. **3**, 127–148 (2010)
56. C.A. van Blitterswijk, D. Stamatialis, H. Unandhar, B. Papenburg, J. Rouwkema, R. Truckenmuller, A. van Apeldoorn, M. Wessling, J. de Boer, Materiomics: dealing with complexity in tissue engineering. Tissue Eng. A **14**(5), 796 (2008)

57. J. de Boer, H. Fernandes, L. Moroni, C. van Blitterswijk, Extracellular matrix and tissue engineering applications. J. Mater. Chem. **19**(31), 5474–5484 (2009)
58. D. Stamatialis, B.J. Papenburg, J. Liu, G.A. Higuera, A.M.C. Barradas, J. de Boer, C.A. van Blitterswijk, M. Wessling, Development and analysis of multi-layer scaffolds for tissue engineering. Biomaterials **30**(31), 6228–6239 (2009)
59. H.V. Unadkat, M. Hulsman, K. Cornelissen, B.J. Papenburg, R.K. Truckenmuller, G.F. Post, M. Uetz, M.J.T. Reinders, D. Stamatialis, C.A. van Blitterswijk, J. de Boer, An algorithm-based topographical biomaterials library to instruct cell fate. Proc. Natl. Acad. Sci. USA **108**(40), 16565–16570 (2011)
60. J. Hafner, C. Wolverton, G. Ceder, Toward computational materials design: the impact of density functional theory on materials research. Mater. Res. Soc. Bull. **31**(9), 659–668 (2006)
61. G. Ceder, B. Kang, Battery materials for ultrafast charging and discharging. Nature **458**(7235), 190–193 (2009)
62. F.J. Ulm, Y. Abousleiman, The nanogranular nature of shale. Acta Geotech. **1**(2), 77–88 (2006)
63. F.J. Ulm, The nanogranular nature of hydrated porous materials: concrete, shale and bone. Poro-Mechanics **IV**, 57–68 (2009)

# Chapter 2
# The Materiome

**Abstract** The goal of materiomics is the complete understanding of the materiome—a holistic characterization of a complex material system. The balance of form and function throughout Nature is well recognized, but the materiome must enhance a basic characterization of complex biological phenomena, to enable the prediction and design of new technologies. Analogous to genomics and other "-omic" fields, there is an obvious difference in scope between a gene or genetic sequence, and the human genome. Here, we establish the scope of the materiome beyond the assembly of material components (*e.g.*, architecture or structure), the fundamental difference between application and function, the concept of material behavior scaling, as well as the challenges (and benefits) imposed by material hierarchies and complexity. Material and structure are no longer distinct, and the assembly of building blocks ranges across all scales from the nano to the macro level.

*The structure of tissues and their functions are two aspects of the same thing. One cannot consider them separately. Each structural detail possesses its functional expression. It is through physiological aptitudes of their anatomical parts that the life of the higher animals is rendered possible... Tissues are endowed with potentialities far greater than those which are apparent.*

*Alexis Carrel, Science, Vol. 73, No. 1890, pp. 297–303 (1931)*

## 2.1 Introduction

The above quote indicates a fundamental principal of materials science central to materiomics: the inherent (and reciprocal) relation between a material's *structure* and material's *function*. Superficially, in many applications—both engineering and biological—one can be directly inferred from the other. For example, a steel cable supporting a suspension bridge is constructed to withstand tension, while the bones in our bodies are relatively stiff to provide our limbs and muscles with structural support. The problem arises, of course, when one wishes to optimize and develop (in other words, *engineer*) both structure and function simultaneously. This, as indicated by Chap. 1: Introduction, is the approach Nature astutely implements through *growth*.

S.W. Cranford, M.J. Buehler, *Biomateriomics*, Springer Series in Materials Science 165, DOI 10.1007/978-94-007-1611-7_2, © Springer Science+Business Media Dordrecht 2012

Growth is a complex process that can be influenced by the external conditions including temperature, mechanical loading, and supply of light, water or nutrition. As such, biological materials are not produced in the same manner as engineering materials—the process is highly adaptive and responsive. Moreover, critical factors may differ not only in time, but in scale—what a material system "sees" at the nanoscale may be quite different than at the macroscale. Organisms must necessarily possess the ability of adaptation to external needs, while possible external influences on a technical system must be typically anticipated in its design. Multiscale response, temporal changes, and the need for robustness (applying "survival of the fittest" principle to materials science) are the primary culprits resulting in complex biological materials and systems. Even if you think you have a complete physiological and mechanistic understanding, the system can evolve. This usually means back to the metaphorical drawing board! Clearly, a biological material differs from common static and benign materials such as aluminum or glass (as everyday examples). For this reason, complex materials (in particular complex biological materials such as tissues and cells) need a more comprehensive and holistic framework to completely understand (and exploit) physical, chemical, biological and mechanical characteristics and properties. Such a holistic framework is known as the *materiome*.

**Materiome**: A holistic characterization of a material system, consisting of the material constituents (elemental building blocks and/or structural units), the cross-scale structure-property-process relations, and the resulting functionalities/requirements across *all* levels of hierarchy, from nano to macro.

An integrated and holistic perspective of biological materials is particularly fascinating for the materials scientist—and has likewise been approached in the past with various motivations. For example, the classical book by D.W. Thompson *On Growth and Form* relates the "form" (or shape) of biological objects to their physiological role [1]—an early insight into the integration of structure and function. Thompson quotes:

> An organism is so complex a thing, and growth so complex a phenomenon, that for growth to be so uniform and constant in all the parts as to keep the whole shape unchanged would indeed be an unlikely and an unusual circumstance. Rates vary, proportions change, and the whole configuration alters accordingly.

A recurrent theme of *On Growth and Form* is that contemporary biologists (circa 1919) overemphasized evolution as the fundamental determinant of the form and structure of living organisms, and underemphasized the roles of physical laws and mechanics. Thompson advocated *structuralism* as an alternative to survival of the fittest in governing the form of species. Yet a purely mechanistic approach is also limited in scope, failing to encompass adaptive physiological and environmental aspects of biological systems. From an integrated perspective, evolution and structuralism go hand-in-hand—they are aspects of the same "development" process, contributing to the materiome.

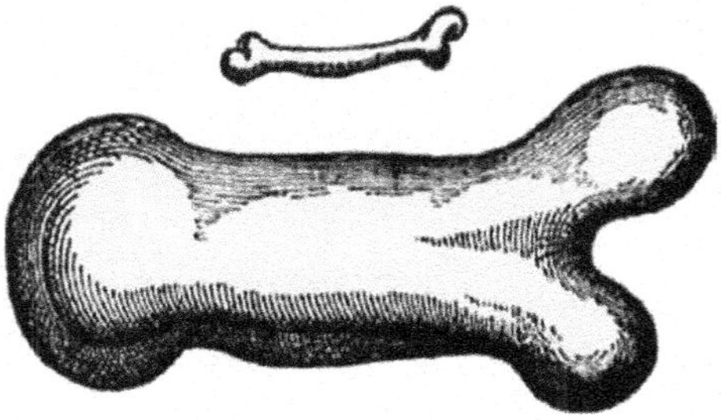

**Fig. 2.1** Galileo's illustration of the same bone (femur) of a human (*top*) and a giant (*bottom*). Whereas the length of the bones differ by about three times, the width of the bones differ ninefold. Typically, "Galileo's bones" are used as an example of scaling of strength and weight (the so-called cube-square law). Here, the bones represent a variation in structure with function—the material (bone) is constant, but the *materiome* varies from human to giant. From *Discorsi e Dimonstrazioni Matematiche intorno a Due Nuove Scienze*, 1638

Even earlier, the relationships between anatomy (*i.e.*, structure) and function of living systems had been explored by Leonardo da Vinci and Galileo Galilei. One of the first to apply fundamental physics to biological systems, Galileo is often considered the father of biomechanics (albeit the modern approach to biomechanics is largely credited to the pioneering work of Y.C. Fung). Among his many other discoveries, Galileo recognized that the shape of an animal's bones are to some extent adapted to its weight. Long bones of larger animals typically have a smaller aspect ratio (see Fig. 2.1).

Galileo's explanation is the basis for the simple cube-square law: the weight of an animal scales with the cube of its linear dimension, while the structural strength of its bones scales with the square. Hence, the aspect ratio of long bones has to decrease with the body weight of the animal. Indeed, for this reason, large animals do not look like small animals: an elephant cannot be mistaken for a mouse scaled-up in size. The bones of an elephant are necessarily proportionately much larger than the bones of a mouse, because they must carry proportionately higher weight. Because of this, the giant animals seen in movies (*e.g.*, Godzilla or King Kong) are unrealistic, as their sheer size would break their bones! While the material properties of bone are similar from animal to animal (the composition of human bones is similar to mouse bones is similar to elephant bones, for example), the structure adapts to function. In other words, the *materiome* changes.

Beyond material properties (*e.g.*, strength or mass), early insights further linked biological processes—such as walking, running, and the flight of birds—using a mechanistic framework (see Fig. 2.2). It was a radical paradigm that all living systems, including the human body, could be described as simple machines ruled by the same mechanical laws. More recently, the works of Steven Vogel [2–6] has introduced the field of comparative biomechanics—application of biomechanics to non-human organisms and reducing the biomechanical universe to a set of simple tubes, surfaces, flows, beams, and levers, all amenable to simple calculation and estimation. In simplest terms, quantifying how living things stack up against non-biological physical reality. Biomechanics is closely related to engineering, because

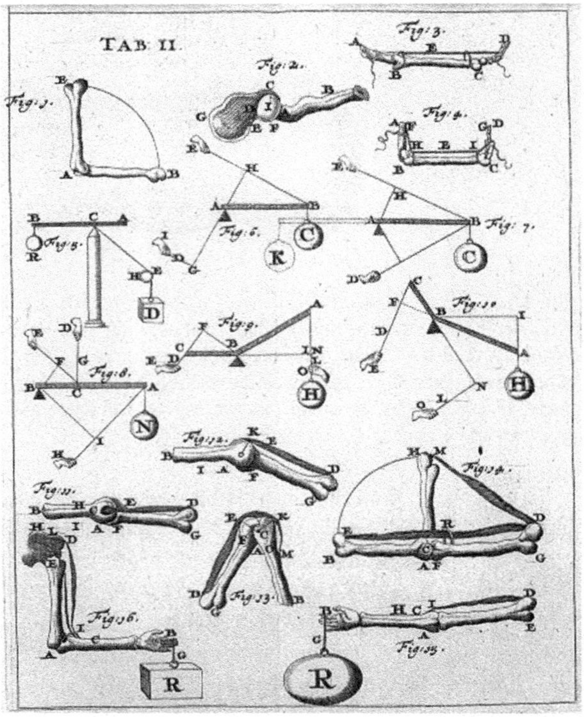


**Fig. 2.2** Biology provides more than just materials, but functional mechanical systems. Illustration is a page of one of the first works of biomechanics, *De Motu Animalium* by Giovanni Alfonso Borelli (1608–1679). Borelli studied walking, running, jumping, the flight of birds, the swimming of fish, and even the piston action of the heart within a mechanical framework. Influenced by the work of Galileo, he had an intuitive understanding of static equilibrium in various joints of the human body well before Newton published the laws of motion

it often uses physical engineering sciences to analyze biological systems. Currently, the study of biomechanics ranges from the inner workings of a cell to the movement and development of limbs, to the mechanical properties of soft tissue, and bones, and the transport of fluids and nutrients *via* cellular mechanisms and large-scale vasculature.

In a similar manner, the challenges of understanding the relationship between material system, function, and adaptation are the focus of materiomics. The goal is not only to understanding such complex biological phenomena, but to enable the prediction and design of new technologies. Materiomics is a unifying field, attempting to merge the efforts of biologists, material scientists, and engineers alike. Indeed, many of the overarching themes and principles have been investigated under different motivations. In this chapter, we introduce and discuss the fundamental concepts of materiomics, such as the *materiome* (depicted schematically in Fig. 2.3), as well as guiding principles and nuances, such as the fundamental difference between function and application, structure-property-process relations, the prevalence of hierarchies and complexity, and the concept of material behavior scaling.

## 2.2 Motivation and Scope

Materiomics is defined as the systematic study of the complete material system and the effect on the macroscopic function and failure in their mechanical context,

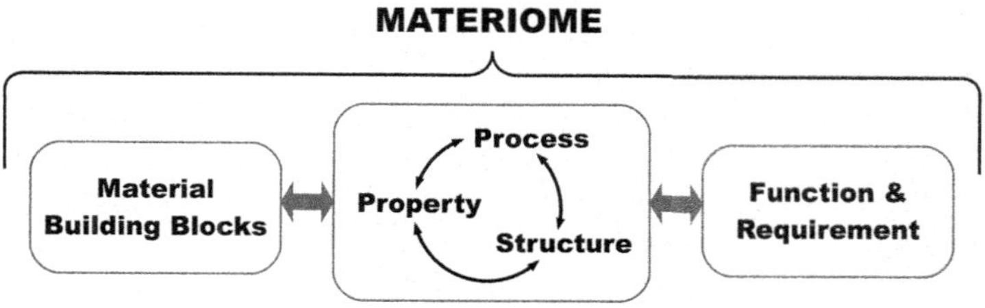


**Fig. 2.3** Schematic representation of materiomic information, consisting of the material constituents (elemental building blocks and/or structural units), the cross-scale structure-property-process relations (structural geometry, stress and strain transfer, failure mechanisms), and the resulting functionalities/requirements (strength, robustness, toughness, and other mechanical properties)

linking processes, structure and properties at multiple scales, from nano to macro, through a materials science perspective, integrating experimental, theoretical, and computational methods. The term has been coined in analogy to genomics—the study of an organism's entire genome—where, indeed, the suffix "omics" refers to "all constituents considered collectively". The development of genomics is a direct result of the *Human Genome Project* [7, 8]—an ongoing international scientific research project with a primary goal of to identify and map approximately three billion base pairs that form the chemical rungs in DNA's signature double-helical shape, from both a physical and functional standpoint. It is a gross oversimplification to state that the aim of the project is to simply map the nucleotides contained in a human genome in a sort of extensive database. Rather, the ultimate goal is to understand the human genome—detailing a genetic instruction set, finding the genetic roots of disease, and providing a scaffold for future work. A common idiom states that the "whole is greater than the sum of its parts"(originating from Aristotle[1]), and the Human Genome Project is the attempt to discover the "whole" of our genetic make-up, assigning meaning to:

...ACCGTAAATGGGCTGATCATGCTTAAACCCTGTGCATCCTACTG...,

beyond a seemingly random sequences of nucleobases (where A refers to adenine, C to cytosine, G to guanine, and T to thymine, the four base pairs of DNA). Moreover, the effort to create a comprehensive map of the human genetic sequence was more than just a breakthrough for geneticists. Genomics marked the launch of a new era of "-omic"-based research [9, 10]. The focus was shifted from individual parts within a system, to the system itself (a holistic approach). In the field of systems theory, this integrative view is sometimes referred to as *emergence* [11]—the way complex systems and patterns arise out of a multiplicity of relatively simple interactions. Biological systems consist of a large number of mutually interacting and interwoven parts, but complex and adaptive systems are not limited to biology. Examples

[1] "...the totality is not, as it were, a mere heap, but the whole is something besides the parts.", *Aristotle, Metaphysics*, Book H, 1045a:8–10.

of non-biological systems in which emergence plays a role include social networks, ancient and modern cultures, languages and writing systems, and economies and ecosystems. Recently, the "omic" approach has found a particularly fitting niche within the field of integrative biology—the study and research of biological systems from a biological science, engineering and physical science perspective. Specific research may involve genetics or environmental studies, yet the goal is always to solve a larger, biological problem—from the smallest, molecular studies to broader studies of the biosphere [12, 13].

It is within this realm that materiomics can make a natural tangent—borrowing from the data collecting and consolidation of genomics and proteomics (among others; see Table 2.1), with the system-level perspective of integrative biology. Whereas genomics delineates genetic sequences, materiomics categorizes structural components; proteomics encompasses protein expression, materiomics can elucidate a material's behavior and function.

Although inspired by genomics, the forerunner of all contemporary "omics disciplines", the recent explosion and adoption of many omics by researchers (see Greenbaum *et al.* [14] for example) gives rise to clarification of the intention of defining materiomics. Traditionally, "omics" is a general term for a broad discipline of science and engineering for analyzing the interactions of biological systems in particular. Such fields are typically characterized by general systems (such as genomics for genes or proteomics for proteins) or processes (for example interactomics for cellular interactions or mechanomics for stress transfer). Indeed, even the term "Omics" (we use capitalization to denote the field rather the suffix) itself can refer the encompassment of all such bioinformatics research fields to understand all the biological information processing phenomena. Table 2.1 presents some common "omics" with their associated focus and scope. However, without prudence, the value of a new "omic" could be viewed as self-serving and counterproductive.

While the intent of omics in general is the collection of knowledge and information *via* holistic understanding and integration, the introduction of too many subfields and specialties can promote separation and reductionism of systems and processes under investigations. This is not to devalue the subfields of bioinformatics, where system complexity warrants specialization (*e.g.*, characterizing DNA through genomics compared to RNA through ribonomics). Many such fields can be viewed as a hierarchical approach to genomic research (see Fig. 2.4 for an illustrative example of both genomic and materiomic "hierarchies"). To be meaningful beyond a label, new omics should be unifying rather than segregating. With this standpoint, materiomics is neither a subdiscipline of biomaterial engineering, materials science or mechanics, nor intended to be applied solely to biological systems. Unpresumptuously, materiomics is not *introducing* a new field of science, but rather *encapsulating* many fields under a common banner. Just as genomics has motivated research to elucidate biological processes ranging from molecular interactions to complete organisms, it is hoped that the field of materiomics will stimulate extensive research, establishing a hierarchical apex shared between many disciplines promoting integration and collaboration. The use of a materials science approach to studying biological materials may have broader impact beyond the areas of biological protein materials and biomimetic systems.

**Table 2.1** Some common "-omics" with corresponding focus and scope; brief definitions meant to provide illustrative descriptions only

| Omic | Focus | Scope |
|---|---|---|
| Omics | Analyzing the interactions of biological information in various "omes" | Applied research paradigm to produce knowledge *en masse* from networks of information *via* holistic principles and methods. |
| Genomics | An organisms' entire hereditary information; genome | Determination of entire DNA sequences of organisms, fine-scale genetic mapping including genes, regulatory and non-coding sequences. |
| Proteomics | Protein characterization; protein-coding regions of the genome; proteome | The entire complement of proteins produced by an organism or system, including protein structure, function, and expression. |
| Metabolomics | Metabolites and metabolic networks; metabolome | All the small molecules present in a cell in a specific physiological state; systematic study of the unique chemical fingerprints that specific cellular processes leave behind. |
| Transcriptomics | RNA transcripts produced by the genome at any one time; transcriptome | Examines the expression level of RNA in a given cell population, which vary with external environmental conditions, including mRNA, rRNA, tRNA, and non-coding RNA. |
| Interactomics | Interactions between all macromolecules in a cell; interactome | Analyses and characterization of gene-gene, protein-protein, or protein-ligand interactions; development of molecular interaction maps/networks. |
| Mechanomics | Mechanical systems and processes within an organism; mechanome | General role of force, stress transfer, mechanics, and molecular machinery in biology, encompassing biological motors, mechanical structures, and processes. |
| Materiomics | Material characterization through components, structure, and function; materiome | Analysis of material systems though constitutive components, hierarchical structure-property-process relations, cross-scale interactions, and effects on functionality. |

Indeed, within the biological sciences, the field of genomics has advanced our knowledge base through the successful sequencing of entire genomes. In recent years however, extensive efforts have been initiated to move beyond genomics, where fields such as systems biology provide explanation to mechanisms of how genes affect phenotypes and biological function. Here, materiomics refers to the general study of a material system's *materiome*—the integrated view of materials cross-scale interactions that collectively define the material's properties, function, and purpose.

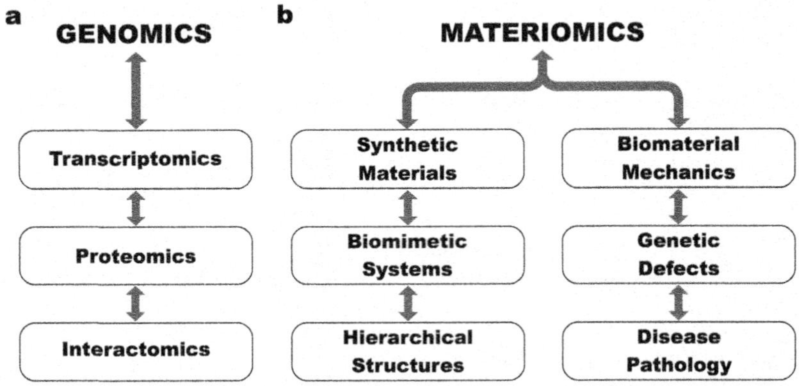


**Fig. 2.4** Example flow of information under genomics and materiomics frameworks. (**a**) Genomics encompasses the entire genetic sequence, which includes specific DNA sequences transcribed to RNA molecules (*transcriptomics*), in turn, mRNA from a DNA templates carry the coding information required for protein synthesis and expression (proteomics), finally, the mapping of protein-protein interactions networks can be characterized by interactomics. It is noted that this is merely one possible flow of information under genomics, with many interactions possible between subdisciplines. (**b**) Two potential *paths* are given for materiomics. First, of all classes of synthetic materials being developed, a subset may find inspiration from biological materials. From these bio-inspired or biomimetic materials, the motivation may arise from multiscale hierarchical structures, such as those found in spider silk, wood, or bone. Materiomics provides a potential framework for the development of such *de novo* materials. Second, there is an advancing knowledge base on the mechanical behavior and properties of biomaterials, both at the molecular and system levels (*e.g.*, cellular mechanics or soft tissue behavior). At the molecular level, genetic point defects (*i.e.*, mutations) can lead to mechanical changes expressed at the macro-scale. Such pathology can be quantified and analyzed, leading to new diagnostic and treatment methods for certain diseases. The diverse aims of biomimetic material design and disease pathology can be unified under a materiomic paradigm through the understanding of material systems and functionalities

## 2.3 Material Versus Materiome

Materiomics, as currently presented, is the study of a system's materiome, or the "complete" material system—its constituents and structure, properties and processes, function, failure, and behavior—in its entirety. The goal is to link the disparate nature of the physical description of a material (*i.e.*, components and structure) with the related phenomenological functionalities (*i.e.*, strength and robustness). The approach is partially motivated by macro-scale engineering techniques such as structural analysis. For example, it is a rather trivial analysis procedure to determine the flexural behavior of a steel joist girder if the geometry and material properties of the truss members are known (see Fig. 2.5). The behavior of an individual joist, in turn, affects the behavior of the system in which it is contained (such as a simple roofing system). If we consider the truss arrangement as the first hierarchy, it is apparent that the mechanical properties of the material used to build the truss, as well as the structure of the truss itself, ultimately affect the mechanical properties and failure of the system. However, at the macro-scale, by convention, there is a distinct differentiation between the "material" and the "structural system".

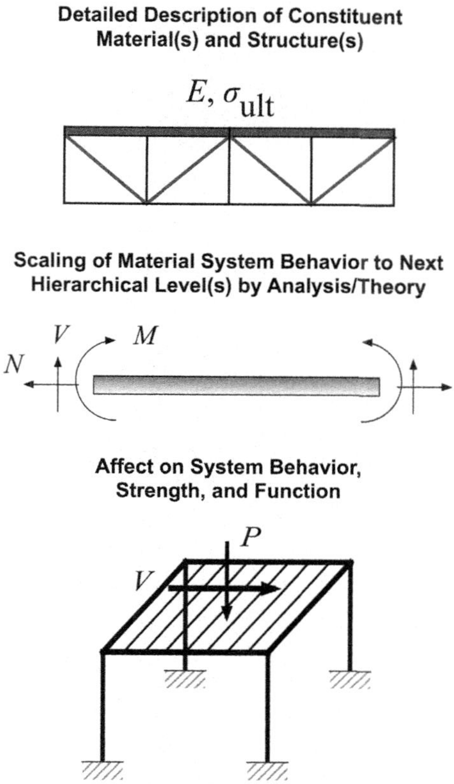


**Fig. 2.5** Schematic of the material to system relationship in a typical engineered structure. The material building block is chosen (in this case, steel joist girder with known member properties such as ultimate stress, Young's modulus, etc., and geometry). The girder is fundamentally a flexural beam element with known behavior and load response from analysis of girder members. The behavior of this scale level (*i.e.*, hierarchy) is predictable and designed accordingly. Finally, these theoretical beam elements are used in analysis of structural frame. The material properties are directly expressed at the system-level behavior. Biological materials, however, neither have an analogue for "theoretical beam elements", nor have consistent properties from scale to scale

Typically, one would not associate conventional material properties such as Young's modulus and Poisson's ratio to a roofing frame. Moreover, the relation between the "material" and the "structure" is well-defined and predictable—a building made of steel is inherently stronger than a building made of wood (perhaps not by weight, but that is another issue). At the nanoscale, however, the "material/structure" distinction is not as clear, nor is the predictability of system-level properties from known building blocks. Knowledge of one does not infer the other.

Consider the hierarchical nature of bone or tendon as an example (depicted in Fig. 2.6), for which there has been ample research at multiple hierarchical levels [15, 17], including the triple-helical polypeptide structure, the subsequent formation of collagen fibrils, and the ultimate macro-scale system of bone/tendon through both

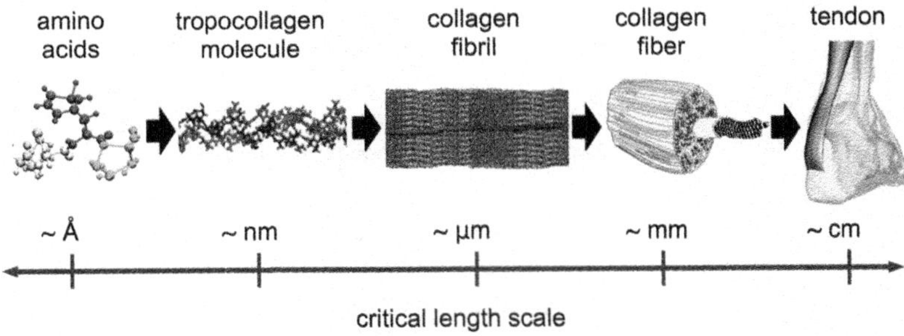


**Fig. 2.6** Hierarchical structure of collagen protein materials [15]. Like all protein-based materials, the fundamental building block is a sequence of amino acids (on the order of *Ångstrom*). Each collagen molecule is made of three peptide chains that form the ~300 nm long triple helical collagen molecule (*nanometer scale*). Collections of collagen molecules aggregate both in lateral and longitudinal directions to form fibrils (*micrometer scale*). Fibrils include tiny hydroxyapatite crystals in bone tissue, which provide stiffness and compressive load resistance. In tendons and ligaments, multiple fibrils make up collagen fiber (*millimeter scale*), formed with the aid of proteoglycans. Figure adapted from [16]

atomistic [16, 18], computational [19, 20] and experimental methods [21–23] (discussed further in Chap. 8: Unlocking Nature: Case Studies). Unlike a steel frame, however, the system-level (bone/tendon) properties are not reduced to the mechanical properties of the first hierarchical level (protein triple-helices). In fact, a literature review results in different Young's moduli defined for tropocollagen triple helices [16, 18], collagen fibrils [16, 24], and bone [25]. It is apparent that, as the scale of the material is decreased, the distinction between what exactly is labeled the "material" and the "structure" is simply a matter of perspective. Within the context of the materiome, the material and structural information is considered at *all* hierarchical levels. While a material description of bone can be defined by the macroscale properties such as Young's modulus or fracture strength, the complete materiome of bone necessarily includes the material information at all levels of hierarchy, the structure-property relations between hierarchies, and the associated functionalities of the system. In this context, Figure 2.3 only summarizes the information contained within *each* hierarchical level of the materiome—it is incomplete.

From a cursory perspective, the materiome may be considered merely a "multiscale snapshot" of a material system, that is, simply a catalogue of material properties and functions throughout different scales. What such a simple snapshot would lack, however, is the communication and cross-scale interactions that define the functionality of complex materiomic systems. In other words, the materiome provides not only the answer to what the material system is in terms of components, structure, and properties, but also to why the system is the way it is and how it is and/or can be manipulated. For example, knowledge of the spatial relationships and interactions of genes and regulatory elements in the cell nucleus are revealing an extensive network of communication within and between chromosomes [26]. Such interactions are, not surprisingly, inherently multiscale with nano-scale details exhibited throughout hierarchical levels [27]. A simple material description of the chromosome nucleotides and structure is unable to construe such information, as the local environment and material requirement effect gene expression. Such gene adaptations continuously change the material but maintain a constant materiome, a clear differentiation of the two concepts. A crucial unresolved issue is the extent to which this organization affects gene function, rather than just reflecting it. By unlocking the complete materiomic information, efforts have been made to utilize gene regulation in the self-assembly and organization of structural DNA materials [28–30], discussed further in Chap. 10: Synthesis and Design. Such applications are only possible through the integration of multi-scale feedback, chemical interactions, and structural-property relations, which are central to the field of materiomics.

Inspired by biological materials hierarchical structures, decentralized processes, material properties and environmental requirements, materiomics amalgamates the combined effects to mutual completion. In contrast to the traditional paradigm in materials science, relations between "external" functions/requirements and "internal" properties exist on several scales resulting in multi-functionality. Though, as requirements are consistently changing (*e.g.*, changing loads, changing environment) on several time and length scales, in addition to multi-functionality, robust feedback loops are required and enable decentralized self-organization and self-optimization (see Fig. 2.7).

The consideration of the complete materiome of a material system allows a fundamental bottom-up design of purpose-specific materials from the atomistic to the continuum levels. Granted, the understanding of the materiome is still at its infancy, where the role of the relationship between processes, structures and properties of materials in biological organisms is thus far only partially explored and understood. Approaches in studying the materiome include multi-scale experiments (*e.g.*, AFM, optical tweezers, etc., discussed further in Chap. 5: Experimental Approaches), multi-scale simulation methods (*e.g.*, molecular dynamics, finite element analysis, see Chap. 6: Computational Approaches and Simulation), as well as high-throughput methods based on combination of these techniques. The objective is to ultimately bridge hierarchical levels and piece together not only material properties and structures at the nano- and micro-scales, but the ultimate effects on both the mechanical properties and function of the entire material system. A complete understanding of the materiome elucidates not only the cross-scale relations between hierarchies and mechanical properties, but also offers clues how to assemble new materials with disparate and mechanical properties from few constituent building blocks, and to identify novel approaches in designing materials that evolve autonomously to adapt to changes in environmental conditions.

## 2.4 Functionality Through Architecture

As will be discussed in subsequent chapters, biological materials have the capacity to integrate multiple, and at times, disparate properties, unaccounted for by a simple

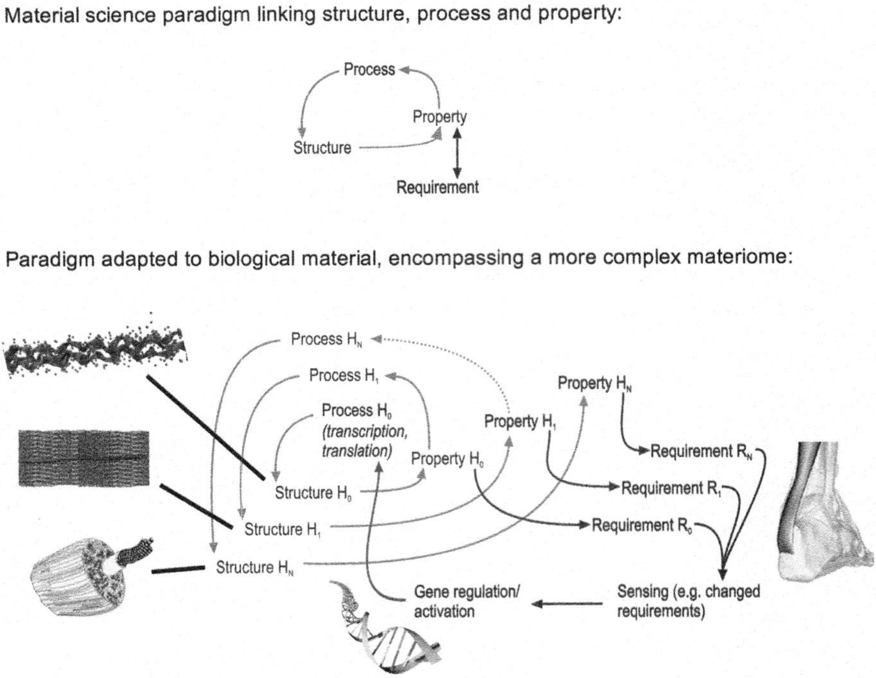


**Fig. 2.7** Alternative schematic representation of materiomic information across scales, for a complex biological material such as tendon. Biological materials integrate hierarchical structures, decentralized processes, material properties and environmental requirements. A materials science paradigm applied to the hierarchical structure of protein materials is necessarily more complex (here, $H_i$ refers to hierarchy levels $i = 0 \ldots N$, and $R_i$ refers to material property requirements at hierarchy levels $i = 0 \ldots N$). The cycle initiates at $H_0$ (process $H_0$, the only level at which protein expression occurs) to form protein constituents (structure $H_0$). Their properties (property $H_0$) control the association at the next hierarchical level (process $H_1$ leading to structure $H_1$). This cycle continues through all hierarchical levels $i = 0 \ldots N$, where process and structure $H_1$ and beyond denote protein assembly stages. At each stage, the properties of the structure control the assembly at the next level. Overall, properties at different hierarchical levels (properties $H_i$) are regulated by corresponding physiologic demands (requirements $R_i$), which are sensed and transduced intracellularly to activate genetic regulation, resulting in changes to process $H_0$. Adapted from [31]

combination of material properties or "*rules of mixture*". The materiomic function is frequently a function of the system architecture and geometry (often across scales, or *hierarchical*). The addition of hierarchies is fruitless without consideration of hierarchical function—the structure at each hierarchical level is associated with a specific property and essentially compartmentalized at that scale. The goal is the definition and systematic characterization of structure-property-process (SPP) relation (*i.e.*, how the manipulation of one influences the other, and how to derive functionality intended combinations rather than *ad hoc* permutations). The ultimate functionality of the *materiome* is differentiated from that of the constituent *material* by the SPP relations.

**Structure-Property-Process (SPP) Relations**: The interplay and underlying correlation between a material system's *structure* (geometry and material components), resulting *properties* (stiffness, strength, stability, etc.), and mechanistic *processes* (including fabrication, synthesis, stress transfer, deformation, and eventual failure). The ultimate functionality of the *materiome* is differentiated from that of the constituent *material* by the SPP relations.

To exemplify, there has been much study on the structure-property relationships of numerous mollusk shell nacreous microstructures [32–34], in which it is generally concluded the amplification of mechanical properties (increase in strength and toughness) exhibited by biological composites compared to their individual constituent materials is beyond simple rule of mixture formulations.

The remarkable performance of biological materials is a consequence of their hierarchical structure, the specific design at each level of organization (such as the skeleton of a sea sponge, discussed in Chap. 1: Introduction). Another example is found in the cuticle of large crustaceans such as the lobster *Homarus americanus* [35], a chitin-based material containing proteins, mineral nanoparticles (mostly amorphous calcium carbonate with some crystalline calcite) and water. The hierarchical organization of the cuticle illustrates several generic design principles of crustacean exoskeletons (see Fig. 2.8). Synthetic structural materials that take advantage of the hierarchical structure-property relationships of such composite systems are increasingly being realized [36, 37].

Another recent study of a deep-sea hydrothermal vent gastropod [38] demonstrates the material properties of the multi-layered shell are dependent on the specific combination of different materials (building blocks), the microstructures, interfacial geometries, gradation and layering (structure-property-process), which are advantageous for penetration resistance, energy dissipation, mitigation of fracture and crack arrest, reduction of deflections, and resistance to bending and tensile loads (function and requirement). Such investigations help elucidate the gastropod shell materiome, integrating the effect of contrasting material components (compliant organic layers and stiff mineralized platelets) with the mechanical properties of the entire material system. Indeed, each layer of the shell is responsible for distinct and multifunctional roles in mechanical protection (see Fig. 2.9).

We can compare the "compartmentalized function" of a gastropod's shell with a more common macro-scale system, a reinforced concrete flexural member (depicted schematically in Fig. 2.10). Here, we consider the materiome of the composite concrete and steel system. We no longer consider the beam a heterogeneous material system composed of concrete and steel, but rather the beam holistically as a functional system. Indeed, as discussed in Chap. 1: Introduction, the concrete itself can be thought of as a complex hierarchical system, and micromechanical models of steel elucidate complex phenomena in steel such as dislocations and grain boundaries—yet typically both materials are treated with bulk macroscale properties. Through engineering of the cross-section, the structure and specific placement

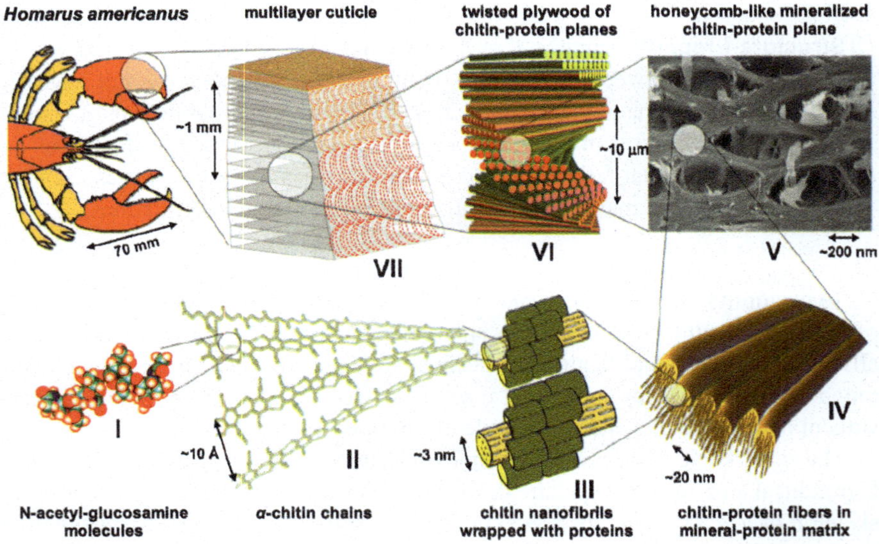


**Fig. 2.8** Hierarchical structure of the lobster cuticle: **I**) N-acetyl-glucosamine molecules, **II**) antiparallel chains of $\alpha$-chitin, **III**) chitin-protein nanofibrils, **IV**) chitin-protein fibers in a mineral-protein matrix accommodating and stiffening the chitin network (not shown), **V**) cuticle with pore canal system (the well-developed, honeycomb-like system of pore canals facilitates biomineralization by water-mediated transport of calcium and other ions across the cuticle), **VI**) chitin-protein fibers organized in a twisted plywood structure for reinforcement, and **VII**) three-layered cuticle at the macroscopic level. From Raabe *et al.* [35], used with permission, copyright © 2010 WILEY-VCH Verlag

of the material components serve a distinct mechanical role. The concrete serves as the primary load-bearing medium while protecting the steel elements from corrosion. The longitudinal steel reinforcement is designed to carry tensile stress, while the concrete carries compressive stress. Indeed, the amount of steel ultimately dictates the failure mechanism of the beam (brittle or ductile failure). Finally, the transverse steel stirrups increase the shear capacity of the member while limiting crack propagation. An integrated view of the flexural member's materiome provides complete view of the materials (concrete and steel), the structure-property-process relations (cross-sectional geometry, stress distributions, failure mechanisms), and the ultimate function (flexure). Of course, a concrete cross-section is a rather trivial example, not requiring a materiomics perspective. However, it illustrates the subtle interplay between structure-property-process relations and function. For example, based on the geometry of the cross-section alone (e.g., placement of the steel reinforcement), one could deduce the function of the member—as a beam subject to a positive bending moment. Alternate functions (such as a cantilever member, or as a column) would require variations in the structure, but could (potentially) implement the same material components. The structure—not the material—determines the function.

Functionally is ultimately obtained though material architecture. In the case of the concrete beam, an engineer, through analysis of structural requirements,

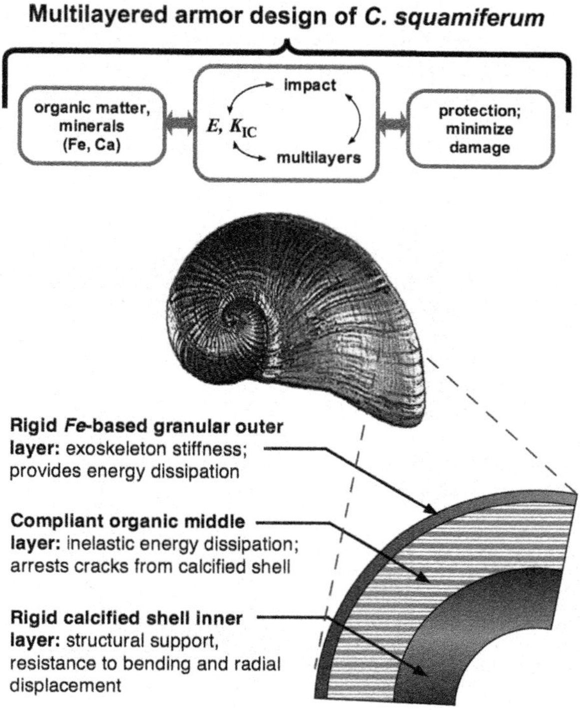


**Fig. 2.9** Schematic representation of the materiome of a multilayered armor of *C. squamiferum*; adapted from Yao *et al.* [38]. Each material layer (structural component) serves distinct functional roles, contributing to the overall mechanical behavior of the shell. Variation of geometric parameters such as layer order of thickness would change the shell performance

determined macro-scale structure. However, the architecture of the gastropod shell, adequately summarized at the micro-scale as a composite of organic layers and mineralized platelets, explicates only the highest level of hierarchy. Hierarchical sub-levels including the protein-based composure of the organic layer(s) [39, 40] and the properties of the organic-mineral interfaces [41] are still being investigated in nacreous materials, and are ultimately required for a complete description of the materiome and potential design of *de novo* synthetic materials [42].

## 2.5 Function Versus Application

It is apparent that the concept of materiomics is a subtle shift in the way we think about material systems. Materiomics attempts to ask more than simply what are the material components and properties of a system, but rather ask *why*. It shares a common goal with materials science in the fundamental understanding of materials, their structure and characterization, but extends breadth to multi-scale, multi-component attributes of the materiome. The overlap in concept between a material and materiome is discerning, and can easily muddle the distinction between the two. As such, a didactic example can be useful.

To facilitate the discussion, we consider the typical components of the materiome: (1) material components (building blocks); (2) structure-property-process

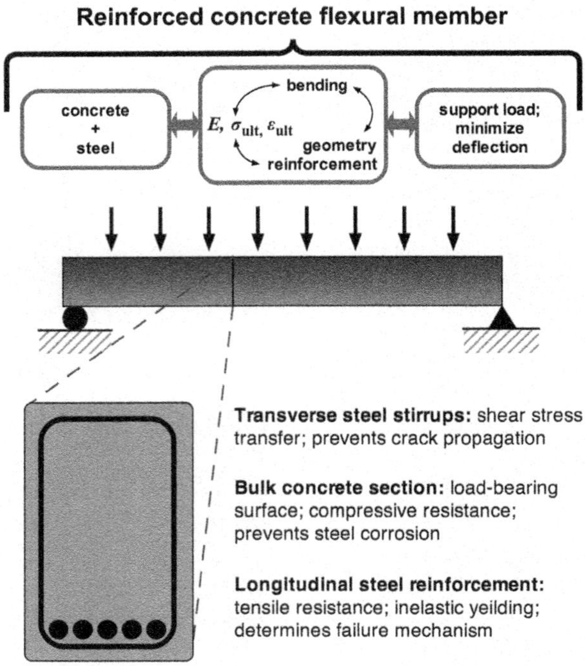


**Fig. 2.10** Schematic representation of the materiome of a reinforced concrete flexural member as used in civil engineering. Each structural component is utilized for distinct purpose, the combination of which determines the properties and ultimate function of the flexural member. For example, relocation of the longitudinal steel reinforcement to the upper region of the cross-section is more suitable for cantilever beams—the same components can be assembled to serve difference functions

characteristics, and; (3) the function/requirement (depicted in general in Fig. 2.3; examples in Fig. 2.9 and 2.10). First we consider a typical material taught to many undergraduate students: copper. The critical question we ask is: *Do we need a materiomics approach to investigate bulk copper*?

We can certainly define the material components simply as copper atoms. We can even expand this definition to include the electron density attributed to each atom (which is critical to the electrical properties of copper). The structure is known to be "face-centered-cubic", of "FCC" for short, thereby describing the repeating lattice structure. From this structure and properties of the constituent atoms, we can theoretically determine material properties such as Young's modulus and shear modulus. From this we know that copper is ductile and malleable, and can investigate the onset of yielding, plasticity, and dislocations and the ultimate effects on material behavior and properties. At this point, copper has a well-defined materiome, however, copper, in itself, has no function or requirement. Copper, albeit, has many applications, including copper wires, piping, circuits, etc. that take advantage of the properties of copper, but these applications are not beget by the material itself, but rather from the ingenuity and foresight of a material scientist or engineer—copper is *chosen* as a component because its material *characteristics match the needs* (schematic of bulk material to application for copper depicted in Fig. 2.11). It has neither evolved nor been grown for function in the natural environment. It seems that a materiomics

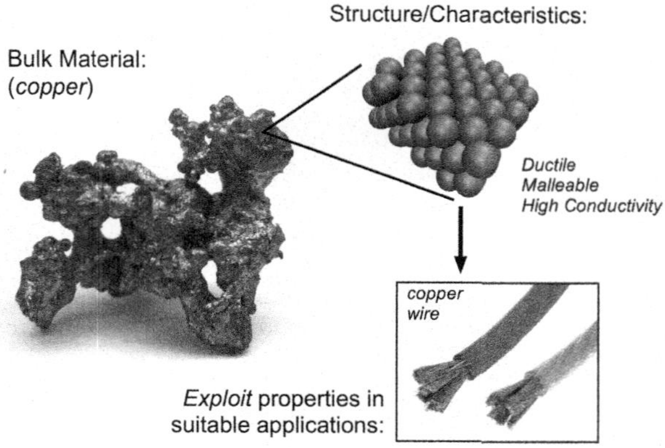


**Fig. 2.11** Schematic of typical (non-materiomic) approach in determining material applications. A bulk material (in this case copper) is investigated to determine structural features, chemical make-up, and associated characteristics (mechanical and electrical properties, for example). From this checklist of features, suitable applications can be found that exploit the known properties—the malleability and high conductivity of copper, for instance, make it an ideal candidate for electrical wiring. The understanding we gain from copper allows us to look for other similar materials for similar applications (such as using gold, which has a similar crystalline structure), but copper—unlike biological materials—has no intrinsic *function* for which it has evolved or has been grown

approach—aside from offering a simple example—is superfluous in the description of copper.[2]

We can also consider a more "recent" material, such as carbon nanotubes [46]. Carbon nanotubes (CNTs) are among the most widely studied nanomaterials, with many potential applications that take advantage of their unique mechanical, electrical, thermal, and optical properties. There are many concurrent investigations involving carbon nanotubes, ranging from experimental synthesis to atomistic and continuum modeling with focus on a variety of properties, behaviors, and applications [47]. The superior mechanical properties of carbon nanotubes are appealing for their potential use in novel nanomaterials. For instance, the Young's modulus of a single-walled nanotube approaches a terapascal ($10^{12}$ Pa) [48], with an ultimate stress approaching 60 GPa [49] implying that CNTs are one of the strongest known synthesized materials in terms of elastic stiffness and ultimate tensile strength [50]. Since their discovery over a decade ago [46], carbon nanotubes have been continuously studied, modeled, probed, and investigated for their potential applications in

[2]That is not to say, however, that we cannot improve on the properties of current widely used materials such as copper. Recent approaches have successfully enhanced the yield strength and ductility copper nanowires and films through a process of nanotwinning [43–45], exploiting the behavior of grain boundaries at the molecular scale. Improvements are possible, but such improvements can only enhance the intrinsic properties—new applications for copper may indeed emerge in electronics, biochips, NEMS, and many others, but if and only if the "new" enhanced properties satisfy the requirements of the chosen application.

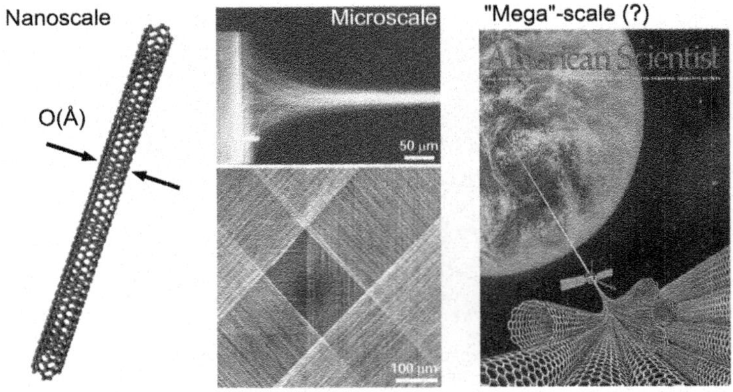


**Fig. 2.12** Why can we not exploit the exemplary properties of carbon nanotubes for macroscale applications? At the nanoscale, carbon nanotubes are one of the strongest materials ever tested in terms of stiffness and ultimate tensile strength (*left*: molecular model of a (5, 5) single-walled carbon nanotube; length of 10 nm). Yet, when we combine multiple nanotubes together, either in sheets or in bundles, the intrinsic strength is lost (*center*: SEM micrographs showing the co-operative rotation of multi-walled nanotubes in a forest to form a sheet and a two-dimensionally reinforced structure fabricated by overlaying four nanotube sheets with shifts in orientation; printed with permission from American Association for the Advancement of Science, *Science*, [51] © 2005). Currently, the use of such high-strength carbon nanotube cables as a mega-scale "space elevator" from the Earth to a geostationary orbital tether is still the realm of science fiction (*right*: cover of American Scientists depicting an artistic interpretation of a carbon nanotube-based space elevator)

nanotechnological engineering. *Surely, the carbon nanotube is an ideal candidate for a materiomic description*?

Defining the material component is trivial: by definition, the building blocks carbon nanotubes are carbon atoms, specifically an allotrope of carbon consisting of covalent $sp^2$ bonds formed between the individual carbon atoms (similar to graphite). The structure is that of a cylinder, which can be defined by the length, diameter, number of nanotubes (*i.e.*, single-, double-, or multi-walled), and chirality (essentially the orientation of the rolled carbon structure). Further, there are also many known structure-property relations for carbon nanotubes, such as the buckling behavior of single- and multi-walled nanotubes [52, 53], the relation of chirality (*e.g.*, armchair or zigzag) to electrical conductivity [54, 55], or the effects of defects on thermal conductivity [56]. Indeed, one could easily claim that carbon nanotubes are perhaps the most thoroughly studied and investigated nanomaterial in recent years.

Amongst non-biological approaches, carbon structures such as carbon nanotubes offer exciting possibilities. Many potential applications have been proposed for carbon nanotubes, including conductive and high-strength composites; energy storage and energy conversion devices; sensors; field emission displays and radiation sources; hydrogen storage media; and nanometer-sized semiconductor devices, probes, and interconnects [47]. The pending question, or course, is what are benefits of a materiomics framework if it does not contribute to the understanding of a material such as carbon nanotubes? Why should we bother with defining the materiome at all?

With all the interesting characteristics and extensive knowledge base developed for carbon nanotubes, they are not (currently) a prominent engineering material—a number of difficulties are associated with the large-scale implementation of such nanotubes. Indeed, with all the potential functions based on their unique properties, a practical *application* remains elusive (see Fig. 2.12). It seems carbon nanotubes, with all their potential, are still awaiting a purpose. Again, as with copper, the materiome is lacking intrinsic function.

While individual carbon nanotubes are like minute bits of string, and many trillions of these invisible strings must be assembled to make useful macroscopic articles. The question arises, to what extent are the extraordinary properties of CNTs scalable? One of the more exotic proposals for carbon nanotubes is the construction of a space elevator—a non-rocket spacelaunch structure which acts essentially as a tethered guideline from Earth to orbit. If the cable is long enough, centrifugal forces exceed gravitational forces and the cable will work under tension. The cable would represent the largest engineering structure, hierarchically designed from the nanoscale (single nanotube with length of the order of a hundred nanometers) to the megascale (space elevator cable with length of the order of a thousands of kilometers). A space elevator would revolutionize the methodology for carrying payloads into space at low cost, but its design is very challenging. The most critical component in the space elevator design is undoubtedly the cable [57], which requires a material with very high strength and low density—properties possessed by carbon nanotubes. While relegated to mere science fiction for over a hundred years, researchers are recently claiming that the feasibility of space elevator cable is now realistic, thanks to carbon nanotube technology, proposing its realization within a decade [58].

If we consider a space elevator cable with constant cross-section, the maximum stress reached at the geosynchronous orbit has been approximate to be on the order of is 60 to 70 GPa (if low carbon density is assumed for the cable). It is only recently, after the discovery of carbon nanotubes, that such large failure stresses have been measured experimentally, during tensile tests on ropes composed of single-walled [59] or multiwalled [60] carbon nanotubes, both of which were expected to have an ideal strength of $>100$ GPa, compared to steel and Kevlar at under $\sim 5$ GPa.

However, the current view of basing the design of the megacable on the theoretical strength of a single carbon nanotube has also been considered naïve [61, 62]. Indeed, the role of thermodynamically unavoidable atomistic defects (effecting fracture, fatigue and elasticity) is detrimental for scaling of nanotubes and nanotube bundles. In particular, the first *ab initio* statistical prediction for megacable strength suggest that a megacable of carbon nanotubes would have a strength lower than 45 GPa [62], and thus not strong enough for (current) designs of a space elevator. This result may not be surprising—inevitable defects in the nanotubes mean that nanoscale strength can never be realized. *Can we expect such scaling behavior in all materials*?

If we shift our focus to Nature, we find a "natural" tension cables in the form of spider silk (specifically, the dragline silk implemented as structural components

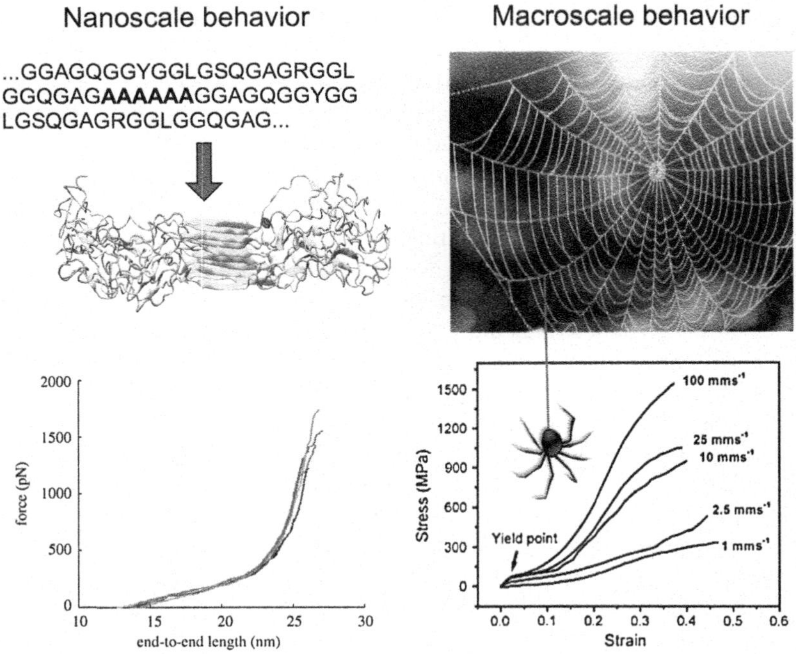


**Fig. 2.13** Comparison of computational characterization of the molecular structure of dragline spider silk with experimental measurements of silk threads. Atomistic model constructed based on the known protein sequence, and the force-displacement behavior of the macromolecular structure can be investigated *via* simulation (plot from [64]). Converting the force-displacement behavior to stress and strain results in material behavior that closely resembles the results of macroscale experiments on silk threads, shown here for various spinning speed (plot from Du *et al.* [65], used with permission, Elsevier © 2006). Unlike ordered and synthetic nanotubes, the macroscale strength observed in silk threads is almost equivalent to the nanoscale strength observed in the protein structure (both on the order of ~1 GPa) (see stress-strain graph in Fig. 8.8)!

in spider webs [63]). Spider silk will be discussed in more detail in Chap. 8: Unlocking Nature: Case Studies, but here we mention it for one remarkable fact: the macroscale strength observed in silk threads is almost equivalent to the nanoscale strength observed in the protein structure (both on the order of ~1 GPa). Moreover, silk it is not even an ordered structure, like carbon nanotubes or crystalline metals, yet achieves strength by local, nanoscale crystalline regions of aligned $\beta$-strand protein segments (see Fig. 2.13).

While webs are obviously not on the scale of a space elevator, they are significantly large structures to the perspective of a spider, and yet the spider exploits the strength of silk at the molecular level in web construction. Is spider silk defect free? This is highly unlikely. Then how does Nature solve the scaling problem? Perhaps one key is the disorder of silk—it has been known for decades that crystalline materials are sensitive to defects, and that they show a clear drop in strength with increasing size. Crystalline structures typically lead to stress concentrations and brittle fracture induced by the symmetric surfaces and slip planes, features avoided by more amorphous arrangements of materials, exemplified by silk.

While silk is characterized by the combination of disordered, semi-amorphous regions peppered with small sections of crystalline $\beta$-strands (similar to cross-linked polymers), silk is also composed of atomistic to macro-scale hierarchies, in a complex arrangement still being investigated. The hierarchies elicit flaw-tolerant behavior, inherently optimizing the performance of silk. The structure and robust properties of silk are attained by relatively few chemical constituents (amino acids), with interactions much weaker than those found in metals and nanotubes (*i.e.*, hydrogen bonding), yet silk accomplishes a feat we have yet to engineering—expressing nanoscale strength at macroscale functionality.

What can we learn from the complex materiome of silk? What are the underlying principles for flaw tolerance? Can we apply the understanding of silk to the use of carbon nanotubes in a space elevator? Perhaps the introduction of disorder, making the strength of individual nanotubes less crucial, and the introduction more hierarchies for a megacable? Quoting N. Pugno: "*... is the space elevator out of order? Our opinion is: at present, yes; but never say never*" [62]. Perhaps the key lies in Nature, and the understanding the materiome of silk, and what it may share with other material systems.

What kind of properties does materiomics discover? Unfortunately, the nomenclature and prospective metrics of materiomics are still being developed. Definitions and measures for material structure-property relations, hierarchical effects, links between functionality and requirement, and others, are crucial. Cooperativity, hierarchical degrees of freedom, multiscale interactions, flaw tolerance, and other concepts have qualitative meanings, but are more difficult to quantify and compare than traditional material properties such as mechanical stiffness of thermal conductance. Appropriate terminology for cross-scale relations such as scale separation, integration and interaction must be defined, encompassing all material components and effect on system properties and behavior. The field of materiomics will stimulate extensive research in these directions, beyond the investigation of one-off material applications, to the cohesive extension and design of materiomic function from nano to macro, and develop a distinct lexicon.

## 2.6 Material Behavior Scaling: Multiscale Perspectives

The previous section illustrated a critical problem in material selection: just because a material has exceptional properties at one scale (such as carbon nanotubes; Fig. 2.12), it does not mean we can exploit those properties at larger scales (yet accomplished by silk; Fig. 2.13). Perhaps we can use silk as a component of our space elevator? Even if we could train millions of spiders to spin our space elevator, there is another problem—silk, while extremely strong in terms of ultimate stress, is also extremely extensible (*i.e.*, stretchy). Steel, for example, has a yield strain of about 0.5 % whereas dragline silk can sustain strains of over 30 %. While the combination

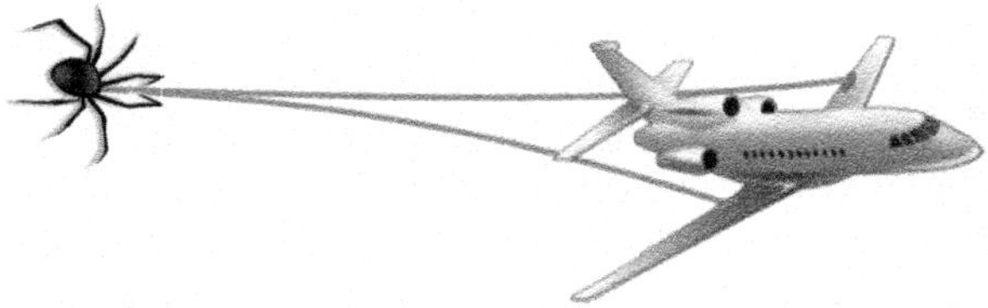

**Fig. 2.14** Can we use spider silk to stop a plane in flight? Mechanically, the answer is yes. A one centimeter thick thread of silk can stop a plane traveling ~200 mph. Unfortunately, this is impractical for two reasons: (1) the thread would have to be over thirty kilometers long and (2) due to the extensibility of silk, the plane would need another nine kilometers to come to a complete stop. The full strength of silk is only realized at such high extension. It seems the properties of natural silk are not suitable for large scale applications. Can we improve over Nature's design?

of strength and extensibility enhances the toughness of silk (a good property if you want to catch flies in your web), practical problems arise if you try and envision macroscale structures made from silk. A popular statement about the strength of spider silk is:

> A pencil thick spider's silk thread is capable of stopping a Boeing-747 in full flight.

While mechanically true (Fig. 2.14), it would require a silk thread approximately thirty kilometers in length, which would stretch another nine kilometers before bring the plane traveling at approximately 200 mph to a stop [66]. The extensibility of silk is disadvantageous as the application gets larger—for example, a suspension bridge supported by silk would greatly sag in the center (where load is the greatest). If we assume Spiderman swings over Manhattan using natural silk, he would hit the streets more often due to the extreme stretchiness (of course, the radioactive spider that bit Peter Parker may spin silks with different properties).

A pertinent question immediately arises: How does the function of silk relate to its material properties? This question is currently being investigated by biologists and materials scientists alike to unlock the "blueprint" (or materiome) of silk across multiple scales. Perhaps in the future we *can* design bridges with silk-like cables or robust and tough silk-like textiles and armors. Can we likewise integrate carbon nanotubes in silk-like configurations? The key lies in the material scaling behavior of the constituent "building blocks", and how those blocks are assembled. This requires a multiscale perspective from the molecules up.

Even limited knowledge of the nanoscale structure combined with molecular (or atomistic) interactions or a material can potentially shed great insight into the mechanical properties and potential material applications (or functions) of a material. Again, we can examine a simple crystalline material (such as copper) as an illustrative example where this approach works extremely well. Due to the known homogeneous repeating crystalline structure, combined with a rudimentary (in terms of potential parameters) atomistic interactions, relatively accurate approximations of macroscale mechanical and bulk elastic properties can be deduced from a nanoscale perspective (see Fig. 2.15).

Such a "molecular mechanics" approach is encompassed by such techniques as the Cauchy-Born rule (in which atomistic potentials are equated with continuum

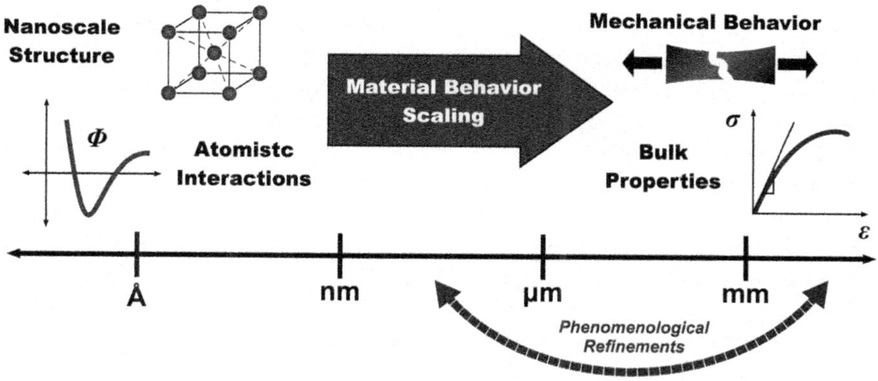


**Fig. 2.15** Schematic of material behavior scaling, wherein known behavior at a fine-scale (such as atomistic) is adequate to describe large-scale behavior and properties. Here, we depict the scaling of a crystalline material (such as FCC based polycrystalline copper) wherein the atomistic interactions can be used to describe mechanical properties such as bulk modulus and fracture toughness through relatively simple approximations (*e.g.*, Cauchy-Born rule). Refinement is sometimes necessary to account for phenomenological effects, such as the presence of grain boundaries, inclusions or other defects

theories) as well as atomistic-scaled finite element formulations. In essence, it is assumed *a priori* that the behavior and properties between atoms can be easily (and suitably) extended to the micro- and macro-scales. The Cauchy-Born rule relates microscopic deformation fields with macroscopic stresses and strains [67, 68]. The central assumption of the Cauchy-Born rule is to express the energy of an atomic system as a function of an applied strain tensor, $\underline{\underline{\varepsilon}}$, and it is assumed such continuum fields can be mapped to the atomic scale (see Chap. 7: Mechanical Characterization in Molecular Simulation for further discussion). An atomic unit cell is selected as a representative volumetric element (RVE) to formulate the strain energy density, $\phi(\underline{\underline{\varepsilon}})$. In crystal lattices with pair-wise interactions, the strain energy density can be formulated as a some over all the bonds a unit cell, or:

$$\phi(\underline{\underline{\varepsilon}}) = \frac{1}{\Omega}\sum \phi_i(r_i) \tag{2.1}$$

The interatomic potential, $\phi_i(r_i)$ to represent crystalline metals, for example, can be the relatively simple Morse potential [69, 70], which is defined by three parameters, D, $\alpha$, and $r_0$, or:

$$\phi(r) = D\left(1 - e^{-\alpha(r-r_0)}\right) \tag{2.2}$$

From the simple assumption of the Cauchy-Born rule, we invoke the atomic potential and structural information, and can directly derive material constants such as the bulk modulus, shear modulus, elastic modulus and Poisson ratio. For example, the bulk modulus can be determined as:

$$K = V_0 \left.\frac{d^2\phi}{dV^2}\right|_{r=r_0} \tag{2.3}$$

With some further refinement, such scale-bridging extension has proven successful for many systems, especially crystalline and metallic systems with homogeneous repeating (*e.g.*, polycrystalline) structures (common materials subjected to micromechanical modeling; see Fig. 2.15). It would be trite to label such complex systems as "simple" and "completely described by the atomistic potentials" as there exists a multitude of complex phenomenon that require refinements and nontrivial model formulation for accurate predictions of such models including (but certainly not limited to) dislocations, grain boundary effects, defects and voids, composite material behavior, viscoelastic effects, or thermal transitions (*i.e.*, brittle to ductile behavior).

The point, of course, is that knowledge of the nanoscale structure and interactions of copper atoms versus carbon atoms, for example, gives some indication (subject to refinement) of macroscale properties. One may infer, without any physical specimens to test, that carbon, with a more robust crystalline structure and stronger interatomic bonds, results in the stronger, stiffer material (*i.e.*, diamond) than bulk copper. As you may have guessed by now, such is not the case for biological protein materials. Hierarchical multiscale structures, seemingly disordered ordered arrangements (similar to polymer blends), complicated convoluted structures (*i.e.*, the quaternary structures of folded proteins), preferred interactions (such as protein "lock-and-key" mechanisms), and environmental influences (including solvent pH, temperature, etc.) make any kind of material property extrapolation from the nanoscale to the macroscale naive. On the one hand, just because we can investigate how actin[3] behaves at the molecular level [71, 72] does not mean we can deduce the behavior of the cytoskeleton [73] (see Fig. 2.16). Be that as it may, even limited understanding of keratin[4]—specifically the disulfide bond bridges at the molecular level—resulted in the popular "perm" hairstyle, which uses heat and chemicals to break the disulfide links in hair. It seems some molecular characteristics *are* expressed at the macroscale, for some protein-based materials. How do we know what we can predict and control at the macroscale?

Again, whereas there was an explicit dichotomy between materials with complex hierarchical structures (such as silk, collagen (bone or tendon), actin (cytoskeleton), keratin (hair), etc.) and homogeneous crystalline structures (such as copper or diamond) we also note a discrepancy in the scaling of the properties between material systems. To overcome this scaling issue, smaller increments of scaling must be introduced to bridge hierarchical levels. Thus, we introduce the concept of *material behavior scaling* (MBS) to our materiomics framework.

[3] Actin is a type of globular protein found in all eukaryotic cells in species as diverse as algae and humans, and is one of the three major components of the cytoskeleton. Actin participates in many important cellular processes including muscle contraction, cell motility, cell division and cytokinesis, vesicle and organelle movement, cell signaling, and the establishment and maintenance of cell junctions and cell shape.

[4] Keratin refers to a family of fibrous structural proteins, the key structural material making up the outer layer of human skin and the key structural component of hair and nails. It is part of the family of intermediate filament proteins.

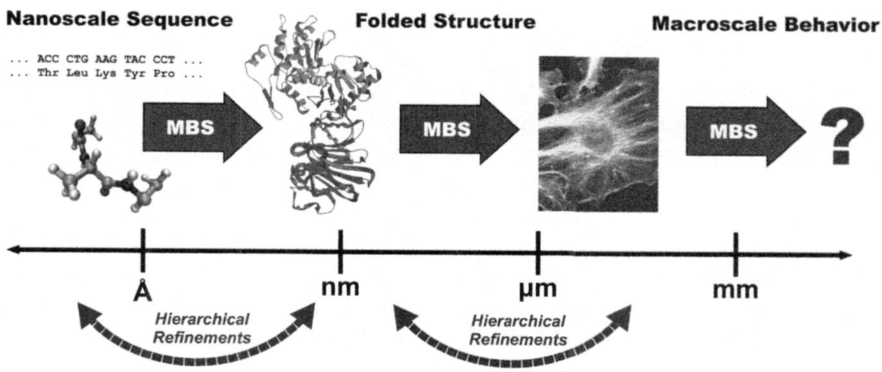


**Fig. 2.16** Schematic of material behavior scaling for more complex material systems. Here, known behavior at a fine-scale (such as atomistic) is adequate to describe behavior and properties at the next subsequent scale only, where the ensuing structure is known. Thus, to predict ultimate macroscale behavior and properties, a hierarchical approach is required, wherein knowledge at each scale is combined in a materiomic framework. Here, we depict the scaling of actin, a protein material found in the cytoskeleton. Atomistic interactions can be used to approximate the folded structure, for example, but at this scale, interactions with other proteins (such as myosin) and the cellular environment must be taken into account to fully describe the cytoskeleton behavior affected by actin

> **Material Behavior Scaling**: The extent in which a fine-scale description of the mechanical properties and behavior of a material adequately describes a larger scale. An *ideal scaling material* is one in which a small-scale description—considered here the *ab initio* material description—is sufficient to describe the material properties and behavior at *all* scale levels, transcending Ångstroms to meters. A *nonscaling material* exhibits disparate properties from one scale to another.

A complex materiome has, by definition, limited material behavior scaling—knowledge of the properties and behavior at any (arbitrary) scale gives little indication of properties at higher or lower scales. By extension, without consideration of the entire multi-scale system—bridging atomistic structure to intended functionality—analysis, investigation and understanding of a single (limited) scale is moot. This is not a new perspective attributed to materiomics. Indeed, it is a well known that modifications of ultra-small or nanoscale features in materials can lead to significant changes in the macroscopic behavior. Similar mechanisms have been known for a long time in the physics or mathematics communities, who have been studying the dynamical properties of nonlinear (or chaotic) systems. In fact, many optical or magnetic properties are based on such microscopic collective mechanisms, leading to dramatic changes at the macroscopic scale.

Basic concepts of mechanics of materials typically assert a "bulk" perspective to material systems. This is insufficient for complex materiomic systems in which a multiscale perspective is necessary. Classical theories of the strength of solids, such as fracture mechanics or those based on the maximum stress, assume a continuum. Even if such a continuum hypothesis can be shown to work at the nanoscale for elastic calculations (such as in our example of copper), it has to be revised for computing the strength of nanostructures or nanostructured materials (such as carbon nanotubes). Biological materials, on the other hand, are particularly complex, which requires the analysis of mechanisms across multiple scales. For mechanics of materials, such concepts have thus far not systematically been exploited, and are often poorly understood. One reason is that in the past, structural changes at the nanoscale have not been possible, which made it difficult to confirm theoretical explanations and concepts that link nanoscale and macroscale quantitatively. However, such atomic-scale manipulations of matter is now reality.

For biological materials, and in particular those based on proteins, genetic engineering has become an important tool to address precisely these questions. These techniques enable scientists and engineers to change specific genes, leading to mutations in the sequence of proteins, at the scale of several Ångstrom. This has significant impact in understanding diseases, which often origin from such mutations. Concurrently, genetic manipulation also allows the creation of new materials. Nature can be exploited to produce new biological materials—known as biocomposites—derived from known structures modified to feature improved properties [74, 75]. Biocomposites are structural "green" materials made from renewable resources that biodegrade in an anaerobic environment after their useful service life to produce a fuel or feedstock to produce a biopolymer for a new generation of composites. Some of these, such as cellulose and starch are actively used in products today, while many others remain underutilized. Promising candidates for such biocomposites range from a diverse set of polysaccharides, proteins, lipids, polyphenols, and specialty polymers produced by bacteria, fungi, plants and animals [76]. Such materials are being researched and developed to replace less eco-friendly structural and non-structural materials used in the construction industry [77, 78]. The key for such material development and rapid advancement lies in the understanding of fundamental biosynthetic pathways and options to modulate or tailor these pathways through genetic manipulations—a process than necessitates a holistic bottom-up perspective. Such biologically produced systems are intriguing examples for a systematic microscopic changes in building block growth (resulting from, for example, genetically modified bacteria)—which can actually be performed, with extremely high accuracy—and whose affect on macroscale behavior can be measured. Complete understanding of interactions and growth factors across scales remains elusive. Yet, it is the key to decipher the origins of diseases, mechanisms of cellular and tissue processes, and the development of new biologically inspired materials.

## 2.7 Extending the Materiome: Hierarchies and Complexity

Clearly, from the discussions of bone and tendon (Fig. 2.6), lobster cuticles (Fig. 2.8), actin (Fig. 2.16), and spider silk (Fig. 2.13), as well as the deficiencies of scaling carbon nanotubes (Fig. 2.12), and the simplicity of a material like copper (Fig. 2.11), a materiomic framework is most beneficial to describe complex, multiscale hierarchical materials—such materials common to biological systems and tissues. In this context, the term "hierarchical" is used loosely to indicate a material system with at least a single distinct differentiation between constituent material components and global system structure. Collagen fibrils found in tendon, for example, are composed of as tropocollagen molecules, while the tropocollagen fibrils are themselves composed of a hierarchical arrangement of polypeptide chains, which are also composed of constituent amino acids. Depending on the level of scrutiny, the defined "lowest level of hierarchy" need not be the fundamental building blocks of the system.

**Hierarchical System:** A system composed of stable, observable sub-elements that are unified by a super ordinate relation. Thereby, lower level details can affect higher levels and thus the overall system behavior. A common characteristic of biological materials.

The materiome must adequately encompass both the hierarchical structure and intrinsic complexity beyond a simple catalogue of scale-by-scale structures, properties and interactions. A holistic approach must integrate the information inherent to the materiome (consisting of the material constituents, the cross-scale SPP relations, and the resulting functionalities/requirements) across all-scales, as hinted by Fig. 2.7. Conceptually, this is depicted in Fig. 2.17. Note that each "hierarchy" and reduction in scale can represent multiple branches and material components that require investigation and characterization. For example, actin is only one of three primary components of the cytoskeleton, the other two being intermediate filaments and microtubules. Each structure can be characterized at the molecular scale, but the interactions must be collectively considered at the cytoskeleton scale and beyond.

The combinatorial nature of these building blocks results in the emergence of complexity within the materiome. Again, we see that "whole is greater than the sum of its parts". Even if we had a full molecular understanding of actin, intermediate filaments, and microtubules, we would not be able to predict the behavior of the cytoskeleton by simple additive (or similar) relations—the system is considered nonlinear. The system is more complex, with each protein material affecting the behavior and properties of the other in ways unpredictable from inspection of a single element. Note that this is different than the common structural engineering approach depicted in Fig. 2.5—we know how a steel truss will behave regardless if is in a system of other steel members, concrete walls, or wooden columns. Such inability to predict behavior limits potential scaling of such material components.

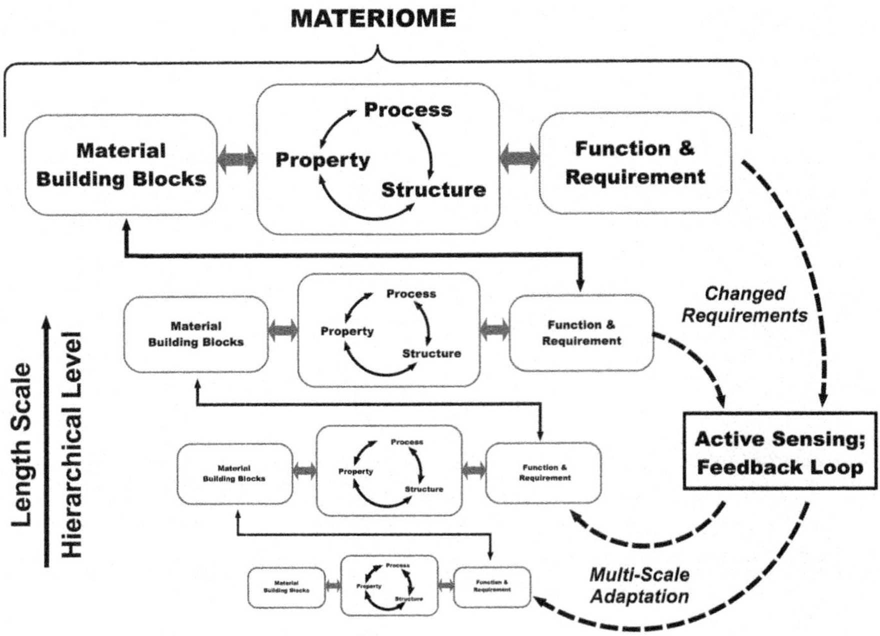


**Fig. 2.17** Schematic of *extended* materiomic information, consisting of the material constituents, the cross-scale SPP relations, and the resulting functionalities/requirements across *all* levels of hierarchy. Commonly, biological materials exhibit hierarchical structures in which material properties and environmental requirements are brought together in mutual completion transcending multiple scales (*i.e.*, subtle changes at the molecular scale due to macroscale environmental influences). In contrast to a traditional materials science paradigm, relations between "external" functions/requirements and "internal" properties exist on several scales resulting in multifunctionality

**Complexity**: The existence of many interacting components and leads to emerging nonlinear behavior of a system. Complexity in a material system (*i.e.*, a complex *materiome*) necessitates the quantification of cross-scale interactions and mechanisms, which cannot be deduced from general scaling relations.

This complexity can be likened to the mathematical field of chaos. Chaos theory is a field of mathematics, with applications in several disciplines including physics, economics, biology, and philosophy. In simplest terms, chaos theory studies the behavior of dynamical systems that are highly sensitive to initial conditions—an effect which is popularly referred to as the butterfly effect.[5]

[5]The butterfly effect is the sensitive dependence on initial conditions where a small change at one point in a nonlinear system can result in large differences to a later state. The effect is coined after a thought experiment, where a butterfly flapping its wings in Japan can directly lead to creation of a hurricane in Florida.

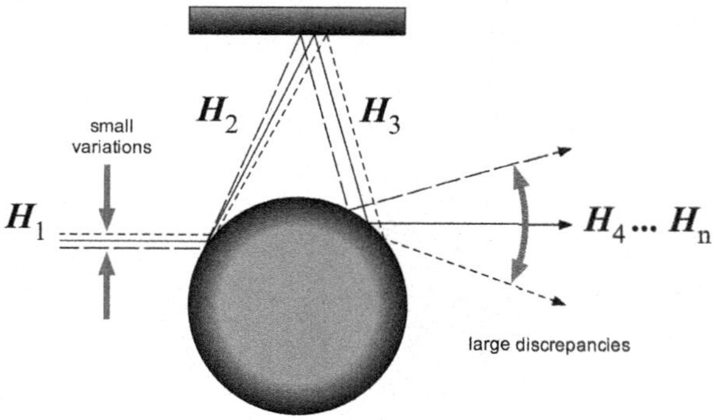


**Fig. 2.18** Simple depiction of effect of variations in a complex hierarchical system (chaotic behavior). Slight perturbations in input (lowest level of hierarchy) propagate across all scales, resulting in large discrepancies in system-level (highest level of hierarchy) properties and behavior. Single-scale analysis and characterizations are rendered inadequate without consideration of system level effects

Chaotic behavior can be observed in many systems, such as the weather or financial markets. Small differences in initial conditions (such as those due to rounding errors in numerical computation) yield widely diverging outcomes for chaotic systems, rendering long-term prediction impossible in general. This happens even though these systems are deterministic, meaning that their future behavior is fully determined by their initial conditions, with no random elements involved. In other words, the deterministic nature of these systems does not make them predictable. This behavior is known as deterministic chaos, or simply chaos. From a materiomic perspective, small differences in initial conditions refers to a change in protein sequence, or a small geometrical re-arrangement in structure. Unlike the weather of financial markets, however, the chaos (or complexity) of a materiome arises due to multiscale interactions and bridging of hierarchies. Such chaotic behavior across hierarchies can be visualized by the path of a multiple reflected light beam (Fig. 2.18), where each reflection corresponds to a higher scale.

This may present a problem in the development and understanding of complex materiomic systems. If small perturbations result in divergent properties, any behavior or property observed in Nature, however, would be idiosyncratic to each individual system. Obviously, this is not the case. The self-assembly (or growth) of Nature's materials is not an ordered atom-by-atom (or even protein-by-protein) construction—it is a highly stochastic process driven by varying factors and environmental conditions which preclude systematic "quality controls". Yet the final systems are remarkable consistent in terms of behavior and functionality. If such systems are complex, chaotic, and unpredictable, *how does Nature achieve such dependable materials and systems*?

Nature, it seems, has developed multiple safeguards or buffers to ensure slight deviations in input do not result catastrophic changes in macroscale system behavior. They are robust by design, without resorting to the redundancies or backup systems typically introduced to engineered systems to prevent complete failure. This

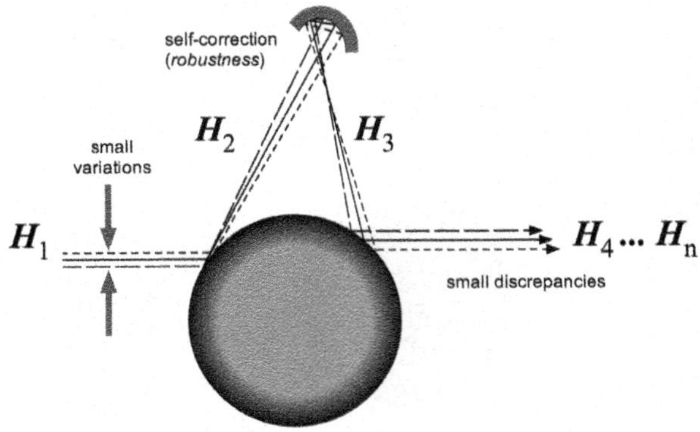


**Fig. 2.19** Simple depiction of effect of variations in a complex hierarchical system with intrinsic robustness. Here, there is a natural "safeguard" mitigating divergent behavior from the first to second hierarchies ($H_1$ to $H_2$). Small variations at the lowest level have marginal effects at larger scales. Such safeguards can come in the form of geometrical arrangement, material combinations, etc., that reduce the chaotic behavior of biological materials and systems, without resorting to the redundancies or backup systems typically introduced to engineered systems

too can be represented by the reflecting light analogy—while deviations in input cause divergent behavior across hierarchies, natural safeguards at specific hierarchies reverse, reduce, or eliminate the divergent behavior. The result is such that the variations at lower scales have marginal effects at larger scales (see Fig. 2.19).

The means by which such safeguards are accomplished is highly dependent on system function and scale. For some systems, such as bone, elimination of divergent behavior can be seen in the geometric assembly, such as the characteristic length scale of collagen fibrils, which increases flaw tolerance, enhancing toughness. Unavoidable flaws such as voids or cracks in the mineralized hydroxyapatite, for example, do not propagate throughout they system and bone maintains its inherent strength (the toughening mechanisms in bone are discussed further in Chap. 8: Unlocking Nature: Case Studies). The same flaws would significantly decrease the strength of carbon nanotubes. At a larger scale, the composite shells of gastropods or lobster cuticles combine materials (soft protein-based materials with stiffer mineralized materials) that serve to arrest cracks and dissipate energy. Small structural defects are not detrimental to the system. Even the cytoskeleton—with a seemingly random arrangement of actin, intermediate filaments, and microtubules—exploits the mechanical properties of each in a synergistic manner (intermediate filaments provide tensional strength, while microtubles are compressive members, in a biological tensegrity structure, for example). The combined behavior within the cellular membrane is consistent, even if there are deviations amongst the structural members in spite of the seemingly random arrangement.

Nature effectively avoids the divergent aspects of chaotic systems, but does not eliminate the inherent complexity. Note, however, that Nature has not (and cannot) prevent the effect of all minute changes in initial (or molecular) conditions. For instance, single point mutations (a simple swapping of amino acids) within the tropocollagen structure results in *osteogenesis imperfecta* (more commonly known as brittle bone disease), a disease discussed further in Chap. 9: Pathological Materiomics. Can such conditions be predicted? More importantly, can such conditions be avoided by understanding what kind of mutations propagate across the hierarchical scales? Which mutations are inherently "chaotic"? This are critical questions materiomics attempts to answer.

## 2.8 Summary

The preceding chapter has introduced a fundamental concept central to materiomics: the materiome. The materiome is a holistic perspective of a material system,—"the whole is greater than the sum of its parts"—that attempts to integrate properties, behaviors, and interactions beyond a simple scale-by-scale characterization. Like the genome, it will associate meaning with the constituent building blocks, and allow the decomposition of complex phenomena found in biological materials. Just as proteomics elucidates protein-protein interactions, materiomics can investigate the structure-property relationships of nutrient transport in tissue engineering, for example.

As discussed, a materiomic framework is necessary when the distinction between structure and material is difficult to define, and may change according to scale, and when functionality emerges from the materiome itself (not selected for an *ad hoc* application based on desired/required properties). Defining the materiome is not without challenges—characterized by multiscale hierarchies, complexity, and (sometimes) chaotic relationships inherent in biological materials. Such systems require new approaches to investigate, and warrant new metrics and heuristic techniques to characterize. Can abstract terms such as hierarchies and complexity be quantified? Can the system level effects of atomistic changes even be predictable? Are there universal motifs of structure and geometry within biological systems? What material combinations can possible enhance tissue growth? Can such design principles even be adapted to inorganic construction materials (such as carbon nanotubes)? Many of these questions are being investigated in one form or another, but are separated by disparate disciplines, motivations, and goals. They are all, however, defining the materiome. Therein lies the opportunity to exploit the designs of Nature.

## References

1. D.W. Thompson, *On Growth and Form* (Dover, New York, 1992)
2. S. Vogel, *Life in Moving Fluids: The Physical Biology of Flow* (W. Grant Press, Boston, 1981)
3. S. Vogel, *Life's Devices: The Physical World of Animals and Plants* (Princeton University Press, Princeton, 1988)
4. S. Vogel, J.G. Vogel, Copying life's devices. Curr. Sci. **78**(12), 1424–1430 (2000)

5. S. Vogel, *Comparative Biomechanics: Life's Physical World* (Princeton University Press, Princeton, 2003)
6. S. Vogel, The emergence of comparative biomechanics. Integr. Comp. Biol. **47**(1), 13–15 (2007)
7. J.C. Venter, et al., The sequence of the human genome. Science **291**(5507), 1304–1351 (2001)
8. F.S. Collins, M. Morgan, A. Patrinos, The human genome project: lessons from large-scale biology. Science **300**(5617), 286–290 (2003)
9. J. Lederberg, A. McCray, 'ome sweet' omics—a genealogical treasury of words. The Scientist **15**(7), 8–9 (2001)
10. E. Kolker, Editorial. OMICS: J. Integr. Biol. **6**(1), 1 (2002)
11. J. Goldstein, Emergence as a construct: history and issues. Emergence: Complexity and Organization **1**(1), 49–72 (1999)
12. E.J. Alm, B.J. Shapiro, Comparing patterns of natural selection across species using selective signatures. PLoS Genet. **4**(2) (2008)
13. C. Fraser, E.J. Alm, M.F. Polz, B.G. Spratt, W.P. Hanage, The bacterial species challenge: making sense of genetic and ecological diversity. Science **323**(5915), 741–746 (2009)
14. D. Greenbaum, N.M. Luscombe, R. Jansen, J. Qian, M. Gerstein, Interrelating different types of genomic data, from proteome to secretome: 'oming in on function. Genome Res. **11**(9), 1463–1468 (2001)
15. P. Fratzel, *Collagen: Structure and Mechanics* (Springer, New York, 2008)
16. A. Gautieri, S. Vesentini, A. Redaelli, M.J. Buehler, Hierarchical structure and nanomechanics of collagen microfibrils from the atomistic scale up. Nano Lett. **11**(2), 757–766 (2011)
17. J.D. Currey, *Bones: Structure and Mechanics* (Princeton University Press, Princeton, 2002)
18. M.J. Buehler, Atomistic and continuum modeling of mechanical properties of collagen: elasticity, fracture and self-assembly. J. Mater. Res. **21**(8), 1947–1961 (2006)
19. S.C. Cowin, A.M. Sadegh, G.M. Luo, An evolutionary Wolff law for trabecular architecture. J. Biomech. Eng. Trans. ASME **114**(1), 129–136 (1992)
20. S.C. Cowin, Bone poroelasticity. J. Biomech. **32**(3), 217–238 (1999)
21. N. Sasaki, S. Odajima, Elongation mechanism of collagen fibrils and force-strain relations of tendon at each level of structural hierarchy. J. Biomech. **29**(9), 1131–1136 (1996)
22. S.J. Eppell, B.N. Smith, H. Kahn, R. Ballarini, Nano measurements with micro-devices: mechanical properties of hydrated collagen fibrils. J. R. Soc. Interface **3**(6), 117–121 (2006)
23. J.A.J. van der Rijt, K.O. van der Werf, M.L. Bennick, P.J. Dijkstra, J. Feijen, Micromechanical testing of individual collagen fibrils. Macromol. Biosci. **6**, 697–702 (2006)
24. M.J. Buehler, Nanomechanics of collagen fibrils under varying cross-link densities: atomistic and continuum studies. J. Mech. Behav. Biomed. Mater. **1**, 59–67 (2008)
25. H.-C. Spatz, E.J. O'Leary, J.F.V. Vincent, Young's moduli and shear moduli in cortical bone. Proc. Biol. Sci. **263**(1368), 287–294 (1996)
26. E. Fedorova, D. Zink, Nuclear genome organization: common themes and individual patterns. Curr. Opin. Genet. Dev. **19**(2), 166–171 (2009)
27. N. Kepper, D. Foethke, R. Stehr, G. Wedemann, K. Rippe, Nucleosome geometry and internucleosomal interactions control the chromatin fiber conformation. Biophys. J. **95**, 3692–3705 (2008)
28. N.C. Seeman, An overview of structural DNA nanotechnology. Mol. Biotechnol. **37**(3), 246–257 (2007)
29. E.S. Andersen, M. Dong, M.M. Nielsen, K. Jahn, R. Subramani, W. Mamdouh, M.M. Golas, B. Sander, H. Stark, C.L.P. Oliveira, J.S. Pedersen, V. Birkedal, F. Besenbacher, K.V. Gothelf, J. Kjems, Self-assembly of a nanoscale DNA box with a controllable lid. Nature **459**(7243), 73–75 (2009)
30. C. Lin, Y. Liu, H. Yan, Designer DNA nanoarchitectures. Biochemistry **48**(8), 1663–1674 (2009)
31. M.J. Buehler, Y.C. Yung, Deformation and failure of protein materials in physiologically extreme conditions and disease. Nat. Mater. **8**(3), 175–188 (2009)

32. B.L. Smith, T.E. Schaffer, M. Viani, J.B. Thompson, N.A. Frederick, J. Kindt, A. Belcher, G.D. Stucky, D.E. Morse, P.K. Hansma, Molecular mechanistic origin of the toughness of natural adhesives, fibres and composites. Nature **399**, 761–763 (1999)
33. R.Z. Wang, Z. Suo, A.G. Evans, N. Yao, I.A. Aksay, Deformation mechanisms in nacre. J. Mater. Res. **16**(9), 2485–2493 (2001)
34. F. Barthelat, C.-M. Li, C. Comi, H.D. Espinosa, Mechanical properties of nacre constituents and their impact on mechanical performance. J. Mater. Res. **21**(8), 1977–1986 (2006)
35. S. Nikolov, M. Petrov, L. Lymperakis, M. Frik, C. Sachs, H.-O. Fabritius, D. Raabe, J. Neugebauer, Revealing the design principles of high-performance biological composites using ab initio and multiscale simulations: the example of lobster cuticle. Adv. Mater. **22**(4), 519–526 (2010)
36. E. Munch, M.E. Launey, D.H. Alsem, E. Saiz, A.P. Tomsia, R.O. Ritchie, Tough, bio-inspired hybrid materials. Science **322**, 1516–1520 (2008)
37. C. Ortiz, M.C. Boyce, Bioinspired structural materials. Science **319**(5866), 1053–1054 (2008)
38. H. Yao, M. Dao, T. Imholt, J. Huang, K. Wheeler, A. Bonilla, S. Suresh, C. Ortiz, Protection mechanisms of the iron-plated armor of a deep-sea hydrothermal vent gastropod. Proc. Natl. Acad. Sci. USA **107**(3), 987–992 (2010)
39. M.A. Meyers, A.Y.M. Lin, P.Y. Chen, J. Muyco, Mechanical strength of abalone nacre: role of the soft organic layer. J. Mech. Behav. Biomed. Mater. **1**(1), 76–85 (2008)
40. F. Song, A.K. Soh, Y.L. Bai, Structural and mechanical properties of the organic matrix layers of nacre. Biomaterials **24**, 3623–3631 (2009)
41. N. Yao, A.K. Epstein, W.W. Liu, F. Sauer, N. Yang, Organic-inorganic interfaces and spiral growth in nacre. J. R. Soc. Interface **6**, 367–376 (2009)
42. Z. Tang, N.A. Kotov, S. Magono, B. Ozturk, Nanostructured artificial nacre. Nat. Mater. **2**, 413–418 (2003)
43. L. Lu, X. Chen, X. Huang, K. Lu, Revealing the maximum strength in nanotwinned copper. Science **323**(5914), 607–610 (2009)
44. L. Li, N.M. Ghoniem, Twin-size effects on the deformation of nanotwinned copper. Phys. Rev. B **79**, 075444 (2009)
45. D. Xu, W.L. Kwan, K. Chen, X. Zhang, V. Ozolins, K.N. Tu, Nanotwin formation in copper thin films by stress/strain relaxation in pulse electrodeposition. Appl. Phys. Lett. **91**, 254105 (2007)
46. S. Iijima, Helical microtubules of graphitic carbon. Nature **354**(6348), 56–58 (1991)
47. R.H. Baughman, A.A. Zakhidov, W.A. de Heer, Carbon nanotubes—the route toward applications. Science **297**, 787–792 (2002)
48. M.M.J. Treacy, T.W. Ebbesen, J.M. Gibson, Exceptionally high Young's modulus observed for individual carbon nanotubes. Nature **381**, 678–680 (1996)
49. M.-F. Yu, O. Lourie, M.J. Dyer, K. Moloni, T.F. Kelly, R.S. Ruoff, Strength and breaking mechanism of multiwalled carbon nanotubes under tensile load. Science **287**(5453), 637–640 (2000)
50. F. Li, H.M. Cheng, S. Bai, G. Su, M.S. Dresselhaus, Tensile strength of single-walled carbon nanotubes directly measured from their macroscopic ropes. Appl. Phys. Lett. **77**(20), 3161–3163 (2000)
51. R.H. Baughman, M. Zhang, S.L. Fang, A.A. Zakhidov, S.B. Lee, A.E. Aliev, C.D. Williams, K.R. Atkinson, Strong, transparent, multifunctional, carbon nanotube sheets. Science **309**(5738), 1215–1219 (2005)
52. T.C. Chang, W.L. Guo, X.M. Guo, Buckling of multiwalled carbon nanotubes under axial compression and bending via a molecular mechanics model. Phys. Rev. B **72**(6) (2005)
53. A. Sears, R.C. Batra, Buckling of multiwalled carbon nanotubes under axial compression. Phys. Rev. B **73**(8) (2006)
54. T.W. Odom, J.L. Huang, P. Kim, C.M. Lieber, Atomic structure and electronic properties of single-walled carbon nanotubes. Nature **391**(6662), 62–64 (1998)
55. L.C. Venema, V. Meunier, P. Lambin, C. Dekker, Atomic structure of carbon nanotubes from scanning tunneling microscopy. Phys. Rev. B **61**(4), 2991–2996 (2000)

56. N. Mingo, D.A. Stewart, D.A. Broido, D. Srivastava, Phonon transmission through defects in carbon nanotubes from first principles. Phys. Rev. B **77**(3) (2008)
57. N.M. Pugno, On the strength of the carbon nanotube-based space elevator cable: from nanomechanics to megamechanics. J. Phys., Condens. Matter **18**(33), S1971–S1990 (2006)
58. B.C. Edwards, Design and deployment of a space elevator. Acta Astronaut. **47**(10), 735–744 (2000)
59. R.S. Ruoff, M.F. Yu, B.S. Files, S. Arepalli, Tensile loading of ropes of single wall carbon nanotubes and their mechanical properties. Phys. Rev. Lett. **84**(24), 5552–5555 (2000)
60. R.S. Ruoff, M.F. Yu, O. Lourie, M.J. Dyer, K. Moloni, T.F. Kelly, Strength and breaking mechanism of multiwalled carbon nanotubes under tensile load. Science **287**(5453), 637–640 (2000)
61. N.M. Pugno, The role of defects in the design of space elevator cable: from nanotube to megatube. Acta Mater. **55**(15), 5269–5279 (2007)
62. N.M. Pugno, Space elevator: out of order? Nano Today **2**(6), 44–47 (2007)
63. J.M. Gosline, P.A. Guerette, C.S. Ortlepp, K.N. Savage, The mechanical design of spider silks: from fibroin sequence to mechanical function. J. Exp. Biol. **202**(23), 3295–3303 (1999)
64. S. Keten, M.J. Buehler, Nanostructure and molecular mechanics of spider dragline silk protein assemblies. J. R. Soc. Interface (2010)
65. N. Du, X.Y. Liu, J. Narayanan, L.A. Li, M.L.M. Lim, D.Q. Li, Design of superior spider silk: from nanostructure to mechanical properties. Biophys. J. **91**(12), 4528–4535 (2006)
66. E. Nieuwenhuys, How thick should a spider silk thread be to stop a Boeing-747 in full flight? (online). http://ednieuw.home.xs4all.nl/Spiders/Info/SilkBoeing.html (2007)
67. J.H. Weiner, Hellmann-Feynman theorem, elastic moduli, and the Cauchy relations. Phys. Rev. B **24**(2), 845–848 (1981)
68. E.B. Tadmor, M. Ortiz, R. Phillips, Quasicontinuum analysis of defects in solids. Philos. Mag. A **73**(6), 1529–1563 (1996)
69. P.M. Morse, Diatomic molecules according to the wave mechanics. ii. vibrational levels. Phys. Rev. **34**(1) (1929)
70. M.J. Buehler, *Atomistic Modeling of Materials Failure* (Springer, Berlin, 2008)
71. K.C. Holmes, D. Popp, W. Gebhard, W. Kabsch, Atomic model of the actin filament. Nature **347**(6288), 44–49 (1990)
72. T.D. Pollard, G.G. Borisy, Cellular motility driven by assembly and disassembly of actin filaments. Cell **112**(4), 453–465 (2003)
73. E. Frixione, Recurring views on the structure and function of the cytoskeleton: a 300-year epic. Cell Motil. Cytoskelet. **46**(2), 73–94 (2000)
74. A.K. Mohanty, M. Misra, L.T. Drzal, *Natural Fibers, Biopolymers, and Biocomposites* (Taylor & Francis, Boca Raton, 2005)
75. R.P. Wool, X.S. Sun, *Bio-Based Polymers and Composites* (Elsevier/Academic Press, Amsterdam/Boston, 2005)
76. D.L. Kaplan, *Biopolymers from Renewable Resources* (Springer, New York, 1998)
77. S. Thomas, M.J. John, Biofibres and biocomposites. Carbohydr. Polym. **71**(3), 343–364 (2008)
78. G. Hinrichsen, A.K. Mohanty, M. Misra, Biofibres, biodegradable polymers and biocomposites: an overview. Macromol. Mater. Eng. **276**(3–4), 1–24 (2000)

# Chapter 3
# The Challenges of Biological Materials

**Abstract** The challenges of biological materials and systems—the inherent complexity—is a result of the diversity of Nature itself. The evolution of structure in biological materials is guided by the ever-changing requirements of the external environment and have resulted in materials with desirable engineering properties, such as high strength, toughness, adaptability, flaw tolerance, self-healing, mutability and multifunctionality, all with a limited set of—and frequently inferior—building materials. As a result of such constraints, Nature implements a flexible material composed of a limited set of molecular components: soft, deformable, highly convoluted proteins, composed of a minute set of amino acids. Providing a common base, protein form and function has evolved intimately, such that even the prediction of folded structure from a known peptide sequence is a technological challenge. Potential functionality is extended through the use of structural hierarchies—resulting in system robustness, efficiency, and design tolerance—while decreasing the efficacy of any single-scale analysis. At the microscale, the complexity manifests as functional, growing, adaptable materials—producing "shaky" platforms for tissue engineering and growth. While biological and chemical cues may change across scales, mechanical insights can provide a common basis regardless of scale or analytical progression.

*Look deep into Nature, and then you will understand everything better.*

*Albert Einstein (1879–1955)*

## 3.1 Introduction

In the development of new materials, researchers more frequently rely on Nature as a muse (commonly referred to as "bio-inspiration"). *Buy why*? Such materials are the building blocks of life, forming a diverse group of biological materials, ranging from spider silk to bone, tendon to the skin, all which play an important role in providing key functions to biological systems [1–7]. Nature, through meticulous trial-and-error and centuries of optimization, has intricately combined material

S.W. Cranford, M.J. Buehler, *Biomateriomics*, Springer Series in Materials Science 165,
DOI 10.1007/978-94-007-1611-7_3, © Springer Science+Business Media Dordrecht 2012 

structure, properties and functionality to the point in which they are indistinguishable. Structure and function are so intimately linked that substitution of even the most commonly found materials is not possible—you obviously cannot swap out bone and replace it with wood! This is fundamentally different than engineering materials—you can design a building using steel, for example, but substituting reinforced concrete can serve the same purpose.

Commonly, biological protein materials feature hierarchical structures to constitute a diverse range of physiologic materials. In addition, one of the defining characteristics of biological systems is the ability of even the most intricate of its component molecular structures to self-assemble with precision and fidelity. The complex hierarchies arise naturally across scales, typically under limited external guides or controls. As such, biological materials, by their very essence, require a holistic view of the *materiome*. The analysis of protein materials is integral to the field of *biomateriomics*, utilizing insight based on structure-process-property (SPP) relations in its *biological* context, to probe deformation and failure phenomena at the molecular and microscopic level. Such analysis requires the integration of advanced experimental, computational and theoretical methods (discussed in subsequent chapters of Part II: Methods and Tools), utilized to assess, monitor and predict mechanisms associated with failure of protein materials and structures composed of them. Simply put, if we can *fully* understand complex biological materials, more benign synthetic materials should be relatively simple.

The natural evolution of structure in biological materials is guided by the interactions between these materials and their environment. As a result, materials in biology, specifically those made of proteins, realize a diversity of functional properties such as catalysis, signal transmission, material transport, gene regulation and protein formation, structural support, and movement generation—many of them simultaneously—resulting in multi-functional materials [1]. Significant evidence has recently emerged that proteins are organized in functional networks, resulting in multi-functional structures that span many hierarchical scales [7–13]. However, a major frontier in protein science is our lack of understanding as to how the exceptionally complex functionality found in natural biological systems is created despite [14]: (i) a limited number of building blocks (*e.g.*, 20 amino acids); (ii) constraints in available material volume and energy for synthesis (*e.g.*, low temperature and pressure), and; (iii) only a handful of simple chemical interactions. To explicate, the same library of amino acid building blocks creates materials as diverse as spider silk, tendon, cornea, blood vessels, and cells, each of which displays greatly variegated functions. Moreover, biological materials feature an intriguing set of properties that would be considered sophisticated advancements (*e.g.*, self-healing capabilities) for many traditional materials (listed in Table 3.1).

## 3.2 Proteins, Proteins... Everywhere!

Among the many building blocks of biological materials, the structure and behavior of proteins, as well as the materials that are based on proteins, stand out. The

**Table 3.1** Definition of important properties of biological materials. Adapted from [15, 16]

| Term | Definition |
|---|---|
| Strength | The maximum applied force or stress (pressure) at which failure of a system occurs (*e.g.*, through fracture, tissue break down, etc.). |
| Robustness | Measures the ability of a system to tolerate flaws and defects, that is, still being capable of providing the required function under the presence of flaws and defects. A possible definition of robustness related to strength properties is the ratio of strength of a flawed structure divided by the strength of a perfect structure. |
| Adaptability | Ability of system to respond to changes in the environment (such as formation of defects due to injuries, or changes in physiological requirements, or due to the formation of fractures, etc.). |
| Flaw (defects) | Deviation of the structure of a system from its perfect, ideal or reference configuration. Examples for defects include cracks, inclusions, protein misfolds, or mutations in the amino acid sequence. |
| Failure | Sudden, typically uncontrolled and irreversible loss of the functional properties of a system. An example is the breakdown of tissue due to injuries under very large applied forces, or the collapse of a building. |
| Self-healing | Ability of a system to reform from a perturbed structure to its reference configuration (re-assemble). May involve, for example, the curing of flaws and defects such as cracks or voids, or the replacement or the addition of tissue. |
| Mutability | Formation of distinct (sometimes preprogrammed) structures with different properties, which can be controlled (and reversed) by external cues. Examples include the existence of multiple conformations of proteins based on pH, or applied forces. |
| Multifunctionality | Ability of a system to provide multiple properties to satisfy a set of target properties. An example is the combination of strength and robustness. |
| Evolvability | Ability of a system to evolve over generations of synthesis. In contrast to adaptability, evolvability reflects a change of structural makeup and/or properties over generations of synthesis. |

properties of materials are of fundamental importance to biological tissues, organs, organisms and species, and are crucial to providing functional properties to all living systems. Example properties that are crucial in biology include the strength toughness of bone [17–20] the elasticity of spider silk [21–23], or the stretchiness of the skin and other soft tissues [24–27]. Most fibers, tissues, organs and organisms found in nature are the result of a highly hierarchical and organized protein structure, where features are found at all scales, ranging from individual molecules ($\sim$ 50 Å), assemblies ($\sim$ 1 nm to 10 nm), fibrils and fibers ($\sim$ 10 nm to 100 μm), to cells ($\sim$ 50 μm), and to tissues and organs (1000 μm and much more) [1, 7, 28, 29]. By definition, a protein is simply a polypeptide: a single linear polymer chain of amino acids. The universality of such amino acids is astounding upon first glance, as Francis Crick noted in his 1962 Nobel Lecture:

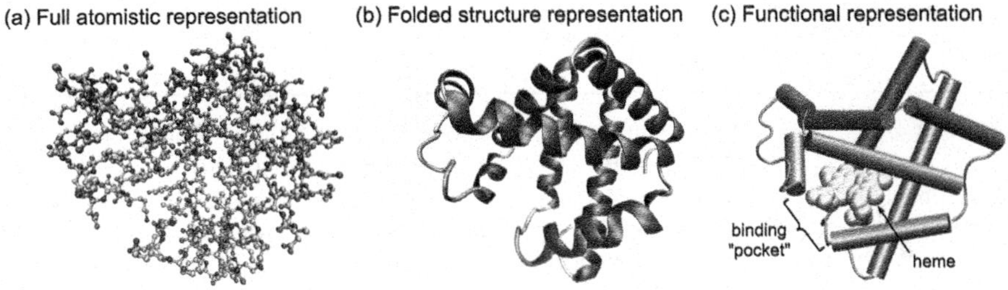


**Fig. 3.1** From *polypeptide* to *protein* to *function*: more than a collection of amino acids. (**a**) Full atomistic representation of the globular protein myoglobin—a single-chain protein composed of approximately 150 amino acids that provides oxygen to muscle tissue. The protein structure is difficult to determine from an atomistic view. (**b**) Common "ribbon" representation of the protein alpha-helices formed by the folded polypeptide. The particular arrangement of alpha-helices exposes a molecular "pocket". (**c**) Representation of protein structure indicating function, wherein the alpha-helices allow the secure binding of a *heme* molecule (carrying iron). Myoglobin was the first protein to have such a three-dimensional structure defined (by high-resolution X-ray crystallography in 1958 [31], work that led to a share of the Nobel Prize in Chemistry in 1962)

> It is one of the more striking generalizations of biochemistry—which surprisingly is hardly ever mentioned in the biochemical textbooks—that the twenty amino acids and the four bases, are, with minor reservations, the same throughout Nature.

Protein molecules are essentially long polymer chains, constructed of these limited amino acid monomers, and synthesized *via* a translation process that converts the genetic information transcribed from DNA into RNA into a polypeptide sequence. Protein synthesis is initiated at precise sites during mRNA translation and terminated when signalled by the presence of encoded stop codons (*i.e.*, messages). Once a protein molecule is constituted, the polypeptide chain folds into its unique three-dimensional conformation, sometimes with the aid of chaperone proteins [1]. Functional activation of the protein structure often involves binding with other molecules to form the protein's intended functional structure [1], with specific protein shapes and structures having affinity for one another (*e.g.*, lock-and-key mechanisms, or complementary binding domains [30]). For example, we can consider the protein myoglobin: an iron- and oxygen-binding protein found in the muscle tissue of almost all mammals. Myoglobin is a globular protein composed of a single-chain polypeptide with about 150 amino acids (see Fig. 3.1). The polypeptide characteristically folds into eight alpha-helices (*i.e.*, molecular coils), which form a structure to bind oxygen and supply muscle tissue this critical fuel. Indeed, high concentrations of myoglobin in muscle cells allow organisms to hold their breaths longer—mammals such as whales and seals have muscles with particularly high amounts of myoglobin. Can this function be predicted from the original sequence of amino acids? Unfortunately, such predictions are currently arduous and limited in scope. The difficulty in predicting such functions and interactions lies in the *complexity* of protein folding—identical peptide sequences forming different geometrical configurations may behave in different ways.

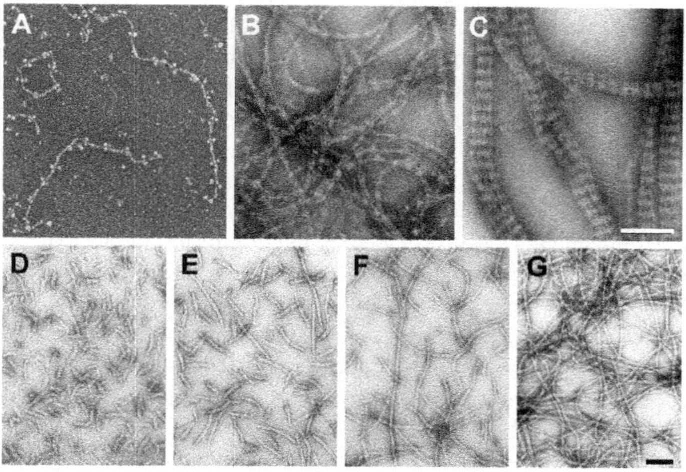


**Fig. 3.2** Protein material self-assembly: axial growth of lamin fibrils. Electron microscopic images of assembly stages from (**A** to **C**) rat liver lamin A/C, and (**D** to **G**) human vimentin. (**A**) Head-to-tail fibers, (**B**) beaded filaments that (**C**) extensively laterally associate to eventually yield paracrystals. (**D** to **G**) Assembly of recombinant human vimentin with similar hierarchical structuring. From Herrmann and Aebi [33], used with permission, copyright © 2004 Annual Review of Biochemistry

An initial peptide sequence, however, does not always directly translate to a particular or intended function. *In vivo*, many protein materials obtain their specific functionality by post-translational modifications, which include hydroxylations (*e.g.*, during formation of tropocollagen's hydroxyproline residues), phosphorylations, glycosilations (*e.g.*, during formation of cartilage tissues), as well as enzymatic cross-linking. These modifications are particularly important for material properties of tissues, as they control the interaction between proteins and with other material components (*e.g.*, inorganic materials, sugar-based components) as well as their bioactive properties. These modifications are particularly difficult to mimic *ex vivo* or through synthetic approaches, posing a major challenge in structure prediction (and design) and the development of biomimicking and biocompatible materials.

The resultant individual protein components self-assemble at different timescales to form a protein structure with hierarchical geometrical entities. Assembly mechanisms have been investigated based on the combination of imaging tools with controlled assembly conditions (*e.g.*, pH, temperature, solvent). For example, this approach has elucidated the mechanism of lamin assembly [32, 33]. In this process, lamin dimer association and axial growth is followed by radial growth to form larger fibrils, resulting in formation of a network. Figure 3.2 displays the direct experimental visualization of these assembly processes. The biological process of *in vivo* assembly of many structural protein materials involves a dynamic change of physical and chemical conditions, as has been shown for the synthesis of spider silk [21, 34], to be discussed further in Chap. 8: Unlocking Nature: Case Studies.

In addition to studies of naturally occurring biological protein materials, recent research led to the development of techniques that enables one to change their structural makeup, and to design and manufacture synthetic analogs [35–39] through recombinant DNA techniques, RNAi knockdowns, or sequence insertions. Two primary routes of development pursued include *ex vivo* assembly (*e.g.*, self-assembling peptide systems [35, 37, 39]) and *in vivo* expression of protein materials (*e.g.*, through bacterial hosts [36, 40]). The ability to control the DNA sequence information at a fundamental level provides us the ability to engineer the structure of protein materials at the molecular (amino acid) scale. For example, the ionic peptide EAK16 belongs to an interesting class of self-assembled peptides that constitutes a material platform for a variety of biological, biomimetic and nanotechnology applications [41, 42]. The knowledge about the details of the *in vitro* or *in vivo* processes enables one to mimic these processes *ex vivo*, as it has been demonstrated for spider silk fibres in microfluidic devices [21].

Proteins undoubtedly constitute the critical building blocks of life, and the specific sequence of amino acids within protein molecules form the basic constituents of this group of materials. While there are (only) 20 amino acids, sequences consist of thousands of such molecules (for example, the largest known proteins are the titins, a component of the muscle sarcomere, with a total length of almost 27,000 amino acids [43]), and the sequence subsequently determines the function through a three-dimensional hierarchical structure. One small change in amino acid sequence (*i.e.*, a defect or mutation) can result in catastrophic consequences (discussed further in Chap. 9: Pathological Materiomics). These materials are distinct from the conventional paradigms of structure and material, as they connote the merger of these two concepts through hierarchical formation of structural elements that range from the nanoscale to macroscale.

Protein materials play a crucial role in the biological function of all cells and tissues within organisms. Many such materials with mechanical function form structural filaments, trusses or fibers (such as collagen), while others retain the globular structure of their protein constituents (such as myoglobin). Biological protein materials are commonly characterized by where they reside with respect to their associated tissue, to either serve as extracellular or intracellular protein materials [1]. Intracellular protein materials provide architecture to a cell's structure and include vimentin, microtubules, actin (proteins from a cell's cytoskeleton) and lamins (forming the cell's nuclear envelope) [1, 2, 44]. Extracellular protein materials, secreted by cells into the surrounding microenvironment, include elastin and collagen [1, 3, 4]. Other protein materials such as fibrin [5, 6, 45, 46] appear in biological processes such as the clotting of blood. Complicating the matter, distinct protein materials in organisms can not be considered isolated. For instance, lamins are connected to extracellular protein materials (*e.g.*, collagen) *via* cytoskeletal proteins (*e.g.*, actin and nesprin).

Materiomics provides a means to define the structural components of biological protein materials from a materials science perspective, and their role in mechanical materials phenomena, materials failure, and in diseased or altered physiologic conditions. As previously discussed, significant advances and convergence in experimental, theoretical and computational materials science have enabled a deeper understanding through the linking of material structure-process-property and material functionality. For a complete understanding of the materiome, concurrent with natural performance, the understanding of failure in biological systems (broadly defined as the loss of its ability to provide an intended physiologic function) must be considered in the context of material breakdown due to a range of causes, including altered chemical/physical boundary conditions (*e.g.*, extreme forces), weakening of tissues due to structural flaws (such as genetic defects), or the inability of a tissue to provide its function due to the interaction with an ectopic or foreign material. As such, materials science and mechanistic approaches provide critical insight to append the growing knowledge base of biological sciences. This study of material properties of hierarchical protein structures and their effect on molecular and microscopic properties, by utilizing mechanistic insight based on structure-process-property relations in its biological context, provides a basis for understanding such complex systems *via* translating material concepts from biology. However, we reiterate that biological systems are not directly comparable to more traditional engineering materials such as metals and polymers. As such, the investigation of biological materials and related systems give way to a host of materials science challenges and opportunities that need to be addressed and overcome (summarized in Table 3.2).

One of the apparent advantages of exploiting biological materials lies in the fact that the formation of protein materials in natural systems occurs through self-assembly of protein molecules whose structure is encoded in an organism's DNA. Ideally, one could construct an amino acid or DNA "template", and a complex hierarchical system would emerge from the initial, molecular components. Such templates, like building blueprints or mechanical schematics, can then be engineered and compiled for multifarious purposes, and many systems produced with the same base "ingredients". Such approaches are currently being explored, and are discussed further in Chap. 10: Synthesis and Design. The ability to control self-assembly synthetically and genetically provides a powerful framework for the study of the links between structure, property and process. The approach of adding and deleting molecular domains with distinct chemical functionality has enabled the detailed study of the mechanisms of self-assembly processes (*e.g.*, identification of essential protein domains) [47–50]. This allows one to determine how changes to the protein structure alters biological function and disease properties.

The evolution of protein materials through genetic selection and structural alterations has resulted in a specific set of protein building blocks that define their structure. Protein materials exist in an abundant variety, and the need exists to formulate a widely applicable model to systematically categorize all such materials, in order to establish a fundamental understanding, and to exploit the use of hierarchical structural building blocks to develop a new generation of advanced nanomaterials [15, 51, 52]. Biological materials exploit a common set of structural building blocks (*e.g.*, alpha-helices, beta-sheets, random coils) during their formation and function, and the process or mechanism of use of this material (*e.g.*, synthesis, breakdown, self-assembly). The phenomenon of *universality* exists ubiquitously in biology, where certain protocols are commonly found in all protein materials (such as the use of hierarchical levels of building blocks: DNA nucleotides, DNA double helical structure, alpha-helices, beta-sheets), and the process of transcrip-

**Table 3.2** Key materials science challenges associated with the study of biological protein materials, examples and opportunities for further development of methods (experiment, theory, models). Adapted from [7]

| Materials science challenge | Example | Significance | Need for model development |
|---|---|---|---|
| Confinement effects | Change of material properties at different length- and time-scales (*e.g.*, size effects, rate dependence) | Adaptation of universal yet diverse protein structures, biomaterials | Constitutive material models with explicit consideration of hierarchies |
| Environmental effects | Effects of chemicals, solvent (*e.g.*, pH), temperature, etc., on formation, function and material breakdown | Tissue remodelling, protein filament adaptation/remodelling, mechanotransduction (chemomechanical signalling) | Chemomechanical models of tissues, atomistic-level descriptions that integrate mechanics and chemistry |
| Material failure | Material degradation in genetic and infectious diseases, failure due to impact and stress in injuries and/or trauma | Mechanisms of genetic diseases (*e.g.*, *osteogenesis imperfecta*, *progeria*, muscle dystrophies, Alzheimer's), nanomedicine | Constitutive strength models with explicit consideration of hierarchies, plasticity and failure models (*e.g.*, rupture of H-bonds), disease models (*e.g.*, role of molecular defects and misfolds) |
| Multi-scale synthesis and fabrication | Structure control, from nano to macro, how to control implementation of hierarchies | Synthesis of synthetic biofibres (*e.g.*, spider silk) and bioinspired or biomimetic materials (scaffolding protein materials for tissue engineering, interfaces between microdevices and biological systems) | Integration of universal and diverse elements in hierarchical structures, translation from biology to nanotechnology (carbon nanotubes, nanowires, peptides) |
| Multi-scale characterization | Structure measurement at various time- and length-scales (*e.g.*, assembly stages), characterization of mechanisms of deformation | Application of AFM, optical or magnetic tweezers, microfluidic devices, TEM or SEM for material characterization, integration of measuring and imaging approaches across multiple scales to aid model development | Statistical analysis, high-throughput methods, error/uncertainty analysis |

tion/translation, protein synthesis). However, other protocols are highly specialized (such as the use of specific DNA sequences for a particular protein structure, the resultant protein motifs of tendon fascicles, lattice-like lamin structure, etc.), thus representing *diversity*. Protocols can therefore be classified as either *universal* or *diverse*. This paradigm, long known in the biological sciences, will be re-visited in Chap. 4: Universality-Diversity Paradigm: Music, Materiomics, and Category Theory.

## 3.3 Soft Matter, Entropy, and Folding for Function

While materials in biology often involve non-proteineous components such as crystal platelets (*e.g.* in bone, dentin, or nacre), which have important effects on their mechanical properties, the primary component is the polymer-like polypeptide chains. Such materials are typically considered *soft matter* (as opposed to condensed matter), or "matter that is squishy" at room temperature. More specific, soft matter refers to substances whose molecules can be made to move significant distances (deformation) by the application of comparatively weak forces, including thermal stresses or thermal fluctuations, external stresses, or electromagnetic fields. The field of soft matter consists of liquids, colloids, polymers, foams, and gels. Material such as plastic and rubber, which are synthetic polymers, are common examples of soft matter. But more abundant (and more important!) are the soft matter biological materials that compose your cells, tissues, and organs, and allow the functioning necessary for life. Unfortunately, they present more of a challenge.

If we consider synthetic polymers, they typically have repeating segments where each unit is identical. For example, nylon, like a protein, is a polyamide containing monomers of amides joined by peptide bonds. Nylon can be described by the simple chemical formula: $[\text{–NH–}[CH_2]_n\text{–CO–}]_x$, where the variable "$n$" indicates the number of repeat $CH_2$ units between each peptide bond and "$x$" the number of repeat sequences. Kevlar, known for its high tensile strength-to-weight ratio, is another synthetic polymer fiber, characterized by a para-aramid (aromatic polyamide) chemistry, and can be described by the slightly more complicated formula: $[\text{–CO–}C_6H_4\text{–CO–NH–}C_6H_4\text{–NH–}]_n$, where the aromatic groups are represented by $C_6H_4$. A protein, on the other hand, is described by the specific chain of amides, resulting, for example, in the following chemical description of myoglobin (the sequence depicted in Fig. 3.1):

GLY LEU SER ASP GLY GLU TRP GLN LEU VAL LEU ASN VAL TRP GLY LYS VAL GLU ALA ASP VAL ALA GLY HIS GLY GLN GLU VAL LEU ILE ARG LEU PHE LYS GLY HIS PRO GLU THR LEU GLU LYS PHE ASP LYS PHE LYS HIS LEU LYS SER GLU ASP GLU MET LYS ALA SER GLU ASP LEU LYS LYS HIS GLY ASN THR VAL LEU THR ALA LEU GLY GLY ILE LEU LYS LYS LYS GLY HIS HIS GLU ALA GLU LEU THR PRO LEU ALA GLN SER HIS ALA THR LYS HIS LYS ILE PRO VAL LYS TYR LEU GLU PHE ILE SER GLU ALA ILE ILE GLN VAL LEU GLN SER LYS HIS PRO GLY ASP PHE GLY ALA ASP ALA GLN GLY ALA MET SER LYS ALA LEU GLU LEU PHE ARG ASN ASP MET ALA ALA LYS TYR LYS GLU LEU GLY PHE GLN GLY

Again, myoglobin is a relatively small protein (only 153 amino acids). In Nature, the position of each amino acid affects the resulting protein. Indeed, the specificity of each amino acid and their placement and organization begets infinitely variable properties, forms, and functions. The repeating structures of nylon and Kevlar may appear "dull" by comparison! It remains a major challenge to understand the fascinating properties of such complex soft matter from first principles, *i.e.*, by deriving its properties from the microscopic interactions.

We emphasize the importance of entropic contributions to elasticity, and the interplay of different chemical bonding on different strengths and behaviors at different length- and time-scales. Due to the relative flexibility of protein molecules, it is vital to consider various types of chemical interactions that compete with thermal fluctuations, including the covalent bonds of the peptide chains (due to overlap of electron orbitals) and weak, inter-chain interactions such as ionic (electrostatic; columbic interactions), hydrogen-bonds (H-bonds), and dispersive, or van der Waals (vdW) interactions. In general, the elastic properties of materials can be expressed as the partial derivatives of the free energy density with respect to the strain tensor that characterizes the deformation. Based on the free energy:

$$A \equiv U - TS \tag{3.1}$$

where $A$ is the Helmholtz free energy, $U$ is the internal energy of the system, $T$ is the absolute temperature (kelvins), and $S$ is the entropy. We can define the free energy density (*e.g.*, per volume) as:

$$\Phi = \frac{A}{V} = \frac{1}{V}(U - TS) \tag{3.2}$$

From which the elastic modulus (a measure of material stiffness) can be defined as:

$$E = \frac{\partial^2 \Phi}{\partial \varepsilon^2} = \frac{\partial^2}{\partial \varepsilon^2}\left(\frac{A}{V}\right) = \frac{\partial^2}{\partial \varepsilon^2}\left(\frac{1}{V}(U - TS)\right) \tag{3.3}$$

The eccentric behavior of soft matter and biological materials can be traced to that remaining (and irksome!) entropic term, $TS$. Whereas the elasticity of crystalline materials is primarily controlled by energetic changes of the internal energy ($U \gg TS$), in natural and biological materials, elasticity is also significantly influenced by entropic contributions ($S$). This is because in many crystalline materials, the entropic terms can be neglected, such that $A \simeq U$. However, in biopolymers and polypeptides composing proteins, entropic contributions can dominate ($TS \gg U$) the elasticity, in particular for small deformation, such that $A \simeq TS$.

Theoretical descriptions such as the Neo-Hookean constitutive equation are derived based directly from these concepts, albeit often using empirical models. Being said, the introduction of entropic terms (*i.e.*, thermal fluctuations) into constitutive relations and continuum theory (such as finite element analysis) has been successful in biological applications, such as genomic analysis (mapping) of DNA [54, 55], the mechanics of fibrin networks [53, 56] related to blood clots (see Fig. 3.3), and the behavior of actin networks [57] found in the cell's cytoskeleton.

Entropically driven systems (such as polymers and biological materials) are typically characterized by seemingly amorphous (or disordered) structures. A common analogy of describing a polymer melt, for example, is the entanglement of cooked spaghetti. By definition, entropy and thermal fluctuations can induce many geometric configurations in a soft matter system, and as such, theories have been developed

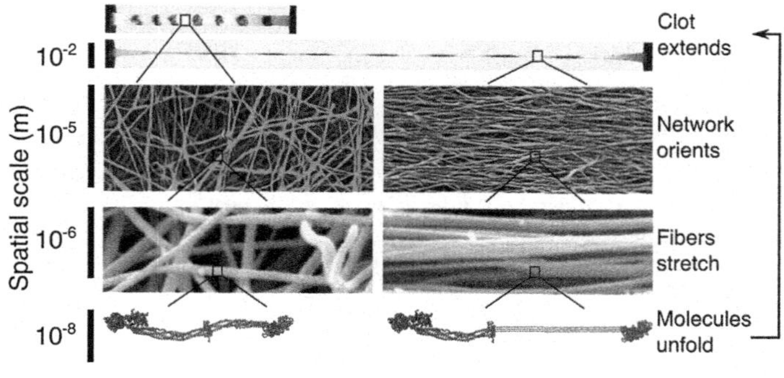


**Fig. 3.3** Entropic contributions to biological materials. Blood clots are highly extensible supramolecular protein polymers formed from well-separated, relatively straight and stiff fibers $\approx 200$ nm in diameter. When stretched, the fiber network aligns in the direction of the applied strain (change in orientation) and the individual fibers stretch, forcing the fibrin monomers that make up the fibers to extend. Ultimately, it is this molecular unfolding that allows clots to stretch so far. Thus, understanding fibrin clot mechanics requires knowledge of the mechanical response and the corresponding structural changes spanning from the centimeter scale to the nanometer scale. From Weisel *et al.* [53], used with permission, copyright © 2009 American Association for the Advancement of Science

to describe the behavior of such systems (such as worm-like chain models, or Flory-Huggins solution theory). For proteins, however, the random convolutions of polymers give way to predetermined folded structures—a phenomena known as *protein folding*—wherein entropy and ease of deformation is necessary to induce structural conformations (see Fig. 3.4).

The variable geometric and topological conformations (*i.e.*, folding) of polypeptides is the fundamental basis for creation of the first hierarchies of biological materials. For proteins, these hierarchies are typically referred to as the secondary, tertiary and quaternary structures (the primary structuring being the sequence itself). The introduction of "folded" variations can be though of as a mechanism to extend the possible "design space" (*e.g.*, achievable properties and functions) with a limited number of building blocks (*e.g.*, amino acids).

As a simple though experiment, one could consider a protein-like sequence with chemically amicable domains designated "A" and "B" (see Fig. 3.5). For this purpose, the pairing of such domains is desirable for functionality (such as a catalytic reaction), but the domains are not adjacent to each other. Without folding, there are only two routes to pair the domains: either change the sequence (requiring a change in DNA transcription) or introduce a second sequence (more material). The introduction of folding, however, provides many possible successful pairings. Moreover, the flexibility of the material (*e.g.*, soft matter) facilitate such geometric manipulations with nominal energetic expense. While we disregard the mechanism to induce specific folded structures (for now), this simple example illustrates the potential (and advantage) of allowable folding in extending the combinatorial matching of amino acids and chemical functional groups. Note that this example does not necessarily have to be limited to the scale of the peptide sequence—indeed, the "A" and "B"

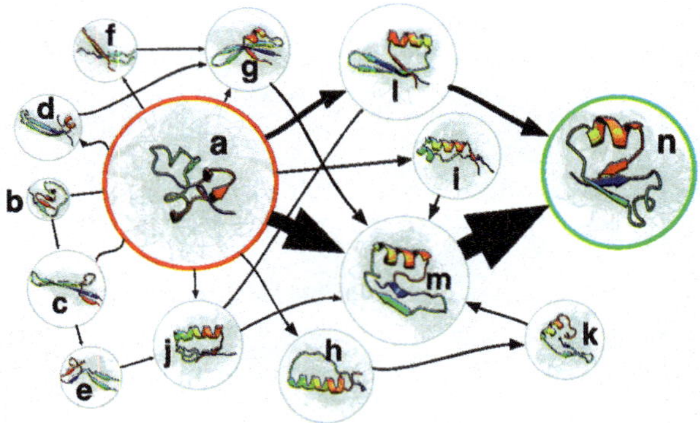


**Fig. 3.4** The complexity of the protein folding problem (also referred to as protein structure prediction). A complete understanding of how proteins fold, *i.e.* self-assemble to their biologically relevant "native state", remains an unattained goal. Not only does the final structure have to be determined, but also the energetic pathway to the final configuration along with any possible intermediate states. The figure depicts a sampling of the global free energy landscapes of folding using statistical Markov State Model (MSM). Shown is the superposition of the top 10 folding states (molecular structures) and associated pathways (*arrows*) labeled (**a**) to (**n**). The visual size of each state is proportional to its free energy, and *arrow* size is proportional to the interstate flux. Thus, the most "likely" folding pathway (and intermediate structures) is given by (**a**) → (**m**) → (**n**). From Pande *et al.* [58], used with permission, copyright © 2010 American Chemical Society

domains could represent higher-order (hierarchical) structures, but the analogy is the same.

The intrinsic flexibility and entropically governed nature of protein materials permit multiple variations in structure (*qua* function) with nominal required changes in sequence. Considering the origin of the sequence (*via* DNA transcription), exploitation of geometric and topological space provides a more efficient means to attain desired results, opposed to complete changes in sequence requiring significant changes to DNA and possibly re-direction of available resources (*i.e.*, raw chemical materials).

While the concept of folding is quite simple, the implications of such behavior is complex. From a molecular perspective, the problem (or challenge) lies in the *prediction* of secondary and higher order folded protein structures [59, 60]. In addition, several neurodegenerative and other diseases are believed to result from the misfolding of proteins (further discussed in Chap. 9: Pathological Materiomics), providing a practical impetus to understanding the folding phenomena. In broad terms, the protein folding problem is to determine how proteins so consistently fold into a stable state. Ingenious efforts have developed approaches and techniques to explore the stable configurations of proteins, ultimately attempting to correlate a given sequence with a deterministic structure (and indirectly, linking the sequence with function). Examples include the `Folding@home` project developed at Stanford University[1]—a distributed computing project launched in 2000 where people

[1] http://folding.stanford.edu.

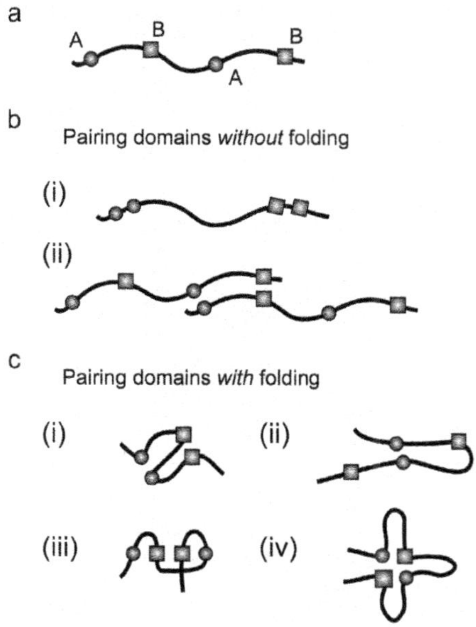


**Fig. 3.5** Folding as mechanism to extend the possible "design space". (**a**) A representative protein-like sequence with chemically amicable domains designated "A" and "B" (*circles* and *squares* respectively). (**b**) Without folding (*i.e.*, a rigid sequence), paring domains can only be accomplished *via* (i) a fundamental change of sequence (*i.e.*, change of building blocks) or (ii) introduction of another sequence (*i.e.*, more material resources). Both means are energetically inefficient. (**c**) Allowing folding introduces a plethora of new configurations and possible pairings of domains, including (i) A-A and B-B pairing, (ii) A-A pairing while isolating B domains, (iii) sequential A-B-B-A assembly, (iv) parallel or concurrent matching of A-A, A-B, and B-B domains (*e.g.*, 3D folding)

throughout the world download and run software to effectively make one of the largest supercomputers in the world [61]. Researchers from the University of Washington[2] involved with another distributed computer project, `Rosetta@home`, have developed an experimental video game about protein folding: *Foldit*—a multiplayer online game that engages non-scientists in solving hard prediction problems [62] (screenshot depicted in Fig. 3.6).

Foldit is an attempt to apply a human's natural three-dimensional pattern matching abilities to protein folding problem. Puzzles are based on well-understood proteins and by analyzing the ways in which humans intuitively approach these structures, researchers hope to improve the algorithms employed by existing protein folding software. Such efforts are ultimately mechanistic in nature, encompassing the systematic and parametric exploration of the free energy landscapes of macromolecules to determine the local minima (sometimes referred to as the principal of minimal frustration [63]). Assumption that such a *native* folded state exists led to

[2]http://boinc.bakerlab.org/rosetta.

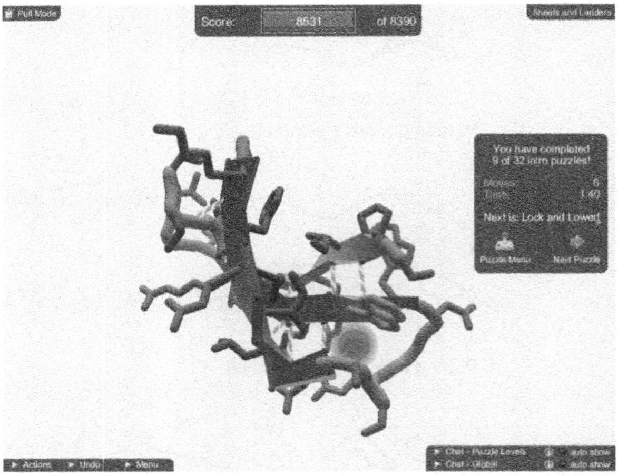


**Fig. 3.6** Screenshot of the game "Foldit"—an experimental crowdsourcing video game about protein folding, developed as a collaboration between the University of Washington's departments of Computer Science and Engineering and Biochemistry [62]. Determining how the primary structure of a protein turns into a functioning three-dimensional structure—*i.e.*, how the molecule "folds"—is difficult; the general mechanisms are known, but predicting protein structures is computationally demanding. Foldit is an attempt to apply the human brain's natural three-dimensional pattern matching abilities to this problem. By analyzing the ways in which humans intuitively approach such puzzles, researchers can improve the algorithms employed by existing protein folding software

the thermodynamic hypothesis known as *Anfinsen's Dogma*. Simply put, Anfinsen's Dogma states that the native folded structure of a protein is determined only by the protein's amino acid sequence. The minimum of the free energy of the native structure is (1) unique (the sequence does not have any other configuration with a comparable free energy), (2) stable (small perturbations cannot give rise to changes in the minimum configuration) and (3) kinetically accessible (the energetic path must be reasonably smooth) at the environmental conditions (temperature, solvent concentration and composition, etc.) at which folding occurs [64].

One-to-one understanding of sequence to final structure, however, is only part of the problem. The other challenge lies in determining the folding *pathway* [65]. Thus, countering Anfinsen's Dogma is *Levinthal's paradox*—a thought experiment in the theory of protein folding. The Levinthal paradox states that, due to the very large number of degrees of freedom in an unfolded polypeptide chain, the number of possible conformations available to a given protein is astronomically large. For example, even a polypeptide of a mere 20 residues will have 19 peptide bonds, and therefore 38 different *phi* and *psi* bond angles. If each of these bond angles can be in one of three stable conformations, the protein may misfold into a maximum of $3^{38}$ different conformations (including any possible folding redundancy). Therefore if a protein were to attain its correctly folded configuration by sequentially sampling all the possible conformations (say, one per *nanosecond*), it would require over forty years to arrive at its correct native conformation. That is only considering a mere

20 residues. The protein hemoglobin, for example, contains *over 500* residues. The "paradox" is that most small proteins fold spontaneously on a millisecond or even microsecond time scale [66]. The paradox also makes computational prediction of protein structures under the same basis unfeasible if not impossible (the number of all possible conformations of a polypeptide chain is too large to be sampled exhaustively). The presupposition of environmental conditions further complicates the folding problem, as variations in solvent type, concentration of salts, pH, temperature, and molecular chaperones can serve to actively guide the folding pathway.

Folding to achieve and manipulate form and functionality is currently being explored outside the domain of proteins and biological material. Successful assembly of constituent elements which interact and rearrange intelligently can lead to the development of "programmable matter"—materials whose properties can be programmed to achieve specific shapes or stiffnesses upon command. A synthetic analog to proteins, programmable matter could form different shapes autonomously by folding. Past approaches to creating transforming machines have been limited by the small feature sizes, the large number of components, and the associated complexity of communication among the units, but recent attempts mitigate such difficulties through the unique concept of self-folding origami with universal crease patterns. Like biological materials, such systems are capable of achieving multiple shapes for multiple functions.

The theoretical basis of self-folding programmable matter arises from the field of computational origami [68, 69]. One early result in this field [70] states that every polyhedral surface can be folded from a sufficiently large square of paper. The concept of folding and creating almost any "structure" or shape just from a pliable piece of material is akin to biological materials where universal building blocks create diversity of function. Folding—as opposed to complete reorganization of chemical bonds and structure—can be though of as an energetically favorable means to sample many potential configurations (and optimize function). Just as a sheet of paper can be manipulated to resemble a "crane" or a "boat" (as depicted in Fig. 3.7), a single polypeptide can be stable in multiple conformations. Moreover, due to the noncovalent nature of the folded protein, most structures are reversible (some proteins, such as myosin, use transitional states as a type of protein "motor" [71]). Nominal changes in folded structure may be driven by the need for new functionalities of variations in environmental conditions, and can be attained without "redesigning" the material platform but rather a simple shifting of the required folds, creases, and coils.

Protein structure prediction (the "protein folding problem") is ultimately a molecular level problem, characterized by amino acid interactions and atomistic states. Although the folding of proteins is an important element that affects most materials in biology, it does not provide the information necessary to characterize all relevant material properties and functions across scales. There also exists larger-scale structural features, which influence the overall mechanical properties, that typically arise in distinct *hierarchical* arrangements critical to the understanding of the materiome.

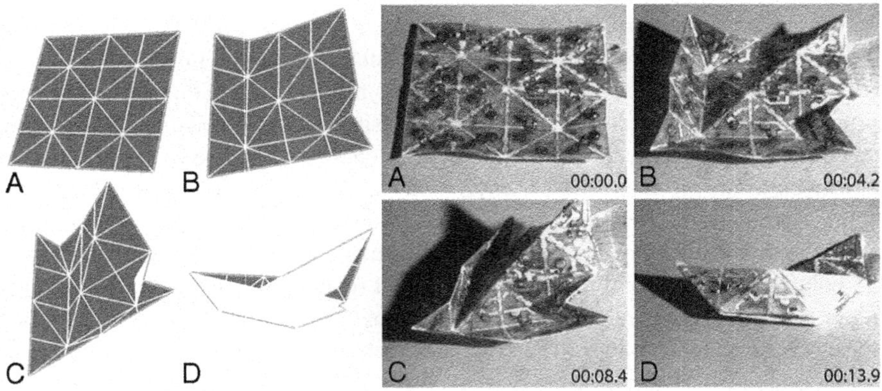


**Fig. 3.7** Programmable matter by folding... the protein way? Programmable matter is a material whose properties can be programmed to achieve specific shapes or stiffnesses upon command. This concept requires constituent elements to interact and rearrange intelligently in order to meet the goal. The sheet shown can fold with a certain crease pattern to create multiple three-dimensional shapes, depending on which creases fold, in which direction, and in which order. Simulation (*left*) and experiments (*right*, with time shown in *lower right*—mm:ss.s) of a self-folding "boat". (**A**) Initial sheet. (**B**) All actuators receiving current. (**C**) Immediately before magnetic closures engage. (**D**) Finished boat on side. From Wood *et al.* [67], copyright © 2010 National Academy of Sciences

## 3.4 Nature's Hierarchies

The limited number of constituent amino acids combined with multiple folded states cannot account for the diverse materials and functional systems found in Nature. Whereas metals and ceramics often show a regular structure of polycrystalline nature (yet draw from a greater variety of elements), many biological materials show a much more complex hierarchical and rather heterogeneous structure. Functionality arises across scales, wherein the same common building blocks can be assembled in different hierarchical arrangements (*e.g.*, molecules, filaments, mesoscale structures) to attain different purposes at the macroscale and control a material's properties. Moreover, while diverse protocols are distributed heterogeneously across hierarchical levels, common (and successful) structural arrangements are found in many diverse systems.

For example, the four DNA nucleotides (adenine (A), cytosine (C), guanine (G) and thymine (T)) represent a universal protocol common to all protein materials (forming complementary hydrogen-bonded base pairs), where their arrangements in diverse patterns form the immense variety of genetic sequences found in biology. Through the universal process of transcription and translation, protein molecules are synthesized into a one-dimensional sequence of the universal twenty amino acid building blocks. Genetic sequences are universally encoded in double-helical DNA, regardless of the specific nucleotide sequence. While DNA serves a particular biological role, other hydrogen-bond stabilized helical structures can be found in the alpha-helices of intermediate filament dimers within cellular membranes (a single helix), topocollagen fibrils composing bone (a triple helix), and superhelical keratin

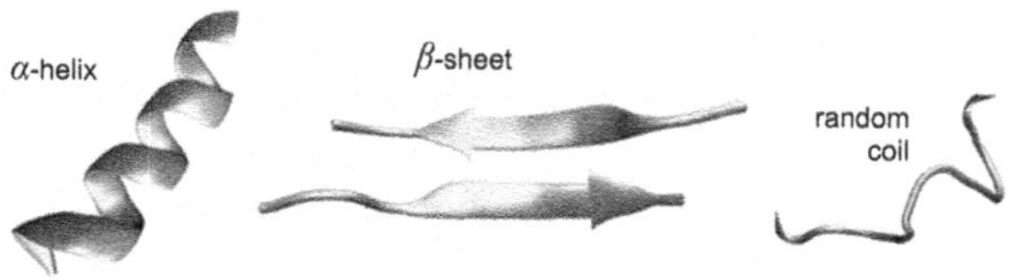


**Fig. 3.8** The secondary structural "building blocks" of protein materials: $\alpha$-helices, $\beta$-sheets and random coils. These universal motifs arrange into unique, diverse larger-scale protein structures. Note that each structure depicted consists of approximately the same number of atoms/residues

fibrils found in hair (helically coiled helices). Moreover, the secondary structures of virtually all proteins consist of a distribution of one or more of the universally found motifs: $\alpha$-helices, $\beta$-sheets and random coils (as seen in Fig. 3.8). These universal motifs arrange into unique, diverse larger-scale protein structures (*e.g.*, enzymes, fibres, filaments).

Sometimes, as for example in the case of elastin [72] or spider silk [73], the microstructure at the molecular or atomistic level is not completely known—consisting of random domains that have, thus far, largely thwarted any attempt to decipher their structure. This also applies to some natural composite materials such as cement [74, 75] for which not even the basic building blocks or their structure is fully known (as previously discussed in Chap. 1: Introduction). Similarly, the grain boundaries of polycrystalline metals and alloy (such as Ni, Co, and Cu and binary alloys such as Ni-Fe or Ni-W), while well-observed and defined [76], depict little order or structuring across-scales (that is not to say however, that such grain boundaries depict characteristic scales for particular materials). Within the field of soft matter, many polymeric systems, composites, and rubbers lack any long-range order, yet have well defined "hierarchies" such as layered striation [77], colloidal hollows or inclusions [78], and enhanced crystallinity [79]. Many such materials—considered chemically complex—share critical features with many biological composites and the computational and experimental techniques discussed within this text—a materiomic perspective—may also be broadly applicable to these materials.

In Nature, a greater diversity of protocols is generally found at higher levels, suggesting that biological functionality is associated with structural diversity. However, in the case of biological materials, simple mechanical behavior is complicated, and the "function" of hierarchical features unknown. As a contrasting example, a decrease in size of grain boundaries serves as a strengthening mechanism in many polycrystalline materials (the so-called Hall-Petch effect). In simple terms, grain boundaries impede dislocation movement, and changing grain size can effectively toughen a material by influencing dislocation propagation associated with plastic deformation and thereby increase yield strength (note that this is not universal). As a direct result, material processing techniques such mechanical alloying and compaction, severe plastic deformation, gas-phase condensation of particulates and consolidation, and electrodeposition have been developed to specifically alter grain sizes and exploit the Hall-Petch effect [80]. Complicating the case of biological materials, due to the heterogeneous and hierarchical design, mechanical behavior may be different at different scales. For example, bone is composed of two phases: single crystal platelets of hydroxyapatite, and a matrix of collagen [81–83]. Whereas hydroxyapatite itself is extremely fragile (think classroom chalk), collagen is extremely compliant. However, the combination of these two phases in a nanostructural composite makes a material that is though [81], and the interaction between soft protein materials and stiff mineral is only recently being investigated. Thus, in order to understand the mechanics of hierarchical materials, it is critical to investigate the responses at different scales, including a treatment of the individual phases and the interaction of different phases. Few (if any) "Hall-Petch" (or "inverse Hall-Petch")-type relations[3] have been formulated for biological materials, and fewer still "material processing" procedures to exploit them systematically.

Consider a complex material subject to a macroscopic defect such as a crack. Due to a heterogeneous structure of a material, the response of the individual building blocks needs to be considered. The individual building blocks may respond in various ways, including energetic elasticity, fracture, or entropic elasticity, or by a mixture of different types of elastic and plastic behavior. The size of the defect relative to the size of the building blocks (or hierarchical scale) is important to characterize the response of the material. A fundamental difference between engineered materials and naturally formed biological materials is that functionality in biology is created by arranging universal building blocks in different patterns, rather than by inventing new types of building blocks. The formation of hierarchical arrangements provides the structural basis to enable the existence of universality and diversity within a single material. This combination of concepts may explain how protein materials are capable of combining disparate material properties, such as high strength and high robustness, together with multi-functionality.

Biological functionality must be understood at varying scales, and as a complex interaction of mechanisms that occur simultaneously at multiple levels. While biochemistry focuses on biological functionality at molecular scales, the potential expression of varying molecular structures may be lost at higher levels. The mesoscale that encompasses length-scales that range from nanometers to micrometers and time-scales of nanoseconds to microseconds is a particularly important level necessary to understand how specific protein materials derive their unique properties and what role they play (both in physiologic and pathologic phenomena). The mesoscale science of protein materials, through the linking of molecular properties to properties of protein materials at the microscale, thus represents an important frontier of materials research with great potential for fundamental contributions to biology, medicine as well as for the *de novo* synthesis of engineered materials. The engineering question that ultimately arises is simple: Can an arbitrary set of universal building blocks be assembled to attain a desired and unique functionality?

Such a paradigm is not limited to molecular components at the nanoscale—the approach of utilizing universal building blocks to create diverse multifunctional

[3]The Hall-Petch relation explains how reducing the grain size in polycrystalline materials (*e.g.* metals, ceramics) leads to a strengthening; the inverse Hall-Petch effect refers to a weakening of the material below a maximum strength.

hierarchical structures has been successfully applied in current macroscale engineering paradigms. For example, in the design of structures such as buildings or bridges, universal constituents (bricks, cement, steel trusses, glass) are combined to create multi-functionality (structural support, living space, thermal properties, light harvesting) at larger length-scales. It is quite obvious that a steel bridge serves a different purpose than a steel building or a steel ship, yet each structure can be constructed *via* a selection of cleverly assembled prefabricated sections. A steel beam or a plate has specific mechanical properties, responses to loads, and performance (found in numerous construction handbooks and design specifications). From a top-down perspective, it is only the hierarchical arrangement of steel that differentiates the *Golden Gate Bridge* from the *RMS Queen Mary 2*.

The challenge of utilizing similar concepts that span to the nanoscale—as exemplified in biological protein materials—by merging structure and material could enable the emergence of novel technological concepts. At this time, we do not have a catalogue of "prefabricated protein sections" with detailed properties and specifications. Moreover, another key obstacle in the development of new materials lies in our inability to directly control the structure formation at multiple hierarchical levels—we cannot "weld" or "bolt" together a biological system a single alpha-helix at a time. However, the concept of universality and diversity and the knowledge gained from how to characterize these materials at different hierarchical levels greatly contributes to addressing these challenges. Nature's use of a limited number of universal building blocks, arranged diversely in a variety of ways, is a limitation as well as a strength of biological systems. For example, although the performance of structural tissues in our body is poor compared with most engineered materials (such as steel or concrete), their performance is remarkably good considering their inferior building blocks. Understanding these material concepts and the translation to the design of synthetic materials could provide us with new concepts for materials design based on inexpensive, abundant and environmentally benign constituents.

Figure 3.9 shows a summary of the structural makeup of two example protein materials, intermediate filaments and amyloids, revealing their hierarchical structures that range from nano to macro. The fact that virtually all biological structures feature nanostructural elements at the most fundamental level suggests that new insight into nanoscience and nanotechnology can be derived from the study of these materials. The hierarchical makeup of biological materials is elementary to their ability to provide *specific* biological functions. However, at this point it remains poorly understood what the role of these distinct hierarchical structures are, how they regulate the growth and function of biological systems, and what the driving forces are for their formation. While hierarchial material structures an undoubtedly common amongst biological systems, less clear is the inherent benefit (if any) to such ordered and multi-scale arrangements. Turning once again to more traditional analogs, hierarchies are frequently employed with various motivations and advantages. Some such feasible benefits of common hierarchical systems are presented in Table 3.3, but, while hinting the possible gains of natural hierarchies, direct conferral to biological systems is currently unwarranted and merely conjecture.

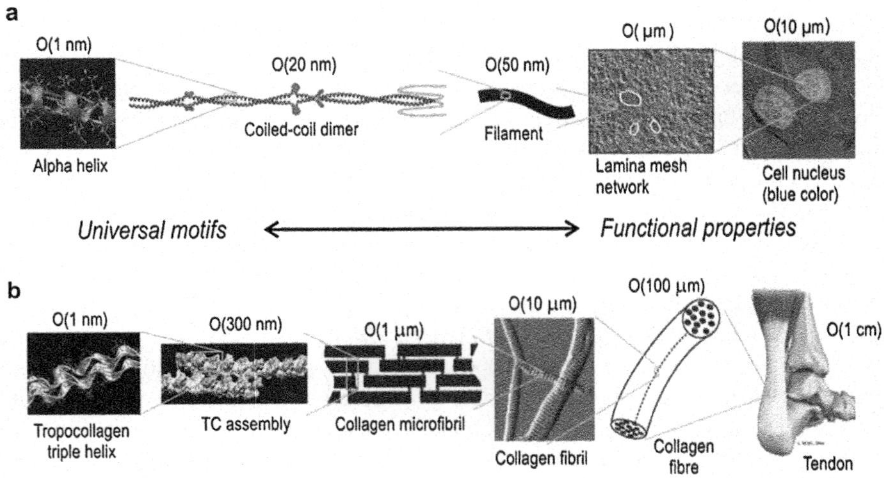


**Fig. 3.9** Example hierarchical multi-scale structures of biological protein materials: lamin intermediate filaments (lamins), collagenous tissues (tendon). (**a**) Structure of lamins. Alpha helical protein domains assembly into dimers, which form filaments that define a lattice-like lamin network of the cell's nucleus (lamina mesh network snapshot adapted and reprinted with permission from Macmillan Publishers Ltd. from [84]). (**b**) Structure of collagenous tissues, from nanoscale to macroscale. Tropocollagen molecules assemble to form fibrils, which form fibres that provide the structural basis of collagen tissues such as tendon (experimental picture of collagen fibril reprinted with permission from Elsevier Ltd. from [85]; eOrthopod image of Achilles tendon provided as a courtesy of Medical Multimedia Group, http://www.medicalmultimediagroup.com/)

The requirement of natural systems to support multiple and diverse functions explains the formation of hierarchical structures with defined length-scales for key protein domains that are, as a consequence, found as universal features throughout biology. In fact, the formation of hierarchical structures facilitates the emergence of diverse properties created by universal building blocks, a concept discussed in subsequent chapters and referred to as the Universality-Diversity Paradigm (or UDP). Previously it was discussed that Nature has developed multiple safeguards or buffers to ensure slight deviations in input do not result catastrophic changes in macroscale system behavior—a kind of natural design tolerance or quality control. Most small variations do not result in divergent system-level properties. However, the inherent complexity of potential cross-scale interactions introduces nonlinear multiplicative effects, at times detrimental to the entire system. While the favoring of robustness over precision enables functionality and imparts flaw tolerance, some small variations sometimes result in significant system changes. Such changes are typified by mutations (such as the glutamic acid to valine substitution in hemoglobin that results in sickle cell disease), wherein a single amino acid substitution can be result in complete loss of system function. The same structure that imparts adaptability, robustness and tolerance can utterly fail with minimal perturbations—a metaphorical *Achilles' heel.* The complexity associated with such hierarchical systems are difficult to predict, and one sequence substitution may be benign while swapping the

**Table 3.3** Feasible benefits and advantages of engineered hierarchical systems

| Hierarchical advantage/benefit | Description | Common or engineering analog |
|---|---|---|
| Multiplicative components (efficiency) | Efficient use of designed small-scale subsystems (components or building blocks), employed numerous times, to efficiently bridge larger scales in the formation of a superstructure or supersystem. | Standard selection of codified roof trusses to erect a roofing system. The "subsystems" can be arranged many ways for many different structures without a change in design or behavior at the component-scale. |
| System robustness (redundancy) | Failure restricted to small-scale or single-phase components such that system operation and functionality is maintained in spite of sustained damage due to multiple or repetitive features that localize failure, mitigate structural damage, or prevent catastrophic propagation. | The hub and spoke model of airline transportation or a communication network. Failure of a single hub can be mitigated by re-routing of traffic through other pathways whereas failure of a single or multiple spokes is localized and non-damaging to the entire system. |
| Resource dissemination (supply) | Facilitation of distribution and material interactions across multiple scales, allowing function, potential efficacy and possible growth at many scales within the same system, inaccessible to single-scale design. | The distribution of electrical power, where large transmission lines transfer electricity from generation to a target area (such as a city or town) and smaller scale transformers and power lines feed individual consumers, ranging from a large skyscraper to a single traffic light. |
| Extended design space (combinatorial) | Functionality and required performance can be created by arranging common component building blocks assembled in different patterns, arrangements, and structures across scales, rather than by inventing new types of building blocks, thereby exploiting the available and/or limited materials. | The organization of an musical orchestra, an ensemble of various string, brass, woodwind, and percussion instruments. The composed music can range from Mozart to Bach with only re-arrangement of not only of the instruments (structure), but the notes they play (function). |
| Design tolerance (quality control) | A hierarchical structure facilitates scale homogenization without a need for precision at the component level. While each component is similar, exact and accurate construction at small-scales is unnecessary if functionality is maintained at pertinent scales. | The hierarchical organization of military units (*i.e.*, chain of command). Each individual soldier need not perform ideally, but the platoon must act as a cohesive whole for the company to be successful, each company cohesive for the battalion, each battalion for the regiment, etc. |

very next amino acid may be fatal. A more complete understanding of the mechanisms and hierarchical functions is necessary to enable the prediction (and possible prevention) of such system flaws.

The origin of how naturally occurring biological protein materials are capable of unifying disparate mechanical properties such as strength, robustness and adaptability is of significant interest for both biological and engineering science. Strength and robustness are properties of fundamental importance to biological materials and structures, and are crucial to providing functional properties to living systems. *Strength* is defined as the maximum force (or pressure) a material can withstand before breaking. *Robustness* can be considered the ability of a material to tolerate flaws and defects in its structural makeup while maintaining its ability to provide functionality. These properties are crucial for materials in biology (such as skin, bone, spider silk, or cells), which either must provide reliable structural support themselves (such as the skeleton formed by bone), or must reliably withstand mechanical deformation under normal physiologic conditions (such as cells and tissue associated with blood vessels that are exposed to the pressure of the blood).

A fundamental question that arises in whether hierarchical structuring is advantageous (in terms of strength and robustness) in general, independent of material, biological, and physiological factors. To explore, we consider a simple case study of hierarchical assemblies of alpha-helical protein domains—a known natural "building block" for complex biological systems—and compare the strength-robustness properties of composed filaments. The basic building block for all filaments considered is an alpha-helical protein domain as shown in Fig. 3.10, a common secondary protein structure stabilized by three to four H-bonds per turn (an alpha-helical turn has an average of ~3.6 H-bonds). The question examined is simple: *is it possible to build larger-scale protein material structures out of individual protein domains that maintain high levels of strength and robustness, a property that is crucial for physiological function (survival)*? In other words, can we improve performance by hierarchies? Note for this analysis, the concept of robustness ($R$) is defined as the strength of a filament in which one element (here, one alpha-helix) is missing at the smallest level divided by the strength of an intact filament (*e.g.*, the ratio of strength of a defected system, $F(i-1)$ and an intact system, $F(i)$; a measure of fault insensitivity), according to the following equation:

$$R(i) = \frac{F(i-1)}{F(i)} \tag{3.4}$$

Figure 3.10 depicts the geometries and results for permutations of eight alpha-helices (the definition of sub-elements and their arrangement are those shown in the inlay of the figure). The analysis shows that even though no additional material is used, the mechanical performance changes significantly as the hierarchical arrangement of the structure is varied (see caption of figure for details regarding the nomenclature). The {8} structure provide very high levels of robustness, albeit at low strength. In contrast, the {4, 2} structure provide high strength, albeit at low

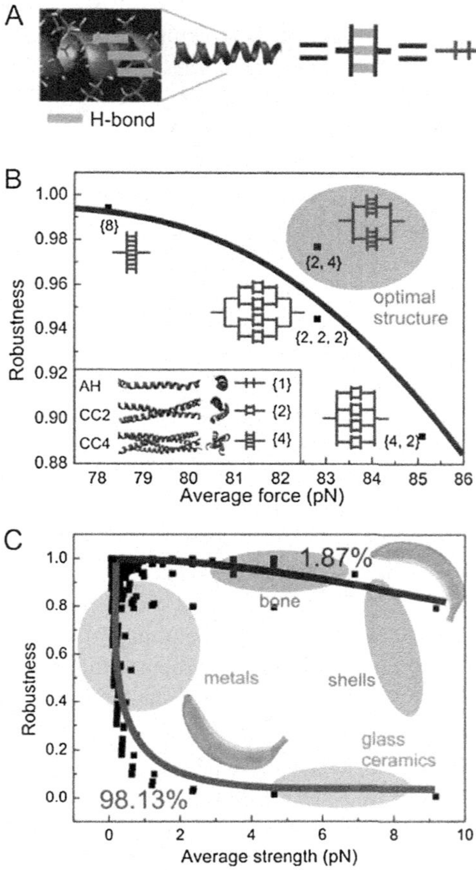


**Fig. 3.10** Strength-robustness relation for alpha-helix based protein filaments, arranged in different hierarchical structures to assess their respective performance (results adapted from [15, 86–88]). *Panel A*: geometry of a single alpha-helix, composed of 3–4 H-bonds per turn. We study how the performance in the strength-robustness domain changes if several alpha-helices are assembled in different hierarchical patterns. *Panel B*: results for eight sub-elements in the protein filament arranged in all possible hierarchical patterns. The definition of sub-elements and their arrangement are shown, differentiated by the number of elements that are found in parallel with each other at a particular hierarchical level. A single alpha-helix (AH) is characterized by {1}, a bundle of two alpha-helices {2} resembles a coiled-coil structure (CC2), and a bundle of four alpha-helices {4} resembles a four-fold coiled coil structure (CC4); see *insets*. The {8} structure represents a single bundle of eight alpha-helices in parallel; the {2, 4} structure represents a fiber composed of two bundles of four alpha-helices; the {2, 2, 2} structure represents a fiber composed of two bundles of two bundles of two alpha-helices each; the {4, 2} structure represents a fiber composed of four bundles of two alpha-helices. *Panel C*: results for 16,384 sub-elements in the protein filament. An analysis of the distribution of structures and their performance shows that most data points ($> 98\ \%$) fall onto the so-called "banana curve". Less than 2 % of all structures lead to high strength and high robustness. This analysis shows how high-performance materials can be made out of relatively weak constituents such as alpha-helices that are bonded by mechanically inferior H-bonds [86–88]

robustness. However, there are some structures that provide an optimal combination of both properties, the {2, 2, 2} and {2, 4} structures. Among these, the {2, 4} structure is the best performer as it provides the highest levels of strength and robustness. The {2, 4} structure represents a fiber composed of two bundles of four-fold coiled coil alpha-helices (abbreviated as CC4). The results suggest that biological materials may have developed to yield materials with multiple objectives, such as high strength and high robustness, a trait that is achieved with relatively weak materials by utilization of hierarchical structures.

A key result from this analysis is that the particular arrangement of constituents at the nanoscale matters. Specifically, a particular design is required in order to observe the unique mechanical traits founds in biological materials; here, high strength at high robustness. This observation explains the relatively poor mechanical performance of engineered materials. These materials typically contain random arrangements of nanostructural elements (due to a lack of hierarchical structures, *e.g.* the microstructure of grains in polycrystalline metals are typically randomly arranged). For the filaments considered, for a large number of constitutive elements (>16,000) most "random" hierarchical structures fall onto a so-called *banana curve* on which strength and robustness remain exclusive. Less than 2 % of all structures provide the inverse behavior, typified by natural materials, such as bone.

Clearly there is a relationship between a materials robustness, strength and level of hierarchical complexity, exemplified throughout a range of biological materials and explicitly shown in the investigation of alpha-helical assemblies. A possible conceptual explanation is the increase of system entropy with an increase in hierarchies (a kind of "hierarchical degree of freedom"). By using an approach that multiple hierarchies inherently increase the number of microstates resulting in global system macrostates, the entropy is also increased. The hierarchical structure provides additional complexity to the energy landscape of the system, increasing the total number of possible energy states for a given loading. This can be compared to structural dynamics, where increasing modal degrees of freedom increases possible modal shapes, frequencies of the system, and energy dissipation capacity. Likewise, the number of "hierarchical degree of freedom" explicitly changes the entropy of the system and affects strength and robustness [88].

Studies of the effects of hierarchical structures on fracture toughness have recently been extended to silica-based materials [89] (motivated by the usage of silica in biological systems, such as sea sponges and diatoms as discussed in Chap. 1: Introduction). The study reported a systematic investigation of the effect of hierarchical structures on toughness and defect-tolerance based on a single and mechanically inferior brittle base material, silica, using a bottom-up approach rooted in atomistic modeling. Figure 3.11 shows some of these results, illustrating how introducing hierarchical structures in a simple silica material yields a significant increase in the toughening behavior. What is notable is that drastic changes in the material crack-propagation resistance (R-curve) emerge solely due to the introduction of hierarchical structures, and not due to the introduction of additional materials (building blocks). The two fundamental building blocks are created from an

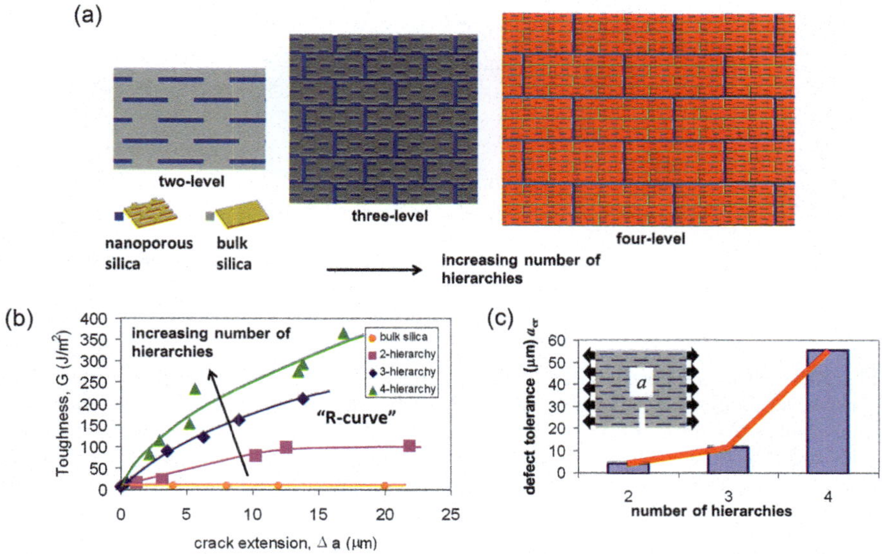


**Fig. 3.11** Effect of hierarchical structures on fracture toughness (R-curve) and defect tolerance (it can be regarded as a generic model for mineralized biological materials, such as nacre or bone). (**a**) shows three different hierarchical geometries, all generated from the same material—silica. Two fundamental building blocks are created by introducing holes into the material ("nanoporous silica") and another one resembles bulk silica. (**b**) shows R-curves, plotting the fracture toughness for growing cracks. The increasing number of hierarchies lead to a significant toughening behavior, much beyond the very low and constant fracture toughness of bulk silica. (**c**) depicts the damage tolerance capacity of the different structures considered, demonstrating that solely through the introduction of hierarchical structures, cracks on the order of 50 micrometers can be tolerated. The *inlay* in (**c**) illustrates the fracture toughness test carried out. Adapted from [89]

identical base material, silica. One building block resembles nanoporous silica (by introducing nanoscale holes) and another one resembles bulk silica. The structures and resulting stress-strain curves are shown in Fig. 3.12, obtained using reactive molecular dynamics [89–91]. These two building blocks were then assembled into different composite structures. Over a range of up to four hierarchy levels, an exponential increase in the defect-tolerance is observed, approaching hundred micrometers (Fig. 3.11(b) and (c)). This presents a significant departure from the defect-tolerance of the base material, silica, which is brittle and highly sensitive even to extremely small nanometer-scale defects. Mechanistically, large loads are dissipated in multiple-hierarchy materials through the initiation and arrest of cracks that occur at different length scales. This is in contrast to single hierarchy materials, in particular for the case of a brittle material like silica, that would shatter with a single crack propagating through the material. If the load does not increase to levels that would lead to unstable crack propagation, this mechanism provides a tissue or an organism the necessary time to repair the material in order to restore its full load carrying capacity.

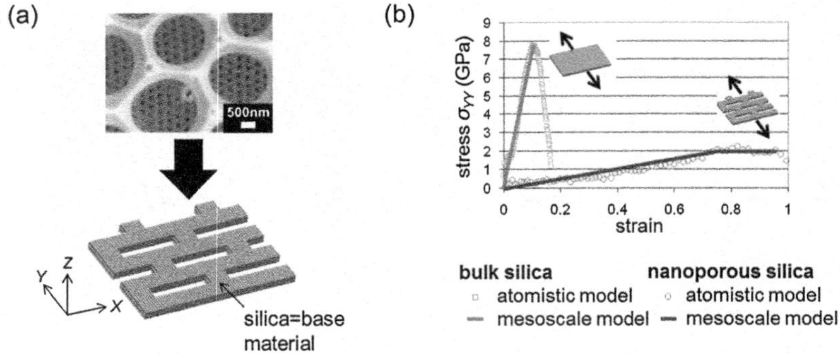


**Fig. 3.12** Illustration how two distinct material building blocks are created from a single base material, silica, by modulating the nanostructure. The nanoporous structure of silica is inspired from the ultrastructure of diatoms as shown in (**a**). Panel (**b**) shows the stress-strain curves for bulk silica and nanoporous silica, revealing a distinct material behavior. The introduction of nanoscale holes in silica yields a more ductile, highly dissipative response, as shown in reactive atomistic modeling [90, 91]. Adapted from [89] and [90]

## 3.5 Mechanistic Insights

Through what means can we hope to understand the complex hierarchies of Nature's materials? Advances in genomics, molecular and biological chemistry—while successful in describing the behavior of proteins, carbohydrates, lipids, nucleic acids and other biomolecules—are restricted to the description of biochemical signaling and metabolic processes such as the chemistry of enzyme-catalyzed reactions. Such information—necessary for a complete description of the materiome—is difficult to "upscale" across hierarchical levels. For example, knowledge of the enzymatic cleavage of collagen (through enzyme matrix metalloproteinase-1, or MMP-1; see Fig. 3.13) is not indicative of the performance and behavior of cells and tissues. However, a direct linkage between microscale mechanical forces and local material remodeling undoubtedly has unknown consequences in cell, developmental, and cancer biology [92]. A consistent materiomic perspective requires a *scale invariant* approach: descriptors and heuristics at one hierarchical level should be equally applicable across all levels. The common language of mechanics—the physics concerned with the behavior of materials when subjected to *forces* or *displacements*—provides such a view.[4] To decipher the fundamental mechanisms of biological materials systematic mechanical studies of deformation, failure, structural dynamics and mechanochemical transduction in living cells and biomolecules are needed [93].

The mechanical behavior, stability, and failure mechanisms of biological protein materials play an integral part in defining their functional role in biological systems.

[4]We note that a purely mechanistic framework is not sufficient, and requires the integration of biology and chemistry to fully understand processes and systems across scale. Mechanics, however, provides a common basis regardless of scale or analytical progression (*e.g.*, top-down or bottom-up).

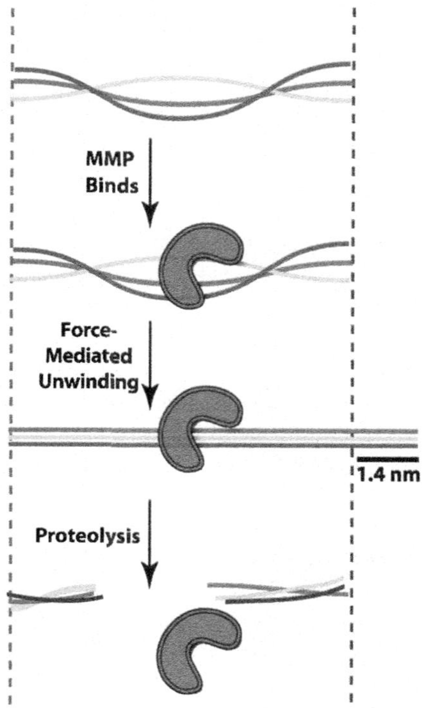


**Fig. 3.13** Example of mechanical influences on biological processes, here for the example of collagen fibers (an abundant building material in biology). Proposed mechanism of collagen proteolysis (cleavage) through enzyme matrix metalloproteinase-1, or MMP-1. Applied load stabilizes a stretched, proteolytically accessible collagen conformation (from Dunn *et al.* [92], copyright © 2011 American Chemical Society). While the applied mechanical force is shown to increase the rate of collagen cleavage at a the molecular scale, unresolved is what hierarchical effects could trigger and potentially modulate such a mechanical load at the system level?

In biological protein materials, each hierarchical level potentially has its own defects and failure mechanisms. Over multiple hierarchical levels, there are two primary types of response. First, if a material is able to tolerate defects and failure at specific scales, it retains its overall integrity since the presence of multiple hierarchical levels decrease the overall probability of catastrophic failure. Second, some defects and failure mechanisms lead to a catastrophic breakdown of the material (*e.g.*, in genetic disease, where proteins are misfolded due to errors in the DNA code). Many physiologic conditions exist in the realm where the material is able to tolerate defects and failure mechanisms (*e.g.*, microcracks in bone [20, 81], discussed further in Chap. 8: Unlocking Nature: Case Studies), while many diseases originate due to the material's inability to mitigate material defects and breakdown. In both cases, the basic mechanisms of failure can be exhibited through the breaking of H-bonds, protein unfolding, sliding of molecules against each other, and breaking of cross-links. Failure occurs because the intramolecular bonding (*e.g.*, H-bonds) is crucial to defining the structure of protein building blocks. Similarly, adhesion forces due to a charged surface or hydrophobic forces between molecules, fibrils, fibres and filaments are elementary to providing function at larger length-scales. Often, these failure mechanisms are competing and depend on loading rate, pH, density of cross-links or molecular geometry [94–97]. These biological failure mechanisms can be analogous to failure in engineering materials, which is mediated by defects such as cracks, dislocations or mass transport along grain boundaries. The quantitative analysis of multi-scale failure mechanisms provides great potential in explaining the role of materials in their respective biological context.

Much effort has recently gone into understanding the nano- and micromechanics of living organisms, as for example cells [93]. Early mechanical analysis of cells focused on simple network models to describe the cellular filaments (using simple components, such as rods and ropes), leading to molecular modeling at the scale of spectrin networks, for example. More recently, the mechanics of red blood cells within the context of malaria uncovered that the disease has a dramatic effect of making the cell much stiffer, leading to spreading of the disease (see Chap. 9: Pathological Materiomics). The quantitative, precise analysis revealed that this stiffness effect due to malaria is much more severe than previously assumed, leading to stiffening up to ten times compared to the original stiffness of the cell. To what gain are such investigations? It is clear that biological systems are similar to a mechanical system or device: if the performance or behavior of a single component is compromised, the system is subject to undue stress, wear, and potential catastrophic failure.

Proteins—with their purpose built functionalities—exemplify such natural types of "nano devices", suggesting an unmistakable promise in the development of biomimetic synthetic nanomachines based on the polypeptide building blocks. Rather than starting from scratch, natural protein systems provide a clear platform and practical approach to design functional materials at the nanoscale. The biological role of a protein is determined by its function, which is largely dependent on its structure. A clear gap in our understanding of protein "nano devices" is the direct the linking of the convoluted, hierarchical structure of folded proteins and the eventual function. Mechanical analysis is required to accurately model protein molecules as functional devices rather than biological processes, facilitating understanding of their function or ability to design and manipulate protein based materials and systems. Predictive mechanistic knowledge and exploitation of such links would allow the vital ability to design and manufacture protein based nano "machines" that perform specific engineering functions. If successful, such research will provide the ability to develop devices of increasing sophistication at the nano level, capable of acting as sensors, or carrying out more complex tasks. *Why re-design the assembly lines Nature has provided and successfully trial tested*? Such an understanding would open new doors in the science and engineering world of tomorrow.

Moreover, the mechanical performance of biological materials commonly surpasses that of traditional, engineered materials, while implementing relatively weaker structural building blocks. Silk, for example, displays impressive toughness and relative strength (see Fig. 3.14). Toughness can be thought of as the energy required to break a material, and determined by calculating the area under a stress-strain curve. While a material such as aramid (a synthetic fiber used in ballistic rated body armor) is stronger in terms of ultimate stress, and materials such as rubber is more extensible, silk balances these properties and is one of the toughest known materials. Moreover, in terms of strength-to-weight ratio, the spider silk strength (1.1 GPa) is about equivalent to high-tensile engineering steel (1.3 GPa), yet spider silk has a relative density of 1.3 compared with that of steel (7.8) [98]. In terms of toughness, spider silk is $165 \pm 30$ kJ per kilogram, which is substantially higher

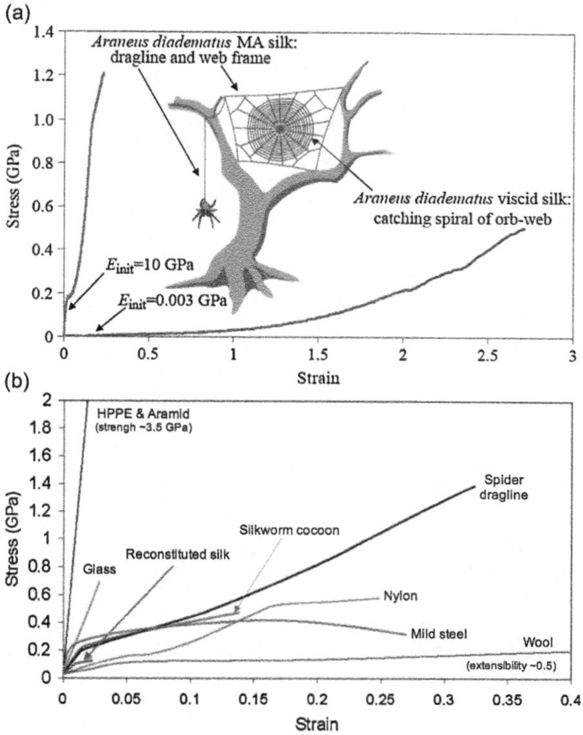


(c)

Tensile mechanical properties of spider silks and other materials

| Material | Stiffness (GPa) | Strength (GPa) | Extensibility (%) | Toughness (MJ/m³) |
|---|---|---|---|---|
| *Araneus* MA silk | 10 | 1.1 | 27 | 160 |
| *Araneus* viscid silk | 0.003 | 0.5 | 270 | 150 |
| *Bombyx mori* cocoon silk | 7 | 0.6 | 18 | 70 |
| Tendon collagen | 1.5 | 0.15 | 12 | 7.5 |
| Bone | 20 | 0.16 | 3 | 4 |
| Wool | 0.5 | 0.2 | 50 | 60 |
| Elastin | 0.001 | 0.002 | 150 | 2 |
| Resilin | 0.002 | 0.003 | 190 | 4 |
| Synthetic rubber | 0.001 | 0.05 | 850 | 100 |
| Nylon fibre | 5 | 0.95 | 18 | 80 |
| Kevlar 49 fibre | 130 | 3.6 | 2.7 | 50 |
| Carbon fiber | 300 | 4 | 1.3 | 25 |
| High-tensile steel | 200 | 1.5 | 0.8 | 6 |

**Fig. 3.14** Mechanical properties of silk. (**a**) Stress-strain curves for major ampullate (MA) gland silk (structural web components) and viscid silks (prey capture) from the spider *Araneus diadematus*. The parameter $E_{init}$ indicates the initial stiffness. Composed of similar protein building blocks, the properties range from a relatively stiff and strong dragline silk with moderate strain-to-failure all the way to capture silks that have reduced strength but very high toughness through huge strains-to-failure. From Gosline *et al.* [98], copyright © 1999 The Company of Biologists Ltd. (**b**) Representative experimental stress-strain plots for a range of different materials compared with silk. Adapted from Vollrath and Porter [100], used with permission, copyright © 2009 John Wiley and Sons, Inc. (**c**) Silk displays impressive toughness and relative strength in comparison to a compilation of material data from multiple sources (specific references omitted as they are intended to indicate the relative magnitude rather than exact values); based on data from [98]

than that of Kevlar (33 kJ per kilogram) [99]. This suggests that energy absorption and toughness may be the crucial properties that will lead us to an understanding of the function of silk [100]. Silk is an extremely useful reference materials that allows valuable insights into the structure-property-function relationships of a multitude of important structural proteins.

A hierarchical paradigm exemplifies how functional diversity is achieved through the reliance of structural variation of few and simple building blocks at distinct length-scales, rather than through a great diversity of building blocks. While proteins composing such materials is intrinsically "soft", a vast range of stiffness

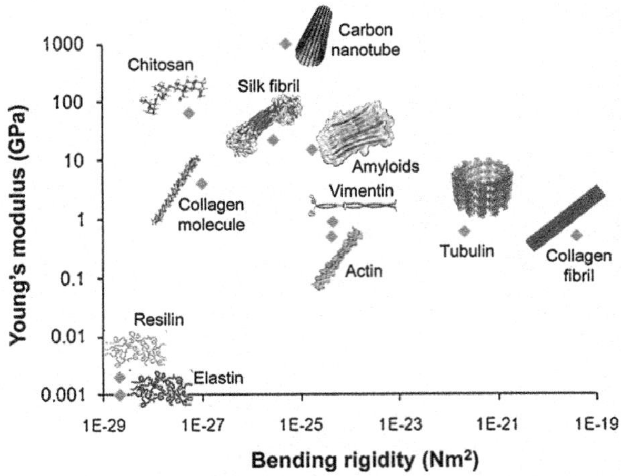


**Fig. 3.15** Mechanical properties of materials at the nanoscale, comparing both biological and synthetic materials. Biological fibrils and fibers present a vast range of mechanical properties in terms of Young's modulus and bending rigidity. However, most of the protein materials feature Young's moduli in the range of 100 MPa to 10 GPa, well below the stiffness of many synthetic nanostructured material such as carbon nanotubes. On the other hand, bending rigidity shows a much greater variability. Collagen microfibrils and fibrils, for example, present a significantly enhanced bending rigidity with only a relatively small decrease in the Young's modulus compared to a single molecule (from Gautieri *et al.* [101], copyright © 2011 American Chemical Society)

(Young's modulus) and bending rigidity can be achieved for a variety of biological and synthetic fibers (depicted in Fig. 3.15).

Biological materials feature mechanical properties that are characteristic for this entire class of materials, including: robustness, adaptability and multifunctionality [3, 102]. Robustness is defined as the degree of separation between stability and failure, and is crucial to understand the role of protein materials in biology, for both physiologic and pathologic conditions (*e.g.*, varying pH, forces, structural changes). Adaptability refers to the ability of a material to cope with environmental changes by changing its structural arrangement to better cope with changed conditions (*e.g.*, formation of actin stress fibres in cell adhesion, tissue remodeling in bone). For example, tissue remodeling in bone plays a crucial role to improve the material's damage tolerance through its intrinsic ability to repair itself [20]. The underlying mechanism is that small cracks, formed due to physiologic mechanical load, are detected by so-called bone remodeling units and removed from the tissue before they reach a critical size at which catastrophic failure would occur. Such mechanisms show the intricate connections between tissue formation and overall failure properties.

Many protein materials exhibit graceful degradation, reflecting the material's ability to induce a controlled breakdown of a system's function after damage without leading to catastrophic failure and without affecting materials in the environment. The hierarchical structure may be the basis to these unique properties, distinguishing them from many engineered materials. The study of biological materials and structures could impact the advancement of nanotechnology through the "mechanistic translation" of design rules. Recently developed nanomaterials are often found to have material properties superior to conventional engineering materials (*e.g.*, nanomaterials such as carbon nanotubes and graphene represent some of the strongest materials known to date). However—as discussed in Chap. 2: The Materiome—we are currently unable to fully exploit and translate these superior nanoscale properties to larger length-scales (micrometers and beyond), which is necessary for technological applications.

For a variety of applications, hierarchical multiscale effects will be critically important as we push the limits of what we can see, and how small and how effective we can design. For efficiency and conservation of (finite) resources, novel multiscale modeling methods will be required that enable us to explore the full design space, from nano to macro. We may then be able to utilize these material concepts to impact the field by exploiting the full potential (higher strength, higher robustness, or properties such as adaptability or mutability) of multi-scale engineered materials. This could lead to the design of novel forms of construction materials (such as a new generation cement and concrete), adhesion systems or cables. A more robust mechanically understanding of biological materials could also have significant impact in several areas of biomedical research. For example, atherosclerosis (hardening of blood vessels due to plaque formation), or blood clots in large vessels (*e.g.*, carotid artery), and other blood vessel diseased states are related to a complex interplay of materials-cell interactions at multiple scales, including the nanoscale. Similar considerations hold for many other diseases, including: cancer, infectious diseases, genetic diseases, and injuries. For example, injuries and genetic diseases are often caused by structural changes in protein materials (defects), resulting in failure of the material's intended function. Materiomics approaches enable us for example to probe how mutations in structure alter the mechanical properties of protein materials. The long-term impact of biomateriomics could be used to predict diseases in the context of diagnostic tools by measuring material properties and mechanical markers rather than focusing on symptomatic chemical readings alone, providing new strategies for treatment options. Indeed, nanomedicine is now emerging as a new field of study, in which the links between nanoscience and biomedical applications are investigated.

## 3.6 'Shaky' Foundations of Biological Materials

The folding of polypeptide chains combined with the multiscale hierarchies of common biological materials hinders a simple description of such systems. From another perspective, even a material such as water seems to become more complex when we inspect the molecular level details such as H-bonding and polarization of $H_2O$ molecules. Yet, such "nano-complexity" does not limit the description of macroscale behavior (such as the differences between laminar and turbulent flow). In Chap. 1: Introduction, we discussed the nanoscale hierarchies and structure of

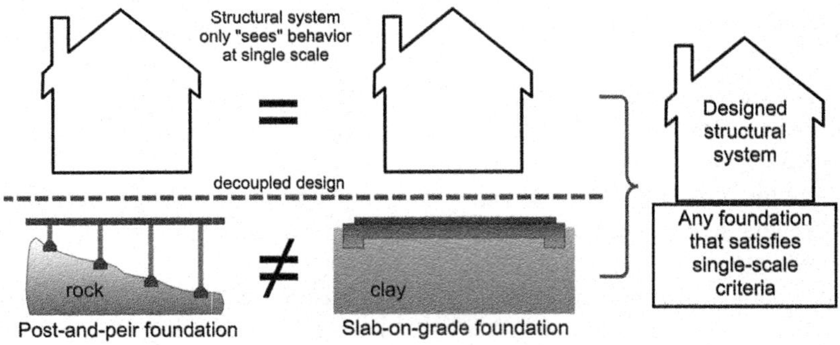


**Fig. 3.16** Schematic of macro-scale, structural engineering in which the designed structural system is decoupled from the foundation. The overlaying structure only "sees" foundation behavior at a single-scale, encompassing critical properties such as settlement and bearing capacity, and any foundation that satisfies such properties (*e.g.*, post-and-pier versus slab-on-grade) can support equivalent functional structures. More detailed "small scale" properties (such as material type or architecture) have no external affect on the borne structure

concrete and cement are only now being fully understood—but that has not prevented the successful use of concrete (the most common structural material in the world). For many common, well-understood and "everyday" systems, the details and behavior are adequately homogenized with an increase in scale. Surely such a homogenization occurs with biological materials? Is knowledge of macroscale behaviour sufficient in practice? Unfortunately (and predictably) the answer to both questions is *no*. To illustrate the reason, we turn to another engineering analogy: the foundations of constructed structures [103].

The design and assembly of any safe structure requires a thorough knowledge of the foundation. No structural engineer would approve the construction of a skyscraper without a complete geotechnical report, where the stability of the foundation below supports the building above. Yet, at the macro-scale, the groundwork and the structure are sufficiently decoupled. As long as the foundation satisfies minimum requirements, there is no adverse effect on the overlying structure. What the building "sees" is limited to a single scale and characterized by the design parameters of the foundation, including settlement restrictions, bearing capacity, and other such phenomena at the same scale as the structural system. Assuming that the critical parameters are met, more detailed and smaller scale properties (such as material type or microstructure) have no external affect on the borne structure, where different engineered foundations can support equivalent functional structures (see Fig. 3.16). One can design and construct a building independent of the structural foundation, as long as it is assumed *a priori* that the foundation has satisfied the required design parameters such as maximum settlement and bearing capacity. On a more complex level, one may also consider seismic and slope stability, pore water pressure variation, and other parameters. The point being is that the foundation is designed independently of the structure—one need not know the details of either for a successful design, only the design requirements. In the development of biological materials, systems, and tissues, however, this fundamental relationship changes. The foundation for nano- and micro-scale assembly—the material substrate, scaffold, or matrix that will support cellular processes and mechanical requirements—is no longer disassociated from the assembled system; they interact intimately. Cellular processes recognize the foundation properties at multiple levels, and they are inextricably linked as the details of the foundation define what tissue grows (see Fig. 3.17).

Indeed, different substrates induce different affects, are not transferable, and are more complex. The complexity (and hence challenge) arises from the multi-scale and hierarchical nature of this relation. Changes can easily be made for system-level properties (such as stiffness, porosity, etc.), but the changes have effects that subsequently cascade downward from the system-level to the fundamental molecular interactions at the nanoscale. From a cursory perspective, they are uncertain—or *shaky*—foundations.

Currently, there is no single set of "design parameters" that will satisfy more than the most rudimentary system. As discussed previously, tissues and biological materials are commonly hierarchical as there is underlying structure and function at a multitude of diverse scales [3]. As a result, slight variations at a lower scale, such as scaffold topology or material choice, can propagate and express at larger scales, typically to the detriment of the growth or assembly of the system. This is in contrast to macroscale engineered systems, in which structural details, such as spacing of piers for example, do not affect the above structure (assuming alternate designs have equivalent mechanical performance). Likewise, material details, such as the choice of reinforced concrete or structural steel components, do not affect the borne structure. The overlaying structure does not "care" what supports it—it will function the same. Thus, a fundamental challenge of tissue engineering and biological material synthesis lies within the understanding of material-substrate interactions across all scales, from individual atoms, to molecules, to subsequent tissues and entire organisms. Continuing the analogy with foundation engineering, not only do you have to analyze properties of the soil, in the case of biological tissues you have to know the behavior of each individual rock, mineral, and grain.

Certainly, for the construction of a building, the foundation is of little importance to the overlying structure—a post-and-pier foundation on bedrock provides a similar platform to a slab-on-grade over clay, and a home built atop does not change functionality based on the foundation below. On the other hand, it is well known in tissue engineering that different substrates can result in different materials. One important difference is that the tissue grows itself whereas a building is "grown" by the external means (typically construction personnel). But it is in the case of biological tissues—the same cells can develop into different tissues depending on the substrate (*i.e.* foundation) they live. For construction, a structure can be designed independently from the foundation, whereas in biology, the foundation may dictate the structure! For example, a promising candidate for a tissue engineering scaffold is the use of extracellular matrix (ECM), which is a key component in the natural regeneration and maintenance of tissues and organs [105]. Methods of producing ECM-inspired tissue platforms have been successful in replicating the required physiochemical properties and structural features of their natural analogs, but, in most cases, do not match the mechanical properties of the tissue to be regenerated.

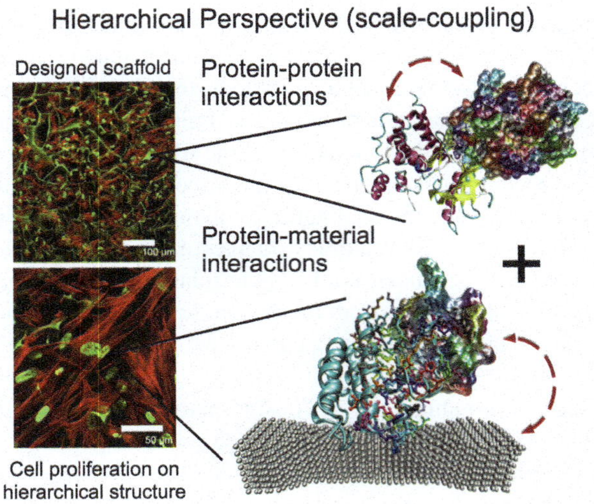


**Fig. 3.17** What does a biological tissue or material "see"? Biological systems are intrinsically hierarchical, integrating cross-scale interactions from nano to macro. Successful (predictive) scaffold design requires a bottom-up, hierarchical perspective. Tissue growth is rooted in the discrete molecular-scale interactions (such as protein-protein and protein-material) that express themselves throughout a multitude of scales. Designed hierarchical scaffold represented by 3D silk fibroin structure depicting cultured human primary dermal fibroblast cells (SEM images from Mandal and Kundu [104], reproduced with permission, copyright © 2009 Elsevier Ltd.). Figure adapted from Cranford and Buehler [103]

Yet, the elasticity of the matrix can determine stem cell differentiation: soft matrices are neurogenic, stiffer matrices are myogenic, and rigid matrices are osteogenic [106, 107]. This example focused on a simple and single parameter, stiffness, clearly shows that the properties of the foundation affect the resulting structure.

We must also account for the geometry of the foundation. At the microscale, for example, the advance of rapid prototyping techniques has significantly improved control over the pore network architecture (*e.g.*, pore size, channel geometry) of tissue engineering scaffolds, which are known to influence the signal expression and subsequent differentiation of a transplanted cell population [108, 109]. Indeed, the interconnectivity of pores, permeability, confinement, and other geometric properties have been shown to affect the transport of oxygen and nutrients throughout tissue scaffolds [109]. There is a multitude of known design parameters considered important to achieve a successful synergy between material and substrate (cell and scaffold), including porosity, interconnectivity, surface properties, mechanical strength, the amounts and types of filler material, cell seeding density, and other exogenous growth factors [110–112]. The common aspect of such design parameters is that they are typically considered at a *single scale*. While the results of such property variations are known, the underlying protein/substrate interactions are not fully understood. Be that as it may, such *ad hoc* approaches are quite advanced and ingenious, and have been successful in delineating appropriate substrates and scaffolds for particular tissues (such as collagen or bone [113]) and biological macromolecular structures (such as amyloid films [114]). However, the specific molecular mechanisms resulting in successful tissue generation remain largely unknown. Often, synthesis is achieved by experimental trials and high-throughput screening, and the continuous refinement of previous insights. Such steps are necessary for the progression and immediate application to tissue engineering.

The fundamental challenge lies in the complex hierarchical structure of the tissues and materials, where changes at the molecular level propagate and are expressed in unpredictable ways. What a cell (with a diameter of 10 to 30 micrometers) in a collagen tissue "sees" can be very different in a natural system than in most currently used tissue engineering scaffolding biomaterials. The structure and properties of the implemented scaffold are typically designed on the scale of tissue assembly, whereas the critical scales may be orders of magnitude smaller. A successful and predictive synthesis and assembly of biological tissues must incorporate these natural hierarchical structures when we design materials in which tissues are supposed to assemble or grow. One natural solution is to mimic the complexity of such materials by the development of hierarchical foundations, where the form and function of the substrate are specifically designed at each scale. The caveat, of course, is that this requires intimate knowledge of the specific molecular, protein, cellular, and tissue interactions across all scales.

Hierarchical structures can be of great advantage for tissue engineering application as they provide a more natural environment for cells to grow and develop into tissues. For example, the multiscale complexity of bone necessitates hierarchical features of a scaffold/matrix such as commonly implemented electrospun porous nanofiber networks [113, 115]. This presents a conundrum: we require thorough understanding and synthesis of complex hierarchical substrates to facilitate the synthesis and growth of complex hierarchical tissues. Moreover, *in lieu* of molecule-by-molecule constructions, self-assembly provides the only practical route to synthesize such hierarchical substrates. On what platform should such assembly occur? It seems we have arrived at a Catch-22! Currently, we do not have the technologies available to tailor-make a hierarchical substrate, but recent investigations of novel hierarchical substrate materials, *e.g.*, as injectable cell carriers for *in vivo* tissue regeneration [110] are beginning to fill the necessary gaps (see Fig. 3.18).

The complex hierarchical structure and interactions of biological materials presents fundamental challenges in the development and prediction of successful material foundations. However, even if we are able to solve this "multi-scale problem", there is another challenge in the design of robust platforms, the *temporal* variations in system behavior and properties. Biological materials feature selectively tailored molecular assemblies and interfaces that elicit specific form and functionality, which can readily change and adapt to their environment. Natural systems often vary the stiffness, hierarchical structure, and other parameters overtime to control how cells grow. A theoretically perfect scaffold for one stage of differentiation may be completely inadequate to *sustain* growth. Indeed, engineered tissues must not only grow to fill a defect and integrate with the host tissue, often they must also grow and thrive subject to the changing needs of a varying biological environment. Tissues capable of adapting with time could be engineered by supplying stimulus

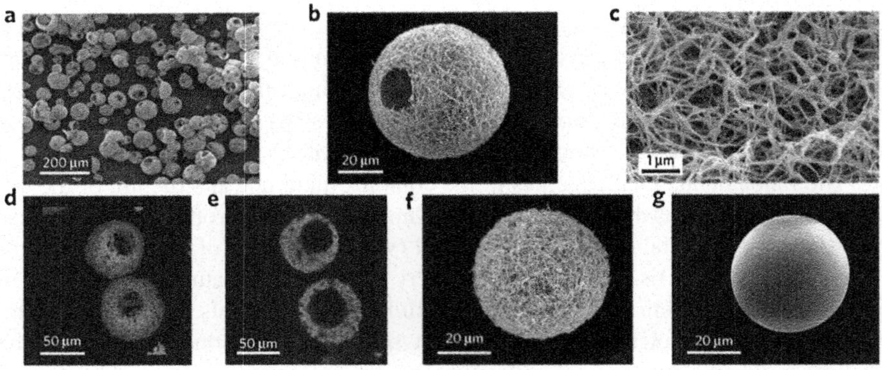


**Fig. 3.18** Hierarchical foundations for tissue regeneration through nanofibrous microspheres and solid-interior microspheres. (**a**) SEM image of nanofibrous hollow microspheres—the acting foundation for tissue growth and the largest hierarchical level considered—showing that almost every microsphere had one or more open hole(s) on the shell. (**b**) SEM image of a representative nanofibrous hollow microsphere (a lower hierarchical level), showing the nanofibrous architecture and a hole of approximately 20 µm on the microsphere shell. (**c**) A high-magnification image of the microsphere in (**b**), showing the nanofibre network (another level of hierarchical structuring), which have an average diameter of about 160 nm. (**d**) A 3D reconstruction of nanofibrous hollow microspheres. (**e**) A 2D cross-section confocal image of the nanofibrous hollow microspheres, confirming the hollow structure of the microsphere. (**f**) SEM image of a representative nanofibrous microsphere, showing the nanofibrous architecture on the microsphere surface. (**g**) SEM image of a representative solid-interior microsphere, showing the smooth surface of the microsphere. From Ma *et al.*, [110], used with permission, copyright © 2011 Nature Publishing Group

signals to cells from the biomaterial or scaffold used [116]. Mimicking the target systems, a possible solution is to make the properties of the substrate dynamic and controllable *via* external stimuli or internal feedback, a concept known as *mutability* (see Table 3.1). Mutable materials are found widely in biology, characterized by a material's capacity to change its properties under external cues based on directed structural changes at specific material levels. Mutable materials are also inherently hierarchical, where property changes are often driving by interactions and processes at the molecular level. Through monitoring of self-assembly and growth or by internal feedback (*e.g.*, mechanical or geometric cues) mutable materials could potentially optimize according to the needs of the system.

Potential candidates for such mutable materials are stimuli-responsive polymer nanostructures. Scaffolds currently used in tissue engineering and cell therapy are mostly passive in that they deliver biological agents mainly through mechanisms involving molecular diffusion, material degradation, and cell migration, which do not allow for dynamic external regulations. Yet, *in vivo*, the structures of protein materials self-maintain or adapt *via* feedback loops by translating spontaneous demands in the microenvironment (*via* intracellular signalling) to regulate gene activation/deactivation to ultimately alter the material's structural makeup in order to better suit the local physiologic needs. Responsive polymer systems exhibit similar features as biological materials, and are capable of conformational and chemical

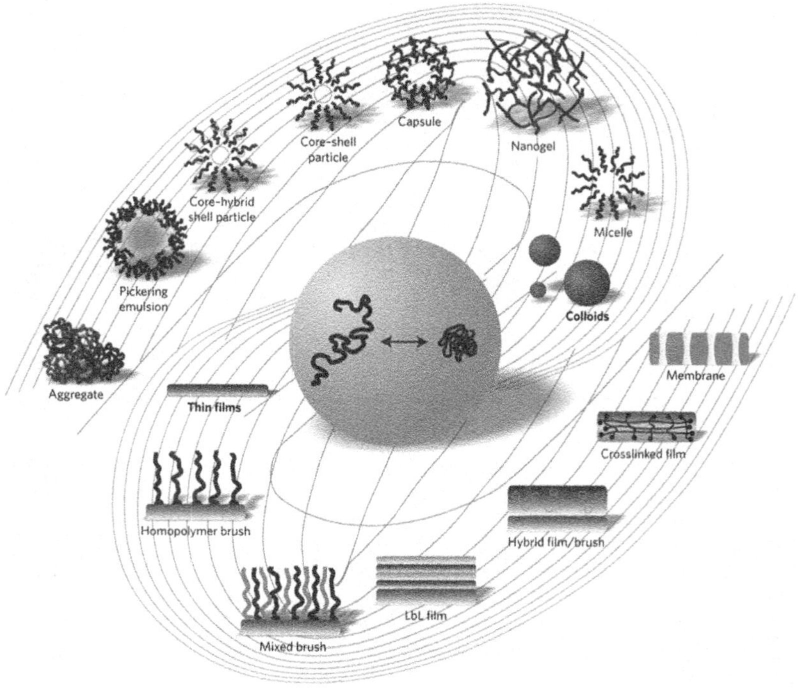


**Fig. 3.19** Schematic of possible nanostructured stimuli-responsive polymer materials. Biological materials are highly dynamic, and thus successful foundations and material interfaces must actively change via external or internal mechanisms. Such materials rely on the phase behavior of macromolecule assemblies in thin films (polymer brushes, multilayered films made of different polymers, hybrid systems that combine polymers and particles, thin films of polymer networks, and membranes that are thin films with channels/pores), and nanoparticles (micelles, nanogels, capsules and vesicles, core–shell particles, hybrid particle-in-particle structures, and their assemblies in solutions and at interfaces in emulsions and foams). From Luzinov and Minko *et al.*, [117], used with permission, copyright © 2010 Nature Publishing Group

changes on receiving an external signal. Such materials can adapt to surrounding environments, regulate transport of ions and molecules, change wettability and adhesion of different species on external stimuli, or convert chemical and biochemical signals into optical, electrical, thermal and mechanical signals, and *vice versa* [117]. Examples of these types of materials are given in Fig. 3.19.

The concept of mutability provides a paradigm shift and exciting opportunity in the area of cell growth—namely, dynamic control of the self-assembly process. Like their biological analogues, a holistic materiomic approach is necessary to predict the structure and mutability of large-scale material properties from the nano-scale up. Synthetic polymer systems with desired characteristics are currently being developed for a multitude of biological applications, such as responsive biointerfaces that are functionally similar to natural surfaces [118], coatings that are capable of interacting with and responding to their environment [119], and composite materials that

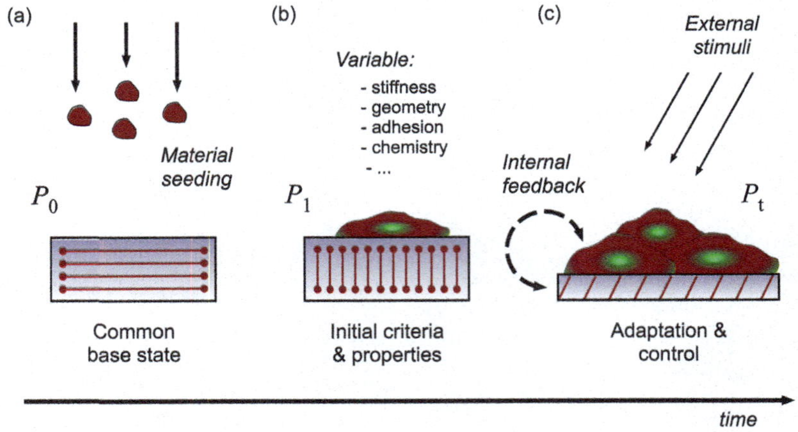


**Fig. 3.20** The potential of implementing mutable/responsive materials within biological material scaffolds to enhance/optimize growth over time. (**a**) A generic but mutable material scaffold with an initial (or neutral) property set ($P_0$; common base state) could be varied depending on the desired tissue (neurogenic or osteogenic, for example). (**b**) Range of relevant properties (such as stiffness, geometry, chemistry, etc.) can be modified accordingly for predictable differentiation ($P_1$; initial criteria). (**c**) Once differentiation is achieved, sustained growth can be enhanced by either internal feedback (mechanical or chemical cues) or by external stimuli (based on system monitoring) allowing dynamic adaptation properties ($P_t$) and precise control of tissue growth. Adapted from [103]

actuate and mimic the action of muscles [120]. *Do such responsive polymer systems offer a suitable foundation for biological material assembly*?

The effect of mechanical cues on the stimulation of cellular signal expression can exploit materials such as photo-crosslinking polymer composites [108, 121] or pH-responsive systems [122, 123]. Material properties (stiffness) and geometry (pore architecture and connectivity), can be tuned on a system-by-system basis to investigate the effect on cell growth. Other studies have been undertaken exploring potential spatial patterning [124] and temporal variations [125] in cross-linked polymer systems, resulting in the coupling between inherent responsive material properties and geometry. Such a material can offer scaffolds with dynamic, tunable architectures and bulk properties, triggered at the molecular level. The responsive properties of reconstructable polymer systems are relevant to many biotechnological and biomedical applications [119, 126] because these materials can undergo dynamic changes in accord with changes in living systems (see Fig. 3.20). The possibility of tuning and switching adhesion between stimuli-responsive materials and proteins and cells has been explored for the control of cell and protein adhesion [127], as well as exposing and masking potential biointerfaces and manipulation of cellular signals, protein interactions, and growth factors [128]. Moreover, precise control of the permeation of chemicals, nanoparticles, and ions through nanoporous membranes and 3D scaffolds offers a unique opportunity for control of assembly and growth process [129–132].

One important aspect of responsive material systems is the coupling that exists between the chemical and molecular scales. The challenge is to understand at each molecular species with as much atomistic and chemical detail as possible, leading to the rational design of mutable and hierarchical scaffolds. Prediction and understanding of thermodynamic, chemical, and structural properties is crucial, incorporating many of the different interactions (such hydrogen bonding [133] and chemical reactions [134]) present in these systems. The resulting increased functionality of tissue-engineering materials may rival the complexity of the tissue itself. Responsive polymer systems can be used for a variety of applications, and biomaterials and tissue engineering is just an example of important areas that will benefit greatly from further development of tunable responsive materials. The critical feature, which could potentially be exploited for other material systems, lies in the mutability and dynamic control of properties and behavior. In fact, the challenge is to develop complex systems that are responsive to biochemical signals during tissue growth (internal feedback) that mimics biological response. Such systems need a complex, hierarchical organization of the responsive chemical and molecular components to adapt to potential environmental factors and responses.

A fundamental understanding of cross-scale interactions and mechanisms in self-assembly and tissue growth can be used to exploit the process for both biological and synthetic materials. If assembly and growth is dictated by material/substrate interactions, an ability to dynamically tune substrate properties provides vast potential for control across all scales. With increasing complexity, such systems start to resemble their biological counterparts (*e.g.*, adaptation to their surrounding environment), mimicking the concepts natural systems have been relying on for millions of years. However, practical technological application has so far been severely hindered due to lack of understanding of how to link the atomistic scale with material structure and device properties and function. The exploitation of hierarchical interactions provides a novel paradigm to make progress in tissue engineering and unpredictability can be eliminated. Such an objective can be attained by the combination of bottom-up, multiscale investigations and top-down synthetic approaches, and ultimately, the stabilization of a shaky foundation. This shaky attribute is a result of a single-scale perspective in the current design paradigm and one that can potentially be remedied if we had the fundamental knowledge to exploit such complex interactions (both in theory and in practice). The implementation of system specific foundations can lend insight to the molecular foundations of disease and disease pathology rooted at the molecular scale (*i.e.*, single amino acid mutations)—potentially tracking variations at the nanoscale and effects and expression at the tissue level. Ultimately, mechanistic understanding can complement robust screening methods to reverse engineer the complexity of biological materials and tissue growth. The combination of multi-scale structural control and integration of living and non-living systems into technologies and interfaces may lead to the development of new technologies that utilize the advantages of both micro and nanotechnology with the principles of biology and provide a new foundation for biological materials.

## 3.7 Summary

In contrast with most material systems, the mechanical behavior of biological materials (especially those in living organisms) cannot be characterized simply in terms of fixed "properties"—the material structure is a dynamic system that adapts to its local mechanochemical environment. Mechanistic understanding of the relationships among environment, structure and function, however, requires concurrent investigation of each component. Despite the sophistication of experimental and computational approaches in cell and molecular biology, the mechanisms which couple biochemical and mechanical processes are poorly understood.

Along with biomedical applications, understanding the material concepts in protein materials could also be used to advance nanomaterials for engineering applications. Protein based materials hold great potential as biomaterials or new synthetic polymers with unique mechanical, optical, and electrical properties. In recent years, proteins have also been proposed as basis of new materials for technological applications [36]. Genetically engineered biopolymers based on recombinant DNA technologies haven been developed for various applications, including pH sensitive hydrogels [35]. This is an example for the vanishing borderline between technology and biology, enabling new critical breakthroughs for novel material concepts, allowing to translate Nature's structural concepts into engineered materials and opportunities to exploit knowledge gained from biological protein materials.

Material properties of nanomaterials are often found to be superior to conventional engineering materials. For instance, carbon nanotubes are one of the strongest materials known, but have rarely been employed in structural materials due to our inability to utilize the properties of this material at larger length-scales. Understanding protein materials could provide knowledge that can eventually be translated to connecting disparate material scales, from nano to macro. We may then be able to utilize these material concepts to impact the field by exploiting the full potential (higher strength, higher robustness) of multi-scale engineered materials. An intriguing application may be the development of a new material platform based on similar concepts as found in biological protein materials, but rather than using the universal protein constituents (such as alpha-helices, beta-sheets, random coils, or tropocollagen, etc.) we could instead use building blocks such as carbon nanotubes, nanowires and self-assembled peptides. Some steps towards this goal have been taken in recent studies, where conducting metallic nanowires were grown through use of amyloids as templates [135].

Biology utilizes hierarchical structures to create multifunctional materials. Even though biochemical and image-based diagnostics will remain important, the integration of scales, as well as the mixing of physical, biological and chemical concepts into novel engineering designs could complement the current practice of disease diagnosis and treatment, as well as the design of new materials, and thereby unfold numerous opportunities for technological innovations. The wide impact of the use of materials science approaches in biology and biomedical sciences has yet to be fully demonstrated. In some specific areas such as bone (*e.g.*, osteoporosis), cartilage (*e.g.*, arthritis), cardiovascular aspects and, particularly, in the context of tissue engineering and regenerative medicine, materials science approaches have started to play an important role. In addition to mechanical deformation and failure mechanisms of protein materials, the investigation of other material properties could be of great interest. Optical properties (*e.g.*, in the eye's cornea and lens), photoelectric properties (*e.g.*, photosynthesis in plants), electrical (*e.g.*, links between neuron cells—synapses), motility (*e.g.*, in muscle tissues, cilia), or thermal properties (*e.g.* thermal management, heat shock proteins) are important issues that could be studied using a materials science approach.

## 3.8 Suggested Readings

The following texts provide extended discussion of biological materials across scales with varying perspectives. Edited texts:

- D. Kaplan, K. McGrath (eds.), *Protein-Based Materials (Bioengineering of Materials)* (Birkhäuser, 1997)
- M. Elices (ed.), *Structural Biological Materials: Design and Structure-Property Relationships* (Elsevier, 2000)
- V. Renugopalakrishnan, R.V. Lewis (eds.), *Bionanotechnology: Proteins to Nanodevices* (Springer, 2006)
- T. Vo-Dinh (ed.), *Nanotechnology in Biology and Medicine: Methods, Devices, and Applications* (Taylor & Francis, 2007)
- F. Mollica, L. Preziosi, K.R. Rajagopal (eds.), *Modeling of Biological Materials* (Springer, 2007)
- K. Eom (ed.), *Simulations in Nanobiotechnology* (CRC Press, 2011)

Informal pedagogical texts regarding biology, nanotechnology, and biomaterials:

- R.H. Carlson, *Biology Is Technology: The Promise, Peril, and New Business of Engineering Life* (Harvard University Press, 2011)
- T. Sargent, *The Dance of the Molecules: How Nanotechnology is Changing Our Lives* (Avalon, 2006)
- D.S. Goodsell, *Bionanotechnology: Lessons from Nature* (Wiley, 2004)
- R.A.L. Jones, *Soft Machines: Nanotechnology and Life* (Oxford University Press, 2004)

Readers are referred to cited works for further information.

## References


1. B. Alberts, A. Johnson, J. Lewis, M. Raff, K. Roberts, P. Walter, *Molecular Biology of the Cell* (Taylor & Francis, New York, 2002)
2. N. Wang, D. Stamenovic, Mechanics of vimentin intermediate filaments. J. Muscle Res. Cell Motil. **23**(5–6), 535–540 (2002)

3. P. Fratzl, R. Weinkamer, Nature's hierarchical materials. Prog. Mater. Sci. **52**, 1263–1334 (2007)
4. K. Gelse, E. Poschl, T. Aigner, Collagens—structure, function, and biosynthesis. Adv. Drug Deliv. Rev. **55**(12), 1531–1546 (2003)
5. W. Liu, L.M. Jawerth, E.A. Sparks, M.R. Falvo, R.R. Hantgan, R. Superfine, S.T. Lord, M. Guthold, Fibrin fibers have extraordinary extensibility and elasticity. Science **313**(5787), 634 (2006)
6. J.W. Weisel, Biophysics—enigmas of blood clot elasticity. Science **320**(5875), 456–457 (2008)
7. M.J. Buehler, Y.C. Yung, Deformation and failure of protein materials in physiologically extreme conditions and disease. Nat. Mater. **8**(3), 175–188 (2009)
8. R. Lakes, Materials with structural hierarchy. Nature **361**(6412), 511–515 (1993)
9. R.C. Strohman, The coming kuhnian revolution in biology—commentary. Nat. Biotechnol. **15**(3), 194 (1997)
10. P. Nurse, *Reductionism and Explanation in Cell Biology*. Limits of Reductionism in Biology (Wiley, New York, 1998)
11. E.M. Marcotte, Detecting protein function and protein-protein interactions from genome sequences. Science **285**, 751–753 (1999)
12. D. Eisenberg, E.M. Marcotte, I. Xenarios, T.O. Yeates, Protein function in the post-genomic era. Nature **405**(6788), 823–826 (2000)
13. G.B. Olson, Designing a new material world. Science **288**(5468), 993 (2000)
14. M.J. Buehler, Tu(r)ning weakness to strength. Nano Today **5**(5), 379–383 (2010)
15. T. Ackbarow, M.J. Buehler, Hierarchical coexistence of universality and diversity controls robustness and multi-functionality in protein materials. J. Comput. Theor. Nanosci. **5**(7), 1193–1204 (2008)
16. M.J. Buehler, Y.C. Yung, How protein materials balance strength, robustness, and adaptability. HFSP J. **4**(1), 26–40 (2010)
17. G.E. Fantner, T. Hassenkam, J.H. Kindt, J.C. Weaver, H. Birkedal, L. Pechenik, J.A. Cutroni, G.A.G. Cidade, G.D. Stucky, D.E. Morse, P.K. Hansma, Sacrificial bonds and hidden length dissipate energy as mineralized fibrils separate during bone fracture. Nat. Mater. **4**(8), 612–616 (2005)
18. T. Hassenkam, G.E. Fantner, J.A. Cutroni, J.C. Weaver, D.E. Morse, P.K. Hansma, High-resolution afm imaging of intact and fractured trabecular bone. Bone **35**(1), 4–10 (2004)
19. R.O. Ritchie, M.J. Buehler, P. Hansma, Plasticity and toughness in bone. Phys. Today **62**(6), 41–47 (2009)
20. D. Taylor, J.G. Hazenberg, T.C. Lee, Living with cracks: damage and repair in human bone. Nat. Mater. **6**(4), 263–268 (2007)
21. S. Rammensee, U. Slotta, T. Scheibel, A.R. Bausch, Assembly mechanism of recombinant spider silk proteins. Proc. Natl. Acad. Sci. USA **105**(18), 6590–6595 (2008)
22. F. Vollrath, D. Porter, Spider silk as archetypal protein elastomer. Soft Matter **2**(5), 377–385 (2006)
23. N. Du, X.Y. Liu, J. Narayanan, L.A. Li, M.L.M. Lim, D.Q. Li, Design of superior spider silk: from nanostructure to mechanical properties. Biophys. J. **91**(12), 4528–4535 (2006)
24. T. Ackbarow, D. Sen, C. Thaulow, M.J. Buehler, Alpha-helical protein networks are self-protective and flaw-tolerant. PLoS ONE **4**(6), e6015 (2009)
25. Z. Qin, L. Kreplak, M.J. Buehler, Hierarchical structure controls nanomechanical properties of vimentin intermediate filaments. PLoS ONE **4**(10), e12115 (2009)
26. S.G. Wise, S.M. Mithieux, A.S. Weiss, Engineered tropoelastin and elastin-based biomaterials. Adv. Protein Chem. Struct. Biol. **78**, 1–24 (2009)
27. A.S. Weiss, J.F. Almine, D.V. Bax, S.M. Mithieux, L. Nivison-Smith, J. Rnjak, A. Waterhouse, S.G. Wise, Elastin-based materials. Chem. Soc. Rev. **39**(9), 3371–3379 (2010)
28. M.J. Buehler, T. Ackbarow, Hierarchical coexistence of universality and diversity controls robustness and multi-functionality in protein materials. J. Comput. Theor. Nanosci. **5**(7), 1193–1204 (2008)

29. M.J. Buehler, Computational and theoretical materiomics: properties of biological and de novo bioinspired materials. J. Comput. Theor. Nanosci. **7**(7), 1203–1209 (2010)
30. J.L. Morrison, R. Breitling, D.J. Higham, D.R. Gilbert, A lock-and-key model for protein-protein interactions. Bioinformatics **22**(16), 2012–2019 (2006)
31. J.C. Kendrew, G. Bodo, H.M. Dintzis, R.G. Parrish, H. Wyckoff, D.C. Phillips, A three-dimensional model of the myoglobin molecule obtained by x-ray analysis. Nature **181**(4610), 662–666 (1958)
32. P.A. Janmey, J.F. Leterrier, H. Herrmann, Assembly and structure of neurofilaments. Curr. Opin. Colloid Interface Sci. **8**(1), 40–47 (2003)
33. H. Herrmann, U. Aebi, Intermediate filaments: molecular structure, assembly mechanism, and integration into functionally distinct intracellular scaffolds. Annu. Rev. Biochem. **73**, 749–789 (2004)
34. E. Bini, D.P. Knight, D.L. Kaplan, Mapping domain structures in silks from insects and spiders related to protein assembly. J. Mol. Biol. **335**(1), 27–40 (2004)
35. W.A. Petka, J.L. Harden, K.P. McGrath, D. Wirtz, D.A. Tirrell, Reversible hydrogels from self-assembling artificial proteins. Science **281**(5375), 389–392 (1998)
36. R. Langer, D.A. Tirrell, Designing materials for biology and medicine. Nature **428**(6982), 487–492 (2004)
37. A. Mershin, B. Cook, L. Kaiser, S.G. Zhang, A classic assembly of nanobiomaterials. Nat. Biotechnol. **23**(11), 1379–1380 (2005)
38. J.M. Smeenk, M.B.J. Otten, J. Thies, D.A. Tirrell, H.G. Stunnenberg, J.C.M. van Hest, Controlled assembly of macromolecular beta-sheet fibrils. Angew. Chem., Int. Ed. **44**(13), 1968–1971 (2005)
39. X.J. Zhao, S.G. Zhang, Molecular designer self-assembling peptides. Chem. Soc. Rev. **35**(11), 1105–1110 (2006)
40. J.C.M. van Hest, D.A. Tirrell, Protein-based materials, toward a new level of structural control. Chem. Commun. **19**, 1897–1904 (2001)
41. S.G. Zhang, C. Lockshin, R. Cook, A. Rich, Unusually stable beta-sheet formation in an ionic self-complementary oligopeptide. Biopolymers **34**(5), 663–672 (1994)
42. P. Chen, S. Jun, Y. Hong, H. Imamura, B.Y. Ha, J. Bechhoefer, Self-assembly of the ionic peptide eak16: the effect of charge distributions on self-assembly. Biophys. J. **87**(2), 1249–1259 (2004)
43. A.B. Fulton, W.B. Isaacs, Titin, a huge, elastic sarcomeric protein with a probable role in morphogenesis. BioEssays **13**(4), 157–161 (1991)
44. D.E. Ingber, L. Dike, L. Hansen, S. Karp, H. Liley, A. Maniotis, H. McNamee, D. Mooney, G. Plopper, J. Sims, N. Wang, *Cellular Tensegrity—Exploring How Mechanical Changes in the Cytoskeleton Regulate Cell-Growth, Migration, and Tissue Pattern During Morphogenesis*. International Review of Cytology—a Survey of Cell Biology, vol. 150 (Academic Press, San Diego, 1994), pp. 173–224
45. A.E.X. Brown, R.I. Litvinov, D.E. Discher, J.W. Weisel, Forced unfolding of the coiled-coils of fibrinogen by single molecule afm. Biophys. J., 524a–524a (2007)
46. B.B.C. Lim, E.H. Lee, M. Sotomayor, K. Schulten, Molecular basis of fibrin clot elasticity. Structure **16**(3), 449–459 (2008)
47. L. Kreplak, U. Aebi, H. Herrmann, Molecular mechanisms underlying the assembly of intermediate filaments. Exp. Cell Res. **301**(1), 77–83 (2004)
48. S.V. Strelkov, J. Schumacher, P. Burkhard, U. Aebi, H. Herrmann, Crystal structure of the human lamin a coil 2b dimer: implications for the head-to-tail association of nuclear lamins. J. Mol. Biol. **343**(4), 1067–1080 (2004)
49. R. Schietke, D. Brohl, T. Wedig, N. Mucke, H. Herrmann, T.M. Magin, Mutations in vimentin disrupt the cytoskeleton in fibroblasts and delay execution of apoptosis. Eur. J. Cell Biol. **85**(1), 1–10 (2006)
50. A.V. Sokolova, L. Kreplak, T. Wedig, N. Mucke, D.I. Svergun, H. Herrmann, U. Aebi, S.V. Strelkov, Monitoring intermediate filament assembly by small-angle x-ray scattering reveals

the molecular architecture of assembly intermediates. Proc. Natl. Acad. Sci. USA **103**(44), 16206–16211 (2006)
51. M.E. Csete, J.C. Doyle, Reverse engineering of biological complexity. Science **295**(5560), 1664–1669 (2002)
52. U. Alon, Simplicity in biology. Nature **446**(7135), 497 (2007)
53. J.W. Weisel, A.E.X. Brown, R.I. Litvinov, D.E. Discher, P.K. Purohit, Multiscale mechanics of fibrin polymer: gel stretching with protein unfolding and loss of water. Science **325**(5941), 741–744 (2009)
54. R. Phillips, P.K. Purohit, J. Kondev, Mechanics of DNA packaging in viruses. Proc. Natl. Acad. Sci. USA **100**(6), 3173–3178 (2003)
55. T.X. Su, S.K. Das, M. Xiao, P.K. Purohit, Transition between two regimes describing internal fluctuation of DNA in a nanochannel. PLoS ONE **6**(3) (2011)
56. J.W. Weisel, P.K. Purohit, R.I. Litvinov, A.E.X. Brown, D.E. Discher, Protein unfolding accounts for the unusual mechanical behavior of fibrin networks. Acta Biomater. **7**(6), 2374–2383 (2011)
57. J.S. Palmer, M.C. Boyce, Constitutive modeling of the stress-strain behavior of f-actin filament networks. Acta Biomater. **4**(3), 597–612 (2008)
58. V.S. Pande, V.A. Voelz, G.R. Bowman, K. Beauchamp, Molecular simulation of *ab initio* protein folding for a millisecond folder *ntl*9(1-39). J. Am. Chem. Soc. **132**(5), 1526–1528 (2010)
59. S.E. Jackson, How do small single-domain proteins fold? Fold. Des. **3**(4), 81–91 (1998)
60. A. Annila, V. Sharma, V.R.I. Kaila, Protein folding as an evolutionary process. Physica A, Stat. Mech. Appl. **388**(6), 851–862 (2009)
61. M. Shirts, V.S. Pande, Computing—screen savers of the world unite! Science **290**(5498), 1903–1904 (2000)
62. Z. Popovic, S. Cooper, F. Khatib, A. Treuille, J. Barbero, J. Lee, M. Beenen, A. Leaver-Fay, D. Baker, F. Players, Predicting protein structures with a multiplayer online game. Nature **466**(7307), 756–760 (2010)
63. J.D. Bryngelson, J.N. Onuchic, N.D. Socci, P.G. Wolynes, Funnels, pathways, and the energy landscape of protein-folding—a synthesis. Protein. Struct. Funct. Genet. **21**(3), 167–195 (1995)
64. C.B. Anfinsen, The formation and stabilization of protein structure. Biochem. J. **128**(4), 737–749 (1972)
65. V.S. Pande, A.Y. Grosberg, T. Tanaka, D.S. Rokhsar, Pathways for protein folding: is a new view needed? Curr. Opin. Struct. Biol. **8**(1), 68–79 (1998)
66. C. Levinthal, Are there pathways for protein folding? J. Chim. Phys. PhysicoChimie Biol. **65**(1), 44–45 (1968)
67. E. Hawkes, B. An, N.M. Benbernou, H. Tanaka, S. Kim, E.D. Demaine, D. Rus, R.J. Wood, Programmable matter by folding. Proc. Natl. Acad. Sci. USA **107**(28), 12441–12445 (2010)
68. E.D. Demaine, M.L. Demaine, Recent results in computational origami. Origami3 (2002), pp. 3–16
69. E.D. Demaine, J. O'Rourke, *Geometric Folding Algorithms: Linkages, Origami, Polyhedra* (Cambridge University Press, Cambridge, 2008)
70. E.D. Demaine, M.L. Demaine, J.S.B. Mitchell, Folding flat silhouettes and wrapping polyhedral packages: new results in computational origami. Comput. Geom. Theory Appl. **16**(1), 3–21 (2000)
71. R.D. Vale, The molecular motor toolbox for intracellular transport. Cell **112**(4), 467–480 (2003)
72. V. Daggett, B. Li, Molecular basis for the extensibility of elastin. J. Muscle Res. Cell Motil. **23**(5–6), 561–573 (2002)
73. S. Keten, M.J. Buehler, Nanostructure and molecular mechanics of spider dragline silk protein assemblies. J. R. Soc. Interface **7**(53), 1709–1721 (2010)
74. R. Pellenq, A. Kushima, R. Shahsavari, K. Van Vliet, M.J. Buehler, S. Yip, F.-J. Ulm, A realistic molecular model of cement hydrates. Proc. Natl. Acad. Sci. USA **106**(38), 16102–

16107 (2009)
75. F.J. Ulm, Chemomechanics of concrete at finer scales. Mater. Struct. **36**(261), 426–438 (2003)
76. E. Arzt, Size effects in materials due to microstructural and dimensional constraints: a comparative review. Acta Mater. **46**(16), 5611–5626 (1998)
77. G. Decher, Fuzzy nanoassemblies: toward layered polymeric multicomposites. Science **277**(5330), 1232–1237 (1997)
78. F. Caruso, R.A. Caruso, H. Mohwald, Nanoengineering of inorganic and hybrid hollow spheres by colloidal templating. Science **282**(5391), 1111–1114 (1998)
79. G. Chen, S. Shen, A. Henry, J. Tong, R.T. Zheng, Polyethylene nanofibres with very high thermal conductivities. Nat. Nanotechnol. **5**(4), 251–255 (2010)
80. K.S. Kumar, H. Van Swygenhoven, S. Suresh, Mechanical behavior of nanocrystalline metals and alloys. Acta Mater. **51**(19), 5743–5774 (2003)
81. H. Gao, B. Ji, I.L. Jger, E. Arzt, P. Fratzl, Materials become insensitive to flaws at nanoscale: lessons from nature. Proc. Natl. Acad. Sci. USA **100**(10), 5597–5600 (2003)
82. P. Fratzl, H.S. Gupta, E.P. Paschalis, P. Roschger, Structure and mechanical quality of the collagen-mineral nano-composite in bone. J. Mater. Chem. **14**(14), 2115–2123 (2004)
83. P. Fratzl, H. Peterlik, P. Roschger, K. Klaushofer, From brittle to ductile fracture of bone. Nat. Mater. **5**(1), 52–55 (2006)
84. U. Aebi, J. Cohn, L. Buhle, L. Gerace, The nuclear lamina is a meshwork of intermediate-type filaments. Nature **323**(6088), 560–564 (1986)
85. J.K. Rainey, C.K. Wen, M.C. Goh, Hierarchical assembly and the onset of banding in fibrous long spacing collagen revealed by atomic force microscopy. Matrix Biol. **21**(8), 647–660 (2002)
86. S. Keten, M.J. Buehler, Strength limit of entropic elasticity in beta-sheet protein domains. Phys. Rev. E **78**, 061913 (2008)
87. S. Keten, M.J. Buehler, Geometric confinement governs the rupture strength of h-bond assemblies at a critical length scale. Nano Lett. **8**(2), 743–748 (2008)
88. Z. Qin, S. Cranford, T. Ackbarow, M.J. Buehler, Robustness-strength performance of hierarchical alpha-helical protein filaments. Int. J. Appl. Mech. **1**(1), 85–112 (2009)
89. D. Sen, M.J. Buehler, Structural hierarchies define toughness and defect-tolerance despite simple and mechanically inferior brittle building blocks. Sci. Rep. **1**, 35 (2011)
90. A. Garcia, D. Sen, M.J.B. Buehler, Hierarchical silica nanostructures inspired by diatom algae yield superior deformability, toughness and strength. Metall. Mater. Trans. A **42**(13), 3889–3897 (2011)
91. D. Sen, A. Garcia, M.J. Buehler, Mechanics of nano-honeycomb silica structures: a size-dependent brittle-to-ductile transition. J. Nanomech. Micromech. (ASCE) **1**(4) (2011)
92. A.R. Dunn, A.S. Adhikari, J. Chai, Mechanical load induces a 100-fold increase in the rate of collagen proteolysis by mmp-1. J. Am. Chem. Soc. **133**(6), 1686–1689 (2011)
93. S. Suresh, G. Bao, Cell and molecular mechanics of biological materials. Nat. Mater. **2**(11), 715–725 (2003)
94. M.J. Buehler, Nature designs tough collagen: explaining the nanostructure of collagen fibrils. Proc. Natl. Acad. Sci. USA **103**(33), 12285–12290 (2006)
95. T. Ackbarow, X. Chen, S. Keten, M.J. Buehler, Hierarchies, multiple energy barriers and robustness govern the fracture mechanics of alpha-helical and beta-sheet protein domains. Proc. Natl. Acad. Sci. USA **104**, 16410–16415 (2007)
96. M.J. Buehler, S. Keten, T. Ackbarow, Theoretical and computational hierarchical nanomechanics of protein materials: deformation and fracture. Prog. Mater. Sci. **53**, 1101–1241 (2008)
97. J.M. Ferrer, H. Lee, J. Chen, B. Pelz, F. Nakamura, R.D. Kamm, M.J. Lang, Measuring molecular rupture forces between single actin filaments and actin-binding proteins. Proc. Natl. Acad. Sci. USA **105**(27), 9221–9226 (2008)
98. J.M. Gosline, P.A. Guerette, C.S. Ortlepp, K.N. Savage, The mechanical design of spider silks: from fibroin sequence to mechanical function. J. Exp. Biol. **202**(23), 3295–3303

(1999)
99. D.L. Kaplan, F.G. Omenetto, New opportunities for an ancient material. Science **329**(5991), 528–531 (2010)
100. F. Vollrath, D. Porter, Silk as a biomimetic ideal for structural polymers. Adv. Mater. **21**(4), 487–492 (2009)
101. A. Gautieri, S. Vesentini, A. Redaelli, M.J.B. Buehler, Hierarchical structure and nanomechanics of collagen microfibrils from the atomistic scale up. Nano Lett. **11**(2), 757–766 (2011)
102. H. Kitano, Computational systems biology. Nature **420**(6912), 206–210 (2002)
103. S.W. Cranford, M.J. Buehler, Shaky foundations of hierarchical biological materials. Nano Today **6**(4), 332–338 (2011)
104. B.B. Mandal, S.C. Kundu, Cell proliferation and migration in silk fibroin 3d scaffolds. Biomaterials **30**(15), 2956–2965 (2009)
105. H. Fernandes, L. Moroni, C. van Blitterswijk, J. de Boer, Extracellular matrix and tissue engineering applications. J. Mater. Chem. **19**(31), 5474–5484 (2009)
106. A.J. Engler, S. Sen, H.L. Sweeney, D.E. Discher, Matrix elasticity directs stem cell lineage specification. Cell **126**(4), 677–689 (2006)
107. D.E. Discher, P. Janmey, Y.L. Wang, Tissue cells feel and respond to the stiffness of their substrate. Science **310**(5751), 1139–1143 (2005)
108. K. Kim, D. Dean, J. Wallace, R. Breithaupt, A.G. Mikos, J.P. Fisher, The influence of stereolithographic scaffold architecture and composition on osteogenic signal expression with rat bone marrow stromal cells. Biomaterials **32**(15), 3750–3763 (2011)
109. F.P.W. Melchels, A.M.C. Barradas, C.A. van Blitterswijk, J. de Boer, J. Feijen, D.W. Grijpma, Effects of the architecture of tissue engineering scaffolds on cell seeding and culturing. Acta Biomater. **6**(11), 4208–4217 (2010)
110. X. Liu, X. Jin, P.X. Ma, Nanofibrous hollow microspheres self-assembled from star-shaped polymers as injectable cell carriers for knee repair. Nat. Mater. **10**, 398–406 (2011)
111. K. Kim, A. Yeatts, D. Dean, J.P. Fisher, Stereolithographic bone scaffold design parameters: osteogenic differentiation and signal expression. Tissue Eng. B Rev. **16**(5), 523–539 (2010)
112. J.L. Drury, D.J. Mooney, Hydrogels for tissue engineering: scaffold design variables and applications. Biomaterials **24**(24), 4337–4351 (2003)
113. A. Nandakumar, L. Yang, P. Habibovic, C. van Blitterswijk, Calcium phosphate coated electrospun fiber matrices as scaffolds for bone tissue engineering. Langmuir **26**(10), 7380–7387 (2010)
114. T.P.J. Knowles, T.W. Oppenheim, A.K. Buell, D.Y. Chirgadze, M.E. Welland, Nanostructured films from hierarchical self-assembly of amyloidogenic proteins. Nat. Nanotechnol. **5**(3), 204–207 (2010)
115. M. Shin, H. Yoshimoto, J.P. Vacanti, In vivo bone tissue engineering using mesenchymal stem cells on a novel electrospun nanofibrous scaffold. Tissue Eng. **10**(1–2), 33–41 (2004)
116. E. Alsberg, K.W. Anderson, A. Albeiruti, J.A. Rowley, D.J. Mooney, Engineering growing tissues. Proc. Natl. Acad. Sci. USA **99**(19), 12025–12030 (2002)
117. M.A.C. Stuart, W.T.S. Huck, J. Genzer, M. Muller, C. Ober, M. Stamm, G.B. Sukhorukov, I. Szleifer, V.V. Tsukruk, M. Urban, F. Winnik, S. Zauscher, I. Luzinov, S. Minko, Emerging applications of stimuli-responsive polymer materials. Nat. Mater. **9**(2), 101–113 (2010)
118. W. Senaratne, L. Andruzzi, C.K. Ober, Self-assembled monolayers and polymer brushes in biotechnology: current applications and future perspectives. Biomacromolecules **6**(5), 2427–2448 (2005)
119. P.M. Mendes, Stimuli-responsive surfaces for bio-applications. Chem. Soc. Rev. **37**(11), 2512–2529 (2008)
120. Z.S. Liu, P. Calvert, Multilayer hydrogels as muscle-like actuators. Adv. Mater. **12**(4), 288–291 (2000)
121. J.P. Fisher, D. Dean, A.G. Mikos, Photocrosslinking characteristics and mechanical properties of diethyl fumarate/poly(propylene fumarate) biomaterials. Biomaterials **23**(22), 4333–4343 (2002)

122. K.K. Chia, M.F. Rubner, R.E. Cohen, ph-responsive reversibly swellable nanotube arrays. Langmuir **25**(24), 14044–14052 (2009)
123. S.W. Cranford, C. Ortiz, M.J. Buehler, Mechanomutable properties of a paa/pah polyelectrolyte complex: rate dependence and ionization effects on tunable adhesion strength. Soft Matter **6**(17), 4175–4188 (2010)
124. R.A. Marklein, J.A. Burdick, Spatially controlled hydrogel mechanics to modulate stem cell interactions. Soft Matter **6**(1), 136–143 (2010)
125. S.J. Bryant, R.J. Bender, K.L. Durand, K.S. Anseth, Encapsulating chondrocytes in degrading peg hydrogels with high modulus: engineering gel structural changes to facilitate cartilaginous tissue production. Biotechnol. Bioeng. **86**(7), 747–755 (2004)
126. C. Alexander, K.M. Shakesheff, Responsive polymers at the biology/materials science interface. Adv. Mater. **18**(24), 3321–3328 (2006)
127. L. Ionov, N. Houbenov, A. Sidorenko, M. Stamm, S. Minko, Stimuli-responsive command polymer surface for generation of protein gradients. Biointerphases **4**(2), 45–49 (2009)
128. R.W. Sands, D.J. Mooney, Polymers to direct cell fate by controlling the microenvironment. Curr. Opin. Biotechnol. **18**(5), 448–453 (2007)
129. I. Tokarev, M. Orlov, S. Minko, Responsive polyelectrolyte gel membranes. Adv. Mater. **18**(18), 2458 (2006)
130. I. Tokarev, S. Minko, Multiresponsive, hierarchically structured membranes: new, challenging, biomimetic materials for biosensors, controlled release, biochemical gates, and nanoreactors. Adv. Mater. **21**(2), 241–247 (2009)
131. M. Motornov, T.K. Tam, M. Pita, I. Tokarev, E. Katz, S. Minko, Switchable selectivity for gating ion transport with mixed polyelectrolyte brushes: approaching 'smart' drug delivery systems. Nanotechnology **20**(43), 434006 (2009)
132. X.H. Zhao, J. Kim, C.A. Cezar, N. Huebsch, K. Lee, K. Bouhadir, D.J. Mooney, Active scaffolds for on-demand drug and cell delivery. Proc. Natl. Acad. Sci. USA **108**(1), 67–72 (2011)
133. C.L. Ren, R.J. Nap, I. Szleifer, The role of hydrogen bonding in tethered polymer layers. J. Phys. Chem. B **112**(50), 16238–16248 (2008)
134. M. Tagliazucchi, E.J. Calvo, I. Szleifer, Redox and acid base coupling in, ultrathin polyelectrolyte films. Langmuir **24**(6), 2869–2877 (2008)
135. T. Scheibel, R. Parthasarathy, G. Sawicki, X.M. Lin, H. Jaeger, S.L. Lindquist, Conducting nanowires built by controlled self-assembly of amyloid fibers and selective metal deposition. Proc. Natl. Acad. Sci. USA **100**(8), 4527–4532 (2003)

# Chapter 4
# Universality-Diversity Paradigm: Music, Materiomics, and Category Theory

**Abstract** The transition from the material structure to function, or from nanoscale components to the macroscale system, is a challenging proposition. Recognizing how Nature accomplishes such a feat—through universal structural elements, relatively weak building blocks, and self-assembly—is only part of the solution. The complexity bestowed by hierarchical multi-scale structures is not only found in biological materials and systems—it arises naturally within other fields such as music or language, with starkly different functions. If we wish to exploit understanding of the structure of music as it relates to materials, we need to define the relevant properties and functional relations in an abstract sense. One approach may lie in category theory, presented here in the form of ontology logs (ologs), that can transcend the traditional definitions of materials, music, or language, in a consistent and mathematically robust manner.

*A sentence should contain no unnecessary words, a paragraph no unnecessary sentences, for the same reason that a drawing should have no unnecessary lines and a machine no unnecessary parts...*

*William Strunk, Jr. and E.B. White, The Elements of Style (1919)*

## 4.1 Introduction

To solve society's most pressing problems, including medical, energy, and environmental challenges, we will need *transformative*, rather than *evolutionary*, approaches. Many of these depend on finding materials with properties that are substantially improved over existing candidates, and we are increasingly turning to complex materials. As discussed in the previous chapter, biological materials and systems present many challenges before we can "unlock" the secrets of Nature. The bottom-up approach of genetic determinism has proven insufficient to encompass the complexity of biological materials [1, 2]. At the level of coded information in DNA—of replication, inheritance, and decoding of DNA messages—the theory of the gene is elegant in its simplicity, accurately captured by:

$$\text{DNA} \Rightarrow \text{RNA} \Rightarrow \text{Protein(s)}$$

S.W. Cranford, M.J. Buehler, *Biomateriomics*, Springer Series in Materials Science 165, DOI 10.1007/978-94-007-1611-7_4, 

At this level the theory is complete or nearly so. Indeed, from this paradigm, we have successfully been able to "recreate" biological materials from proteins (such as recombinant silks [3, 4], for example), but have yet to replicate their performance or properties. We do not fully understand the following process:

$$\text{DNA} \Rightarrow \text{RNA} \Rightarrow \text{Protein(s)} \Rightarrow \text{Function(s)}$$

This "break in the chain"—the inability to predict function from DNA—has partially motivated the paradigm of *systems biology*. Systems biology can be thought of as a field that focuses on complex interactions in biological systems, through a perspective of holism (*e.g.*, system based) rather than reductionism, the modeling and discovery of emergent properties.[1] A reductionist approach can successfully identify basic components and interactions but, unfortunately, offers no convincing concepts or methods to understand the mechanisms by which system properties emerge. We encounter complex hierarchies, multiscale mechanisms, folded and entangled structures, entropic behaviors, and environmental sensitivity, all coupled with a living, growing organism. Yet, these material perform much better than engineered analogues, exhibiting adaptability, toughness, flaw tolerance, self-healing, and energy efficient production. By what means can we decode such complex materials? In other words, how can we explicitly *define the materiome*?

While biological materials are composed of complex hierarchical structures, the molecular basis is comprised of a restricted set of amino acids—simple nanoscale building blocks. In this chapter, we turn to *music* as a "model system", to teach us how complex and intricate compositions can be created (*e.g.*, composed) from simple building blocks (*i.e.*, sound). Like material characterized by mechanical, thermal, electrical, and optical properties across scale, music can be described by categories such as notes and scales, timbre and pitch, tones and instruments, tempo and style (see Fig. 4.1). We demonstrate how analogies between seemingly different fields enable the understanding of general principles by which functional properties in hierarchical systems emerge, similar to an analogy learning process. The mathematical field of *category theory* [5] provides a necessary abstraction to integrate our understanding of complex materials. Biological science has already introduced the concepts of phenotypes (*i.e.*, observed properties) and genotypes (*i.e.*, hereditary information), similar to the morphisms and structures central to category theory. The extension to materials, mechanical properties, and most importantly functionality, is a consequent extension.

Specifically, natural hierarchical materials (such as spider silk) exhibit properties comparable to other constructs (such as classical music) in terms of their hierarchical structure and function. In this chapter, we demonstrate category theoretic analysis as an abstract means to describe emergence of functionality from first principles, *e.g.*, on the basis of fundamental interactions between universal building blocks. As a comparative tool, we discuss hierarchical ontology logs (also referred to as

[1] Systems biology can be considered an umbrella term, encompassing many of the *-omic* fields previously discussed such as genomics, phenomics, proteomics, interactomics, etc. Indeed, "materiomics" itself is concerned with the material *system*.

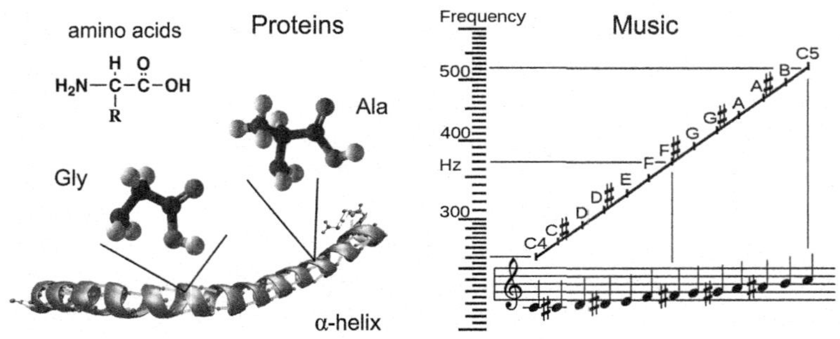


**Fig. 4.1** How are proteins and music similar? Proteins are composed of a sequence of amino acids—general molecules containing an amine group (–$NH_2$), a carboxylic acid group (–COOH) and a side-chain (here, designated as R). Variation in the side-chain results in specific amino acids such as glycine (Gly) or alanine (Ala), which form a sequence that dictate the structure and function of larger protein biomolecules (*e.g.*, an $\alpha$-helix). Music is also composed of a sequence—a sequence of notes, rather than amino acids. Each note is physically related to a frequency (*e.g.*, sound). Variation in the frequency results in specific notes such as $D$ or $A\sharp$. A sequence of notes create a riff or melody. Can we use music to understand complex biomaterials, to design them, and can we create musical expressions of material mechanisms to gain insight into how materials work?

*ologs*) [6, 7] that follow a rigorous mathematical formulation based on category theory [5] to provide an insightful system representation by expressing knowledge in a conceptual map. We explain the process of analogy creation, draw connections at several levels of hierarchy and identify similar patterns that govern the structure of the hierarchical systems and discuss the impact of the derived analogy for nanotechnology. The construction of simple analogies and their accomplishment to collocate a broad picture of materials structure and function can be achieved. Through such representations, similar patterns between systems which trace back to their hierarchical build-up are identified. The broad aim of such constructions is the institution of category theoretic tools as a rigorous and comprehensive means to systematically depict and communicate hierarchical structure-function relationships for applications across fields in engineering, science and the arts—from *music* to *materiomics*.

## 4.2 Universality-Diversity Paradigm

One of the most striking aspects of physics is the simplicity of its laws. Maxwell's equations, Schrödinger's equation, and Hamiltonian mechanics can each be expressed in a few lines [8, 9]. Such simplicity motivates the continuing search for the proverbial "Universe on a T-Shirt", wherein the fundamental equations of Nature can be expressed on a single swath of poly/cotton blend [10]. Everything is simple and neat, except, of course, for biology. Biology is "messy" because it has many different working parts, yet derived from a few fundamental components [11]. Can we use these "limited building blocks" to formulate simple and neat laws? A vision shared by most researchers in complex systems—whether biological, economical, or sociological—is that certain intrinsic, perhaps even universal, features capture fundamental aspects of complexity in a manner that transcends specific domains [12]. The challenge, of course, is choosing the appropriate perspective. For example, most undergraduate students are introduced the concept of polar coordinates (*e.g.*, $r, \theta$) as a means to simplify the analysis of certain problems (such as the equation of motion of a pendulum); using Cartesian coordinates (*e.g.*, $x$, $y$) complicates the analysis. How we *choose* to formulate the problem affects the *solution*.

As mentioned in Chap. 3: The Challenges of Biological Materials, the evolution of protein materials through genetic selection and structural alterations has resulted in a specific set of successful (but not necessarily optimized) protein building blocks that define their structure. While protein materials exist in an abundant variety, the need exists to formulate a widely applicable model to systematically *categorize* all such materials, to reduce the seemingly complexity of biological materials [13]. The approach of utilizing universal building blocks to create diverse multifunctional hierarchical structures has been successfully applied in current macroscale engineering paradigms. Advanced technologies and biology have extremely different physical implementations, but they are far more alike in systems-level organization than is widely appreciated [13].

The *universality-diversity paradigm* (UPD) incorporates the recognition and analysis of biological materials based on the universality and diversity of its fundamental structural elements and functional mechanisms [14, 15]. For example, proteins constitute the elementary building blocks of a vast variety of biological materials such as cells, spider silk or bone, where they create extremely robust, multi-functional materials by self-organization of structures over many length- and time scales, from nano to macro. Examples of such universal building blocks include $\alpha$-helices, $\beta$-sheets or tropocollagen (triple-helical) molecules. In contrast, other features are highly specific to tissue types, such as particular filament assemblies, $\beta$-sheet nanocrystals in spider silk or tendon fascicles. Similarly, cellulosic materials [16], such as wood, grasses, and other green plants, exhibit a wide array of macro-scale mechanical properties dependent on the fiber morphology and structure [17, 18], yet are composed of similar molecular building blocks (various polysaccharides in both crystalline and amorphous phases [19]). Some of the structural features of both protein-based and cellulosic materials are commonly found in a many different systems, that is, they are highly conserved and constructed in similar structural hierarchies. It is apparent that using only a limited number of components, Nature has produced a broad range of materials with diverse properties and biological functions, and created multifunctionality (*diversity*) by changing structural arrangements of few (*universal*) constituents rather than inventing new building blocks. This paradigm is a paradox: *How can a structure be universal and diverse at the same time*? In protein materials, the coexistence of universality and diversity is enabled by utilizing hierarchies of structure, which serve as an expansion of the design space.

**Table 4.1** Intermediate filaments are remarkable due to their diverse appearance in organisms, where they fulfill multiple functions at different hierarchical levels

| IF | Found in | Protein level | Filament level | System level |
|---|---|---|---|---|
| Vimentin [20, 21] | Cell's cytoskeleton | cell signaling mechanisms, associated protein organization | responsible for location, shape and stability of cell organelles, protein targeting processes | "security belt" of the cell |
| Keratin [22, 23] | Hair, nails, hoofs | protein synthesis, cell signaling mechanisms, associated protein organization | cell pigmentation, organization of cell organelles | cell growth, wound healing, locomotion, prey procurement |
| Lamin [24, 25] | Nuclear envelope | signaling mechanisms, mechanotransduction, chromatin positioning | gene regulation and transcription, chromatin positioning | protection of the chromatin, involved in cell mitosis |

**Universality-Diversity Paradigm (UDP)**: The analysis of materials systems based on the recognition of the *universality* of structural elements (building blocks) and potential *diversity* of fundamental functional mechanisms and material behavior. Through the UDP it is possible to improve our understanding of how complex systems are capable of robustly unifying seemingly incompatible features at different hierarchical scales.

The UDP can be exemplified by a particular class of proteins called intermediate filaments (IFs). Intermediate filaments form protein networks in the cytoskeleton of eukaryotic cell, stabilize the nuclear envelope and provide the basis for extra-cellular tissues such as hair or nails. Interestingly, the elementary building block of all kind of intermediate filaments is identical—a universal $\alpha$-helical coiled-coil motif. The three intermediate filaments of interest are described in Table 4.1.

The lowest level of hierarchy encodes the structure of these proteins in the sequence of amino acids. This is reflected by the fact that each intermediate filament type has a distinct amino acid sequence (diversity). Intriguingly, the differences at the lowest hierarchy do not influence the immediately following hierarchical level. This can be verified since all intermediate filaments feature the $\alpha$-helical motif, despite differences at the amino acid sequence level and/or differences at larger scales (universality). However, moderate effects can be observed at the dimer level (two biomolecules). In the dimer structure subtle differences emerge. For example, amino acid inserts in the periodic heptad repeat lead to a local uncoiling of the super helix (creating a "stutter"—a distinct region where the amino acid sequence differs from other parts of the $\alpha$-helical domains), which effects the assembly process as well as the unfolding mechanics [26–28]. Another example is the occurrence of mutations in desmin intermediate filament coiled-coils. It was shown that disease related mutations do not destroy the $\alpha$-helix structure but build additional stutters or stammers in

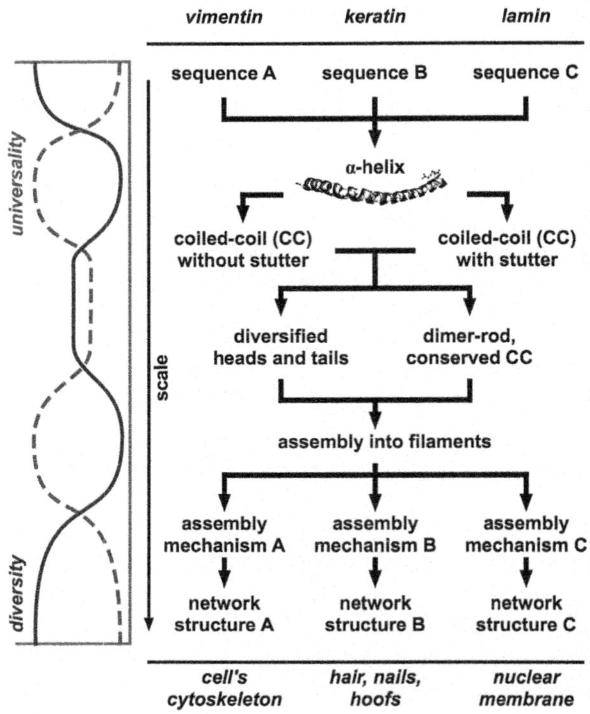


**Fig. 4.2** Hierarchical biological materials, here exemplified for the example of three types of intermediate filaments, are governed through interplay of universal and diverse patterns, which, combined with silencing and activation are unified over multiple hierarchical scales. This enables to forward information that is completely coded at the lowest scale (amino acid sequence), safely by means of silencing through intermediate scales (alpha helix, coiled-coil) up to higher scales, where they are activated in order to fulfill specific requirements. There is a balancing between *universal* features (*e.g.*, α-helical structures, filament assembly) and *diverse* functionality. The *plot on the left* shows the balancing exchange between universality and diversity across scales. Adapted from [14]

the coiled-coil [29, 30]. Even though all types of intermediate filaments commonly show an assembly into filaments, lower scale differences (that is, for instance the amino acid sequence and presence of the stutter) affect the pattern and process of assembly, such as the number of proteins per filament cross-sectional area, or the way dimers associate. The differences on the filament level are of utmost importance, as they influence the properties at the network and the super-structural level, which are dominated by (but not limited to) mechanical functions. The cross-scale universality-diversity of intermediate filaments is depicted in Fig. 4.2.

As shown in this example, nanoscopic modifications do not always influence the properties at the next hierarchical layer, but potentially those of one or more hierarchical layers above. It appears as if specific functional requirements at several higher scales are "forwarded" to lower scales, where modifications are implemented. Through this mechanism biological materials are not only multi-functional but are further continuously adapted to the required scale-specific processes, with the goal to fit the diverse required functions in the best possible way.

This coexistence of universality and diversity as described in the universality-diversity paradigm, is an overarching feature in biological materials and a crucial component of materiomics. Nature has a unique capacity of creating toughness from weak components, capable of balancing multiple, seemingly incompatible properties such as strength, robustness, and adaptability. Materials like bone, being a nano-composite of strong but brittle and soft but ductile materials, illustrate this unification of components with disparate properties within a hierarchical structure. Primarily, the utilization of structural hierarchies enables the coexistence of universality and diversity. Through the UDP it is possible to improve our understanding of how materials are capable of robustly unifying seemingly incompatible features at different hierarchical scales, a fundamental keystone to materiomics. Indeed, material characteristics such as strength and robustness are contradicting properties that cannot be easily combined within a single scale of traditional materials science, requiring a materiomics approach to reconcile such disparate concepts. Further, such a perspective clearly indicates that in structures and processes are amalgamated and cannot be considered alone. Although wide and varying, biology commonly exhibits unlikely harmony within material structures and physiologic functionality.

## 4.3 Tu(r)ning Weakness into Strength

The importance of hierarchies is elusive for many material systems. For the sake of argument, justification of a continued focus on such material hierarchies is prudent. Indeed, one may ask why such complex and redundant hierarchies are even necessary? The simple answer is that hierarchical structures are not required under ideal circumstances (*e.g.*, unlimited resources, both material and energetic), but serve to *extend* the physical design space while limited to a restricted set of constituent building blocks. In other terms, it provides material scientists and engineers more design parameters to manipulate within the same set of building blocks (*i.e.*, structural elements) to realize multiple functions, which together create a highly functional system that is much more than the sum of its parts. Nature implements this trick frequently. A major trait of biological materials is the occurrence of not only hierarchies, but also hierarchies that: (a) contain weak interactions (such as H-bonds in spider silk); (b) exploit abundantly available materials (*e.g.*, silica in some sea creatures), or; (c) implement structurally inferior materials (*e.g.*, extremely brittle crystals in mineralized tissues like nacre or bone). Yet such materials are commonly more robust and tough than those currently designed and engineered. The paradox of a "strong" material being composed of "weak" components is difficult to theoretically reconcile. The entire materiome across multiple scales must be introduced, where universal and diverse patterns are unified hierarchically, and the integrated view of it results in a quantitative understanding of how highly functional materials are created based on inferior building blocks.

Despite this apparent complexity of functional properties, the structural designs of biological materials have evolved under extreme evolutionary pressures to ensure a species' survival, often in adverse environments. As a result, materials are created with the lowest possible energy consumption, under simple processing conditions, and are exquisite as they often form from abundant material constituents. Notably, these abundant material constituents typically represent functionally inferior material building blocks that are exceptionally brittle, such as silica or other minerals, or extremely weak, such as H-bonding or Van der Waals forces. The integration of weak building blocks is not a design *choice*, but rather a natural evolutionary constraint. Moreover, comparative studies show that most biological materials are made up from only a few select universal elements (*e.g.*, C, H, O, N, S...), despite their functional diversity [31]. This is exemplified at different levels, such as in the occurrence of only twenty natural amino acids (which provides the basis for all known protein materials), and a few universal secondary structures such as $\alpha$-helices, $\beta$-sheets or random coils, or more fundamentally, the four DNA or RNA nucleotides that are universal to encode the structure of all protein materials at the genetic level. The definition of these "fundamental building blocks" is also a matter of perspective. In terms biological systems, the primary building blocks of proteins are commonly considered the amino acids (which constitute a unique sequence). However, these acids are also composed of side-chains (*e.g.*, amino acid = amine + carboxylic acid + side-chain), which can also be deconstructed to a few elements and atoms (illustrated by Fig. 4.3). Even the atoms themselves can be considered "hierarchies" of protons, neutrons, and electrons. Down the rabbit hole we go! Practical application requires a "truncation of decomposition", which is dictated by a system under analysis. For certain systems, the behavior of neutrinos and other elementary subatomic particles may be required.

The realization of how biology perpetually uses a limited number of elements to create highly diverse systems poses a fundamental question about the design of biological materials, and specifically, how such a great diversity of functional properties is achieved, despite the apparent inferiority and universality of the building blocks [15, 32]. Recent research suggests that the basis of understanding the remarkable properties of these materials lies within the biological design paradigm where multifunctionality is created not through the use of high quality, or use of a large number of, distinct building blocks. Rather, functionality is created by compiling simple and often inferior elements into assemblies where structures are created at multiple length-scales, resulting in hierarchical material architectures.

It is also within this self-assembly process that the inherent "weakness" of biological materials is a "strength". The flexibility of proteins facilitate folds, coiling, geometric manipulation and compatibility with other proteins. Weak interactions such as H-bonding, ionic cross-linking, and polarization enables the formation of structural units (*e.g.*, dimers and larger quaternary protein structures) with minimal energetic and material expense. One cannot adequately bond carbon nanotubes to a copper substrate without significant chemical and structural modifications, but it is possible to link two proteins to exploit functional groups and assembly mechanisms [33–35] as depicted in Fig. 4.4. Such approaches are further discussed in Chap. 10: Synthesis and Design.

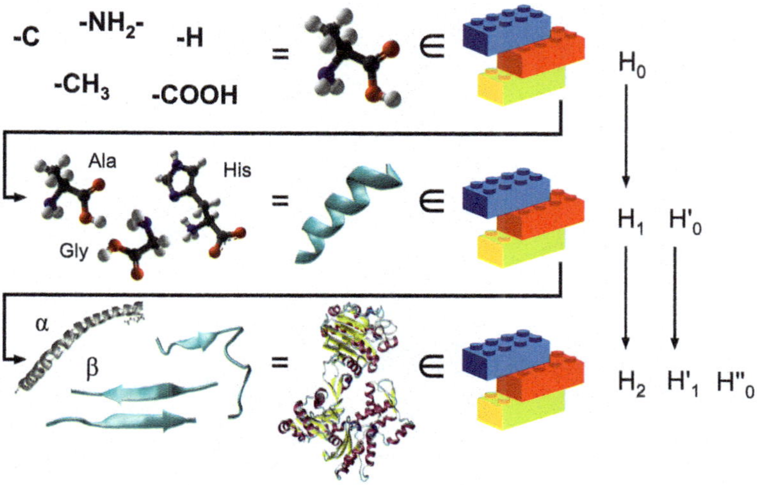


**Fig. 4.3** Building blocks built by building blocks? Hierarchical systems composed of structural units are subject to functional perspective. Here, we assume the molecular groups (amine, carboxylic acid, and a side chain, such as the a methyl group depicted) are the fundamental irreducible building blocks of amino acids (such as alanine), and the set of amino acids provide the building blocks at the primary hierarchy ($H_0$), defined as the first structural unit. The set of amino acids (glycine, alanine, histidine, etc.) can form sequences with characteristic structures (such as $\alpha$-helix), which compose the next hierarchy ($H_1$). Finally, many secondary structures ($\alpha$-helices, $\beta$-sheets, etc.) can form a folded protein (here, actin) at the highest hierarchy considered ($H_3$). At higher scales, proteins themselves form the "set" of building blocks (actin is one of the three major components of the cytoskeleton). Depending on application, amino acids can be considered irreducible building blocks, and thus the secondary structures provide the primary hierarchy ($H_0'$) followed by the protein structures ($H_1'$). Similarly, from a cellular sense, the secondary structures may be considered irreducible, and thus the proteins provide the first structural unit ($H_0''$). Definition of hierarchies and building blocks is dependent on application and the function of each block. Note too that the characterization of secondary structures is a choice—as the amino acid sequence can be considered the building blocks of entire proteins, without the intermediate definition of secondary structures

To fully exploit the "pliability" of proteins as structural building blocks, we must fully understand their potential interactions. Normal biological function requires that a systems's biochemical network be highly interconnected and interactive [36]. The reductionist approach employed by recent biologists is effective in defining the molecular structure and organization of DNA and proteins, but less effective at delineating function, because most biological molecules do not operate independently [37]—they must be considered in the context of all of the other proteins in the cell and their organization [38]. To reconcile this network complexity, the concept of "modular cell biology" was proposed by Hartwell *et al.* [37], grouping interacting molecules, including DNA, RNA, proteins, and smaller molecules, into cellular modules that represent discrete functional entities. Depending on the scale of interest, these modules can be considered the irreducible functional "building blocks", or defined by their finer scale constituents.

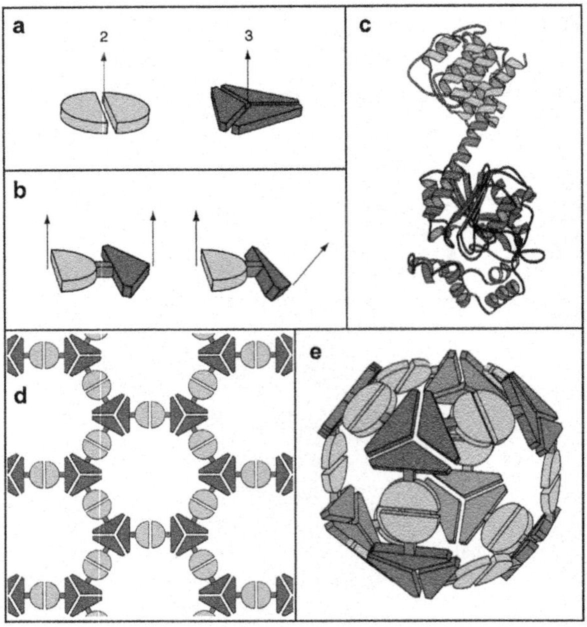


**Fig. 4.4** "Weak" materials can self-assemble: a general strategy for designing proteins that form predictable nanostructures. (**a**) The semicircles represent a natural dimeric protein (*i.e.*, a protein that associates with one other copy of itself), whereas the triangle shape represents a trimeric protein. The symmetry axes of the natural oligomers are shown. (**b**) The two natural proteins are combined by genetic methods into a single protein. Each of the original natural proteins serves as an "oligomerization domain" in the designed fusion protein, with known interaction properties (*i.e.*, function). Two different hypothetical fusion proteins are shown to illustrate that the oligomerization domains can be joined rigidly in different geometries. (**c**) A ribbon diagram of a fusion protein showing one method for joining two oligomerization domains in a relatively rigid fashion. The two are linked by a short stretch of amino acids and the two oligomerization domains are joined physically in a predictable orientation. (**d**) A designed fusion protein self assembles into a particular kind of nanostructure that depends on the geometry of the symmetry axes belonging to its component oligomerization domains. A molecular layer arises from an arrangement like that in (**b**) (*left*). (**e**) A cubic cage arises from an arrangement like the one in (**b**) (*right*). From [33], used with permission, copyright © 2000 National Academy of Sciences

Certain properties of biological modules are recognizable by engineers, such as positive and negative feedback loops, and parallel (redundant) circuits [37]. To exploit a modular building block understanding of cellular function it will be necessary for biologists to understand design principles more familiar to engineers and computer scientists, such as amplifiers, oscillators, and logic circuits [39]. Instead of having to include many of the molecular details of protein-protein or cellular interactions, the functional behavior each module (or system "motif") can be described using mathematical models that have a certain degree of universality. In many cases the models do not require plunging into the details of how every protein works—they only need to include information on whether $X$ activates or inhibits $Y$, and at what concentration (and perhaps a few additional parameters). Such models seem to capture the essential dynamics of protein "circuits", while being, in a sense, insulated from most of the complexity of the proteins themselves [11].

Thereby, delineated levels of resolution (*e.g.*, network module or material building block) provides access to specific *functional* properties, which can be achieved by defining a particular material structure at each hierarchical scale. This paradigm, the formation of distinct structures at multiple length scales, also enables biological materials to overcome the intrinsic weaknesses of the building blocks. This can be illustrated in the use of nanoconfinement that often results in enhanced strength and ductility despite the intrinsic brittleness of the same material in bulk form [40]. In an example relevant for sea creatures such as diatom algae, while silicon and silica is extremely brittle in bulk, the formation of nanostructures results in great ductility and extensibility, where the specific geometry used allows for a continuum of mechanical signatures. Thus, the realization of distinct structural designs provides a means to tune the material to achieve a great diversity of functional properties despite the use of the same building blocks. More generally, if it is possible to alter the material's structure at specific hierarchy levels, independently and during operation of a material, there exists the potential to realize varied material properties depending on functional needs. This is exemplified in echinoderms (*e.g.*, sea urchins and sea cucumbers) that can change their body's modulus manifold through alterations of the cross-linking of constituting collagen molecules. Mutability also occurs in plants that track the direction of sunlight, through a mechanism that involves a change of the plant wall's stiffness exposed to light and resulting in bending towards the softer part (*via* a hypothesized breaking and reforming of H-bonds).

A powerful example to demonstrate the biological material design paradigm is the case of spider silk (discussed in detail in Chap. 8: Unlocking Nature: Case Studies), a remarkable material that-in order to fulfill its biological tasks-must provide extreme levels of strength (1–2 GPa), toughness, and great deformability of up to 50 % tensile strain [41]. Yet, the structural basis of spider silk is extremely simple, and merely consists of a few distinct few amino acids, arranged in long polypeptide chains and that interact only by weak interactions. Moreover, spiders must be able to produce silk quickly, out of a limited stock of solvated protein [42]. The physiologic processing conditions in solvent, at room temperature, and at very short time-scales rules out enzymatic processes and suggests that self-assembly is the primary mechanism to form spider silk under these conditions. This necessitates the use of weak bonding in the fundamental interactions of the silk protein strands. Indeed, spider silk is known to be dominated by H-bonds, one of the weakest chemical bonds known, and also present in liquid water (for example, H-bonds have an energy of $\sim$5 kcal/mol whereas the covalent bond of a $H_2$ molecule has an energy of $\sim$100 kcal/mol). But how is it possible to generate such a mechanically superior material out of clearly inferior constituents and constraints?

The answer is that weak elements in the material, here H-bonds, are arranged geometrically in order to provide maximum strength and toughness (see Fig. 4.5). It has been found that the intrinsic weakness of H-bonds vanishes when grouped into clusters of $\sim$4 H-bonds, which allows them to work cooperatively and thus reach

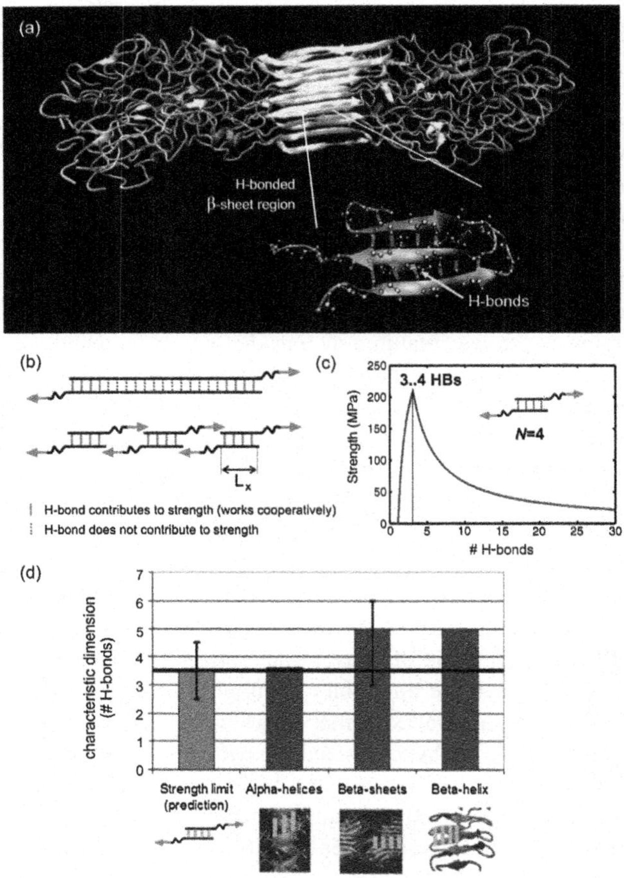


**Fig. 4.5** Clustering of weak H-bonds enhances cooperativity. (**a**) The molecular structure of dragline silk is characterized by a crystalline region of H-bonded $\beta$-sheets [43]—a result of a regular repetition of alanine (poly-Ala) within the silk protein sequence. For natural silk, these poly-Ala regions are typified by six residues resulting in a finite cluster of $\approx$5 H-bonds. *Why not more?* The rupture strength of H-bonds is governed by their cooperativity *via* geometric confinement. (**b**) Schematic illustrating the difference of the strength of a single, long beta strand *vs* a combination of multiple small strands. In the former, only H-bonds at the boundary participate in the rupture process and provide resistance. In the latter, all H-bonds throughout the entire structure contribute to the strength, making the overall structure three times stronger. (**c**) Size effects of the shear strength of $\beta$-sheets. The *plot* depicts the shear strength of $\beta$-sheets as a function of $N$ (*e.g.*, number of H-bonds). The highest shear resistance is found at a characteristic length scale of $\approx$3 to 4 H-bonds. Beyond this length scale, the shear strength drops rapidly (defined as the strength of the $\beta$-strand divided by the sheared area). (**d**) Characteristic dimensions of the size of H-bond clusters in common protein structures: $\alpha$-helices, $\beta$-sheets and $\beta$-helices. Since the theoretical derivation of H-bond cooperativity considers uniform deformation of hydrogen bonds with no particular specificity to geometry, it may also apply to other protein structures where nature utilizes geometric confinement to achieve higher mechanical stability. The fact that 3 to 4 H-bonds per convolution exists on $\alpha$-helices and $\beta$-sheets on the sides of helices occur in clusters of approximately four may be indicative of such a universal biological concept that may be based on the evolutionary driving force to provide maximum strength. Adapted from [44–46]

maximum strength [44, 47]. This induced cooperativity is similar to the contact-splitting mechanism that enhances the adhesion of a gecko's foot—dividing the load among many small interfaces maximizes strength, as opposed to a single, inefficient transfer [48–50]. This phenomenon can be understood by applying basic thermodynamics concepts from fracture mechanics to the protein unfolding problem [45], and explain experimental findings such as DNA unzipping studies or multidirectional loading of complex $\beta$-sheet structures.

Notably, the ability of H-bonds to work cooperatively is also critical to ensure enhanced robustness, where the loss of a single bond does not result in the breakdown of an entire system. Cooperativity is actually facilitated by the weakness of H-bonds, which implies a softness of bonding that endows them with the freedom (entropy) to explore a great variety of structural states such that they can most effectively resist deformation [44]. The assembly of H-bond clusters into geometrically confined $\beta$-sheet nanocrystals, consisting of a pancake-like stack, results in the structural basis for effective cross-linking of multiple polypeptide chains in silk [47]. Because H-bonds can be reformed easily upon breaking, $\beta$-sheet nanocrystals have another highly useful property, toughness, enabled by the ability of H-bonds to self-heal and thereby effectively preventing catastrophic brittle failure as often observed in materials with stronger bonding. Yet, due to the nature of the available building blocks it is not possible to achieve all desired properties at a single material scale. Thus, in order to achieve another functional property of silk, extensibility (while maintaining the great strength facilitated by $\beta$-sheet nanocrystals), the structural design is extended to higher structural scales, and specifically, at the next level through the formation of a nanocomposite achieved by using the same basic material building block (polypeptide) but arranged in a different geometry. Here strong and tough $\beta$-sheet nanocrystals are combined with an additional protein secondary structure that consists predominantly of so-called $3_1$-helices that realize a more disordered phase. Through the provision of extreme amounts of hidden length, this constituent provides the capacity to generate large levels of deformability before the $\beta$-sheet nanocrystal cross-links are deformed and eventually broken [43]. By tuning the relative ratio of the two phases in silk, or the geometric makeup of the $\beta$-sheet nanocrystals, it is possible to achieve a diversity of mechanical signatures (*e.g.*, stiff, soft, extensible, etc.) without a need to define new constituents, a mechanism spiders use to generate different types of silks [41, 51].

The discussion of some of these most salient design features of spider silk strikingly provides an insight into a much broader design paradigm in biology at the nanoscale, pertaining to biology's use of a universal "cement"—H-bonds—in the creation of mechanically stable materials. The formation of confined clusters of H-bonds is indeed observed widely in biology and not only in silk, pointing to a universal design paradigm that enables biological systems to overcome the intrinsic weakness of H-bonds and to form mechanically strong and tough materials. A comparison of the geometric size of H-bond clusters in a diversity of protein found in the Protein Data Bank (http://www.rcsb.org) confirms that H-bonds typically organize in clusters between 4 to 6 in $\beta$-sheets and separated by disordered or differently structured elements, or in clusters of 3 to 4 in the turns of $\alpha$-helices

[44] (see Fig. 4.5(d)). These H-bond clusters represent a geometric feature found across species and highly conserved in biology. Such structures provides simple, yet strikingly effective protocols to achieve mechanical strength out of weak chemical bonding.

There are important insights from the universality of geometrically confined hierarchical structures in protein materials [31]—effectively turning (or tuning) weakness into strength. The application of this concept can impact technological use of abundant, however mechanically inferior (*e.g.*, extremely brittle, or weak) materials such as H-bonded proteins or silica in the design of strong, tough and lightweight materials. The design of hierarchical structures could be the key to overcome their intrinsic weakness or brittleness, properties that currently prevent their widespread technological application. Broadly, this may result in the development of highly functional materials made of abundant, cheap and environmentally friendly constituents, and even materials that are currently considered waste. It suggests that there may exist an enormous technological opportunity that is based on the concept that functional material properties can be design based on almost arbitrary constituents, provided that multiple structural levels are created. An observation made based on the analysis of different materials suggests that universality tends to dominate at smaller scales (*e.g.*, Ångstrom and nanometer), whereas diversity is found predominantly at larger, functional scales.

## 4.4 Music and Nature: Complexity from Common Elements

It is clear that the synergistic interaction of structures and mechanisms at multiple scales, using multiple levels of structure, provides the basis for enhanced functionality of biological materials despite the reliance on few distinct building blocks. The biological paradigm to create diversity of function out of simple, universal elements is not unique to Nature, and can be explained by drawing an analogy to rather far and different fields, from sociology to economics. Here, the importance of multiscale interactions, hierarchical structuring, and multifunctionality is illustrated by using an analogy of *music* [7, 32, 52]. Music, akin to protein materials, is founded by a common basis that can be explained by simple physics: *sound*. Like the combination of elemental building blocks of carbon, hydrogen, and oxygen that constitute protein materials, the phenomenon of music is fundamentally the combination of traveling sound waves with different frequencies (middle C, for example, has a frequency of 261.626 Hz). Such sound waves, however, fail to encapsulate Beethoven's *Symphony No. 9*, just as a listing of atoms fails to convey the function of a protein.

Musical composition is built upon universal elements at the microscale such as basic wave forms, and gathers a small variety of available instruments into hierarchical assemblies to create macroscale functionality, such as a particular orchestral sound (*e.g.*, a symphony). In music, when one considers the synthesis of orchestral music based on universal wave forms, structures at multiple scales are similarly used to arrive at a functional system, which is the resultant assembly of multiple scales—for example in a symphony. The concept is schematically illustrated in Fig. 4.6,

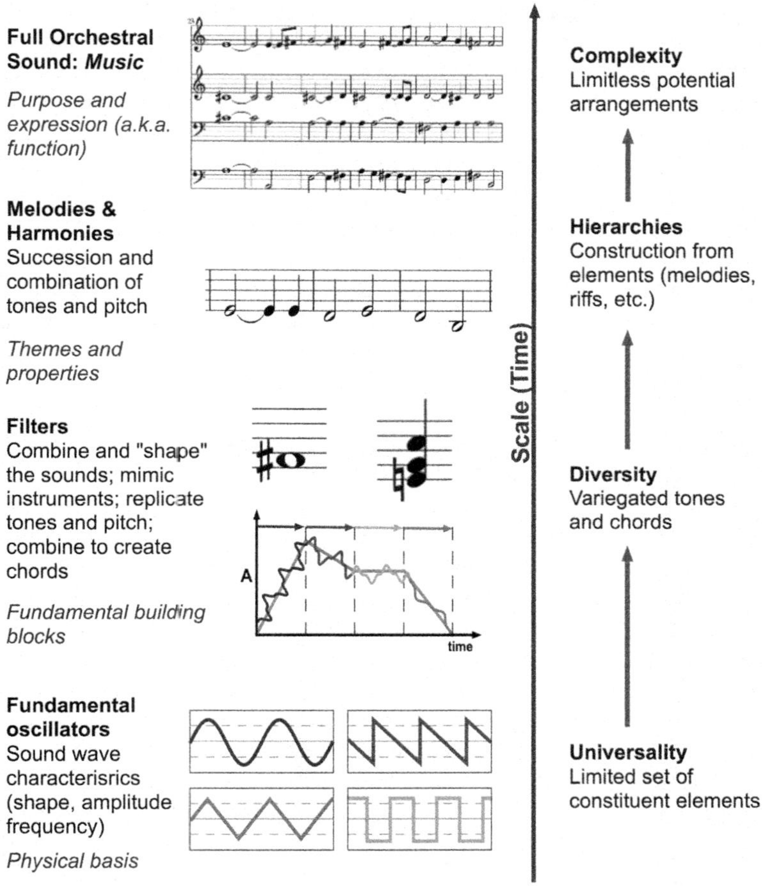


**Fig. 4.6** Music as a complex functional hierarchy. Illustration of multiscale (or cross-scale) interactions in the case of music, here exemplified for the process of synthesizing orchestra-level music based on four fundamental oscillators (a simplistic model). At a fundamental level, four basic oscillators create unique sound waves characterized by physical properties (*i.e.*, shape, amplitude, and frequency). At the next level, these basic sounds are modulated using envelope generators or filters, which shape and mimic the sound of an instrument and construct various tones and pitches (the depicted filter is a schematic representation of an attack-decay-sustain-release envelope). At this level, the fundamental building blocks of music are developed beyond the simple sound waves from which they are composed. The assembly of tones with different duration and pitch over time creates melodies (sometimes referred to as theme or riffs); where all tones come from a universal, limited set of harmonics (assembled in octaves). The theme or type of music is typically dependent on the selection and construction of themes and riffs (*i.e.*, the difference between contemporary jazz and classical baroque, for example), which can be thought of as the musical "properties". Through the combination of multiple instruments, each of which plays characteristic melodies or riffs, a complex orchestral sound is produced at the largest level, fulfilling the intended purpose, or musical "function" that emerges as the hierarchical structures interact synergically across the scales. The example also illustrates how the interplay of diversity and universality provides the fundamental paradigm behind music, resulting in near limitless arrangements from the hierarchical construction of musical elements (figure adapted from [32])

where at a fundamental level, four basic oscillators (chosen here as a fundamental set of constituents) create sine, square, and other wave forms considered raw, unalterable input [53]. *Universality tends to dominate*. At the next level these basic sounds are modulated using envelope generators or filters that change the volume, pitch and duration of the waves over time, employing on-the-fly processing (through Fourier analysis, or band-pass filters, for example), which shape the sound of an electronic instrument (note that similar hierarchies can be formulated for more traditional instruments such as woodwinds, guitars, or pianos).

An assembly of these modulated tones with different duration and pitch, or combinations of several of them into chords, creates melodies or riffs, where all pitches used come from a universal and limited set of harmonics, organized into octaves. Through the combination of multiple instruments, each of which plays characteristic melodies or riffs, a complex orchestral sound is produced at the highest structural level, the *de facto* "functional" scale. Indeed, while four sound waves differ only minimally in terms of physical properties (*e.g.*, frequency, shape, amplitude), the potential for a great diversity of arrangements is vast and manifest at the ultimate functional levels. By analogy, sound waves can represent elements, tones can reflect amino acids, protein sequences embody the melody, and their combination can provide the "music" of protein based structures. Nature has indeed proven to be an adept composer.

The "function" of music, be it the aural aesthetic or emotional expression, is a result of multiscale phenomenon of resonance and dissonance, the creation of chords and harmonies, the choice of classical piano or electric guitar; combined in one way the result is Mozart, while in another the result is *The Rolling Stones*. Variation (mutability) can be achieved by changing any of the levels-leading to variations in rhythm, tones, or melody, which in turn provide a different overall musical piece, or "function". The structure of music and protein material (including how it is changed), and the particular observer (*e.g.*, the audience, or the sensing in a particular physiological environment) are inextricably linked. This is a critical issue in both music and protein materials that is due to the way a particular observer processes and interprets functional properties and how this information is used in feedback loops that can alter the structural makeup at various scales. In biology this may happen through changes in gene expression (at relatively short time-scales) and in evolution (at relatively long time-scales). In music, this may happen through alteration of music during composing (which could involve continuous revisions of a piece), or through changes in the way a particular musical piece is played. In a jazz or rock jam session, music is continuously revised during performance by the interaction between the performers or with an audience (as anyone who has attended a *Phish* concert will surely attest). Similarly, the process of composing music can be regarded as an analogy to the evolutionary process.

While the synthesis of complex sounds from the level of basic oscillators is now possible with modern synthesizers-resembling a bottom-up "nanoscale" paradigm in creating music, composers in ancient days were limited by the availability of certain instruments, such as flutes or harps created from bone. Classical composers (*e.g.*, Bach, Mozart, Beethoven, and others) subsequently used more advanced instruments such as the violin or the piano, whose design was enabled by the materials and technology that became available at the time. Despite the limited set of available instruments (the basic building blocks), composers were able to create music that is considered some of most ingenious of all time (perhaps because it required composers to utilize many levels of structure to achieve a great functionality, which is apparently found appealing by the human brain). On the other hand, the impact of novel approaches to synthesize music from the bottom-up, with the use of complex sounds realized in *de novo* instruments, remains unclear, and music with a classical structure remains preferred by many. Subjective and cultural aspects may likely play a central role in the development and experience of music, an aspect that is evident from distinct types that emerged from different geographical regions and cultures. The construction of music exemplifies how the interplay of diversity and universality provides a powerful design paradigm, which relates directly to that found in biological materials and to what kinds of materials can be designed based on synthetic approaches.

The analogy between protein material design and music illustrates how biology is indeed a proficient composer, and that information can be derived to facilitate the development of more sustainable materials that rely not merely on the strength of building blocks but rather, on the use of structures at multiple levels, to provide heightened functionality. Conversely, humans have found ways to deconstruct and express hierarchical structures in various art forms, such as music, poetry or paintings. The similarity between the design of music and that of materials in biology is intriguing and merits further investigation, perhaps with the potential to unravel more general principles that govern mechanisms by which biological systems achieve functionality. Indeed, it was pointed out that functionality in certain animals (bees, ants, birds, etc.) does not rely on the strength of individual elements, but that functionality is achieved based on collective interactions of elements in clusters (*via* swarming, schooling, flocking, etc.) [54]. This concept is akin to the concept discussed here for the cooperative behavior of H-bonds that turns their weakness into strength. A key lesson reiterated is that in order to create a diversity properties, it is not necessary to rely on strong or numerous building blocks, but rather, the design space can be expanded *via* the formation of hierarchical structures, realized in biology through the merger of the concepts of structure and material and in music through the creation of complex compositions inherent in symphonic pieces.

Although an illustrative analogy, a cursory comparison of music with natural materials (and complex synthetic materials) may appear trite at first glance. Indeed, one could pull many examples where "*the whole is greater than the sum of its parts*" from everyday usage—including recipes for cooking or abstract painting to name a few. Undeniably, a dish of *beef bourgogne* is more than just the summation of beef, wine, and spices; Jackson Pollack's *No. 5, 1948* is more than just drizzles of yellow and brown paint. Being said, music, cooking, and artistic expression do not have obvious objective functions or properties like biological materials. Whether one prefers *The Beatles* or *The Rolling Stones*, chicken or beef, a Picasso or a Renoir, is merely a subjective difference of opinion. This need not be the absolute case.

For example, the book, "*A generative theory of tonal music*" proposes a detailed theory of musical hierarchies [55]. The theory develops a grammar of tonal music based in part on the goals, though not the content, of generative linguistics. The grammar takes the form of explicit rules that assign heard structures from musical surfaces. The term "musical surface" is meant, broadly, as the physical signal (sound waves) of a piece when it is played (the components or building blocks), while the "heard structure" refers to all the structure a listener unconsciously infers when he listens to and understands a piece, above and beyond the data of the physical signal (the function or meaning) [56]. Listeners hear certain structures rather than others. How can these structures be characterized and by what principles does the listener arrive at them? One would ultimately hope to specify those cognitive principles, or "universals" that underlie all musical listening, regardless of musical style or acculturation [57, 58]. Such musical theory attempts to produce formal descriptions in a scientific sense. That is, the goal is not just the description of formal relations, but the "functional form" of music. Thus the theory is *predictive* [59, 60]. In addition to criteria of internal coherence and parsimony, its principles can be verified or falsified by comparing the analysis it generates with one's intuitions about particular pieces of music. In addition, many of its principles can be investigated through laboratory experiment. Even the subjective aesthetic of music can be quantified in a self-consistent manner.

## 4.5 Comprehension by Analogies: Functional Similes and Abstraction

Why even turn to music when our focus is biological materials? Comprehension by analogies is a widely applied concept in science and education [61–64]. When a concept is complex, we tend to simplify our perspectives. Einstein was once quoted as saying "*You do not really understand something unless you can explain it to your grandmother*"—try explaining molecular hierarchies and folded protein structures to Grandma! You may ultimately find yourself describing flexible and folded proteins as cooked spaghetti.

Successful pedagogical strategies comprise features such as "constructivist learning environments" that challenge the view that scientific and mathematical knowledge is static, independent from our minds, and represents a universal truth [65]. In fact, this knowledge serves as a mediator resulting from human inquiry. Children in school become introduced to sciences such as mathematics, physics, chemistry and biology *via* the link to structures and concepts they are more likely to be familiar with. For instance, atom and electron interactions are represented by a model that resembles galaxy structures—the Bohr model—or animal cells are represented by factories [66, 67]. While most people may agree that it does not represent reality even closely, the analogy provides a sufficiently thorough understanding of general mechanisms that take place on the scales of Ångstroms and nanometers. Typically, an abstract comparison can be easily made if the *function* is transferable—white blood cells of your immune system and law enforcement, for example. We can say "white blood cells protect your body like the police" in a kind of *functional simile*, and even a child understands the role of white blood cells (it doesn't matter that our leukocytes are incapable of detective or forensic work).

By systematic abstraction and the deduction of analogical steps, the process of building an analogy itself helps to boost discretion about the important properties and parameters of the inquired system—or at least to ask essential questions [68, 69]. A rigorous methodology to formulate and categorize these analogies can be provided by *ontology logs* (*i.e.*, ologs) based on *category theory* [5, 6]. Category theory is a relatively new branch of mathematics (invented 200 years after the introduction of partial differential equations), designed to connect disparate fields within the larger discipline [5]. It is both a language that captures the essential features of a given subject, and a toolbox of theorems that can be applied quite generally. Category theory originates from a mathematical concept in topology [5] and has recently been used in broader contexts to identify patterns in other fields in a mathematical framework [70–73]. A good overview for non-specialists can be found in Lawvere [74] and Awodey [75].

**Category Theory**: An area of study in mathematics that examines in an abstract way the properties of particular mathematical concepts, by formalizing them as collections of *objects* and *morphisms* (or *functors*), where these collections satisfy certain basic conditions related to the composition of *morphisms*. The most accessible example of a category is the category of sets, where the objects are sets and the morphisms are functions from one set to another. However it is important to note that the objects of a category need not be sets—any way of formalizing a mathematical concept such that it meets the basic conditions on the behavior is a valid category, and all the results of category theory will apply to it.

Quickly after its inception, category theorists realized that its basic ideas were applicable well beyond the borders of mathematics. Category theory has by now been successfully applied in computer science, linguistics, and physics [71]. Whereas the theory of differential equations can be applied throughout science to create *quantitative* models, category theory can be applied throughout science to create *qualitative* models. And once such a qualitative model is formed as a category, its basic structure can be meaningfully compared (again *via* functors) with that of any other category, be it mathematical, linguistic, or other [76]. The use of category theory allows many intricate and subtle mathematical results or functional relations to be stated, and proved, in a much simpler way than without the use of categories. Like a biological system, the basic building blocks of a category are simple, but the networks that can be formed out of them are as complex as mathematics itself. These building blocks are called *objects*, *arrows*, and *composition*: arrows between objects form paths which can be composed into new arrows. It is a wonder that such a simple system can account for the wide variety of forms found in the mathematical universe, but perhaps this is less of a surprise to a biologist who notices the same phenomenon in his or her field.

Categorical algebras consist of objects and arrows which are closed under composition and satisfy conditions typical of the composition of functions [75]. Indeed, category theory provides a bag of concepts (and theorems about those concepts) that form an abstraction of many concrete concepts in diverse branches of mathematics. In a linguistic version, category theory and ologs in particular describe the essential features of a given subject and represent a powerful method to store and share data, knowledge, and insights in structure and functionality.

Equivalent to the learning and understanding process applied in school physics and mathematics, the methodology and hence the above described advantages of analogical thinking can be adapted to the field of materiomics. Recent studies linked for example mechanical properties of protein networks to communication networks [77], active centers in proteins to top predators and top managers [78], and death of living organisms to the strength of solids [79]. Furthermore, the connection between grammar and protein structure has been elaborately studied [80], even in the context of category theory [81].

## 4.6 Category Theoretic Analysis: Linking Hierarchical Structure and Meaning

Our understanding of the synthesis of elements into multi-functional structures remains in its infancy, and is currently limited to specific protein networks or protein materials. For example, mechanistic theories are typically developed for specific systems (including bone [82, 83], cells [84], and spider silk [47]) rather than providing a unified model that is applicable to a variety of distinct materials. A major frontier in the field is thus the extraction of generic principles of how functional properties are derived in functionally diverse systems despite the presence of the same (universal) building blocks, solely by using structure as a design paradigm. The systematic characterization of this knowledge is a keystone to materiomics.

Is it possible to show universality across seemingly disparate biological material systems? That is, by the application of the mathematical field of category theory to protein materials, can the extreme diversity of protein functional properties be described in a unified model that contains only a limited number of universal elements and interaction rules? The successful application of category theory to carry out a qualitative analysis in fields such as linguistics (grammar, syntax, semantics, etc., a toolbox of key concepts that enable the understanding of language [80, 85–88]) and computer science (again modeling syntax and semantics of denotation and operation in programming languages [89–92]), provide promising results. Through the development of such an approach, it may be possible to show that the hierarchical combination of universal elements into multi-level structures enables protein materials to achieve context specific functionalities in an abstract "complexity space"—linking hierarchical structure and function (or meaning) in a logical and self-consistent manner. (See also Fig. 9.15, which shows the hierarchical structure of amyloid materials and an analogy to the structure of language.)

Such an analysis would reveal that in order to create highly functional materials it is not essential to rely on a multitude and a certain quality of building blocks (*e.g.*, with superior qualities, great material volume, strong interactions, etc.). Rather, it is sufficient to use simple interaction rules and simple building blocks—each of which does not need to possess superior qualities—but assembled into hierarchical systems, where the overall structure provides enhanced functionality. Moreover, the universal relations composing the olog would be transferable to a multitude of equivalent systems. For example, the fundamental functional relationships for $\beta$-sheet interactions is not restricted to spider silk. This insight has implications for our understanding of how nanomaterials could be utilized to create macroscopically functional materials, and suggests a paradigm that departs greatly from the current belief in engineering science that material building blocks with superior qualities at the small scale (*e.g.*, carbon nanotubes, graphene, etc.) are critical to reach high performance materials. The implication is that superior functionality can be reached with *any* fundamental building block (or *set* of building blocks), provided that the design space is expanded to incorporate hierarchical structures. Eventually, an understanding of how diverse functional properties can arise out of inferior building blocks could make a profound impact towards the development of environmentally benign and friendly materials, as it would allow manufacturers to use local, abundant, and simple building blocks with overall negative $CO_2$ balance (*e.g.*, wood, plants, silica, water, soy beans) to create highly functional materials and structures. But how can we find a proper mathematical description of these hierarchical mechanisms that generate functional properties? Through category theoretic analysis.

As previously discussed, rather than being selected for a specific application, biological materials evolved to perform specific biological functions [1, 31, 93–96]. The components and connections within a given biological material are analogous to a circuit diagram [38, 39]. But just as it is extremely difficult to deduce the circuitry of a device by experimenting with its inputs and outputs, it is similarly inadequate to describe the higher-level structure of a biological material using only the physical interactions between proteins and some information about gene expression. Instead, we need to take into account additional types of structural information given by the fundamental principles that govern the interactions of the building blocks that define the system and its emerging functionality as these building blocks are connected together, from the molecular to the macro scale. The above considerations are important in any synthetic science; in order to duplicate the functionality of a natural system, we do not need to understand everything about it, only the principles out of which the desired functions arise. As discussed in Chap. 1: Introduction, knowledge is more powerful (and practical) as the representative systems shifts from total mimicry ("stealing" from Nature) to an understanding of the process at its basic level (abstraction). Biological systems contain any number of copies of thousands of different components, each with very specific interactions, and each representing a microscopic device in and of itself—that is, one that works away from local equilibrium. As a result, the microscopic description of a biological system (and materials therein) is intractably complex, unless one moves to a higher level of abstraction in the analysis. In the same manner, one only needs a basic knowledge of lift and drag to understand how a Boeing 777 attains flight, but the a Boeing 777 is fully "fly-by-wire", with 150,000 different subsystem modules, organized *via* elaborate protocols into complex control systems and networks [13]. A 777 is essentially a complex computer network that just happens to fly [13]. If the Orville and Wilber Wright had set out to build and fly a 777, they certainly would have failed.

It is exactly in the face of this complexity that ologs are so appealing. The olog presents us with an opportunity to identify patterns that describe systems and their components, to elucidate possible connections among these components, and to construct isolated functional "modules" by comparing information from many different materials or organisms. That is, by determining fundamental design principles that are simple yet functional, we can not only produce a powerful conceptual model of our system, we also create the possibility of comparing vastly different systems. For the example of a Boeing 777, the functional module that describes the lift of the wings can be equally applied to the wings of an eagle. Indeed, while there is almost nothing physical in common between a protein and a social network, or music and spider silk, we can construct a scenario in which the design principles are well-matched, and thus the systems may be compared. Such a comparison facilitates the transfer of results from other fields (social science, music, linguistics, etc.) to guide us in our study of biological materials of the same structure, and *vice versa.*

If a given study within mathematics (or materiomics) is formalized as a category, it can be connected with other categories that are seemingly far afield, as long as these structures align in the required "functorial" way. Theorems within one branch, like abstract equational algebra, can be applied to a totally different area, like geometric topology. Category theory may not only serve as an alternate foundation to mathematics [97], it unites the various distinct areas within advanced mathematics, formally proving similarities that are not apparent on the surface [98].

## 4.7 Language to Ontology Logs (*ologs*)

Category theoretic analysis and transformations of syntactic structures have been introduced by Chomsky in 1957 [85]. For formal language theory a well-known transformation is for example the left part transformation from non-left-recursive context-free grammars to context-free Greibach normal form where the syntactic structure is preserved during the transformation [99]. Simply put, the grammatical transformation maintains the syntactic meaning, which we can consider (for our purposes), a coherent function or property. These structure preserving transformations are morphisms (or functors) between objects and arrows among categories and constitute the essential operator to form analogies. The linguistic categorical objects are sets, with unique functions between the objects.

In this chapter we use a linguistic version of category theory in which the objects are drawn as text boxes describing some type of thing, like a protein or a genetic code, and where the arrows also have labels describing some functional relationship, as every protein has a genetic code. This notion of category theory has been

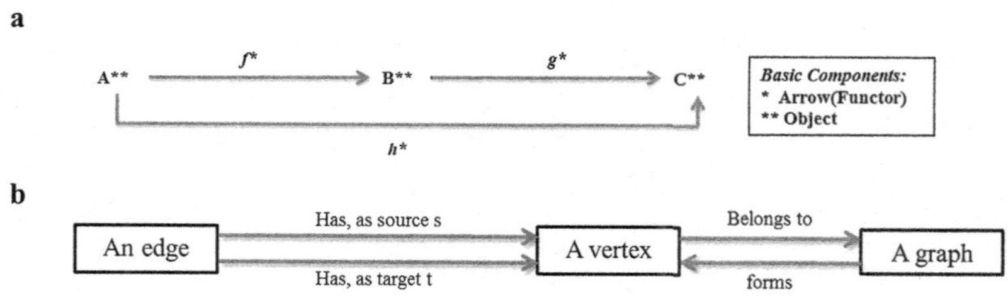


**Fig. 4.7** Basic components of ologs. (**a**) Conceptual diagram of a category theoretic system consisting of objects and arrows. Objects are linked *via arrows* and these paths can again be composed into new paths. Objects in ologs are sets and the *arrows* are unique functions. For example, here, the functor $f$ relates **A** to **B**, or $f : \mathbf{A} \to \mathbf{B}$. Moreover, $f : \mathbf{A} \to \mathbf{B}$ followed by $g : \mathbf{B} \to \mathbf{C}$ is equivalent to $h : \mathbf{A} \to \mathbf{C}$. (**b**) Olog of a graph, where subgraphs can form a new vertex for a graph on a "higher hierarchical level". Each box contains a set of instances which are subject to a functional relationship represented by the *arrow*. From [7]

introduced by Spivak *et al.* [6] and recently applied in analogical categorizations of proteins materials [7, 77], discussed in depth for the remainder of this chapter.

Chains of arrows can be composed, providing a description of how a number of small-scale relationships come together to constitute a single, conceptually simpler, larger-scale relation (like a person's father's sister's daughter is a simply their cousin; an example for "functionality" in the space of linguistics). These linguistic categories are called "ologs," short for "ontology logs" [76] (see Fig. 4.7). Ontology is the study of how or what something is, and ologs are a systematic framework in which to record the results of such as study. Simply put, an olog is a well-defined category presented as a visual abstraction. The term "log" (like a scientist's log book) alludes to the fact that such a study is never really complete, and that a study is only as valuable as it is connected into the network of human understanding. This brings us to the heart of the matter: in order to build a sufficient understanding of hierarchical materials, scientists must integrate their findings more precisely with those of other scientists.

The fact that an olog is fundamentally a category means that such connections can be formulated between ologs with mathematical rigor (see Fig. 4.8), and meaning preserved [76], to facilitate the communication with other fields of science. It is believed that this will ultimately enable the kinds of breakthroughs needed to further our understanding of how functional diversity is achieved despite severe limitations of building blocks. The generation of ologs also allows us to observe the formation of patterns that define certain functionality, and draw connections between disparate fields. A key insight used here is that although patterns of functionality generation can be quite different in the space specific to applications (*e.g.*, proteins, language, music), they are remarkably similar in the space of categories. In other words, we hypothesize it is possible to observe universal patterns of how functionality is created in diverse fields; and that it is possible to generate fundamental laws (similar to PDEs in conventional physics) that describe the emergence of functionality from first principles.

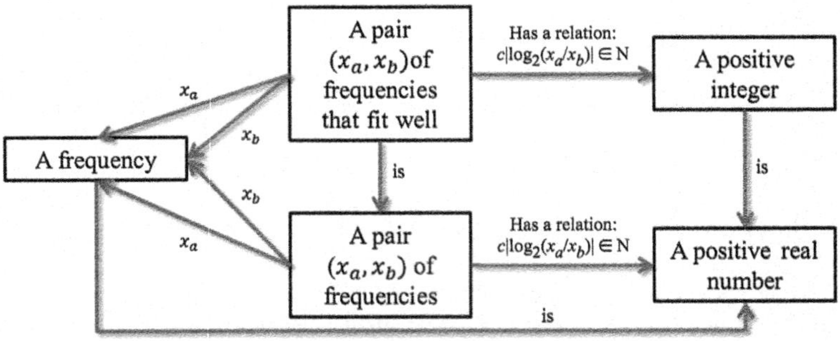


**Fig. 4.8** Special paths in ologs (*i.e.*, fiber products). In case the resulting instance of two distinct paths is equal, two paths commute. These structures define new terms in the olog in terms of already existing structures. For example, consider two arbitrary sound frequencies, $x_a$ and $x_b$, which define "A pair of frequencies". By construction, "A pair of frequencies that fit well" can be defined by the relation $c|\log_2(x_a/x_b)| \in N$ with $c = 12$ [100, 101] being a positive integer, rather than just a positive real number, where positive integers are itself a subset of positive real numbers. All pairs of frequencies have can be formulated by $c|\log_2(x_a/x_b)|$, but not all pairs "fit well". Note that these kind of definitions represent a kind of "worldview" of the author (*e.g.*, extraneous information). From [7]

The actual benefit of an olog is—due to a rigorous mathematical background—its unambiguous way to store and share data, knowledge, and insights in structure and functionality within a single research group and also among many disparate research groups and different fields in science and engineering. As such, ologs can be embedded into a database framework and thus are easily implemented in object based computer languages. Ologs offer means to reveal the origin of the described system property and to connect them to previous results or other topics and fields. For biological materials it is crucial to elucidate the principles from which the superior macroscopic functionality arises in order to define the hierarchical structure-function relationships or even synthesize them. These insights can be gained on the one hand from the category theoretic analysis of protein materials by describing the emergence of functionality from first principles, *e.g.*, on the basis of fundamental interactions between building blocks. On the other hand, the use of systematic analogies with the help of functorial relations supports the researcher in formulating these structure-function relationships in an abstract way, ensuring the ability to connect disparate ologs. Furthermore, by the use of ologs for knowledge creation by shared conceptual models an educational application is feasible [102].

**Ontology Logs (olog)**: The analysis of materials systems based on the recognition of the *universality* of structural elements (building blocks and their interactions), defined with respect to one another in the context space of the olog, and potential *diversity* of fundamental functional mechanisms and material behavior.

The concept of an olog can also be easily extended across hierarchies (not limited to spatial hierarchies, but also encompassing complex functional relations and multi-phenomena coupling). Hierarchical ologs yield similar mathematical features as conventional ologs but are designed to improve the ability to overview the build-up of a hierarchical system by compiling the subunit sets together with sets of superior structural units [7] (see Fig. 4.9). A category theory expert can understand our definition of a hierarchical olog as a category $C$ equipped with a subcategory $H$ with the same set of objects, such that $H$ has the structure of a forest (collection of trees). A morphism $F : (C, H) \to (C', H')$ of hierarchical categories, which we here abuse notation in calling a *functor*, consists of a functor $F : C \to C'$ such that $F(H) \subseteq H'$. A state of a hierarchical olog is just a set-valued functor on $C$; we can denote sequences, graphs, etc. by a simple modification of the olog, which again we abuse notation in eliding. In other words, a hierarchical olog can provide a coarse or fine categorical perspective, depending on the needs of the olog.

The "extended olog" of Fig. 4.9(a) can also be considered hierarchical, as there is no explicit definition or functional relation for amino acids (*e.g.*, molecules containing an amine group, a carboxylic acid group and a side-chain). Such information is not necessary, as it is superfluous to the function of the protein, the amino acid is considered the fundamental constituent. The olog could easily be extended to accommodate the functional build-up of amino acids, but can be encompassed by a single box.

As an example of the typical features found in hierarchical ologs we can also analyze a distinct feature of linguistics, specifically the structure of a sentence which is formed of words (Fig. 4.10). Words consist of phonemes, the smallest pronounceable segments comprising one or more letters [103]. Hence, these phonemes form the "building blocks" for spoken language and are categorized by distinctive articulatory features, *i.e.*, the description of how the sound is mechanically formed in the body [104]. Whether or not a feature is active for a certain phoneme can be indicated by a binary number. The voice laryngeal feature, for instance, determines whether a sound is formed including the vocal folds (1) or not (0) [105, 106]. The "*r*" in "theo*r*y" contains as voice laryngeal feature a (1), while the "e" does not (0).

Each path is constrained to represent a unique function between the instances of the sets. Consequently, each building block can only be uniquely assigned to its higher units by constructing pairs of building blocks and their higher structural units respectively, see Fig. 4.11. Set **A** associates units from set **B** and set **C** together with their positions, an elegant way to maintain functional relationships within the category. The checkmarks indicate commutative paths (*e.g.*, a kind of isomorphism) in the olog where, starting from the same instance of a set, two distinct paths point to the same instance of another set.

We omit further precise mathematical definition of categories and hence ologs, as we will focus on the application of this concept—the discussion will be limited to a general description of ologs, how they are constructed, and focused on the application to describe materials. Hence, we will proceed to describe ologs by example; for a more mathematically precise account of ologs, see Ref. [6]. For further details on isomorphisms and other features such as limits and colimits in ologs see [6, 77].

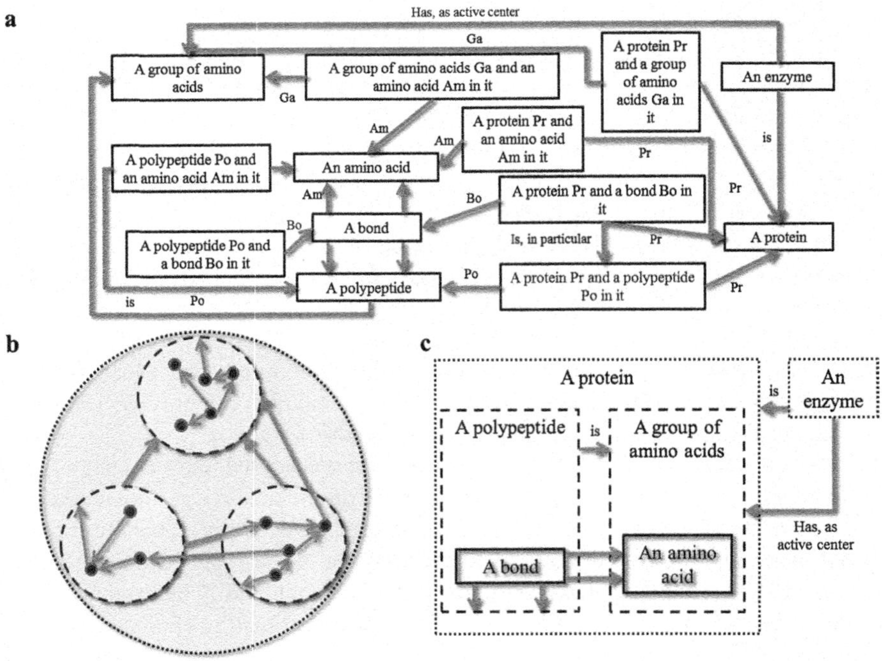


**Fig. 4.9** Development of hierarchical ologs. (**a**) Expanded olog describing the structure of proteins out of their building blocks, amino acids. Each bond has as start point and end point an amino acid/a polypeptide. In order to get a hierarchical level higher a higher structure, *e.g.* the protein, is combined with one of its ingredients of a lower level, for example amino acids or bonds. This assures the unique allocation of lower level elements to superior structures and can be seen as a state or "snapshot" of the system. Apart from the hierarchical levels, a subset of proteins, the enzymes, are included to exemplify the way to include "hierarchical interaction". A function relates a certain property of a higher hierarchical structure, here active centers of enzymes, to an element of a lower level structure, a group of amino acids. A polypeptide is a linear chain of two or more amino acids connected by a bond. (**b**) General graph description of a three level hierarchical structure. Note that each cycle represents a state of its lower level elements and each level may interact (in a functional way) with higher or lower level elements. (**c**) Hierarchical olog depicting the same information included in as in panel (**a**) whereas an overview over the hierarchies is now possible. Analogous to panel (**b**) each box represents a state of its lower level elements. From [7]

The way the olog is represented in Fig. 4.11(a) correlates one-to-one to a computer implementation.

Beneficial to the possibility to overview and clearly identify the underlying hierarchical structure of the system, we introduce hierarchical ologs, Fig. 4.11(b). They yield absolutely identical information but a well-arranged notation. In this context, an increased emphasis on the discovery of the structural makeup of systems and how it relates to the emergence of specific properties can provide an important educational feature. Set **A** of Fig. 4.11(a) is now inherently included and the dashed box represents a state of the system, *i.e.* all words in the former set **B** with their current phonetics and binary information (comparable to a "snapshot"). Often, for

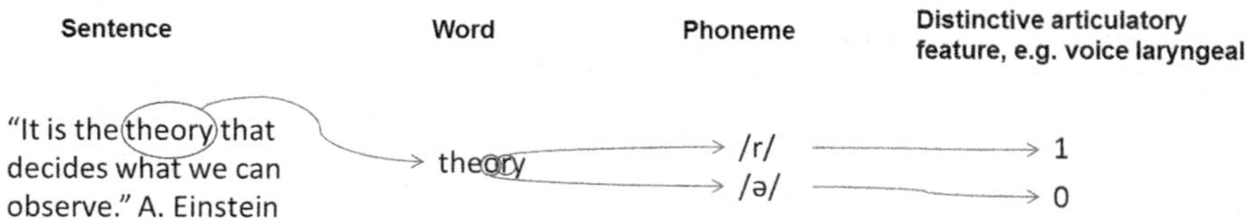


**Fig. 4.10** Hierarchies of linguistic structure. The build-up of linguistics can be seen when analyzing the structure of a sentence. It is formed by words which itself are formed by phonemes, the smallest pronounceable segments of a word. These phonemes are categorized by distinctive articulatory features, *i.e.* the description of how the sound is mechanically formed in the body. Whether or not a feature is active for a certain phoneme is indicated by a binary number. From [7]

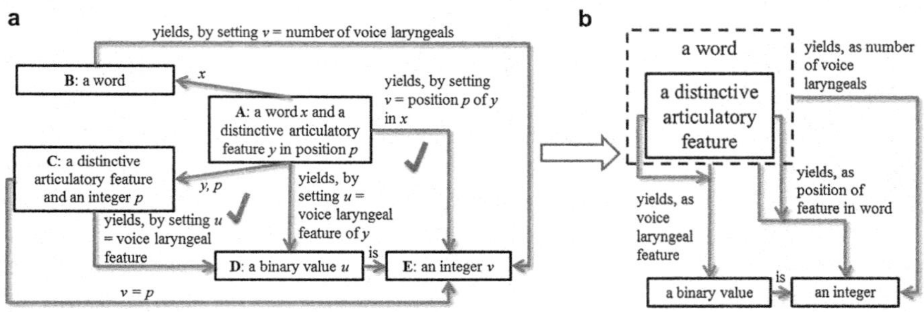


**Fig. 4.11** Linguistic structure to category theory *via* an olog. (**a**) This olog describes the situation where every word consists of phonemes, hence they form the "building blocks" for spoken language. An appropriate way to assign the building blocks to its higher units is a representation of sets of words with associated phonemes (set **A**). The checkmarks indicate commutative paths in the olog where starting from the same instance of a set two distinct paths point to the same instance of another set. The way the olog is represented here correlates one-to-one to a computer implementation. Note, that the paths **A** → **D** and **A** → **C** → **D** as well as **A** → **E** and **A** → **C** → **E** commute as indicated by the checkmarks. No other paths commute. (**b**) A hierarchical olog yields identical information but a better overview of the underlying structure of the problem. From [7]

instance in the case of phonemes in words, a higher structural unit (*e.g.*, the word) is formed by a sequence of subunits (*e.g.*, phonemes). This information is conveyed by a simple dashed "hierarchy-box": the word box surrounds the phoneme box, indicating a hierarchical construction. In other cases, the structure may be arbitrary, but always describable by graph-theoretic tools. This holds true in the case of proteins, which are arrangements of (*i.e.*, hierarchically constructed from) amino acids. Paths combining "inner" and "outer" information, *e.g.*, new "double arrows" and former commutative paths, are automatically commuting and no additional checkmarks are needed.

## 4.8 Proteins and Communication Networks

Let us demonstrate the potential powerful application of category theory. Again, in biology, the same 20 amino acids can have different functions depending on how they are arranged in a sequence as defined by the genes. In other words, the same library of fundamental building blocks can produce different functionality depending on the precise sequence. Just so, an olog serves as a code or formula for a complex structure, but the context in which it is interpreted can lead to different results. Here we ask:

> *Can the same olog represent the structural and functional relationship within a protein filament (such as an $\alpha$-helix or an amyloid fibril) and the relationships between a type of social network?*

In the case of a protein the building blocks are polypeptide fragments with H-bond clusters as glue, whereas in a social network the building blocks are people and communication methods. It is the interplay between *form* and *function* of few universal building blocks that ties biological structuralism and category theory, and which may produce potentially novel approaches to designing engineered systems.

The main ideas will be developed for two examples of protein materials that display a distinct mechanical behavior once exposed to mechanical force [77]. We begin the discussion with a presentation of the proteins and their functional properties, here their mechanical properties under axial extension (realized *via* the application of an axial force applied to the protein filament). The structure, mechanisms and resulting functional properties have been developed in a series of earlier studies based on computational approaches to molecular nanomechanics (for $\alpha$-helices [26] and for amyloids or $\beta$-sheet crystals [47, 107], and we refer the reader to these original papers for further detail into the mechanistic analysis of the nanomechanics).

Figure 4.12 shows the visualization of the two protein materials considered here based on an abstraction of how their mechanical properties can be understood based on the interplay of a set of "building blocks" (Fig. 4.12(a)). Both protein materials considered here resemble a linear arrangement of three available elements: "bricks", "glue", and "lifeline". As a (known) design rule, brick and glue need to alternate in order to achieve a stable structure. Two brick or glue elements immediately next to each other would not stick together. There is a fundamental chemical reason for this constraint as bricks represent the protein's polypeptide backbone and glue represents H-bonding which can only occur between a cluster of amino acid residues in the backbone. The "lifeline" is a third element that is introduced here, reflecting the situation in which there is still a physical connection of bricks even after large force causes the glue to break. Chemically, this resembles the existence of a "hidden" polypeptide length such that there exists a "covalent" link between two brick elements even after the H-bond glue has broken (as present in $\beta$-helices and $\alpha$-helices). This hidden length is not observed as a relevant structural property until the glue breaks, at which point the lifeline comes into play and provides an increasing resistance against further deformation. Thus, although both glue and lifeline can connect neighboring brick elements, they are differentiated in that the lifeline is much stronger than the glue and that its resting extension is roughly the failure extension of the glue.

Although this description of protein filaments is a simplification of how their mechanical properties can be described it enables us to understand the key functional properties based on the interplay of building blocks. Figure 4.12(b) depicts a model

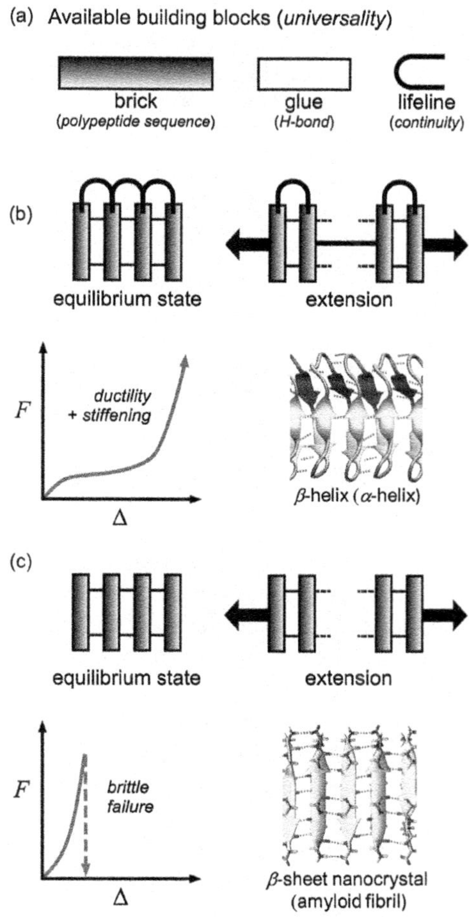


**Fig. 4.12** Visualization of protein materials considered here, and abstraction of how key functional properties (here: mechanical properties under axial extension) can be understood based on the interplay of a set of "building blocks". (**a**) Fundamental building blocks of our protein materials. The protein materials considered here are composed of a linear arrangement of three potential elements, "bricks", "glue", and "lifelines". As a "design rule", brick and glue need to alternate in order to achieve a stable structure. That is, two brick or glue elements immediately next to each other would not stick together—the chemical reason is that bricks represent the protein's polypeptide backbone and glue represents H-bonding which can only occur between residues in the backbone. The "lifeline" is a third element introduced here, reflecting the situation when there is still a physical connection between bricks even after the glue breaks (continuity). (**b**) Model of an $\beta$-helical protein (or, equivalently in terms of structure and behavior, a $\alpha$-helix) under axial loading [26]. This resembles a system with a lifeline, as after breaking of the cluster of H-bonds (glue) that are formed between groups of amino acids (brick) there still exists a physical connection due to the polypeptide backbone (lifeline). The existence of a lifeline has major implications on the functional properties of the overall system, resulting in a stiffening F-$\Delta$ behavior, and enhanced ductility (large extension at failure). (**c**) Model of a $\beta$-sheet crystal protein filament (or similarly, an amyloid fibril) under axial loading [107], resulting in brittle failure. This resembles a system without a lifeline since after breaking of the H-bond cluster (glue) between the layers formed by clusters of polypeptide (brick) no physical connection exists. From [77]

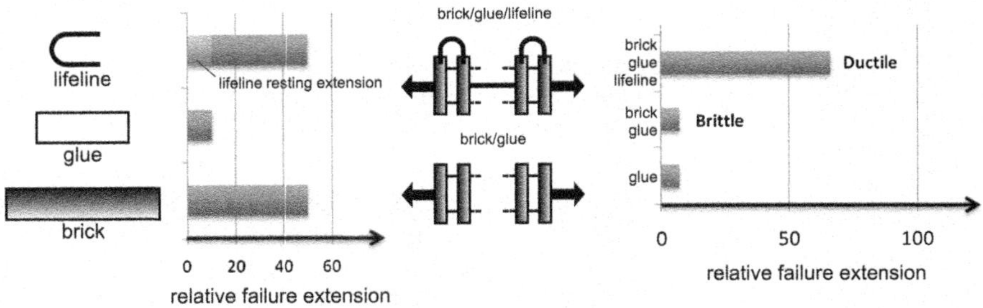


**Fig. 4.13** Mechanical behavior of each of the building blocks characterized by a description of the failure extension. The hidden length of lifelines is reflected in the fact that the resting extension of the lifeline is roughly equal to the failure extension of the glue. Both the brick and the lifeline have large failure extensions relative to the glue. A system with a lifeline shows a *ductile* response, where a connection can be sustained at large extension as compared to the glue alone. A system without a lifeline shows a *brittle* response, where only a small extension can be sustained until the material breaks (which equals roughly the failure extension of the glue). From [77]

of an $\beta$-helix protein (or similarly, a $\alpha$-helix protein) under axial loading, assembled based on an alternating sequence of bricks (amino acid cluster), glue (cluster of H-bonds) and a lifeline element. The lifeline element is formed by the protein backbone that still exists even after the cluster of H-bonds break after unfolding of one $\alpha$-helical turn [26]; providing a physical connection that allows additional glue elements to break after more axial extension is applied. In fact, at large extensions all glue elements will have broken such that the system's overall failure extension is much larger than the failure extension of the glue, marking a "ductile" behavior. Figure 4.12(c) depicts a model of a $\beta$-sheet nanocrystal (or similarly, an amyloid fibril). The structure is realized by the assembly of on an alternating sequence of bricks (amino acid cluster) and glue (H-bond cluster). Upon the increase of the extension one of the glue elements breaks. Since there is no more physical connection between the two brick elements that were previously connected by the glue element the entire system has failed, and at an extension that is roughly equal to the failure extension of the glue. We define this behavior as "brittle".

The comparison of the distinct mechanical behavior of $\beta$-helices or $\alpha$-helices and $\beta$-sheet nanocrystals or amyloid fibrils was achieved by mapping the key mechanisms that generate their specific properties into the abstract space of interactions between a set of building blocks (such as failure extension; see Fig. 4.13). What was described in words in the preceding paragraphs can be rigorously achieved using ologs, which describe the interactions between building blocks. Through the development of ologs for each system we aim to answer a series of questions:

- What are the components of the system, and how do they interact?
- How do these interactions produce the functionality we observe of the overall system?
- When does functionality break down? *E.g.*, failure of building blocks as the system is pushed to extreme conditions, or the presence of defects.

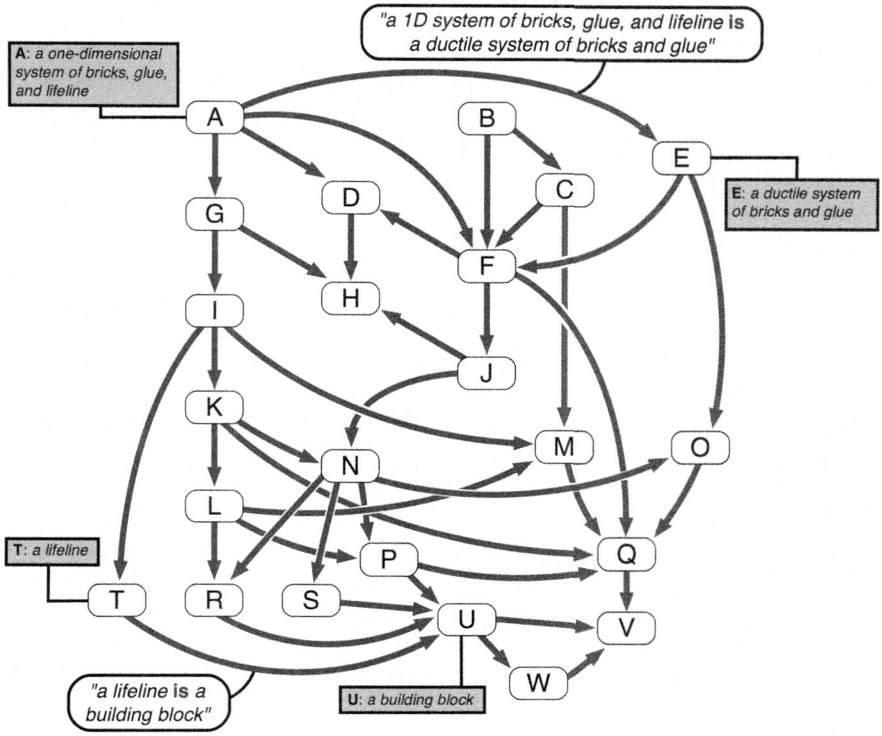


**Fig. 4.14** Olog description of both proteins and a social network in abstract terms of bricks, glue, and lifelines. Each box represents an abstract type, and each arrow represents an aspect (or observable) of that type. Each type refers to a set of intended instances, which we think of as being contained in the box. For example box **E** contains ductile sequences of bricks and glue (like an $\alpha$-helix), whereas box **V** contains real numbers (*e.g.*, 9.228). Each arrow from a source box to a target box refers to an observation one may make on things in the source box, for which the observed result is a thing in the target box. For example, arrow **E** → **O** indicates that one can observe of any ductile material ($S$) a pair of numbers ($R, r$) where $R$ is much greater than $r$. The meaning of these numbers $R$ and $r$ is enforced by a "commutative diagram", illustrated in Table 4.2. Here, the number $R$ must refer to the failure extension of the system $S$ and the number $r$ must refer to the failure extension of its glue. This says that a ductile system fails at a much greater extension than its glue elements do. Each box is meant to contain an intended set of instances and each arrow is meant to functionally relate two such sets. The rest of the olog is recorded in Tables 4.2, 4.3 and 4.4. Some are commutative diagrams which declare two paths through the olog to be equivalent and some are fiber products which define new types in terms of others. From [77]

- A further reaching question may be, by what process did the system come to be constructed, and what selective pressures at the macroscale induce observable changes in the system and at different levels in the structural makeup?

To eventually get us to this point, we will now discuss the components of our brick-and-glue system of proteins, and their integration within the olog presented in Fig. 4.14 which describes both the brittle and ductile protein filaments outlined above. Three universal elements, which we have been calling bricks (b), glue (g), and lifelines (L) are the abstract building blocks composing our systems, and they are defined in relation to one another. Both glue and lifeline are materials that can connect two brick elements. There are two distinctions between them: (i) the failure extension of glue is much less than that of brick, whereas the failure extension of lifeline is roughly equal to that of brick, and (ii) the resting extension of lifeline is roughly equal to the failure extension of glue. These two properties ensure that the lifeline is not detected under axial loading until a glue element breaks and that all the glue elements break long before a lifeline or brick element breaks.

This distinction between one number being roughly equal to another and one number being much greater than another is simple, yet universal in the sciences, and thus we can expect these types (**M** and **O** in the olog) to be quite common in scientific ologs. In fact, we reuse this concept within the olog when we distinguish a ductile system from a brittle one. That is, we characterize a ductile system to be one whose failure extension is much greater than that of its glue element, whereas we characterize a brittle system to be one whose failure extension is roughly equal to that of its glue element. Other common (*i.e.*, universal) patterns that we may find in biological materials is a certain shape (*e.g.*, fibers, helices, spheres), bonds of a certain form (*e.g.*, H-bonds versus covalent bonds, backbone versus side-chain), dimensionality (*e.g.*, 1D, 2D, 3D), and so on. Our olog concentrates on materials whose shape is one-dimensional, a feature we define by the use of mathematical graphs.

Once the fundamental structure of our protein materials and the definition of ductility and brittleness have been defined in the olog, we describe our hypotheses by two arrows, $\mathbf{A} \to \mathbf{E}$ and $\mathbf{B} \to \mathbf{C}$, the first of which hypothesizes that systems with lifelines are *ductile*, and the second of which hypothesizes that systems without lifelines are *brittle*. This hypothesis has now been examined in the paragraphs above, but can be even more carefully explicated using a category theoretic formulation, where each component type and aspect is laid bare. In fact, we have no hope of proving an analogy between this protein setup and the upcoming social network setup without such a formulation. In Fig. 4.14 we show the entire setup as a diagram of boxes and arrows, the precursor to an olog. However, this diagram is not sufficient in the sense that there are mathematical truths present in our system that are not present in the diagram. We include the rest of this information in Tables 4.2, 4.3 and 4.4, which we will describe shortly.

In order to explain what is missing from Fig. 4.14, we should more clearly explain what is there. Each box represents a set. For example box **H**, defined "a graph", represents the set of graphs, whereas box **J**, defined "a system consisting of bricks connected by glue, structured as in graph **G**", represents the set of such systems. Each arrow represents a function from one set to another, and its meaning is clear by reading the definition of the source box, the definition of the arrow, and then the definition of the target box. For example, we read arrow $\mathbf{J} \to \mathbf{H}$ as "a system consisting of bricks connected by glue, structured as in graph **G** is structured as a graph". Thus any element of the set **J** is functionally assigned its structure graph, an element of **H**, by an arrow. Just as the structure graph of a system is an observable of that system, any function from one set to another can be considered an observable of the former.

**Table 4.2** Selection of commutative diagrams in the olog. Each sequence of consecutive arrows through the olog (Fig. 4.14) is called a path, which represents a functional relationship between its starting point and its ending point. Two such paths $\mathbf{A} \rightarrow \mathbf{B}$ may result in the same function. The idea is that given an instance of **A**, each of these paths returns the same instance of type **B**. By having this additional data, we confine the meaning of the label on each box and arrow—they cannot stray far from our intended meaning without "breaking" these path equalities. Thus this table serves as an additional check on our labels. Adapted from [77]

| Start | End | Path 1 | Path 2 | Result |
|---|---|---|---|---|
| **A**: a 1D system (S) of bricks (b), glue (g), and lifelines (L) | **F**: a 1D system (S) of only bricks (b) and glue (g) | $\mathbf{A} \rightarrow \mathbf{E} \rightarrow \mathbf{F}$ | $\mathbf{A} \rightarrow \mathbf{F}$ | Each path simply omits the lifelines; Path 1 provides the additional data of *ductility* |
| **C**: a brittle system (S) of bricks (b) and glue (g) | **Q**: a pair $(x, y)$ of real numbers | $\mathbf{C} \rightarrow \mathbf{F} \rightarrow \mathbf{Q}$ | $\mathbf{C} \rightarrow \mathbf{M} \rightarrow \mathbf{Q}$ | Each path sets the failure extension of the system $(x)$ and glue $(y)$; Path 2 provides the additional data that $x \approx y$. |
| **N**: a pair (b, g) of building blocks, serving as bricks (b) and glue (g) | **Q**: a pair $(x, y)$ of real numbers | $\mathbf{N} \rightarrow \mathbf{P} \rightarrow \mathbf{Q}$ | $\mathbf{N} \rightarrow \mathbf{O} \rightarrow \mathbf{Q}$ | Each path sets the failure extension of the system $(x)$ and glue $(y)$; Path 2 provides the additional data that $x \gg y$. |
| **K**: a threesome (b, g, S) of building blocks, serving as bricks (b), glue (g), and strong-glue (S) | **R**: a brick | $\mathbf{K} \rightarrow \mathbf{N} \rightarrow \mathbf{R}$ | $\mathbf{K} \rightarrow \mathbf{L} \rightarrow \mathbf{R}$ | Each of these paths from **K** to **R** yields the same brick element (b); Path 1 defines a (b, g) pair; Path 2 defines a (b, S) pair. |

A function may be thought of as a "black box" which takes input of one type and returns output of another type. If the output of one function is fed as input to another function and the whole system is imbedded in a black box, it is called the *composition* of functions. Finally, two functions are equal (regardless of the inner workings of their "black boxes") if, upon giving the same input they always return the same output. The first kind of mathematical truth alluded to above that is missing from Fig. 4.14 is a declaration of which compositions of functions in our system are equal. Such equalities of compositions of functions are called commutative diagrams in category theory literature. A selection of such declarations are presented in Table 4.2. These equalities can be considered as checks on our understanding of all the sets and functions in the arrows—declaring them is at the very least "good science".

Table 4.3 describes a certain class of commutative diagrams, called fiber product diagrams. In a fiber product diagram, one set and two observables of it are declared as a kind of "universal solution" to a problem posed by another diagram. In these terms, we consider the diagram $\mathbf{D} \rightarrow \mathbf{H} \leftarrow \mathbf{J}$ as posing a problem, to which

**Table 4.3** Fiber product diagrams in the olog. Some boxes in the olog (Fig. 4.14) are defined in terms of others by use of so-called fiber products. For example, object **A** is defined in terms of three others in relationship, **D** $\rightarrow$ **H** $\leftarrow$ **G**: given a system of bricks, glue, and lifeline (**D**), we observe its structure graph (**H**) and set it equal to a "chain" graph (**G**)—in so doing we define "one-dimensionality" for a system. The notion of one-dimensionality is not up for interpretation, but directly dependent on the other notions in this olog. Thus this table serves to anchor the interpretation(s) of the olog more firmly. Adapted from [77]

| Object | Fiber product object name | Defining attributes | Equated terms | "Idea" |
|---|---|---|---|---|
| **A** | a one-dimensional system of bricks, glue, and lifeline | **D** $\leftarrow$ **A** $\rightarrow$ **G** | **D** $\rightarrow$ **H** $\leftarrow$ **G** | A system of bricks, glue, and lifeline is defined as "1D" if its structure graphs (brick/glue) and (brick/lifeline) are both chains |
| **C** | a brittle system of bricks (b) and glue (g) | **F** $\leftarrow$ **C** $\rightarrow$ **M** | **F** $\rightarrow$ **Q** $\leftarrow$ **M** | A system is defined as "brittle" if its failure extension is roughly equal to the failure extension of its glue |
| **E** | a ductile system of bricks (b) and glue (g) | **F** $\leftarrow$ **E** $\rightarrow$ **O** | **F** $\rightarrow$ **Q** $\leftarrow$ **O** | A system is defined as "ductile" if its failure extension is much greater than the failure extension of its glue |
| **F** | a one-dimensional sequence (S) of bricks (b) and glue (g) | **D** $\leftarrow$ **F** $\rightarrow$ **J** | **D** $\rightarrow$ **H** $\leftarrow$ **J** | A system of bricks and glue is defined as "1D" if its structure graph is a chain |
| **I** | a threesome (b, g, L) of building blocks, serving as bricks, glue, and lifeline | **M** $\leftarrow$ **I** $\rightarrow$ **K** | **M** $\rightarrow$ **Q** $\leftarrow$ **K** | A strong-glue element is defined as "lifeline" if its resting extension is roughly equal to the failure extension of a glue element |
| **K** | a threesome (b, g, S) of building blocks, serving as bricks, glue, and strong-glue | **N** $\leftarrow$ **K** $\rightarrow$ **L** | **N** $\rightarrow$ **R** $\leftarrow$ **L** | A "brick/glue/strong-glue threesome" is defined to be a brick/glue pair and a brick/lifeline pair where the bricks are the same in both instances |
| **L** | a pair (b, S) of building blocks, serving as bricks and strong-glue | **M** $\leftarrow$ **L** $\rightarrow$ **P** | **M** $\rightarrow$ **Q** $\leftarrow$ **P** | Two building blocks, one of which can connect together two instances of the other, are defined as "bricks and strong-glue" if their failure extensions are roughly equal |
| **N** | a pair (b, g) of building blocks, serving as bricks and glue | **O** $\leftarrow$ **N** $\rightarrow$ **P** | **O** $\rightarrow$ **Q** $\leftarrow$ **P** | Two building blocks, one of which can connect together two instances of the other, are defined as "bricks and glue" if the failure extension of the connector is much less than the failure extension of the connectee |

$\mathbf{D} \leftarrow \mathbf{F} \rightarrow \mathbf{J}$ is a solution, as we now explain. The diagram $\mathbf{D} \rightarrow \mathbf{H} \leftarrow \mathbf{J}$ poses the problem "what should we call a system consisting of bricks connected by glue, structured as in graph **G**, where graph **G** is a 'chain' graph?". The declared solution is **F**, "a one-dimensional system (S) of bricks (b) and glue (g)", together with its two observables $\mathbf{F} \rightarrow \mathbf{D}$ and $\mathbf{F} \rightarrow \mathbf{J}$. Thus the second kind of mathematical truth alluded to above that is missing from Fig. 4.14 is that some boxes and attributes have fixed meaning in terms of the others. A list of these is given in Table 4.3, where we see terms such as "one-dimensional", "brittle", "ductile", and "lifeline" defined solely in terms of more basic concepts.

Thus, while it is convenient to think of the olog for our protein systems as the diagram in Fig. 4.14, in fact it is the totality of Fig. 4.14, Table 4.2, Table 4.3 and Table 4.4, which really constitute the olog. Just as in biological materials, the parts of the olog (its boxes and arrows) are not sufficient for the system to act as a whole—the less-obvious inter-relationships between these parts give the system its functionality. It is important to note that ologs can be constructed based on modeling and simulation, experimental studies, or theoretical considerations that essentially result in the understanding necessary to formulate the olog. This has been done for the proteins considered here based on the results from earlier work which provided sufficient information to arrive at the formulation of the problem as shown in Fig. 4.12.

Subsequently, we may construct a simple social network that may appear to some as vastly different than a protein filament, and to others as quite similar. The reason for the discrepancy is that semantically and physically the situations have almost nothing in common, but structurally and functionally they do. In fact, we will prove category-theoretically that they are structurally and functionally isomorphic in the sense that their ologs are identical. We now describe the setting for our simple social network as depicted in Fig. 4.15. Imagine a building with sound-proof rooms labeled 1 through 100, equipped with a controlled wireless communication system connecting each pair of consecutive rooms. In each room a human participant sits on a chair with a simple wireless transceiver that can transmit and receive text messages from the participant to the left (his or her predecessor) or the person to the right (his or her successor). We assume that participants in odd numbered rooms are women and people in even numbered rooms are men, just for pronoun clarity. The goal is to faithfully pass messages (sentences of under ten words, say) from room 1 to room 100 and back the other way as quickly as possible. The woman in room 1 (respectively the man in room 100) receives a message from the experimenter. She then inputs it into her transceiver and sends it to her neighbor (2), who passes it along to his neighbor (3), and on down the line until it is received by the man in room 100, who submits it to the experimenter there. Thus the network has a task of faithfully sending messages from one experimenter to the other; if they fail to successfully transmit at least one message per hour we say that the system has failed.

An obstacle can be added by allowing that the transmission of messages between participants is not always error-free. That is, the experimenters can adjust the amount of "noise" in the system, resulting in messages that could be anywhere from error-free to completely unintelligible. For example, the message "*the party was fun and exciting*" may arrive in the next room as "*tha partu was fon and escitin*". Upon receiving a garbled message, a participant may take the time to "fix

**Table 4.4** Component analogies between protein and social network. Because the olog (Fig. 4.14) is specifically designed to abstract away the particulars of either the protein or the social network, and thus represent *both* cases, this table defines the particulars in each case. Some types, such as "a real number", stand on their own and we merely give examples. Others, such as "a one-dimensional system of bricks, glue, and lifeline" require a bit more description. This table provides the necessary description to connect the concrete formulations in the case of our protein and social network to the abstract formulation given by the olog. Adapted from [77]

| Type | Label | Protein | Social network |
|---|---|---|---|
| **A** | a 1D system (S) of bricks (b), glue (g), and lifelines (L) | $\alpha$-helix, $\beta$-helix, etc. | wireless and physical communication |
| **B** | a 1D system (S) of bricks (b) and glue (g); no lifelines | amyloid, $\beta$-sheet nanocrystal, etc. | wireless communication |
| **C** | a *brittle* system (S) of bricks (b) and glue (g) | brittle protein filament | brittle social network |
| **D** | a "chain" graph $* \rightarrow * \rightarrow * \cdots \rightarrow *$ | chain shape for protein | one-to-one communication |
| **E** | a *ductile* system (S) of bricks (b) and glue (g) | ductile protein filament | ductile social network |
| **F** | a 1D system (S) of bricks (b) and glue (g) | $\alpha$-helix or $\beta$-helix, amyloid or $\beta$-sheet nanocrystal | social network |
| **G** | a system consisting of bricks connected by glue (g) and lifelines (L), both structured as a graph, G | lifeline protein of specified shape | lifeline social network of specified shape |
| **H** | a graph, G | shape of protein | shape of network |
| **I** | a threesome (b, g, L) of building blocks, serving as bricks, glue, and lifelines | amino acid cluster, H-bonds, backbone | transceiver, wireless network, possible physical passage/contact |
| **J** | a system consisting of bricks (b) connected by glue (g), structured as a graph, G | protein of specified shape | social network of specified shape |
| **K** | a threesome (b, g, S) of building blocks, serving as bricks, glue, and strong glue | amino acid cluster, H-bonds, backbone | transceiver, wireless network, possible physical passage/contact |
| **L** | a pair (b, S) of building blocks, serving as bricks and strong glue | amino acid cluster, backbone | transceiver, possible physical passage/contact |
| **M** | a pair $(R, r)$ of real numbers such that $R \approx r$ | *e.g.*, $R = 19.6$, $r = 21.4$ | *e.g.*, $R = 19.6$, $r = 21.4$ |
| **N** | a pair (b, g) of building blocks, serving as bricks and glue | amino acid cluster, H-bonds | transceiver, wireless network |
| **O** | a pair $(R, r)$ of real numbers such that $R \gg r$ | *e.g.*, $R = 89.6$, $r = 2.3$ | *e.g.*, $R = 89.6$, $r = 2.3$ |
| **P** | a pair (B1, B2) of building blocks, such that B2 can connect two instances of B1 | amino acid, backbone | wireless network, transceiver |
| **Q** | a pair $(x, y)$ of real numbers | *e.g.*, $x = 17.4$, $y = 38.7$ | *e.g.*, $x = 17.4$, $y = 38.7$ |

**Table 4.4** (continued)

| Type | Label | Protein | Social network |
|---|---|---|---|
| **R** | a brick (b) | amino acid cluster | transceiver |
| **S** | a glue (g) | H-bond cluster | wireless connection |
| **T** | a lifeline (L) | backbone | possible physical passage/contact |
| **U** | a building block | basic unit of material | basic unit of communication |
| **V** | a real number | *e.g.*, 42.6 | *e.g.*, 42.6 |
| **W** | a resting extension | *e.g.*, 12 Ångstrom | *e.g.*, 1 % error per bit |

it up" before sending it along, thereby helping to ensure that the message can be correctly submitted at the end of the line. We define the "extension" of the system to be the amount of noise, measured as the probability that a transmission error occurs for an arbitrary letter in a message. Given sufficient noise, it may happen that no messages can be transferred successfully through the network. Thus we define the "failure extension" of the network to be the amount of noise present when this occurs. Similarly, the failure extension of a glue element is the amount of noise at which a wireless transmission cannot be successfully sent from one room to the next.

Finally, we can add lifelines to this picture by adding physical passageways between consecutive rooms. Now, in case the noise gets too high, individuals may walk or run through these "lifeline passageways" and transmit a message by voice. During low levels of noise, the doorways will typically not be used to relay information because the text messaging is much faster, and hence the existence of the lifelines will be "hidden". However, once the transmission noise is severe enough to prevent good wireless communication (that is, the glue breaks), these passageways come into effect and save the network from breaking altogether. The three basic building blocks of this social network are shown in Fig. 4.15. For a rigorous analysis we also define a failure extension for bricks and lifeline, and resting extension for lifeline (qualitatively identical to those depicted in Fig. 4.13). We can also define the failure extension of bricks and lifeline to be infinite (because messages existing on a given transceiver or passed *via* voice are unaffected by the noise level), and the functional relation is the same. We also define the resting extension of our lifeline passageways to be the amount of noise at which participants begin to use the passageways (*i.e.*, restricts the wireless communication).

We now analyze the performance of the two types of networks constructed here, without and with a lifeline (see Fig. 4.15). In the system without a lifeline as soon as the noise level is high enough to cause breakdown of one of the glue elements the system fails since no more messages can be transmitted. In the system with lifelines, even though glue elements may break there is still the possibility for signals to travel through the passageway such that a much greater noise level (or extension) can be sustained. A brittle network is one in which the failure extension is roughly the

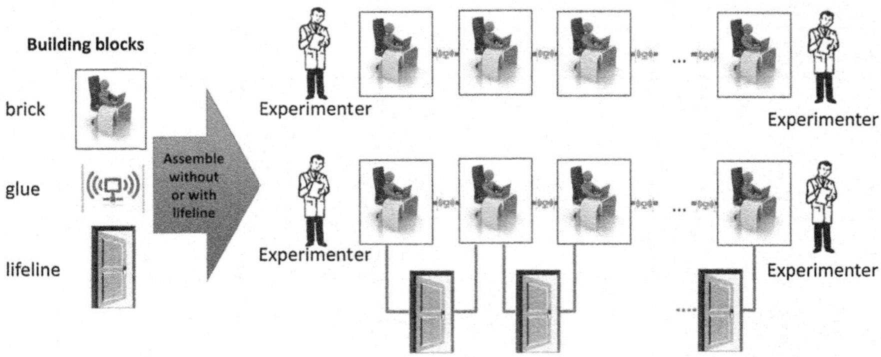


**Fig. 4.15** Visual representation of the idealized social network. The social networks considered here are composed of a linear arrangement of three elements, "bricks", "glue" and in some cases, "lifeline" (*e.g.*, fundamental building blocks). Thereby as a (known) *design rule*, brick and glue need to alternate in order to achieve a stable structure. That is, two brick or glue elements immediately next to each other would not stick together; where the reason is that bricks represent participants with transceivers and glue represents wireless communication that, in our case, can only occur between neighboring participants. The "lifeline" is a third element that is introduced here, reflecting the situation when there is still a physical connection of bricks even after the glue breaks. This reflects the existence of a "hidden" connection in that there exists a physical passageway between two brick elements even after the communication over the wireless connection is no longer feasible. The hidden connection is not "visible" before the glue is actually broken because, for reasons of efficiency, participants will choose to communicate the simple messages wirelessly rather than verbally, as the latter requires much more effort. The hidden length of lifelines is reflected in the fact that the "resting extension" of the lifeline is roughly equal to the failure extension of the glue. In other words, lifeline passageways are used only when wireless communication is no longer feasible. Both the brick and the lifeline have large failure extensions relative to the glue because participants and their verbal communication function perfectly well in the presence of noise on the wireless channels. *Top*: The social network restricts face-to-face interaction (*e.g.*, no lifeline). If noise on the wireless line reaches a critical point, messages can no longer be correctly conveyed. *Bottom*: The social network allows for face-to-face interaction (*e.g.*, possible physical passage/contact), resembling a system with a lifeline. If messages can no longer be conveyed wirelessly, communication can still take place, due to the physical passageways as shown. The existence of a lifeline has major implications on the functional properties of the system. A system with a lifeline shows a *ductile* response, where a connection can be sustained at large displacements as compared to the glue alone. In contrast a system without a lifeline shows a *brittle* response, where only a small displacement can be sustained until the material breaks (roughly the failure extension of the glue). From [77]

same as the failure extension for each glue element. A ductile network is one in which the failure extension is much greater than the failure extension of each glue element. We thus hypothesize that social networks with lifeline passageways will be ductile and that those without lifeline passageways will be brittle. While the above communication network is fairly degenerate as compared with, for example, Facebook,[2] the basic idea is similar. People are connected with a set of "friends"

[2]http://www.facebook.com.

and the basis of this friendship is communication. Communication can be muddled by various kinds of noise, but the use of additional forms of interaction (*e.g.*, talking in face-to-face meetings in addition to using online text messages) increases the probability that the parties understand each other.

We have constructed a system so that the olog describing it is precisely the same as that defining the protein system previously described. The basic layout is in Fig. 4.14, and the tables add "rigidifying information". For example, the participants with their transceivers are the bricks, the wireless communication between neighboring rooms is the glue, the passageways are the lifelines (see Table 4.4 for all definitions). We define brittleness and ductility exactly as we did in the protein case and as described in the previous paragraph; in fact this is forced on us because boxes $C$ and $E$ are fiber products. The fact that the same olog describes our protein materials and our social network should be considered as a rigorous analogy or isomorphism between these two domains.

This is the analogy between the protein strands (*e.g.*, $\alpha$-helix and amyloid, $\beta$-helix and $\beta$-sheet crystal, or any mechanically equivalent protein structures) and the social network experiment: In both cases a network (protein/social) consisting of bricks (amino acid clusters/human participants) connected together by glue (H-bond cluster/wireless communication) is subject to pulling (axial extension/error-producing noise) and eventually reach a breaking point (when the maximum extension is reached/transmission rate drops to zero). Lifelines (additional physical connections *via* covalent links/passageways) serve to increase the ductility (failure extension of network divided by failure extension of individual glue elements) of the network. The two situations can be modeled by precisely the same olog. Thus the olog sets out a space of possible systems that includes everything from proteins to social networks (and potentially many other realizations), any two instances of which must be analogous, at least to the level of detail found in Fig. 4.14 and the associated tables. If one desires additional detail, for example to add a precise meaning for resting extensions, or even real numbers, one would simply expand the olog to capture these ideas. The interpretation of what b, g and L mean in different systems can be distinct (*e.g.*, proteins, polymers, music, etc. can have different physical realizations of these concepts). Yet, their fundamental properties and how they relate to others—other elements, different scales in hierarchies, etc.—are defined properly in the olog, and mathematically expressed not only as a fundamental property but in addition as functors to other elements in the system. For physical systems a key aspect of understanding the interplay of building blocks can for instance be expressed in scaling laws that define properties as a function of ratios of length-scales or energy levels, which fundamentally define how elements behave and interact with others. The general presentation of such relationships in networks is what is missing in current theories, and is where ologs present a powerful paradigm for *de novo* design of biologically inspired systems that span multiple hierarchical levels. This is because ologs can achieve a rigorous description of the synergistic interactions of structures and mechanisms at multiple scales, which provides the basis for enhanced functionality despite the reliance on few distinct building blocks.

Of course, our social network was contrived to fit the olog of the protein, in order to show that two very different domains could have identical conceptual descriptions at a very high level of detail. More detail could show differences between these two domains. For example, an observation we purposely did not include is that the bricks in our social network have need to breathe and eat! It is impossible (and perhaps not desirable in some cases) to include every detail of each system—our goal was to emphasize the essential parameters, and to provide a level of abstraction that emphasizes the key elements that define functional properties. Furthermore, whereas it may be rare for two different scientific studies to result in identical ologs, finding reusable parts should be quite common. In our olog, the notion of bricks being connected together by glue to form the structure of a graph is surely reusable not only within materials science but throughout engineering. The category-theoretic notion of "functors," which formally connect one olog to another, will allow scientists in vastly different fields to share their work by rigorously connecting together their ologs. This opens enormous opportunities for design of novel functional properties by drawing from the understanding gained in diverse fields.

A unique aspect of the equivalence that we describe between two seemingly disparate systems is the rigorous analysis of the conceptual interaction rules in protein materials and the establishment of a direct link to those of a social network *via* the use of category theory. We now turn to another pairing of systems—silk and music—to illustrate the use of hierarchical ologs.

## 4.9 Spider Silk and Music

Returning to a functional simile with music [7], we construct an olog that reflects the hierarchical structure found in protein materials such as spider silk, Fig. 4.16. In order to form the analogy to music we must determine a way to dissect the structure to its basic constituents. A generally advisable approach is the definition of building blocks of the systems first. Depending on the level of abstraction, these building blocks can be of real nature, *e.g.* phonemes, or of abstract nature, *e.g.* the predefined "lifeline" in $\alpha$-helices. Here, proteins assemble out of their building blocks, amino acids, whereas we define the building blocks of music as sound waves (sine, triangular, sawtooth, etc. [108, 109]) that are assumed to assemble *via* stacking, *i.e.* without any additional information about amplitude, frequency or pitch.

In a second step, we define the superior structural units and indicate how they are related to their basic constituents. Bonds affiliate amino acids into groups and thereby a polypeptide is a linear chain of two or more amino acids connected by a (peptide) bond. Each bond within the polypeptide has as starting point and end point an amino acid and hence they represent a subset of amino acid groups which are, in contrast to polypeptides, not necessary a linear chain of amino acids but can assemble in more complex structures (possibly describable *via* a 3D graph). In an analogous way we define the creation of musical structural assemblies where stacked groups of sound waves are called a *tone*.

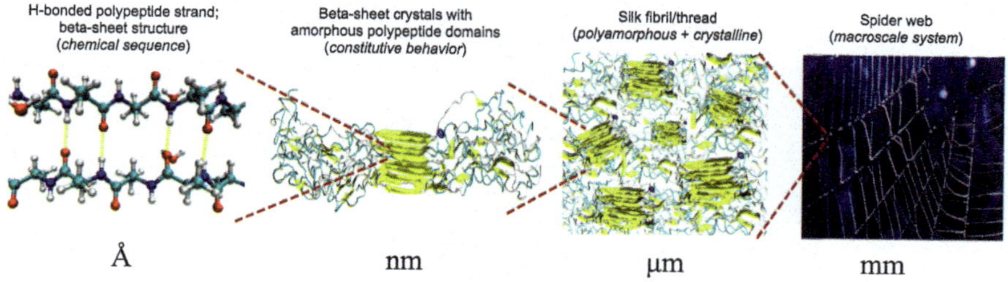


**Fig. 4.16** Hierarchical spider silk structure that ranges from nano (Ångstrom) to macro (millimeters). The image displays key structural features of silk, including the chemical structure found at the level of polypeptide H-bonded $\beta$-strands, the secondary structure $\beta$-sheet nanocrystals embedded in a softer semi-amorphous phase, bulk assembly of poly-crystalline components which assemble into macroscopic silk fibres, and finally the web-structure itself. Further discussion in Chap. 8: Unlocking Nature: Case Studies

So far, the relations only concern structure terminology and the question how functionality can be addressed remains open. Proteins, *i.e.* groups of one or more polypeptides, fold into secondary structures which are crucial to their properties and functionality [31]. Hence, a precursor to the assignment of sequence-structure-function relationships is the sequence-structure identification by experiment and computational studies. Such knowledge-based assignments have already been part of preceding inquiries [110, 111]. The information gathered from these studies, for example the sequence and environmental conditions that lead to distinct structural assemblies, then become data in the olog (Fig. 4.17).

Nanocomposites consist of proteins positioned in certain secondary structures (*e.g.*, $\alpha$-helix, $\beta$-sheet or amorphous phase in spider silk [43, 112]; in any case defined by a graph structure) of a specific size that assemble into higher level networks. The shear strength of these secondary structures, information stored in "a shear strength", is directly related to properties such as size and arrangement [44, 45, 47]. Similarly, the variation of frequency and amplitude of the stacked waves leads to the formation of the functional unit "a note" defined by its property "a pitch" [113]. The pitch corresponds to the audibility which then determines together with duration, loudness and timbre the functional properties of a chord (a grouping of intervals into categories such as thirds, fourths, etc.) which assemble into harmonically stable riff structures [114–118]. Here we identify a major potential of hierarchical ologs. As all chords in the riff are assembled in a weighted graph structure, the information that riffs are made of rhythmic arrangements of chords is inherently included. This designates a novel way of writing music sheets, where chords represent graph nodes connected to their nearest neighbors by edges where the edge length (or weight) directly correlates to the length of the chord.

Apart from the hierarchical levels, a subset of proteins, the enzymes, are included to exemplify the procedure to include "hierarchical interaction". Subsets of higher hierarchical levels contain distinguished members whose functionality is based on lower level architecture. Thus, a function relates a certain property of a higher hierarchical structure, here active centers of enzymes, to an element of a

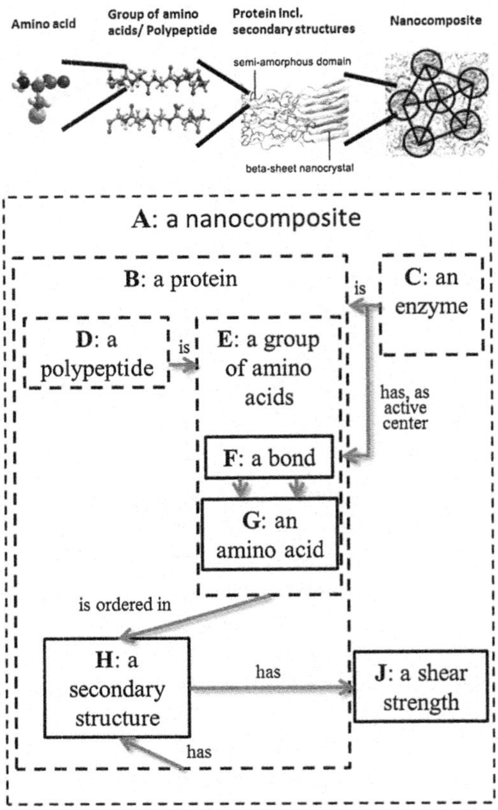


**Fig. 4.17** Olog of silk, describing the hierarchical build-up of proteins out of their building blocks, amino acids. Each bond has as starting point and end point an amino acid. A polypeptide is a linear chain of two or more amino acids connected by bonds. Nanocomposites consist of proteins in certain secondary structures (*e.g.* $\beta$-sheets, amorphous phases) of a specific size that assemble into higher level networks. The geometry of these of the secondary structure directly relates to macroscopic functional properties such as shear strength and extensibility. Each hierarchy-box represents a state or "snapshot" of its inner constituents that are connected in a graph structure, as shown in Fig. 4.11. This assures the unique allocation of lower level elements to superior structures. Apart from the hierarchical levels, a subset of proteins, the enzymes, are included to exemplify the way to include "hierarchical interaction". A function relates a certain property of a higher hierarchical structure, here active centers of enzymes, to an element of a lower level structure, a group of amino acids

lower level structure, a group of amino acids. Similar to the enzyme-protein relation subsets of chords also include group members with distinguished functional meaning [116, 119]. Here, the major chords, as interval special functional class of chords, has as distinguished member the root, the base on which a triadic chord is built. This kind of relation is typical for all kinds of hierarchical organizations, *e.g.* in primate groups [120]. After assembling and relating these insights, a challenge that can be overcome by multiscale studies including graph theoretic tools,

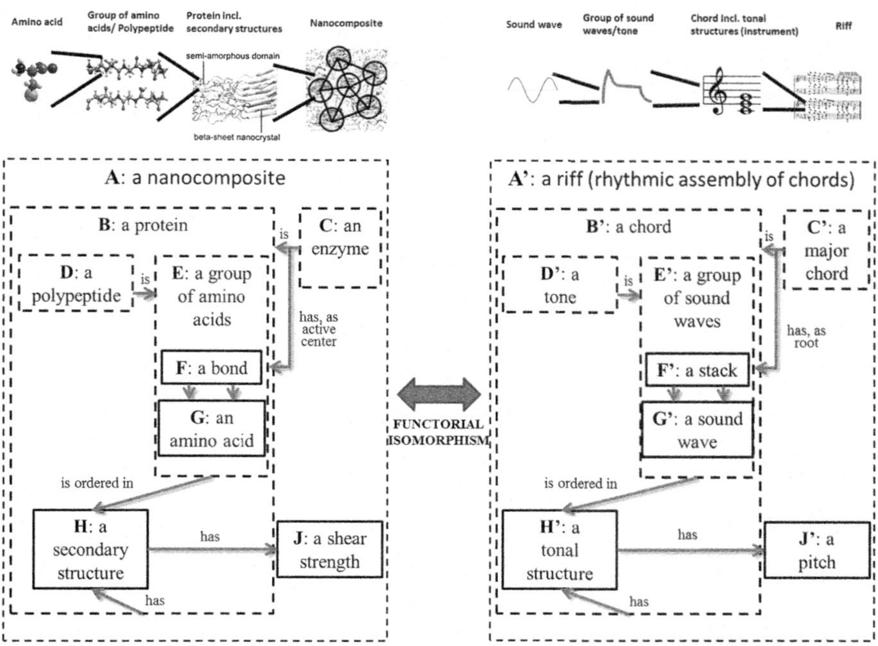


**Fig. 4.18** Functorial analogy of protein materials and musical structure by *hierarchical* ologs I: structure. (*Left*) Olog describing spider silk, identical to Fig. 4.17. (*Right*) A functorial isomorphism relates objects and arrows from the protein network to a music network which shows an equivalent hierarchical buildup. The building blocks in music networks are basic sound waves (sine, triangular, sawtooth, etc.) that assemble *via* stacking. The variation of frequency and amplitude of the stacked waves leads to the formation of the functional unit "a note" defined by its property "a pitch". Similar to the enzyme-protein relation certain subsets of chords, here the major chords, include group members with distinguished functional meaning. From [7]

a deduction to fields which show an equivalent hierarchical build-up by functors is possible (Fig. 4.9). Thereby, the relations and thus the functionality within the category are maintained and the two seemingly disparate fields display their intrinsic connection. In this example the functorial transformation is an isomorphism meaning that the positions of boxes and arrows are the same in both systems; thus it requires no further clarification (Fig. 4.18).

Apart from the simple description of structural details (graph theory can provide potential means), ologs also reveal system properties in a category-theoretic framework. Such a property is for example the H-bond clustering found in protein structures like spider silk [44, 45]. Geometric confinement of protein materials at the nanoscale leads to the rupture of clusters of 3 to 4 H-bonds in the $\beta$-sheet structures and thus to an optimized shear strength. This is shown in Fig. 4.19 where a functional property of the cluster, the shear strength, is related to a structural condition, the geometric confinement. This olog is based on the insights gained from multiscale computational studies [44, 45, 47].

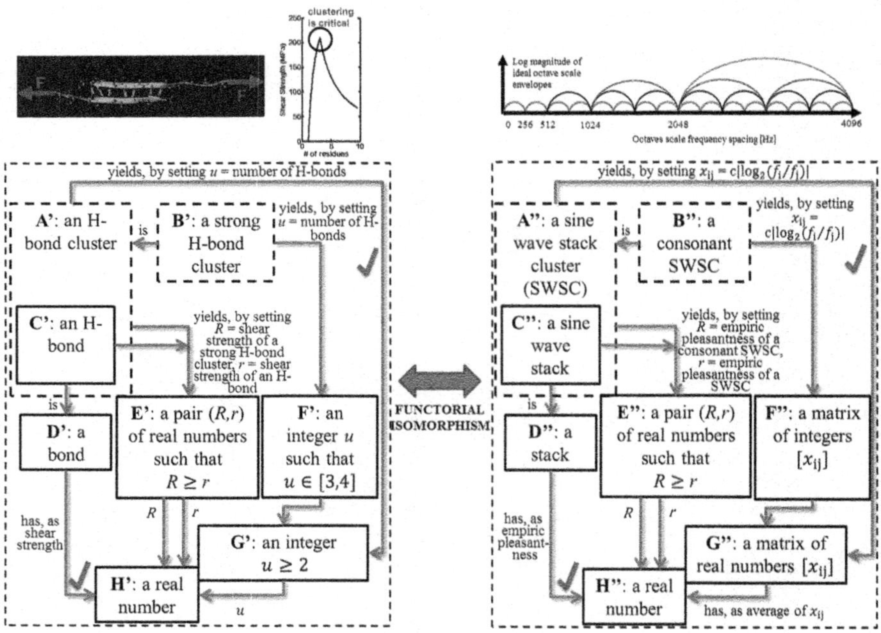


**Fig. 4.19** Functorial analogy of protein materials and musical function by *hierarchical* ologs II: cooperativity. (*Left*) Olog describing the dependence of clustering to obtain functionality. For protein materials such as spider silk H-bonds cluster into groups of 3 to 4 residues. Compared to the strength of a single bond, the shear strength of such a bond cluster becomes significantly higher. In the olog this is typically modeled by relations of real numbers. Therefore, the property must be quantitatively ascertainable. (*Right*) Analogous to this, chords in music form by stacking sound waves on an equal tempered scale, *i.e.* with frequencies that "fit well". Unlike shear strength the benefit of sound wave clustering—in a so-called consonant cluster—is not easily quantifiable but *is* subject to empirical observations, the measured pleasantness. The condition for consonance is given by the condition that all entrances of the frequency matrix are integers. Note, that the paths $\mathbf{C} \to \mathbf{D} \to \mathbf{H}$ and $\mathbf{C} \to \mathbf{E} \to \mathbf{H}$ as well as the paths $\mathbf{B} \to \mathbf{A} \to \mathbf{G}$ and $\mathbf{B} \to \mathbf{F} \to \mathbf{G}$ commute. Both ologs can be adjoined to the ologs shown in Fig. 4.18 simply by connecting it to the set "a bond"/"a stack". From [7]

Similar cluster strategies can be found in music. Chords consisting of "consonant" frequencies are considered to be innately pleasant to humans and even some animals, *i.e.* these frequencies belong to an (equally tempered) linear pitch space where pairs of frequencies $f_i$, $f_j$ follow the approximate relation $c|\log_2(f_i/f_j)| \in N$ with $c = 12$ [100, 101]. This can be thought of more simply as the specific collection of notes that construct a guitar chord (*e.g.*, notes played on adjacent or separate strings), for example. We check this condition for consonance by determining the matrix $[x_{ij}]$ and checking whether the entries have indeed integer values. This physically quantifiable ratio results in the empirically quantifiable sensation revealing emotions, data that can be incorporated in our olog [121, 122]. The major difference between materials sciences and social or artistic sciences is disclosed here: material properties are usually unambiguous (*i.e.*, well-defined) and objective whereas the evaluation of artistic properties is frequently subjective. The present functorial isomorphism allows the correlation of these data and subsequent statistical analysis may reveal additional insights that then lead back to the original system—an advantageous approach to recycle knowledge of well-studied systems such as music for novel applications.

Concerns could relate to the fact that the secondary structure of proteins is often not deterministic, *i.e.* the same protein can fold into more than one structure which endangers the unambiguity of the functional relationships. For our silk, analogous to frequencies that "fit well", we can define two amino acid sequences (in one-letter amino acid codes) such as:

...SQGAGRGGLGGQGAG**AAAAAA**GGAGQGGYGGLGSQG...

and

...GGYGPGQQGPSGPGS**AAAAAAAA**GPGQQGPGGYGPGQQ...

as sequences that "fit will" as the bold poly-Ala sequence constitutes the $\beta$-sheets of the silk [43, 112], but this may not be universally true. The same holds true for music, where for example pitch and timbre are sometimes ambiguous [123]. This has to be addressed by determining the environmental conditions that uniquely specify the protein's secondary structure, thus defining a protein grammar [80] or similar, again by experiment and computational modeling.

Both ologs, Figs. 4.18 and 4.19, are part of a bigger olog which would describe the material system spider silk (or classical music respectively) in total. Addressing the challenge to complete the olog immediately, it is straightforward to start with subunits as presented here. The assembly of a bigger system would work for example with the set "a bond" or "a stack", which is shared in both ologs and hence serves as an attachment point to combine them. Table 4.5 summarizes key structures and functions where connatural hierarchical ologs could be designed to uncover more analogies and then be attached to the existing parts. For instance, higher order structural assemblies such as nanocomposites or musical riffs gain functional importance by pattern building. In proteins this may relate to the repetition of secondary structural units and their overall confinement which ensures macroscale functionality, for example semi-amorphous phase and $\beta$-sheet domains in spider silk that provide a superior toughness by confinement to a fibril size of around 50 nm [124]. The corresponding pattern in music is the formation of chord sequences into riffs and phrases *via* syntactic structures that provide musical tension, an important functional focus in music [125, 126]. Yet another functional commonality of the two systems silk and music is related to the damage tolerance behavior. Localized defects in spider webs do not effect overall mechanical functionality [127] while the deletion of certain chords in a chord sequences do not affect the tonal coherence and hence the functionality [126]. Note that in hierarchical systems functionality is generally obtained by structural arrangement, *e.g.* clustering or stacking, and hence it is mostly impossible to separate structure and function.

**Table 4.5** Overview over similarities between spider silk and music. Beyond the patterns shown in Fig. 4.18 and Fig. 4.19 other structural and functional similarities between spider silk and music can be identified. Adapted from [7]

| General property | Spider silk | Music |
|---|---|---|
| Assembly of building blocks | Amino acids assemble into polypeptides *via* polypeptide bonds | Sound waves are stacked and interfere |
| Assembly of single units | Polypeptides assemble *via* covalent and weak bonds and form secondary structures | Sound waves with different frequency, amplitude and pitch form notes (instrument) |
| Assembly of functional units | Silk protein is formed in a stable structure dependent on solvent condition and ionization state | Sound wave of consonant frequency form chords on the equally tempered scale |
| Assembly of functional structure | Alanine rich repeat units form beta-sheets with high strength whereas glycine rich repeat units form extensible semi-amorphous phases; repetition of functional units creates nanocomposites | Harmonic sequences consist of the three main functions (tonic, sub-dominant and dominant); sequence/repetition of chords creates a melody riff |
| Upscaling of functionality | Nanoconfinement of composite structure ensures functionality (high strength, extensibility and toughness) on the macroscale | Phrases and climaxes within the music ensure musical tension, functional dependency of chord sequence |
| Damage tolerance (robustness) | Localization of deformation upon loading provides spider webs with robustness, damage mitigation, and superior resistance by nonlinear material behavior | According to the dependency structure single chords may be removed from or entered into the sequence without affecting the harmonic function |

Does such a overlap of silk and music provide *practical* engineering value? Due to the repetitive nature of the amino acid sequence of silk and the length of the natural protein, genetic engineering of silk analogs is limited: creating exact replicas of the protein sequence is difficult. Synthetic silks of defined sequence (that act as minimal building blocks) that incorporate structural motifs that represent native sequence elements are therefore highly desirable. In order to explore the idea of creating structures which mimic the function of native materials, one could use music theory to compose music that 'sounds' like silk or natural silks, and variants thereof. This is, in a sense, using Nature to inform music. More importantly, once we establish this equivalency baseline, we can then ask the opposite question—can we design a material according to a musical composition? For example, one could use functional harmony to create a "chord progression", likening a material to a function in musical syntax (*i.e.*, a I-IV-V-I chord progression). Taking this further, one could choose an existing musical composition that is successful and try to follow its path of "sounding like silk". This can then be fed back into our development of the framework of using music to inform materials design through common language and vocabulary of patterning, texture, tone, and expressivity.

This qualitative account will allow us to draw direct analogies to existing models of complex hierarchical structures such as those from social networks, and potentially linguistic theory where similar problems have been studied, and enables the utilization of insights and design paradigms across disparate fields of the science of hierarchical systems. The review of the key concepts presented here provides a generic framework that has the potential to unify existing understanding derived from the myriad existing studies of individual protein materials such as bone, silk, or cells and many others, where a major limitation was that no unifying framework that applies generally to all such materials has yet been proposed. This paradigm and associated design rules, which are applicable to other complex systems such as music, engineered technology and materials, or even food recipes, could emerge as an exciting new field of study and make critical contribution to the field of materiomics for which it serves as a central tool to describe structure-function relationships for hierarchical systems.

## 4.10 Motivating the Abstraction

At first glance, the construction of a complex olog may seem counterproductive. After all, if one requires a thorough understanding of a materiomic system to construct the olog (and all of its relevant connections and relations) what new insight does it bring? Where is the predictive or analytical power? The motivation behind such an abstraction is threefold:

Primarily, it combines a material system's function, structure, and properties as a logical, cohesive (and unique) whole. The complexity of hierarchies and cross-scale interactions can be integrated within the olog to any desired level of precision (consider the "coarse" hierarchical olog of Fig. 4.18 or 4.19 versus the "fine" olog of Fig. 4.14). Definition of the olog embodies the holistic materiomic approach, and supplied a universal means of representation, irrespective of biological material, architectural structure, musical composition, or social network.

Secondly, the ologs provide a common platform of comparison across variegated material systems. They can be considered a kind of complex "Venn diagram" where systems can share similar characteristic and properties unobserved by cursory comparisons. This was demonstrated in the simple example of the social network, wherein wireless communication paths provide the same "transfer function" of H-bonds in protein structures. Discovering the commonalities between complex systems requires both the abstraction and reduction (fundamental building blocks, function, connection, etc.) the ologs represent. The logical reduction from:

$$\textsf{amino acid} \Rightarrow \textsf{sequence} \Rightarrow \textsf{structure} \Rightarrow \alpha\textsf{-helix} \Rightarrow \textsf{H-bond} \Rightarrow \textsf{transfer}$$

is not apparent directly from knowledge of the polypeptide sequence. Delineating the function role of H-bonds, for example, can (a) motivate the implementation in designed, synthetic systems and (b) provide a focus for similar roles in other biological molecular structures. Such investigations are currently being undertaken, such as elucidating the role of H-bonding in stacked graphene oxide systems [128] or the similarities between weak H-bonding and relatively stronger disulfide bonds in human keratin [129, 130], as examples.[3] Is the olog for collagen similar to that of an intermediate filament? Is that of keratin similar to an amyloid fibril? Compilation of this information (*e.g.*, *categorization*) in respective ologs can subsequently reveal subtle similarities and differences between diverse biological materials and systems.

Finally, the abstraction to a reduced form allows an entirely new design space, as yet unexplored (and unexploited). One of the pitfalls of analyzing the material behavior of biological systems is that we resort to common metrics of materials science and engineering. Such tools, successful in describing single scale phenomena are ill-suited in describing a holistic materiome. Undoubtedly, there are differences in stiffness of bone, the stiffness of collagen fibrils, and the stiffness of tropocollagen molecules, yet, they are all within the same complex biological system. Each scale requires a difference method to determine mechanical properties. Integrated computational-experimental approaches must be implemented and further developed to explore the potential and limits of assembling building blocks in hierarchical material design (discussed further in Chap. 5: Experimental Approaches and Chap. 6: Computational Approaches and Simulation). Moreover, even at the same scale, the material can express different mechanical properties (*e.g.*, the stiffness of bone varies depending on the location in the body [132]). While we label "bone" as a material in the traditional sense, with an associated (mechanical) stiffness, we are unintentionally neglecting the hierarchical system composing that material. We overlook the significance of the *materiome*. Would a structural engineering, for example, ever label the "compressive stiffness" of the Eiffel Tower as a single value? Such a property could be formulated (based on the deformation of the tower under load), but would obviously have little value. While such an extreme example is trivial, the determination of the stiffness of bone does have value, such as the case of diagnosing and treatment of genetic diseases such as *osteogenesis imperfecta* (see Chap. 9: Pathological Materiomics). The determined (or assigned) "property" of a system has practical value relative to its function. Again, we see the need for a holistic view, even when we probe benign properties such as "stiffness".

We can also consider the practical aspects of an olog in the *transferability* of system structures and components. Ologs provide a powerful platform to understand the connections between form and function, complex behaviors and simple building blocks. *Solving* a particular olog—that is, determining the governing functional relationships—provides a template to reproduce similar complex systems (see Fig. 4.20). Indeed, through the interaction of bricks, glue, and lifelines, one can design a multitude of synthetic systems to mimic either a protein (such as an $\alpha$-helix)

[3] In graphene oxide, for instance, it has been shown both computationally and experimentally that there exists an optimal water content for a fully formed H-bond network between functionalized graphene layers [131]. Additional water and associated H-bonds serve to weaken the system—similar to the cooperativity of H-bonds in $\beta$-sheets.

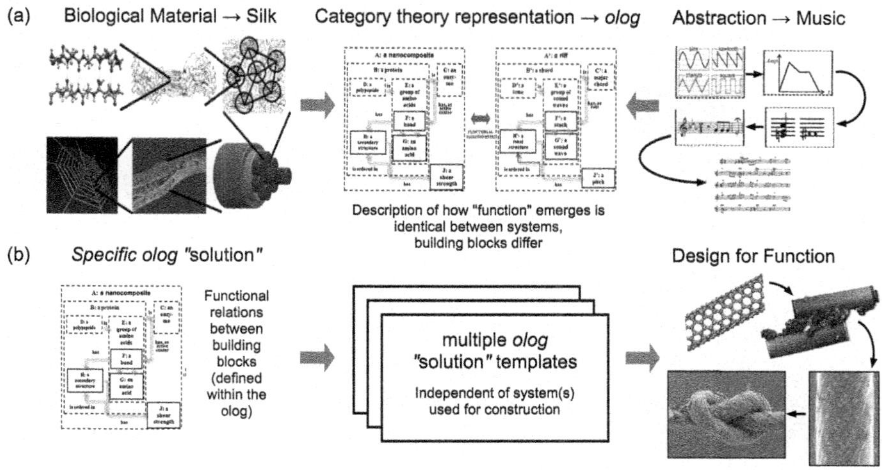


**Fig. 4.20** "Solving" the *olog*—a route to function by design. (**a**) Construction of an olog through the analysis of a complex system. Here, both silk and music can be analyzed to determine the functional relationships between elements, as previously described and depicted in Figs. 4.18 and 4.19. (**b**) Once a solution is obtained (the systems and/or iterations thereof are sufficiently captured by the olog), the olog (and the information contained therein) provides a template for any system that can be defined with the identical abstract elements. The olog is self-contained in the sense that it no longer matters what system was analyzed to initially construct it—be it spider silk or music, the functional relations are the same. A collection of such ologs can be "mixed and matched" for relations that commute—introducing new ways to combine building blocks. The end result is materials with designed functionality. Depicted is possible carbon nanotube "yarn", similar to the threads of spider silk fibers (insets of CNT yarn from [133], used with permission, copyright © 2011 American Association for the Advancement of Science)

or a message-passing communication network. Such "solutions" can be coupled to analysis exploiting analogies between patterns found in the arts with those in materials design, facilitated by the application of category theory as a means to enable a rigorous translation. The realization of category theory as an engineering tool can provide a powerful means to represent and optimize such patterns.

Beginning with proteins and materials we know (such as DNA) we can begin to construct a vast array of materiomic data (*e.g.*, the materiome). Detailed design control allowed *via* genetic templates of protein sequence and length can be used to synthesize and explore the physical design space and performance of functional biomaterials. This *tailorability* and control of chemistry, sequence and size for proteins, all lead to predictable functional material features through the realizing of stable folded configurations. Such relationships in proteins are needed to draw useful analogies to music (and other fields) in a meaningful way in terms of structure and function. For example, we have proposed an olog that links silk to music (see Figs. 4.18 and 4.19). Can we then use that olog as an abstract template to construct new materials? While we construct an olog to delineate the complexity of a known material—spider silk, for example—the principles underlying the exemplary performance of silk

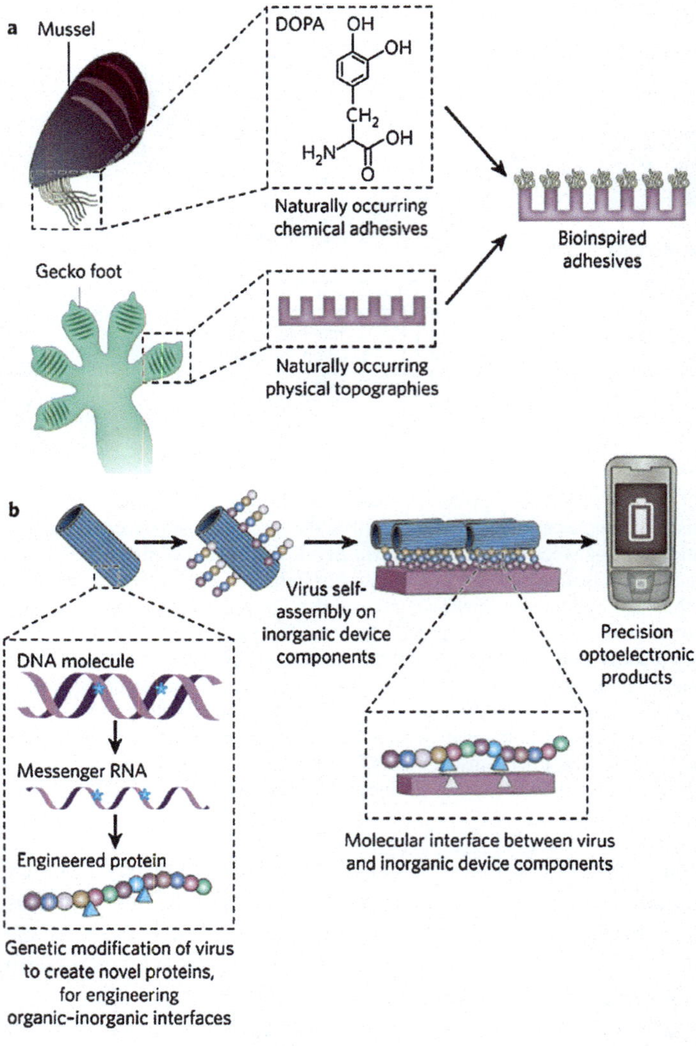


**Fig. 4.21** Using *functional* components of biological organisms and materials in novel applications—the building blocks of engineered materials. (**a**) The chemical and physical properties of materials used by organisms to facilitate surface adhesion can be mimicked, allowing the generation of synthetic coatings that modify surface chemistry or prevent biofouling. For example, $3,4(OH)_2$-phenylalanine (DOPA), a naturally occurring chemical adhesive used to facilitate the adhesion of mussels to surfaces in wet environments, has been combined with the physically patterned nanopillar topography found in the toes of geckos, which facilitates strong adhesion in dry environments, to produce novel adhesives that work in both wet and dry environments. (**b**) The molecular templating of whole viruses allows high-precision, multiscale patterning of electronic devices. Genetic modification of the organism (*left*) is used to engender biomolecular organic–inorganic interactions that lead to the coating of viruses with desired inorganic materials and their macromolecular assembly (*center*). Low-cost, high-precision energy-storage systems (*right*) are one potential application of this concept [134]. From [135], used with permission, copyright © 2009 Macmillan Publishers Limited

can be transferred to other materials (*e.g.*, carbon and other high-performance fibers) and thus help create a synthetic "super-silk" (Fig. 4.20). Exploiting the functional components of biological systems within synthetic devices and novel applications is currently being explored (see Fig. 4.21). Adoption of category theoretic analysis and materiomic descriptions can inform such biomimetic attempts.

*Why is it that the fundamental self-assembly processes for materials manufacturing can draw inspiration from musicians*? Universal commonalities can be found between materials other fields, from music to economics to sociology. Once discovered, the knowledge base can be shared reciprocally. The construction of a complex olog, for example, helps the creator to think about the problem from a new, abstract perspective, and discovery the "missing links" that may exist. The ideas of motifs, structure, form and function, aimed at arriving at enhanced understanding of how functional diversity can be achieved despite a limited set of universal building blocks, can be applied to the design of new mechanism-based materials.

Biological diversity that emerges from combining simple elements to generate structures that span multiple length-scales (from nano to macro) has many parallels to music, dance, language, and other expressions of art. For example, as we have seen in the case of a musical composition, individual structural motifs are assembled to form larger musical structures that emerge from wave forms, leading to instruments, and then to a simple melody and finally to large-scale symphonic works. However, the directed use of multiscale hierarchical assembly in engineering material function remains largely unexplored, and presents an exciting opportunity to realize novel material functions. From one perspective, such disparate fields can be thought of as merely providing new approaches to complex problems, using the successful insights of one field to understand another (yet another instance of scientific *convergence*). From another, bolder, perspective, the analysis of such disparate fields can be thought of as the same, much grander problem.

## 4.11 A New Merger of Science and Art?

Through analogy and abstraction and the use of category theory, we recognize a stunning similarity in the structure of music, language and literature, and materials. Perhaps, all expression of arts are a mere representation of humans' own inner workings to the outside world; without us (or the artists) knowing it. We make use of hierarchical structures (with a selection of forms) to express varied feelings and emotions—such as happiness, sadness, or phenomena such as life and death. While we may not have been able to interpret the arts in such a way in preceding times, we now (since the 20th century or so) have a detailed view into the physical resemblance and mechanisms of our own construction (through computational, experimental, theoretical, efforts as described in this book) so that we can identify the analogy between the structure and function of our own tissues and the structure and function of our expression of art.

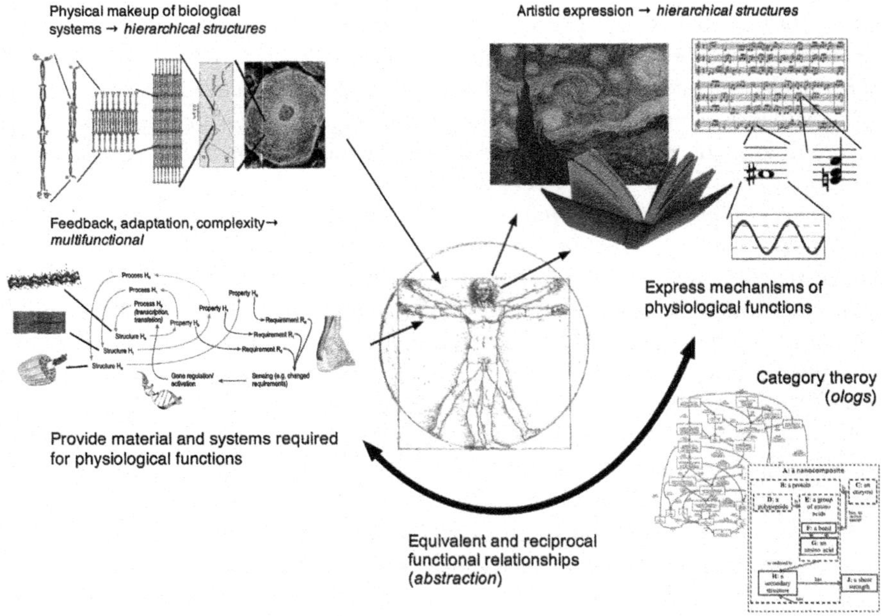


**Fig. 4.22** A new merger of *science* and *art*: linking the materiome to artistic expression through category theory. The biomaterial systems that constitute our living bodies (represented here by Leonardo Da Vinci's *Vitruvian Man*) are composed by hierarchical structures, which results in the necessary feedback and adaptation (and intrinsic complexity) for physiological function (*i.e.*, life), and the emergence of a broad range of material properties including mechanical, electrical, optical, chemical, etc. We also expressive ourselves artistically through mediums that are inherently hierarchical in structure (be it physical or aesthetic), from the composition of a symphony, a painting, a sculpture, or a sonnet. What emerges is a broad range of functions, encompassing emotions, life events, messages, stories, etc. Investigation of biological systems and art reveal universalities—equivalent and reciprocal functional relationships—perhaps as they evolved from the same source (ourselves). Such commonalities can be expressed in category theoretic analysis, and represented by an *olog*, reflecting our understanding of a complex system. Such a theoretical template can be applied to materiomics, or in general to other complex systems, as it reflects our own intrinsic perception

Indeed, the description of happiness, sadness, or life and death are focus areas that reach throughout all forms of arts; and they express who we are as humans and how we function and "fail". It appears as if humans have an innate driving force to express themselves to the outside world, to enable communication and to form social interactions. In this context the way art is created may tell us much more about ourselves than we anticipated. From this angle, the study of the arts may provide a powerful window to learn about the function and failure of biological materials, and biological systems more generally (see Fig. 4.22).

Some of the earlier artists such as Leonardo Da Vinci were both scientists and artists and illustrate the close connection between these two fields. While this connection has been lost over many centuries we may now be able to reinstate and exploit it, and the concepts put forth in a materiomics framework, and with the tools of category theory, can be a powerful means to achieve this for synergistic interactions in both directions. For example, we can create materials that resemble some of the mechanisms seen in the construction of music, as another "microscope" into the inner workings of ourselves; which we can exploit for the design of synthetic materials that share some of the unique characteristics of natural ones.

There is an intrinsic disconnect between description and existence, characterization and phenomena. A material has certain behaviors, of course, subject to the observer. This is always open for interpretation—a song may be "enjoyable", a material may be "stiff". The realization remains that certain materials exist, certain materials behave, and there is some measure of performance. How one attempts to categorize, quantify, or rank such behaviors is ultimately subjective. Categorical analysis (through *ologs* for example) is one means of characterizing a material system in terms of functional relationships (indeed a powerful methodology in terms of self-consistency), but by no means is it the only method. Full understanding (*e.g.*, description, categorization, characterization, representation, modeling, templating, etc.) can take many forms—*ologs* being only one. The materiome can be explicated by an olog, but also any holistic collection of multiscale structure, interactions, and performance (in an extended structure-property-process sense) is adequate. A materiomic perspective—encapsulated by an *olog*—is not dependent on any metric of characterization, but rather, the intrinsic property of "function subject to description". But that is the point. Spider silk will still capture prey, regardless of our knowledge of amino acid sequence, and music can still be enjoyable regardless of tuning (*e.g.*, musical scale). Our knowledge—the description of functional phenomena—is subjective and requires a sort of theoretical mapping. The function (*e.g.*, behavior) and interactions within the materiome—especially in the case of natural materials—exists independent of such mapping.

The use of ologs are a powerful approach to find the level of abstractions to prove these concepts mathematically. While we have used category theory here to bridge the seemingly disparate fields of proteins and social networks, silk and music, we envision that this approach will be generalizable to other systems, thus broadening the impact of this research beyond protein materials. Understanding the inner workings of biological systems, in turn, also allows us to create technologies that are more sustainable as they allow for the continued function of a natural system while providing the kinds of technologies and innovations that improve our lives.

## 4.12 Summary

Here we introduced the concept of hierarchical ologs to describe typical natural hierarchical systems such as language, biological materials and music and draw intrinsic connections between the underlying structures. We showed that in protein materials such as spider silk hierarchical structures identical to classical music can be identified and properly documented by means of hierarchical ologs. Similar to an analogy learning process this method may on the one hand serve as a guide to construct ologs for data and knowledge sharing in research groups and on the other hand as a method to utilize analogies to teach structure and functionality of hierarchical material systems. Conceptually organized like Wikipedia (http://www.wikipedia.org/), but with a substantiated mathematical background, ologs may provide a powerful academic and scientific tool to categorize, organize, relate and share insights gained during research. Specifically, analogy building as a momentous instrument for human understanding and education may be formalized by the use of category theory based ontologies. The ultimate strength of this tool relates to the enforcement of rigor during the analogy building process. Each term and concept defined in one system needs to be precisely related to their analogy counterpart to obtain a structure preserving transformation *via* functors. This characteristic feature of ologs sets the fundament for their superiority over common ontological or heuristic approaches [6].

In general, what ologs can identify for each system is how function emerges (for materials, for example, function can relate to mechanical properties, such as strength, toughness, extension to failure, etc.)—and specifically, how we *understand* these functional properties based on the interplay of the basic building blocks such as molecules or musical notes. Many such systems follow similar—or *universal*—patterns by which diverse function is realized—*e.g.*, through hierarchical structures, size effects, formation of composite structures with certain patterns (*e.g.*, maximum strength for clusters of H-bonds around 3 to 4; complementary frequencies that form chords), etc. While we observe these universalities between diverse biological systems such as, say, silk, diatoms, and bone, there is no current objective mathematical approach to define the similarities between these materials. To resolve this issue we propose the use of category theory—a branch of mathematics, which serves as a powerful tool for us to capture these relationships in an abstract space. When an abstract model is constructed using category theory, the result is a so-called olog, which resembles a graph or network, and encompasses all functional relationships within and between system building blocks.

Ologs can successfully break down—with mathematical rigor—how function emerges based on the interplay of building blocks. All building blocks are defined and self-contained within an olog, *i.e.*, we are free to describe what properties the building blocks have with respect to one another, how they are connected to one another, and how these structures lead to certain material functions and mechanisms. It is important to note that mathematics cannot tell us what these mechanisms are, so we need simulation, theory or experiment to discover these (discussed within Part II: Methods and Tools). This is analogous to the fact that we needed experimentation to eventually confirm the validity of Schrödinger's equation, which serves as the math model to describe the world of quantum mechanics.

The observation of similarities across systems suggest the existence of universal mechanisms by which functional properties emerge, *i.e.*, based on the interplay of the basic building blocks that are arranged in hierarchical structures across vast time-/length-scales, and that hold for a stunning range of natural systems. If such patterns govern much of the natural world, and can be sufficiently defined, the potential for application in engineering the built environment (materials, structures, infrastructure, etc.) are immense. There are several impacts for engineering:

1. Category theory is a tool to compile information about how materials function in a mathematically rigorous way, which allows for comparative analysis between different materials. It also allows us to extract powerful mechanisms from Nature for the use in engineering. One particularly intriguing aspect of natural materials is that function emerges because of hierarchical structures; in spite of or because of the weakness or simplicity of building blocks. These are concepts that we now just begin to understand, and they still need to be translated to engineering.
2. Related to the previous point, we can also ask a question such as: What if we want to replace the building block for a material? *E.g.*, in silk the building block is a $\beta$-sheet protein (possible due to restrictions in available material, which may not be the case in engineering). For a new material, however, we may want to use a carbon nanotubes instead, or a clay mineral, or other naturally occurring structures that may be abundantly available. A successful "solving of the olog" allows us to identify what we need to do be able to use the other building blocks, or how we need to change the structure of these, to create the same function as found in the natural material like silk. What this practically means is that we can learn how to make a material that behaves like silk (*e.g.*, extreme strength, toughness, deformability, etc.) from synthetic building blocks like carbon nanotubes.
3. We can learn about (and from) Nature, and how to design technologies that are more in sync with natural processes—specifically, how function emerges in different systems is of extreme value for engineers. For example, we can learn from patterns in music or language how to make better materials.

Future directions, open research questions, and the impact of an increased understanding of hierarchical protein materials can be discussed at three levels with increasing generality: (i) impact on protein material synthesis (design, engineering and manufacturing, or novel biomaterials); (ii) impact on bioinspired nanoscale material design and assembly (*e.g.*, hierarchical materials such as fibers, yarns or armors); and (iii) impact on macro-scale systems design and engineering (*e.g.*, design of cars, airplanes etc. where the merger of the concepts of structure and material across all the scales provides opportunities for more efficient systems). Immediate future work could be directed towards applying the concept of ologs to specific hierarchical biological materials, such as to silk or bone that show a greater complexity than the simple problems reviewed here. While the resulting ologs are more complex, the basic approach is identical and the main insights discussed here should still hold. Eventually the ologs reviewed in this chapter (Figs. 4.14, 4.18, and 4.19) could be implemented in a computational model, which will open the possibility for design optimization using numerical algorithms or make it easier to reuse existing ologs for the design of new ones.

## 4.13 Key Citations


- D.I. Spivak, R.E. Kent, Ologs: a categorical framework for knowledge representation. PLoS ONE **7**(1), e24274 (2011)

- D.I. Spivak, T. Giesa, E. Wood, M.J. Buehler, Category theory analysis of hierarchical protein materials and social networks. PLoS ONE **6**(9), e23911 (2011)
- T. Giesa, D.I. Spivak, M.J. Buehler, Reoccurring patterns in hierarchical protein materials and music: the power of analogies. BioNanoScience **1**(4), 153–161 (2011)
- T. Giesa, D. Spivak, M.J. Buehler, Category theory based solution for the building block replacement problem in materials design. Adv. Eng. Mater. (2012). doi:10.1002/adem.201200109
- J. Schmalfeldt, *In the Process of Becoming: Analytic and Philosophical Perspectives on Form in Early Nineteenth-Century Music*. Oxford Studies in Music Theory (Oxford University Press, New York, 2011)
- D. Tymoczko, *A Geometry of Music: Harmony and Counterpoint in the Extended Common Practice*. Oxford Studies in Music Theory (Oxford University Press, New York, 2011)
- S. Awodey, *Category Theory* (Oxford University Press, London, 2010)
- M.J. Buehler, Tu(r)ning weakness to strength. Nano Today **5**(5), 379 (2010)
- M.J. Buehler, Y.C. Yung, Deformation and failure of protein materials in physiologically extreme conditions and disease. Nature Materials **8**(3), 175 (2009)
- U. Alon, Simplicity in biology. Nature **446**(7135), 497 (2007)
- R. Kamien, *Music: An Appreciation* (McGraw–Hill Humanities/Social Sciences/Languages, 2007)
- G. Sica, *What is Category Theory*? Advanced Studies in Mathematics and Logic (Polimetrica, Monza, 2006)
- R. Oppenheimer, Analogy in science. American Psychologist **11**, 127–135 (1956).

## References


1. R.C. Strohman, The coming kuhnian revolution in biology—commentary. Nat. Biotechnol. **15**(3), 194 (1997)
2. S. Huang, Complexity—the practical problems of post-genomic biology. Nat. Biotechnol. **18**(5), 471–472 (2000)
3. A. Lazaris, S. Arcidiacono, Y. Huang, J.F. Zhou, F. Duguay, N. Chretien, E.A. Welsh, J.W. Soares, C.N. Karatzas, Spider silk fibers spun from soluble recombinant silk produced in mammalian cells. Science **295**(5554), 472–476 (2002)
4. X.X. Xia, Z.G. Qian, C.S. Ki, Y.H. Park, D.L. Kaplan, S.Y. Lee, Native-sized recombinant spider silk protein produced in metabolically engineered escherichia coli results in a strong fiber. Proc. Natl. Acad. Sci. USA **107**(32), 14059–14063 (2010)
5. S. Eilenberg, S. Maclane, General theory of natural equivalences. Trans. Am. Math. Soc. **58**(Sep), 231–294 (1945)
6. D.I. Spivak, R.E. Kent, Ologs: a categorical framework for knowledge representation. PLoS ONE **7**(1), e24274 (2001)
7. T. Giesa, D.I. Spivak, M.J. Buehler, Reoccurring patterns in hierarchical protein materials and music: the power of analogies. BioNanoScience **1**(4), 153–161 (2011)
8. N. Goldenfeld, L.P. Kadanoff, Simple lessons from complexity. Science **284**(5411), 87–89 (1999)

9. R.B. Laughlin, D. Pines, The theory of everything. Proc. Natl. Acad. Sci. USA **97**(1), 28–31 (2000)
10. D. Falk, *Universe on a T-shirt: The Quest for the Theory of Everything*, 1st edn. (Arcade Publishing, New York, 2004)
11. U. Alon, Simplicity in biology. Nature **446**(7135), 497 (2007)
12. J.M. Carlson, J. Doyle, Complexity and robustness. Proc. Natl. Acad. Sci. USA **99**, 2538–2545 (2002)
13. M.E. Csete, J.C. Doyle, Reverse engineering of biological complexity. Science **295**(5560), 1664–1669 (2002)
14. T. Ackbarow, M.J. Buehler, Hierarchical coexistence of universality and diversity controls robustness and multi-functionality in protein materials. J. Comput. Theor. Nanosci. **5**(7), 1193–1204 (2008)
15. M.J. Buehler, Nanomaterials: strength in numbers. Nat. Nanotechnol. **5**, 172–174 (2010)
16. S.J. Eichhorn, C.A. Baille, N. Zafeiropoulos, L.Y. Mwaikambo, M.P. Ansell, A. Dufresne, K.M. Entwistle, P.J. Herrera-Franco, G.C. Escamilla, L. Groom, M. Hughes, C. Hill, T.G. Rials, P.M. Wild, Current international research into cellulosic fibres and composites. J. Mater. Sci. **36**, 2107–2131 (2001)
17. W. Hamad, *Cellulosic Materials: Fibers, Networks and Composites* (Kluwer Academic, Boston, 2002)
18. K. Abe, H. Yano, Comparison of the characteristics of cellulose microfibril aggregates of wood, rice straw and potato tuber. Cellulose **16**(6) (2009)
19. Y. Nishiyama, Structure and properties of the cellulose microfibril. J. Wood Sci. **55**, 241–249 (2009)
20. H. Herrmann, U. Aebi, Intermediate filaments: molecular structure, assembly mechanism, and integration into functionally distinct intracellular scaffolds. Annu. Rev. Biochem. **73**, 749–789 (2004)
21. N. Mucke, L. Kreplak, R. Kirmse, T. Wedig, H. Herrmann, U. Aebi, J. Langowski, Assessing the flexibility of intermediate filaments by atomic force microscopy. J. Mol. Biol. **335**(5), 1241–1250 (2004)
22. S. Kim, P. Wong, P.A. Coulombe, A keratin cytoskeletal protein regulates protein synthesis and epithelial cell growth. Nature **441**(7091), 362–365 (2006)
23. L.H. Gu, P.A. Coulombe, Keratin function in skin epithelia: a broadening palette with surprising shades. Curr. Opin. Cell Biol. **19**(1), 13–23 (2007)
24. K.N. Dahl, D.E. Discher, K.L. Wilson, The nuclear envelope lamina network has elasticity and a compressibility limit suggestive of a "molecular shock absorber". Mol. Biol. Cell **15**, 119–120 (2004)
25. A.C. Rowat, J. Lammerding, J.H. Ipsen, Mechanical properties of the cell nucleus and the effect of emerin deficiency. Biophys. J. **91**(12), 4649–4664 (2006)
26. T. Ackbarow, X. Chen, S. Keten, M.J. Buehler, Hierarchies, multiple energy barriers and robustness govern the fracture mechanics of alpha-helical and beta-sheet protein domains. Proc. Natl. Acad. Sci. USA **104**, 16410–16415 (2007)
27. S.V. Strelkov, J. Schumacher, P. Burkhard, U. Aebi, H. Herrmann, Crystal structure of the human lamin a coil 2b dimer: implications for the head-to-tail association of nuclear lamins. J. Mol. Biol. **343**(4), 1067–1080 (2004)
28. M. Arslan, Z. Qin, M.J. Buehler, Coiled-coil intermediate filament stutter instability and molecular unfolding. Comput. Methods Biomech. Biomed. Eng. **14**(5), 483–489 (2011)
29. A. Kaminska, S.V. Strelkov, B. Goudeau, M. Olive, A. Dagvadorj, A. Fidzianska, M. Simon-Casteras, A. Shatunov, M.C. Dalakas, I. Ferrer, H. Kwiecinski, P. Vicart, L.G. Goldfarb, Small deletions disturb desmin architecture leading to breakdown of muscle cells and development of skeletal or cardioskeletal myopathy. Human Genet. **114**(3), 306–313 (2004)
30. B. Kiss, A. Karsai, M.S.Z. Kellermayer, Nanomechanical properties of desmin intermediate filaments. J. Struct. Biol. **155**(2), 327–339 (2006)
31. M.J. Buehler, Y.C. Yung, Deformation and failure of protein materials in physiologically extreme conditions and disease. Nat. Mater. **8**(3), 175–188 (2009)

32. M.J. Buehler, Tu(r)ning weakness to strength. Nano Today **5**(5), 379–383 (2010)
33. J.E. Padilla, C. Colovos, T.O. Yeates, Nanohedra: using symmetry to design self assembling protein cages, layers, crystals, and filaments. Proc. Natl. Acad. Sci. USA **98**(5), 2217–2221 (2001)
34. M.M.C. Bastings, T.F.A. de Greef, J.L.J. van Dongen, M. Merkx, E.W. Meijer, Macrocyclization of enzyme-based supramolecular polymers. Chem. Sci. **1**(1), 79–88 (2010)
35. Y. Mori, K. Minamihata, H. Abe, M. Goto, N. Kamiya, Protein assemblies by site-specific avidin-biotin interactions. Organic & Biomolecular Chemistry **9**(16), 5641–5644 (2011)
36. N. Barkai, S. Leibler, Robustness in simple biochemical networks. Nature **387**(6636), 913–917 (1997)
37. L.H. Hartwell, J.J. Hopfield, S. Leibler, A.W. Murray, From molecular to modular cell biology. Nature **402**(6761), C47–C52 (1999)
38. L. Hartwell, Theoretical biology—a robust view of biochemical pathways. Nature **387**(6636), 855 (1997)
39. K.M. Dipple, J.K. Phelan, E.R.B. McCabe, Consequences of complexity within biological networks: robustness and health, or vulnerability and disease. Mol. Genet. Metab. **74**(1–2), 45–50 (2001)
40. H. Gao, B. Ji, I.L. Jager, E. Arzt, P. Fratzl, Materials become insensitive to flaws at nanoscale: lessons from nature. Proc. Natl. Acad. Sci. USA **100**(10), 5597–5600 (2003)
41. F. Vollrath, D. Porter, Spider silk as archetypal protein elastomer. Soft Matter **2**(5), 377–385 (2006)
42. S. Rammensee, U. Slotta, T. Scheibel, A.R. Bausch, Assembly mechanism of recombinant spider silk proteins. Proc. Natl. Acad. Sci. USA **105**(18), 6590–6595 (2008)
43. S. Keten, M.J. Buehler, Nanostructure and molecular mechanics of spider dragline silk protein assemblies. J. R. Soc. Interface **7**(53), 1709–1721 (2010)
44. S. Keten, M.J. Buehler, Geometric confinement governs the rupture strength of h-bond assemblies at a critical length scale. Nano Lett. **8**(2), 743–748 (2008)
45. S. Keten, M.J. Buehler, Asymptotic strength limit of hydrogen-bond assemblies in proteins at vanishing pulling rates. Phys. Rev. Lett. **100**(19), 198301 (2008)
46. M.J. Buehler, S. Keten, Colloquium: failure of molecules, bones, and the earth itself. Rev. Mod. Phys. **82**(2), 1459–1487 (2010)
47. S. Keten, Z. Xu, B. Ihle, M.J. Buehler, Nanoconfinement controls stiffness, strength and mechanical toughness of beta-sheet crystals in silk. Nat. Mater. **9**(4), 359–367 (2010)
48. K. Autumn, Y.A. Liang, S.T. Hsieh, W. Zesch, W.P. Chan, T.W. Kenny, R. Fearing, R.J. Full, Adhesive force of a single gecko foot-hair. Nature **405**(6787), 681–685 (2000)
49. E. Arzt, S. Gorb, R. Spolenak, From micro to nano contacts in biological attachment devices. Proc. Natl. Acad. Sci. USA **100**(19), 10603–10606 (2003)
50. M. Varenberg, N.M. Pugno, S.N. Gorb, Spatulate structures in biological fibrillar adhesion. Soft Matter **6**(14), 3269–3272 (2010)
51. A. Nova, S. Keten, N.M. Pugno, A. Redaelli, M.J. Buehler, Molecular and nanostructural mechanisms of deformation, strength and toughness of spider silk fibrils. Nano Lett. **10**(7), 2626–2634 (2010)
52. R. Kamien, *Music: An Appreciation* (McGraw–Hill Humanities/Social Sciences/Languages, New York, 2007)
53. C. Dodge, T.A. Jerse, *Computer Music: Synthesis, Composition, and Performance* (Cengage Learning, New York, 1997)
54. P. Miller, *The Smart Swarm: How Understanding Flocks, Schools, and Colonies Can Make Us Better at Communicating, Decision Making, and Getting Things Done* (Avery, New York, 2010)
55. F. Lerdahl, R. Jackendoff, *An Overview of Hierarchical Structures in Music* (MIT Press, Cambridge, 1983)
56. F. Lerdahl, R. Jackendoff, An overview of hierarchical structure in music. Music Percept. **1**(2), 229–252 (1983)

57. D.R. Hofstadter, *Metamagical Themas: Questing for the Essence of Mind and Pattern* (Basic Books, New York, 1985)
58. J.S. Bamberger, A. Hernndez, *Developing Musical Intuitions: A Project-Based Introduction to Making and Understanding Music* (Oxford University Press, New York, 2000)
59. D. Lewin, *Generalized Musical Intervals and Transformations* (Oxford University Press, New York, 2007)
60. J. Schmalfeldt, *In the Process of Becoming: Analytic and Philosophical Perspectives on Form in Early Nineteenth-Century Music*. Oxford Studies in Music Theory (Oxford University Press, New York, 2011)
61. R. Oppenheimer, Analogy in science. Am. Psychol. **11**(3), 127–135 (1956)
62. S. Vosniadou, A. Ortony, *Similarity and Analogical Reasoning* (Cambridge University Press, Cambridge, 1989)
63. J. Bransford, *How People Learn: Brain, Mind, Experience, and School*, expanded edn. (National Academy Press, Washington, 2000)
64. D. Gentner, K.J. Holyoak, B.N. Kokinov, *The Analogical Mind: Perspectives from Cognitive Science* (MIT Press, Cambridge, 2001)
65. P.C. Taylor, B.J. Fraser, D.L. Fisher, Monitoring constructivist classroom learning environments. Int. J. Educ. Res. **27**(4), 293–302 (1997)
66. R. Stavy, Using analogy to overcome misconceptions about conservation of matter. J. Res. Sci. Teach. **28**(4), 305–313 (1991)
67. C.-C. Tsai, Overcoming junior high school students' misconceptions about microscopic views of phase change: a study of an analogy activity. J. Sci. Educ. Technol. **8**(1), 83–91 (1999)
68. L.R. Novick, K.J. Holyoak, Mathematical problem-solving by analogy. J. Exp. Psychol. Learn. Mem. Cogn. **17**(3), 398–415 (1991)
69. S. Kaniel, Y. Harpaz-Itay, E. Ben-Amram, Analogy construction versus analogy solution, and their influence on transfer. Learn. Instr. **16**(6), 583–591 (2006)
70. W. Croft, *Typology and Universals*, 2nd edn. Cambridge Textbooks in Linguistics (Cambridge University Press, Cambridge, 2003)
71. G. Sica, *What is Category Theory?* Advanced Studies in Mathematics and Logic (Polimetrica, Monza, 2006)
72. N.C. Ellis, D. Larsen-Freeman, R.C. in language learning (Ann Arbor, Mich.), in *Language as a Complex Adaptive System*. The Best of Language Learning Series (Wiley–Blackwell, Chichester, 2009)
73. W. Croft, Pragmatic functions, semantic classes, and lexical categories. Linguistics **48**(3), 787–796 (2010)
74. F.W. Lawvere, S.H. Schanuel, *Conceptual Mathematics. A First Introduction to Categories* (Cambridge University Press, Cambridge, 2009)
75. S. Awodey, *Category Theory*, 2nd edn. Oxford Logic Guides, vol. 52 (Oxford University Press, London, 2010)
76. R. Brown, T. Porter, Category theory: an abstract setting for analogy and comparison, in *What is Category Theory?* (Polimetrica, Monza, 2006)
77. D.I. Spivak, T. Giesa, E. Wood, M.J. Buehler, Category theoretic analysis of hierarchical protein materials and social networks. PLoS ONE **6**(9), e23911 (2011)
78. P. Csermely, Creative elements: network-based predictions of active centres in proteins and cellular and social networks. Trends Biochem. Sci. **33**(12), 569–576 (2008)
79. N.M. Pugno, A statistical analogy between collapse of solids and death of living organisms: proposal for a 'law of life'. Med. Hypotheses **69**(2), 441–447 (2007)
80. M. Gimona, Protein linguistics—a grammar for modular protein assembly? Nat. Rev. Mol. Cell Biol. **7**(1), 68–73 (2006)
81. S.C. Ji, Isomorphism between cell and human languages: molecular biological, bioinformatic and linguistic implications. Biosystems **44**(1), 17–39 (1997)
82. H. Peterlik, P. Roschger, K. Klaushofer, P. Fratzl, From brittle to ductile fracture of bone. Nat. Mater. **5**(1), 52–55 (2006)

83. D. Taylor, J.G. Hazenberg, T.C. Lee, Living with cracks: damage and repair in human bone. Nat. Mater. **6**(4), 263–268 (2007)
84. D.A. Fletcher, R.D. Mullins, Cell mechanics and the cytoskeleton. Nature **463**(7280), 485–492 (2010)
85. N. Chomsky, *Syntactic Structures*, 2nd edn. (de Gruyter, Berlin, 2002)
86. M. Moortgat, *Categorial Investigations: Logical and Linguistic Aspects of the Lambek Calculus* (Foris Publications, Providence, 1988)
87. M.D. Hauser, N. Chomsky, W.T. Fitch, The faculty of language: what is it, who has it, and how did it evolve? Science **298**(5598), 1569–1579 (2002)
88. D.B. Searls, The language of genes. Nature **420**(6912), 211–217 (2002)
89. E. Moggi, *A Category-Theoretic Account of Program Modules* (Springer, London, 1989)
90. P. Wadler, Monads for functional programming. Math. Struct. Comput. Sci. **2**, 461–493 (1992)
91. M. Barr, C. Wells, *Category Theory for Computing Science* (Prentice Hall, New York, 1995)
92. B.C. Pierce, *Basic Category Theory for Computer Scientists* (MIT Press, Cambridge, 1996)
93. R. Lakes, Materials with structural hierarchy. Nature **361**(6412), 511–515 (1993)
94. E.M. Marcotte, M. Pellegrini, M.J. Thompson, T.O. Yeates, D. Eisenberg, A combined algorithm for genome-wide prediction of protein function. Nature **402**, 83–86 (1999)
95. E.M. Marcotte, Detecting protein function and protein-protein interactions from genome sequences. Science **285**, 751–753 (1999)
96. D. Eisenberg, E.M. Marcotte, I. Xenarios, T.O. Yeates, Protein function in the post-genomic era. Nature **405**(6788), 823–826 (2000)
97. F.W. Lawvere, The category of categories as a foundation for mathematics, in *Proceedings of the Conference on Categorical Algebra* (Springer, New York, 1965)
98. S.M. Lane, *Categories for the Working Mathematician*, 2nd edn. (Springer, New York, 1998)
99. A. Nijholt, From left-regular to Greibach normal form grammars. Inf. Process. Lett. **9**(1), 51–55 (1979)
100. N.C. Maddage, C. Xu, M.S. Kankanhalli, X. Shao, Content-based music structure analysis with applications to music semantics understanding, in *Proceedings of the 12th Annual ACM International Conference on Multimedia* (2004), pp. 112–119
101. D. Tymoczko, *A Geometry of Music: Harmony and Counterpoint in the Extended Common Practice*. Oxford Studies in Music Theory (Oxford University Press, New York, 2011)
102. G. Deline, F. Lin, D. Wen, D. Gagevic, A. Kinshuk, Ontology-driven development of intelligent educational systems, in *2007 IEEE Pacific Rim Conference on Communications, Computers and Signal Processing*, vols. 1 and 2 (2007), pp. 34–37
103. M. Halle, *From Memory to Speech and Back: Papers on Phonetics and Phonology, 1954–2002*. Phonology and Phonetics (de Gruyter, Berlin, 2002)
104. G.N. Clements, The geometry of phonological features. Phonology **2**, 225–252 (1985)
105. A.S. Abramson, Laryngeal timing in consonant distinctions. Phonetica **34**(4), 295–303 (1977)
106. M. Jessen, C. Ringen, Laryngeal features in German. Phonology **19**, 189–218 (2002)
107. R. Paparcone, M.J. Buehler, Failure of a beta(1–40) amyloid fibrils under tensile loading. Biomaterials **32**(13), 3367–3374 (2011)
108. J.E. Cutting, B.S. Rosner, Categories and boundaries in speech and music. Percept. Psychophys. **16**(3), 564–570 (1974)
109. J.A. Moorer, Signal-processing aspects of computer music—survey. Proc. IEEE **65**(8), 1108–1137 (1977)
110. D. Frishman, P. Argos, Knowledge-based protein secondary structure assignment. Protein. Struct. Funct. Genet. **23**(4), 566–579 (1995)
111. Z.R. Sun, S.J. Hua, A novel method of protein secondary structure prediction with high segment overlap measure: support vector machine approach. J. Mol. Biol. **308**(2), 397–407 (2001)
112. S. Keten, M.J. Buehler, Atomistic model of the spider silk nanostructure. Appl. Phys. Lett. **96**(15) (2010)

113. R. Erickson, *Sound Structure in Music* (University of California Press, Berkeley, 1975)
114. D. Deutsch, Music recognition. Psychol. Rev. **76**(3), 300 (1969)
115. J. Bharucha, C.L. Krumhansl, The representation of harmonic structure in music—hierarchies of stability as a function of context. Cognition **13**(1), 63–102 (1983)
116. B. Pardo, W.P. Birmingham, Algorithms for chordal analysis. Comput. Music J. **26**(2), 27–49 (2002)
117. D.M. Randel, *The Harvard Dictionary of Music*, 4th edn. (Belknap Press of Harvard University Press, Cambridge, 2003)
118. K. Jensen, Multiple scale music segmentation using rhythm, timbre, and harmony. EURASIP J. Adv. Signal Process. (2007)
119. C.L. Krumhansl, R.N. Shepard, Quantification of the hierarchy of tonal functions within a diatonic context. J. Exp. Psychol. Hum. Percept. Perform. **5**(4), 579–594 (1979)
120. P. Izar, R.G. Ferreira, T. Sato, Describing the organization of dominance relationships by dominance-directed tree method. Am. J. Primatol. **68**(2), 189–207 (2006)
121. J.A. Sloboda, Music structure and emotional response: some empirical findings. Psychol. Music **19**, 110–120 (1991)
122. T. Schafer, P. Sedlmeier, From the functions of music to music preference. Psychol. Music **37**(3), 279–300 (2009)
123. W.M. Hartmann, *Signals, Sound, and Sensation*. Modern Acoustics and Signal Processing (American Institute of Physics, Woodbury, 1997)
124. T. Giesa, M. Arslan, N. Pugno, M.J. Buehler, Nanoconfinement of spider silk fibrils begets superior strength, extensibility and toughness. Nano Lett. **11**(11), 5038–5046 (2011)
125. E. Bigand, R. Parncutt, F. Lerdahl, Perception of musical tension in short chord sequences: the influence of harmonic function, sensory dissonance, horizontal motion, and musical training. Percept. Psychophys. **58**(1), 125–141 (1996)
126. M. Rohrmeier, Towards a generative syntax of tonal harmony. J. Math. Music **5**(1), 35–53 (2011)
127. S.W. Cranford, A. Tarakanova, N. Pugno, M.J. Buehler, Nonlinear material behaviour of spider silk yields robust webs. Nature (2012)
128. N.V. Medhekar, A. Ramasubramaniam, R.S. Ruoff, V.B. Shenoy, Hydrogen bond networks in graphene oxide composite paper: structure and mechanical properties. ACS Nano **4**(4), 2300–2306 (2010)
129. J. Alegre-Cebollada, P. Kosuri, J.A. Rivas-Pardo, J.M. Fernndez, Direct observation of disulfide isomerization in a single protein. Nat. Chem. **3**(11), 882–887 (2011)
130. S. Keten, C.-C. Chou, A.C.T. van Duin, M.J. Buehler, Tunable nanomechanics of protein disulfide bonds in redox microenvironments. J. Mech. Behav. Biomed. Mat. **5**(1), 32–40 (2012)
131. O. Compton, S. Cranford, Z. An, K. Putz, C. Brinson, M.J. Buehler, S.B. Ngyen, Tuning the mechanical properties of graphene oxide paper and its associated polymer nanocomposites by controlling cooperative intersheet hydrogen bonding. ACS Nano **6**(3), 2008–2019 (2011)
132. P.K. Zysset, X.E. Guo, C.E. Hoffler, K.E. Moore, S.A. Goldstein, Elastic modulus and hardness of cortical and trabecular bone lamellae measured by nanoindentation in the human femur. J. Biomech. **32**(10), 1005–1012 (1999)
133. M.D. Lima, S.L. Fang, X. Lepro, C. Lewis, R. Ovalle-Robles, J. Carretero-Gonzalez, E. Castillo-Martinez, M.E. Kozlov, J.Y. Oh, N. Rawat, C.S. Haines, M.H. Haque, V. Aare, S. Stoughton, A.A. Zakhidov, R.H. Baughman, Biscrolling nanotube sheets and functional guests into yarns. Science **331**(6013), 51–55 (2011)
134. K.T. Nam, D.W. Kim, P.J. Yoo, C.Y. Chiang, N. Meethong, P.T. Hammond, Y.M. Chiang, A.M. Belcher, Virus-enabled synthesis and assembly of nanowires for lithium ion battery electrodes. Science **312**(5775), 885–888 (2006)
135. N. Huebsch, D.J. Mooney, Inspiration and application in the evolution of biomaterials. Nature **462**(7272), 426–432 (2009)

# Part II
# Methods and Tools

# Chapter 5
# Experimental Approaches

**Abstract** System characterization through experimental means is the most common (and successful) approach to extract mechanical properties, performance, and behavior from complex biological systems. Experiment—as opposed to simulation, modeling, and theory—has the intrinsic advantage of not requiring any assumptions about material structure. Here, we review common experimental techniques that span scales from nano to macro. Multiple scales are discussed, encompassing single molecule assays (*e.g.*, through optical tweezers) that probe molecular mechanics and reaction pathways, to the many uses of atomic force microscopy (such as protein stretching or bending), to microscale techniques applied to cells (*e.g.*, micropipette aspiration) and tissues (*e.g.*, nanoindentation). A well equipped materiomics "toolbox" is necessary to further our understanding of how the mechanical behavior of a material affects its biological function.

*Measure what is measurable, and make measurable what is not so.*

*Galileo Galilei (1564–1642)*

## 5.1 Introduction

In the past, the mechanical behavior of complex biological materials has been studied extensively at the tissue, organ and systems levels independently. Such a piecewise approach (in a kind of scale-by-scale "divide and conquer" manner) was highly successful in delineating the structure and properties of systems at particular scales (*i.e.*, the molecular structure of myoglobin [1], collagen fibril architecture [2], the behavior of red blood cells [3, 4]), but defining (and predicting) functionality across scales is more challenging. Concurrent with computational tools, development of nanotechnologies and the resulting emergence of atomistic-scale resolution experimental tools have enabled multi-scale quantitative studies of deformation of individual cells and biomolecules. For the first time, such approaches allow the exploration of biological processes mediated by mechanical signals, with force and displacement resolutions of 0.1 pN and 0.1 nm, respectively [5]. As a result of these capabilities, it is now possible to establish the structure-function relationships across a

S.W. Cranford, M.J. Buehler, *Biomateriomics*, Springer Series in Materials Science 165,
DOI 10.1007/978-94-007-1611-7_5, 

multitude of scales and material hierarchies within biomaterials, under physiological and pathological conditions, in effect, experimentally probing the entire *materiome*.

In particular, application of external mechanical stimuli can induce biochemical reactions, including the synthesis of new biomolecules and the enhanced interaction among biomolecules, which can generate mechanical forces. A quantitative understanding of the mechanical responses of protein molecules is essential for gaining insight into their biological functions as such materials are subject to mechanical stimuli in their natural function—again, encompassing the field of *biomechanics* (see the canonical work in the field, "*Biomechanics: Mechanical Properties of Living Tissues*", by Y.C. Fung [6]). For example, the muscle protein titin facilitates muscle contraction and relaxation through unfolding and/or refolding driven by hydrogen bonds [7–9]. Likewise, changes in chemical stimuli, including pH, temperature, and biomolecular activity, can alter the structure and mechanical integrity of the cell, even in the absence of mechanical stimuli.

Experimental techniques have gained unparalleled accuracy in both length- and time-scales, as reflected in development and utilization of Atomic Force Microscope (AFM) [10–12] magnetic and optical tweezers [3, 13] or nanoindentation [14, 15] to analyze biological protein materials and biological molecules [5, 16]. Researchers from such diverse disciplines as molecular biology, biophysics, materials science, chemical, mechanical and biomedical engineering have developed an impressive selection of experimental tools that can measure and impose forces as small as a few *femto*newtons ($10^{-15}$ N) and displacements as small as a few Ångstroms ($10^{-10}$ m). Single-molecule and single-cell biomechanics assays are necessary to provide critical information and experimental support to develop theoretical materiomic models. At the same time, multiscale modeling has evolved into a predictive tool that complements experimental analysis at comparable length- and time-scales (to be further discussed in Chap. 6: Computational Approaches and Simulation). Integrated experimental, computational, and theoretical approaches are necessary to reconcile the *materiome*.

## 5.2 Advantages of Experimentation

The discussion of computational approaches will present the benefits of modeling and *in silico* techniques, but they lack the "realism" of experimental assays. In essence, computational modeling provides idealized scenarios and unlimited control over parameters and variables. The "real world" of experimentation provides a different forum for characterization. There is inherently a little less control, but experimentation is advantageous for a number of reasons:

- *Physical time- and length-scales*: Experimental approaches occur in realtime, and thus can capture biological and chemical events as they occur in Nature. While practically limited in terms of size and duration (tests occurring over decades, or a metric ton of biological tissue can be challenging), methods can probe the molecular (*e.g.*, single atom) to macro (*e.g.*, organism) scales, from nanoseconds (*e.g.*, chemical reactions) to lifespans (*e.g.*, bone growth).

- *Material specimens are, by nature, non-ideal*: Biological materials are inherently flawed, from both a geometric (disorder structures) and a material (protein sequence variation) perspective. Models typically choose a homogeneous representation (*e.g.*, repeated known sequence for a molecular study, or identical elements for a finite element model), whereas cells and tissues are extremely heterogeneous both spatially and temporally. The imperfection of such materials is difficult to model, but intrinsically captured in a physical sample.
- *In vivo, ex vivo, and in vitro*: Ingenious experimental methods such as micropipette aspiration can probe the mechanical properties of *living* cells and tissues, a critical condition beyond the realm of any computational model. The complex effect of physiological reactions such as mechanochemical signalling and transduction that occur within a living cell cannot be replicated *in silico*. Beyond live samples, *ex vivo* experiment provides a contrast, allowing experimentation under more controlled conditions that may not be life-sustaining—but still on biologically "grown" materials—and thus isolating critical parameters and relevant behaviors. Finally, *in vitro* facilitates a route to synthetic biology and biomimetic materials, culturing specific cells derived from multicellular organisms (*e.g.*, cell culture or tissue culture), subcellular components (*e.g.*, mitochondria or ribosomes), or purified molecules in the test tube (proteins, DNA, or RNA, either individually or in combination) for characterization.
- *Post-phenomenological perspective*: The investigation and manipulation of physical material samples allows the discovery of behaviors, responses, and phenomena that may have been otherwise unpredictable *a priori*. While computational modeling can also reveal unknown and/or surprise behaviors, a simulation is typically designed to illicit a desired (and hypothesized) result. Moreover, simulated phenomena yield little value until they can be repeated experimentally. Phenomena discovered experimentally has the immediate benefit that it is possible and/or tangible, and can be investigated further *a posteriori*.
- *No disconnect between "model" and "material"*: There is always uncertainty in modeling (either theoretical or computational) that the chosen material description is representative of the real material (*e.g.*, does a carbon-carbon bond behave as a harmonic spring? can cells be modeled as neo-Hookean solids?). The value of such representations lies in the power to interpret results. When the assumptions, however, are the basis of model *parameterization* rather than *interpretation*, they must be justified, requiring additional effort for validation. Experimental approaches eliminate the need for such validation—there is no disconnect between the model and material, as there is no model. Validation is self-imposed (assuming, of course, there are sufficient controls).

The contrast between computational modeling and experimental approaches can be reduced to a simple concept: modeling investigates how materials *should* behave, experimentation investigates how materials *do* behave. Until models can approach the complexity of real world environments, constraints and boundary conditions, uncertainties and "unknown unknowns", they will always be limited to a simplified representation (no matter how sophisticated). Being said, until experimental methods identifies, measures, and achieves precise control of every variable, modeling can serve to illustrate mechanisms and phenomena hidden within empirical data. Ultimately, both modeling and experiment are required for a detailed understanding of the materiome, and will continue to converge in terms of scope, scale, and utility.

## 5.3 Overview of Methodologies

There exist a variety of techniques to manipulate the mechanical environment of biomolecules and complex biological materials. These approaches differ in three important respects: mechanical operating principles (*e.g.*, stretch, indenting) force and displacement maxima and resolutions (*i.e.*, scale), and extent of deformation (*i.e.*, global vs. local, or system vs. component). Recent studies have been conducted to establish connections between the structure, mechanical responses, and function of biological tissues, such as the lung, heart, blood vessels, and cardiac muscle with increasing resolution. This trend has led to the development of tools and techniques for monitoring forces arising from cellular and biomolecular processes, sensitive enough to detect forces in the nanonewton (nN) to piconewton (pN) range. Due to the specificity of some developed techniques, it is obvious that an approach successful for single molecules (such as optical tweezers) may not be suitable for probing cellular behavior. Table 5.1 summarizes some experimental methods and techniques in nanobiomechanics, but is by no means exhaustive.

This chapter concentrates on the mechanical characterization of biological materials and systems for two reasons. First, despite the sophistication of experimental and computational approaches in cell and molecular biology, the mechanisms by which cells sense and respond to mechanical stimuli are poorly understood—the intricate coupling between the biochemical and mechanical processes of the materiome impedes prediction beyond a few simple trial cases. Secondly, mechanical characterization provides a metric that can be defined across scales. Mechanical notions of force, stiffness, and deformation—while rooted in different biological and chemical interactions—apply at the nanoscale as well as the macroscale. The challenge lies in seamlessly linking the behaviors.

## 5.4 One at a Time: Single Molecule Assays

If we consider amino acids as the fundamental building block of biological materials, the molecular structure (*i.e.*, polypeptide sequence) is the first level of hierarchy that begets *functionality*. Thus, a thorough understanding and categorization of such polypeptides—single molecules—provides preliminary characterization and properties necessary for higher order predictions. If we consider a structural engineering analogue, we must understand the properties (and limitations) of each steel member of a truss if we want to design the performance of an entire building. This is one of the advantages of using repeated sections over and over—the behavior is well

**Table 5.1** Summary of experimental methods and techniques in biomechanics. Adapted from [17]. "∗" indicates methods reviewed in this chapter

| Method | Mechanical principle | Approx. force resolution/range |
|---|---|---|
| Optical tweezers* (OT) | Dielectric beads of high refractive index are moved using laser beams. | $\sim 2\ \text{pN} \le F_x \le 600\ \text{pN}$ |
| Atomic force microscopy* (AFM) | Relative deformation of cantilever tip and molecule, substrate, or cell is used to estimate forces. | $\sim 10\ \text{pN} \le F_x \le ?$ |
| Micropipette aspiration* (MA) | Gentle suction is applied to micropipette attached to cell. | $10\text{–}20\ \text{pN} \le F_x \le ?$ |
| Magnetic tweezers/magnetic twisting cytometry | Magnetized ferromagnetic or superparamagnetic beads are moved by weaker directional magnetic fields/gradients. | $2\ \text{pN} \le F_x \le 50\ \text{nN}$ |
| Flow chambers | Enclosed chambers with inlet and outlets for fluid flow are used to subject cells to fluid shear stress. | $30\ \text{Pa} \le P_x \le ?$ |
| Elastic substratum method | Wrinkling patterns developed in artificial flexible sheets are used to infer cell traction forces. | Qualitative (forces *via* applied strain) |
| Stretching devices | Flexible membrane is attached to structures that enable membrane to be stretched. | Qualitative (forces *via* applied strain) |
| Flexible sheets with embedded beads/particles | Displacements of beads within flexible sheets (particle tracking) are used to infer cell traction forces. | $140\ \text{nN} \le F_x \le ?$ |
| Array of vertical microcantilevers | Horizontal deflection of individual vertical microcantilevers is used to infer traction forces. | $50\ \text{pN} \le F_x \le 100\ \text{nN}$ |
| Nanoindentation* (NI) | Applied load with known geometry to indent a surface with a local impression, used to determine material hardness, stiffness, fracture toughness, etc. | $1\ \mu\text{N} \le F_x \le 500\ \text{mN}$ |

known (at least, within an acceptable design margin) and quantified by quality standards and measures. Single molecular assays provide the preliminary "ASTM" protocols for characterization of biological materials, allowing both insight into how specific combinations of amino acids result in certain properties (a kind of *top-down* perspective, wherein a specific sequence manifests a particular structure or property), as well as categorization of repeated, universal structures, such as alpha helices (a *bottom-up* perspective, where groups of molecules, defined by common behavior, can contribute to a larger, hierarchical system).

While there are a selection of single molecular assays designed to extract structural and chemical information (such as Raman spectroscopy [18], nuclear magnetic resonance (NMR) spectroscopy [19, 20], enzymatic tagging [21], or single molecule biomarkers [22]) we restrict our discussion here to techniques that can extract mechanical data from single molecules, such as behavior under load (including effective stiffness, strength, toughness, etc.), deformation, and kinematics (*e.g.*, conformational changes, unfolding pathways, failure modes). For the sake of brevity, we limit our discussion to *optical tweezers*, a common single-molecule testing instrument, reflective of a general range of force and displacement resolutions required at the single molecule scale.

Optical tweezers (originally called "single-beam gradient optical trap") is an experimental approach developed to investigate single biomolecule mechanics based on controlled optical displacement—exploiting the forces of laser radiation pressure to trap small particles that are tethered to a single biomolecule [23, 24], or larger structures, such as bacteria and viruses [25].

Optical tweezers (OT) rely on the conservation of photon momentum, and use a highly focused laser beam to provide an attractive or repulsive force (typically on the order of piconewtons), depending on the refractive index mismatch to physically hold and move microscopic dielectric objects (the tethered particle, and not the biomolecule itself). That is, when transmitted through a dielectric object of high refractive index and of radius much less than the laser wavelength, the photons from a focused laser beam are subject to a change in momentum as a consequence of refraction when the beam enters and exits the object. This change exerts a restoring force $F_{OT}$ on the object toward the known and controllable center of the photon intensity gradient (the focal point of the laser beam). As a result, the dielectric object moves toward the laser focal point under a force $F_{OT}$, and can thus be used to impose force through translation of the focal point or to measure externally applied force through the resulting motion of the trapped object:

$$F_{OT} = k_{OT} \cdot x \tag{5.1}$$

where $k_{OT}$ is the optical trap stiffness and $x$ is the imposed bead displacement, measured *via* image analysis or optical interference patterns, for example. Optical tweezers are very sensitive instruments and are capable of the manipulation and detection of sub-nanometer displacements for sub-micrometer dielectric particles [26]. Note that the scattering force utilized in OT is the difference between the attractive force due to refraction and the repulsive force due to reflection. The value of $k_{OT}$ can therefore be tailored by varying the trap design and is on the order of approximately 50 fN/nm [27]. Further stabilization of the optical trap can be attained *via* the use of two opposing, focused lasers (dual beam), such that reflective forces of the lasers cancel out. In experiments on single molecules, forces on the order of 1–10 pN are sufficient to induce deformation [5].

Due to its high accuracy in force measurement, OT has been used extensively in single biomolecule studies [29]. Upon laser entrapment, the forces and displacements involved can be measured, and the interaction can be perturbed mechanically by moving the trap. A frequent and useful application of OT is the simple stretching of a biomolecule, such as DNA [23]. For DNA, mechanical characterization is necessary, as the flexibility of DNA plays a key role in all of its cellular functions, including folding, packaging, regulation, recombination, replication, and transcription. For proteins, this can be thought of as probing "protein unfolding" and shed

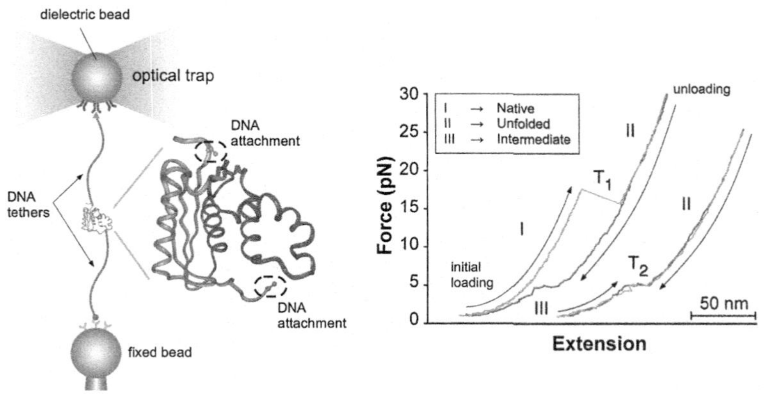


**Fig. 5.1** Protein "unfolding" through optical tweezers. *Left*: Experimental setup. The *Escherichia coli* ribonuclease H (RNase H) contains two unique cysteines suitable for DNA attachment (indicated). The sample was tethered to functionalized beads present at the ends of the DNA linker. One bead is held in place at the end of a micropipette by suction (considered fixed), while the other is held by the optical trap. *Right*: Stretching and relaxation force-extension curves from RNase H display high- and low-force unfolding transitions ($T_1$ and $T_2$ respectively), and three distinct states (I, II, and III). Refolding transitions were observed at the same forces as the low-force unfolding (curves shifted for clarity). Upon initial loading, the native folded state (I) undergoes a transition, $T_1$ to an unfolded state (II) at high force. Upon unloading, the unfolded state reverts to an intermediate state (III). If loading occurs without sufficient time to refold, the II ↔ III transition, $T_2$, is repeated under cyclic loading conditions. Adapted from Cecconi *et al.* [28], used with permission, copyright © 2005 AAAS

light on the transition states of a naturally folding protein molecule. For example, a recent study used optical tweezers to induce complete mechanical unfolding and refolding of individual *Escherichia coli* ribonuclease H (RNase H) molecules [28]. It was previously hypothesized that RNase H folds *via* a hierarchical mechanism, in which the most stable regions of the native structure form first during folding [30, 31]. Using the force-displacement data, it was shown that the protein unfolds in a two-state manner and refolds through an intermediate that correlates with the transient molten globule-like intermediate observed in bulk studies (see Fig. 5.1). The presence of this intermediate displayed unusual mechanical compliance and unfolded at substantially lower forces than the native state. In a narrow range of forces, the molecule hops between the unfolded and intermediate states in real time. The low-force capabilities of OT allow the direct monitoring of the refolding process as well as the equilibrium between the folded and unfolded states. Such studies allow the mapping of energy landscapes of proteins such as RNase H.

On of the most intensive applications of optical tweezers have been studies on the linear motor protein kinesin that transports vesicles and moves along microtubules for up to several nanometers before dissociating [29, 32]. Kinesin represents a type of molecular motor—or mechanoenzyme—capable of converting chemical energy into mechanical work. In muscle fibers, millions of myosin molecules function during contraction, while bending in eukaryotic flagella is produced by thousands of dynein molecules. Such enormous numbers complicate interpretation in terms of

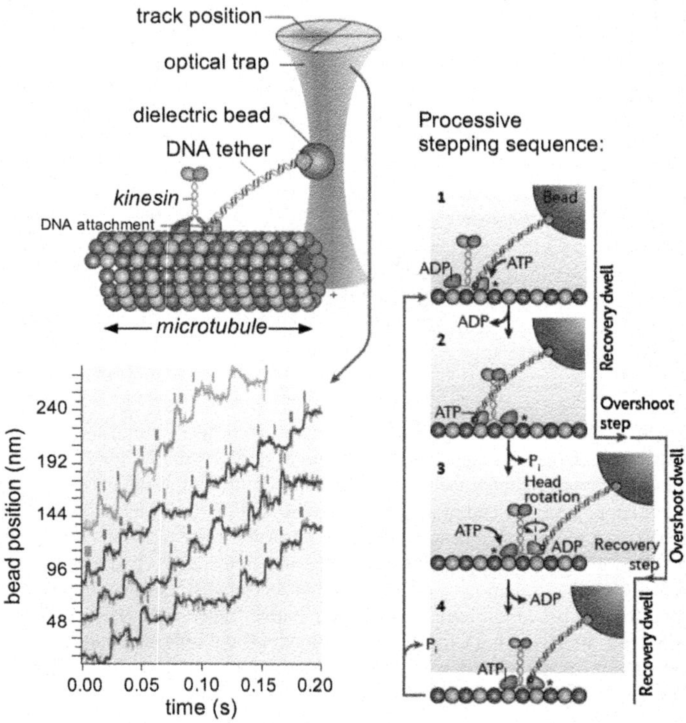


**Fig. 5.2** Optical tweezers to study movement of molecule motors. A single optical trap is shown, to study the "processive" stepping of kinesin, with a load applied to one of the kinesin heads. One head is tethered by a DNA linker to a dielectric bead held in an optical trap. The displacement traces (recorded with an assisting load of ≈2 pN) show the position of the trapped bead and the 8 nm stepping pattern of the molecule. Traces can be interpreted with the help of the schematic sequence of motor states shown in steps **1–4** (*right*). The starting position (step **1**) shows the tethered head without nucleotide (indicated by an asterisk) and strongly bound to the microtubule, just before the binding of ATP. ATP binding causes neck-linker docking and brings the ADP-bound trailing head forward. ADP release from this head follows after binding to the microtubule (step **2**). This all occurs without bead movement. Detachment of the tethered head from the microtubule and its forward movement then actually generates bead displacement 'overshoots' (step **3**), marked by vertical bars above the traces, before the head rebinds to the microtubule in the recovery step (step **4**), marked again by bars above the traces. This demonstrates the "hand-over-hand" stepping mechanism of kinesin. Adapted from [33], used with permission, copyright © 2011 Macmillan Publishers Limited

molecular mechanochemistry, but single molecular assays can shed insight into the process, under relatively well-defined conditions, free of the complex regulatory elements typically present *in vivo*.

Measuring the displacement of kinesin along a microtubule indicated the kinesin molecule moved along a microtubule with regular 8 nm steps (see Fig. 5.2), coinciding with the periodic spacing of an associated protein dimer, and clearly differentiating from steady Brownian motion (*e.g.*, motion due to diffusion) at low adenosine triphosphate (ATP) concentration or at high load [29, 32]. Using OT, other single-molecule studies of kinesin and other mechanoenzymes have substantially improved our understanding of how linear molecular motors function and how the mechanical movement is related to the structural features and chemical reactions [33].

From the displacement data, a relation can be formulated between the reaction rates (ATP binding and dissociating with kinesin, or $k_a$ and $k_b$ respectively), the "step" size ($d \approx 8$ nm), the applied load ($L$), and the velocity of the "motor", $v_{kinesin}$:

$$v_{kinesin} = \frac{d}{\beta_L \cdot [(k_b)^{-1} + (k_a \cdot c)^{-1}]} \tag{5.2}$$

where $c$ is the concentration of ATP, and $\beta_L$ is a fitted load-dependent factor [32]. This exemplifies the inherent coupling of chemical environment (*e.g.*, concentration of ATP) and mechanical conditions (*e.g.*, applied load, $L$). Thus, through constant applied force of the OT, chemical and mechanical information can be aptly linked.

While the most obvious application of OT involve the tethering and direct stretching of molecules for various gains (such as unfolding pathways, and molecular speed), clever design of the experiment can extract more indirect behaviors and properties. A recent study investigated the formation of actin bundles—relevant to cell motility—*via* an OT approach [24]. Cell motility involves the formation of membrane protrusions, which test the extracellular environment, attach to surfaces, and finally retract, pulling the whole cell forward [34, 35]. A proposed mechanism of protrusion formation is based on the formation of actin bundles, which are built up by two or more actin filaments attaching to each other [36, 37]. The goal was to quantify the fundamental interaction between two individual actin filaments forming the smallest possible bundle.

Applying holographic optical tweezers, analysis of the bundling mechanism and measured applied boundary conditions enabled the calculation of the forces between two actin filaments during bundle formation. The proposed mechanism works in a splayed configuration of the filaments placed in front of an obstacle, *e.g.*, the cell membrane (Fig. 5.3). An attractive interaction, originating from depletion forces or molecular interactions with actin binding proteins, results in a bundling force between the two filaments. This force acts perpendicular to the bundling axis, the axis defined by the direction of the already bound filaments. The position at which the filaments come together is moving parallel to the bundling axis toward the free end of the filaments, in the direction of the obstacle (a kind of "zipping" mechanism [36]). During this process the bending of the still free ends of the filaments increases in a buckling manner. Thus, the filaments transduce the bundling force exerted by the attractive interaction toward the obstacle through a buckling force acting parallel to the bundling axis. The forces were found to be dependent on the concentration of bundling agents ($Mg^{2+}$), indicating a possible cellular trigger. In this manner, the OT acted as a sophisticated "load cell", returning the reaction forces and displacements, due to the incurred attractive bundling force.

Beyond single molecules, optical tweezers have also been used to probe the cytoskeleton, measure the visco-elastic properties of biopolymers, and study cell

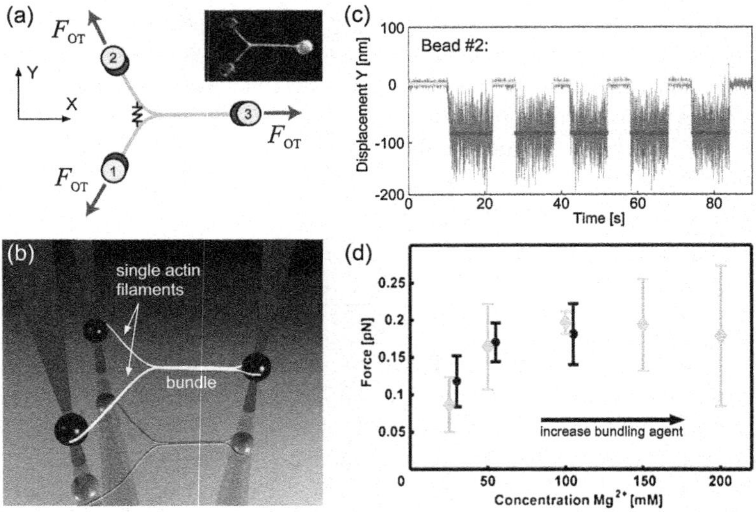


**Fig. 5.3** Optical tweezers are more than just pulling: investigation of actin bundle formation. (**a**) Schematic drawing of the experimental optical trapping geometry in steady state for actin filaments undergoing bundling. Two single actin filaments are attached to three optically trapped microbeads. Bundling is induced by moving the microbead/actin system to a microchannel filled with a buffer containing divalent ions and progresses until the bundling force is balanced by the optical trapping force (fluorescence image of filaments and beads; *inset*). The spring illustrates the foremost binding site which is exerting the bundling force determined from the $y$-components of the optical forces. (**b**) 3D rendering of suspended arrangement in microchannel. (**c**) Steady state, cyclic decrease and increase of the trap stiffness and determination of the displacement of the microbeads. The data show a typical displacement in $y$-direction of microbead #2 during the measurement. (**d**) Forces acting between two single actin filaments determined in steady state and during dynamic bundle formation, as a function of bundling agent, $Mg^{2+}$ (Adapted from [24], used with permission, copyright © 2011 American Chemical Society)

motility. The OT can manipulate relatively large biological samples such as individual cells, as the sample size is determined by the bead diameter (one to several μm sizes are commercially available) or the spot size of the focused laser beam (500 nm or more). OT is useful when low forces and high resolution is desired. A further advantage is that the experimental conditions enable experiments on living cells. The force imposed *via* OT can be as large as approximately 0.50 nN, with force resolution better than 1 pN. Optical tweezers have also been used to impose stretching forces at more than two locations on the cell membrane by proper adhesion of multiple beads at predetermined sites on the cell membrane, in order to approximate loading states such as equibiaxial tension, or by multiplication of a single OT *via* computer generated holograms [38].

A variation on the OT scheme is a magnetic trap or tweezers (MT) [39, 40]. Similar to laser entrapment, here, magnetic beads serve as "grips" which, under electromagnetic field gradients that impose a local magnetic force on these beads, impose displacement. These beads can be attached through specific chemical functionalization to one end of nucleic acids [39, 40] or biomolecules [41], or attached through ligand-specific or aspecific binding to a cell surface for single cell studies [42, 43]. The magnitude of the force is controlled directly by and calibrated according to field intensity (*i.e.*, electromagnetic coil current), and the bead position is maintained in a feedback loop with real-time image analysis, effectively trapping the bead in a potential well. Two advantages of this approach over OT are that out-of-plane rotations and thus torque can be considered (as the magnetic field contains an angular component), and that the potential radiation damage to the molecule is eliminated. Currently, the vertical forces attained by MT are of the highest resolution currently available (10 fN), making it possible to conduct elegant experiments at the single molecule level. However, due to the requirement of real-time image analysis to measure and control bead displacement, the present stiffness of the MT is two orders of magnitude smaller than that of the OT, the maximum vertical/horizontal forces that can be imposed are on the order of 10 pN—at least an order of magnitude smaller than those achievable *via* OT.

If higher forces and displacements are required—either for large biomolecules, inducing failure, or assays at the cellular level—both OT and MT are insufficient. More suitable is perhaps the most ubiquitous experimental approach currently implemented for biological materials: *atomic force microscopy*.

## 5.5 Atomic Force Microscopy: The Molecular Skeleton Key

Atomic force microscopy (AFM) is a specific example of scanning probe microscopy,[1] and was an adaptation of a scanning tunneling microscope that formed images through tunneling current between a probe and a conductive sample [44]. AFM is one of the foremost tools for imaging, measuring, and manipulating matter at the nanoscale, providing piconewton force sensitivity and better than nanometer spatial resolution. In addition, the forces experienced and generated by biological macromolecules can range from the sub-piconewton range up to several nanonewtons [45]—a range accessible by AFM. Information is gathered by "feeling" the surface with a mechanical probe. The AFM consists of a cantilever with a sharp tip (probe) at its end that is used to scan the specimen surface. As the distance between the probe tip and substrate increases, any applied load generates a restoring force that causes the cantilever to bend. This causes deflection of a laser beam directed toward the upper surface of the cantilever, which is measured using a photodetector. The output of the photodetector can be related to the angle of the cantilever and therefore to the applied force, if the elastic properties of the cantilever are known (see Fig. 5.4). The cantilever is typically silicon or silicon nitride with a tip radius of curvature on the order of nanometers [46]. When the tip is brought into proximity

[1] A branch of microscopy that forms images of surfaces using a physical probe that scans the specimen. An image of the surface is obtained by mechanically moving the probe line by line and recording the probe-surface interaction as a function of position.

of a sample surface, forces between the tip and the sample lead to a deflection of the cantilever, again described by the simple relation:

$$F_{AFM} = k_{AFM} \cdot \Delta \tag{5.3}$$

In the simplest form, the effective stiffness, $k_{AFM}$, can be theoretically evaluated considering the cantilever deflection of the AFM tip, $\Delta$, described by classical Euler-Bernoulli beam theory (relating load at the edge of a cantilever beam to maximum tip displacement), such that:

$$k_{AFM} = \frac{3EI}{L^3} \tag{5.4}$$

The stiffness can then be easily manipulated by simple variation in cantilever length, enabling a wide range stiffness with the same probe design. If the cross-section of the AFM cantilever is known, the stiffness can be explicitly calculated [46], such as $k_{AFM} = Ewt^3/4L^3$, for the case of a rectangular cross section of dimensions $w \times t$. In practice, the stiffness of the cantilever is explicitly calibrated to account for any variation in the experimental conditions and environment, such as direct calibration *via* deflection of the cantilever against the intended rigid substrate (*e.g.*, directly measuring applied force and displacement of the cantilever only). At least a dozen different calibration methods can be found in the literature (see work of Tendler *et al.* [47] and the references therein), ranging from elastic formulations of the resonant frequency [48–50], to methods based on the acquisition of the cantilever's thermal distribution spectrum (square of the fluctuations in amplitude as a function of temperature) [51, 52]. For example, assuming a rectangular cross-section, the stiffness can also be expressed as a function of resonant frequency [48]:

$$k_{AFM} = 2\omega(\pi L \nu_k)^3 \sqrt{\frac{\rho^3}{E}} \tag{5.5}$$

where $\omega$ represents the cantilever's width, $L$ its length, $\nu_k$ its resonant frequency, $\rho$ its density and $E$ its elastic modulus. Considering the thermal distribution spectrum, stiffness can be expressed *via* the equipartition theorem (elastic energy must be equal to the thermal energy) for a simple harmonic oscillator [51, 52]:

$$\frac{1}{2}k_B T = \frac{1}{2}k_{AFM}\sqrt{\langle x^2 \rangle} \tag{5.6}$$

where $\langle x^2 \rangle$ is the mean square cantilever tip displacement, $k_B$ is Boltzmann's constant and $T$ is the temperature. Regardless of calibration method implemented, the spring constant $k_{AFM}$ of the AFM cantilever determines the maximum force attainable, and the thermal fluctuations of the cantilever determine the resolution of the instrument. Note that the signal-to-noise ratio does not improve as stiffness decreases, such that softer cantilevers do not provide increased force resolution. As different geometries and base materials are readily available (*e.g.*, rectangular of V-shaped as depicted in Fig. 5.4), $k_{AFM}$ can range from 0.003–10 N/m, such that pN-scale forces and nm-scale deflections are attainable, and the contact-based, force-controlled image can indicate the structure of even living cellular surfaces [53–55].

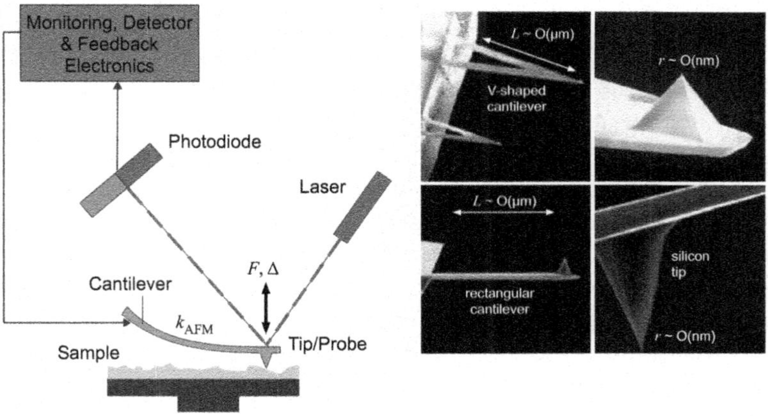


**Fig. 5.4** Schematic of atomic force microscopy (AFM). AFM consists of a cantilever with a sharp tip (probe) at its end that is used to scan the specimen surface. The cantilever is typically silicon or silicon nitride with a tip radius of curvature on the order of nanometers and a cantilever length on the order of micrometers (SEM images depicted right, from [56], used with permission, © 2005 IOP Publishing). When the tip is brought into proximity of a sample surface, forces ($F$) between the tip and the sample lead to a deflection ($\Delta$) of the cantilever with known stiffness, $k_{AFM}$. Depending on the situation, forces that are measured in AFM include mechanical contact force, van der Waals forces, capillary forces, chemical bonding, electrostatic forces, magnetic forces, solvation forces, etc. Along with force, additional quantities may simultaneously be measured through the use of specialized types of probe. Typically, the deflection is measured using a laser spot reflected from the top surface of the cantilever into an array of photodiodes and actively controlled *via* feedback electronics

The most common application of AFM on biomaterials is high-resolution imaging. In AFM, the images are not generated *via* reflection or transmission of photons or electrons, but rather are created *via* local application of mechanical force across a sample surface in one or more spatial dimensions. During a scan, the cantilever is "dragged" across the surface of the sample and the contours of the surface are measured directly using the deflection of the cantilever, exploiting short range repulsive forces and long range attractive forces between the probe tip and the sample to image the surface topography. In the most typical AFM configuration (Fig. 5.4), force ($10^{-12}$ to $10^{-8}$ N) is applied to the sample surface through a sharp tip positioned normal to the free end of the flexible cantilever. Through feedback with the cantilever base transducer, contact mode AFM imaging is attained by maintaining a known voltage—and thus tip deflection and contact force—at a fixed value as the cantilever tip scans the sample surface. The resulting displacement of the cantilever base comprises the height map, or gross features of the surface topography. Since this electronic feedback is not ideal, there are small differences between the desired and the actual tip deflection that comprise an additional image (the error signal image). Tip deflection can be converted to force $F_{AFM}$ *via* Eq. 5.3, *via* the calibrated cantilever stiffness, $k_{AFM}$. Frequently, high resolution imaging is also obtained in the so-called tapping mode, where the AFM cantilever oscillates near its resonance frequency for enhanced sensitivity. High resolution images of many biomaterials such as individual collagen fibrils, cortical bone, living cells, and various proteins have been successfully demonstrated.

While imaging is inherently beneficial to determine structural and topological characteristics, it can also be implemented to deduce mechanical information. For example, the geometric analysis of adhered carbon nanotubes can be used to determine the adhesion energy, *via* analysis of transmission electron microscopy (TEM) images [57], and elastic and interfacial properties of other filamentous assemblies through the analysis of imaged kinks, helices, and rings [58]. A recent study by Mezzenga, Dietler and co-workers [59] implemented single-molecule AFM images of $\beta$-lactoglobulin fibrils (representative of the "misfolding" of globular proteins, a precursor to amyloid aggregation linked to neurological disorders such as Alzheimer's and Huntington's diseases, discussed further in Chap. 9: Pathological Materiomics).

Despite the fact that the individual parameters controlling the aggregation of globular proteins into amyloid fibrils have been well identified, the driving force for such an aggregation process still remains obscure, as well as the possible existence of transition states from protofilament to plaque. Mezzenga, Dietler *et al.* used well-known statistical theory of semi-flexible polymers with simple scaling arguments combined with AFM imaging to elucidate hierarchical structures and aggregation behavior (Fig. 5.5). The AFM images of $\beta$-lactoglobulin fibrils depicted several micrometer-long semiflexible fibrils (see Fig. 5.5(a)), in agreement with previous literature reports [60]. By performing a statistical analysis on AFM images, both contour length and cross-section distribution were able to be determined (see Fig. 5.5(c), for example).

Further, a periodic height fluctuation along the contour of the fibrils was observed, suggesting possible twist. Amyloid fibrils obtained by different aggregation mechanisms and from other proteins do show evidence of helical structures formed by twisting of single filaments around each other [61, 62]. Figure 5.5(b) presents AFM images of fibrils with two different pitches (75 and 100 nm), corresponding to fibril maximum heights of 6 and 8 nm, respectively. The periodicity was found to increase linearly with the height of the fibrils—explained by a proposed hierarchical multi-stranded self-assembly scheme. To further reinforce this self-assembly picture of the aggregation of single filaments into different classes of multi-stranded helical fibrils, the evolution of persistence length with fibril height was analyzed. The persistence length $l_p$ was determined directly from the AFM images by means of the bond correlation function for semiflexible polymers in a two-dimensional conformation:

$$\langle \cos\theta(s) \rangle \cong e^{-\frac{s}{2l_p}} \tag{5.7}$$

where $\theta$ is the angle between the tangent vectors to the chain at two points separated by a contour distance $s$ (Fig. 5.5(d)) and the factor 2 is used to re-scale the exponential decay to account for the two-dimensional nature of protein fibrils absorbed on a substrate [63]. The persistence length $l_p$ is the length above which thermal

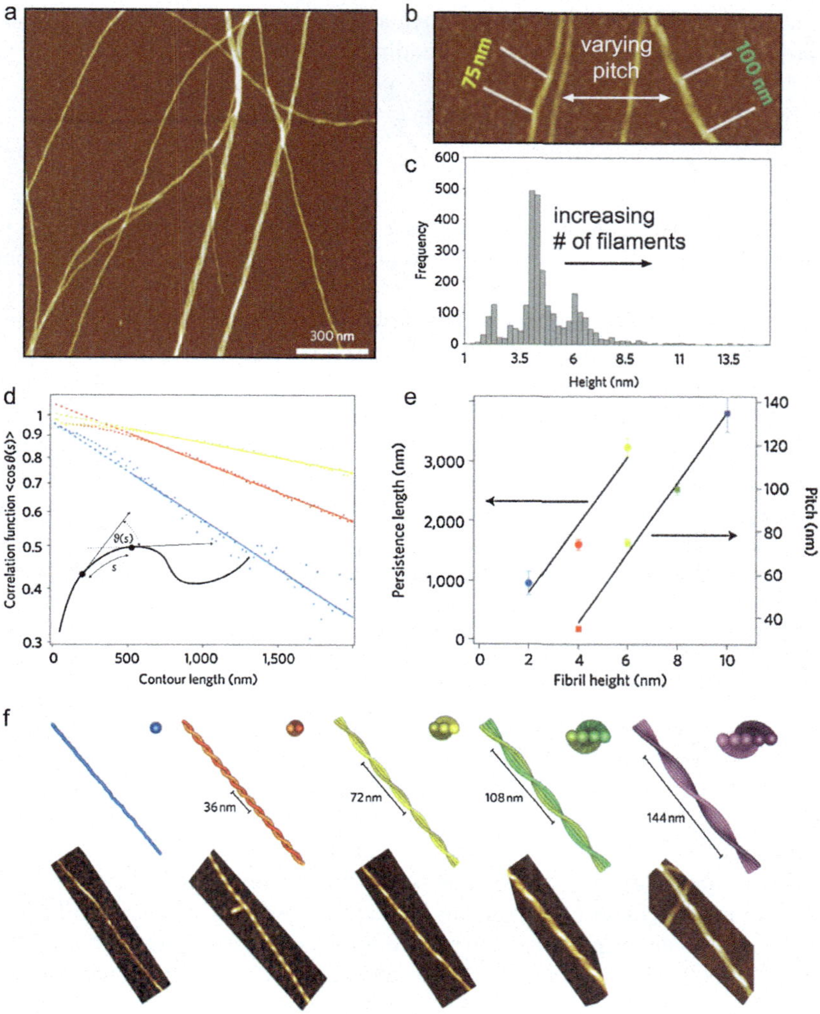


**Fig. 5.5** Mechanical characterization of amyloid fibrils from AFM imaging. (**a**) Tapping-mode AFM height images of $\beta$-lactoglobulin amyloid fibrils. (**b**) Different periods of $\beta$-lactoglobulin fibrils, depicting AFM height image of fibrils with a period of 75 nm (for fibrils with a maximum height of 6 nm) and 100 nm (maximum height, 8 nm). (**c**) Histogram of $\beta$-lactoglobulin fibril maximum heights with three major peaks corresponding to maximum measured heights of approximately 2, 4 and 6 nm, indicating an increase in the number of adhered filaments. (**d**) Decay of $\langle \cos\theta(s) \rangle$ as a function of contour length $s$ for filaments with three varying height ranges: 1.5 to 2.5 nm, 3.5 to 4.5 nm, and 5.5 to 6.4 nm. Fitting with a single exponential decay yields $l_p \cong 968$ nm, 1,620 nm and 3,244 nm respectively. *Inset*: Sketch of a semiflexible fibril chain with contour distance $s$ and angle $\theta$(s) among rigid fibril segments, used to determine the bond correlation function. (**e**) Persistence length and pitch of fibrils plotted against fibril cross-section for the populations of fibrils with varying heights, indicating linear relations of both. (**f**) AFM images and corresponding reconstructions of left-handed helical fibril formation from the twisting of multi-stranded ribbons, with the number of filaments ranging between 1 and 5. Adapted from Mezzenga, Dietler *et al.* [59], used with permission, copyright © 2010 Macmillan Publishers Limited

fluctuations can buckle a rod-like polymer, and it is therefore a measure of the elastic properties of a molecule [64]. The bending rigidity, $EI$ can be related to $l_p$ by:

$$l_p \cong \frac{EI}{k_B T} \tag{5.8}$$

where $E$ is the apparent Young's modulus of the fibril, $I$ the area moment of inertia of the cross-section, $k_B$ the Boltzmann constant, and $T$ the temperature. The statistical analysis clearly indicates a linear increase of the persistence length with fibril cross-section (Fig. 5.5(e)). But what indicates a multi-stranded ribbon configuration?

For a multi-stranded fibril formed by a twisting ribbon of $n$ filaments, the cross-section can be approximated by a rectangle of height $2r_0$ and width $n2r_0$, where $r_0$ is the radius of the single filament. In this case, the observable radius of the fibril, and thus the maximum height attained by AFM, are simply proportional to the number of filaments ($r \approx nr_0$). Furthermore, because the geometry is no longer axisymmetric, *two* area moments of inertia exist, scaling respectively as $I_1 \sim r_0 \cdot (nr_0)^3$ and $I_2 \sim (r_0)^3 \cdot nr_0$. Because the persistence length is proportional to the energy needed to bend the fibril, it essentially reflects the *lowest* energy bending mode, that is, the one with the lowest area moment of inertia, $I_2$, leading to a scaling behavior of the persistence length of the form $l_p \sim nr_0^4 E/k_B T$. Thus, for a multi-stranded helical fibril with twisting ribbon-like cross-section, both the scaling behavior of the maximum height and the persistence length have a linear relationship with the number of filaments in the fibril, as shown in Fig. 5.5(e).

The AFM analysis, through imaging alone, supported by theoretical arguments, revealed that the fibrils have a multi-stranded helical shape with twisted ribbon-like structures (Fig. 5.5(f)). Such results indicate a possible general model for amyloid fibril assembly, and illustrate the potential of this approach for investigating other fibrillar systems, exploiting the statistical sample provided by AFM.

Another major application of AFM (other than imaging) is *force spectroscopy*, the direct measurement of tip-sample interaction forces as a function of the gap between the tip and sample (the direct result of this measurement is the AFM tip force-distance, or F-D, curve). For this method, the AFM tip is extended towards and retracted from the surface as the deflection of the cantilever is monitored as a function of piezoelectric displacement. These measurements have been used to measure nanoscale contacts, atomic bonding, van der Waals forces, dissolution forces in liquids and single molecule stretching and rupture forces [65]. AFM can thus be utilized to study single-molecule biomechanics, including the unbinding of antibody-antigen, ligand-receptor, and DNA-protein pairs, and the folding and unfolding of proteins [66–69]. In such a mode, a single molecule can be stretched and relaxed between the probe tip and an anchoring substrate. Again, we see that understanding the mechanisms determining the folding, stability and function of proteins is one of the fundamental challenges in life sciences—approached by both computational and experimental means. Figure 5.6 shows an example of a such experiment for the protein ubiquitin.

Analysis of such force-displacement curves not only provides information about the strength of interactions established but also about the location of force-induced

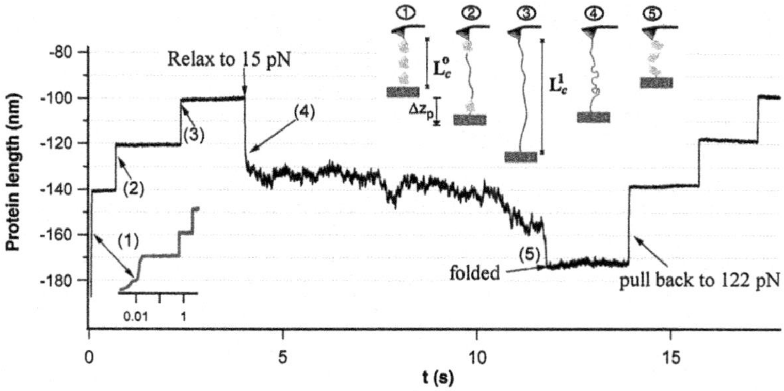


**Fig. 5.6** The unfolding and folding pathway of ubiquitin *via* AFM force spectroscopy. Plot displays the end-to-end length of a protein as a function of time. The *inset* shows a schematic of the events that occur at different times during the stretch-relaxation cycle (numbered from **1** to **5**), where $\Delta z_p$ is piezoelectric actuator displacement and $L_c$ the contour length. The length of the protein (in nanometers) evolves in time as it first extends by unfolding at a constant stretching force (≈122 pN). This stage is characterized by step increases in length of 20 nm each, marking each ubiquitin unfolding event, numbered **1** to **3**. The first unfolding event (**1**) occurred very close to the beginning of the recording and therefore it is magnified with a logarithmic time scale (*inset*) with the length dimension plotted at half scale. Upon reducing the force (to ≈15 pN), the protein spontaneously contracted, first in a steplike manner resulting from the elastic recoil of the unfolded polymer (**4**), and then by a continuous collapse as the protein folds (**5**). The complex time course of this collapse in the protein's length reflects the folding trajectory of ubiquitin at a low stretching force. To confirm that the polyubiquitin had indeed folded, it was stretched again, repeating the initial steplike unfolding extension. From [69], used with permission, copyright © 2004 American Association for the Advancement of Science

structural transitions within the molecule. As discussed in Chap. 3: The Challenges of Biological Materials, when a polymer is relaxed, it forms a coiled structure because this maximizes the entropy of its segments ($TS \gg U$). Extension of the polymer generates an opposing force due to the reduction in entropy ($U \gg TS$). This phenomenon, referred to as *entropic elasticity*, suggests that small extensions require little force but that the resistance to extension rises rapidly as the polymer approaches its full length. To account for this effect, the behavior of polymers under mechanical stress can be described by the worm-like chain (WLC) model of elasticity [70, 71]. This model describes a polypeptide as a continuous string of a given total (or contour) length, and persistence length (reflective of polymer flexibility). The smaller the persistence length, the greater the entropy of the polymer and the greater the resistance to extension (due to the "hidden length"). Peaks in the AFM F-D curve can be fitted using the WLC model, which is accurate up to stretching forces of several hundred piconewtons [23, 66, 71]:

$$F(x) = \frac{k_B T}{l_p}\left[\frac{1}{4}\left(1 - \frac{x}{l_c}\right)^{-2} - \frac{1}{4} + \frac{x}{l_c}\right] \tag{5.9}$$

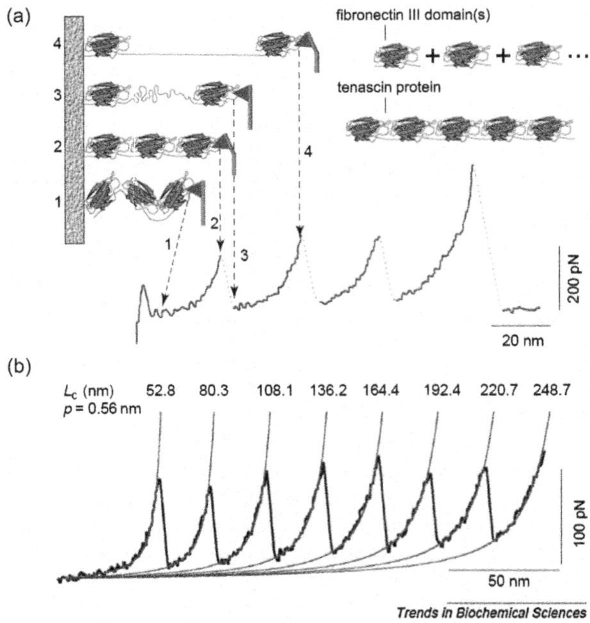


**Fig. 5.7** The entropic elasticity of proteins and domain unfolding. The entropic elasticity of proteins can be described by the WLC (worm-like chain) model (Eq. 5.9). (**a**) AFM force spectroscopy of the extra-cellular matrix protein tenascin, composed of multiple repeats of the fibronectin III (FN-III) domain (*inset*). The saw-tooth pattern of peaks that is observed when force is applied to extend the protein corresponds to sequential unravelling of individual domains of a modular protein. As the distance between substrate and cantilever increases (from state **1** to state **2**) the protein elongates, generating a restoring force that bends the cantilever. When a domain unfolds (state **3**) the free length of the protein increases, drastically decreasing the force on the cantilever. Further extension again results in force on the cantilever (state **4**). (**b**) Consecutive unfolding peaks of recombinant human tenascin were fitted using the WLC model. The contour length for each of the fits is shown, the persistence length was fixed at 0.56 nm, the mean force at which the domains unfold is 137 pN, and the mean interval between peaks is 24.8 ± 2.3 nm [72]. From [74], used with permission, copyright © 1999 Elsevier Science Ltd.

expressing the relationship between force ($F$) and extension ($x$) of a protein using its persistence length ($l_p$) and its contour length ($l_c$). Again, $k_B$ is Boltzmann's constant and $T$ is the absolute temperature.

But why are there multiple "peaks" in the F-D curve? This dramatic deviation from entropic elasticity is a result of the extension of *multidomain* proteins—a continuous polypeptide with localized units of structural order (*i.e.*, domains). The force-extension curves of these proteins show peaks that correspond to the unfolding of single domains (Fig. 5.7). As these proteins are elongated, the restoring force increases. At a certain force, however, one of the domains in the chain unfolds. Like the freeing of a tangle in a rope, this unraveling suddenly adds to the effective length of the protein and allows the force on the AFM cantilever to fall. Further extension is resisted again by entropic forces until a second domain in the chain unravels. The force-extension curve typically displays a characteristic saw-tooth pattern with the number of peaks corresponding to the number of domains stretched between the substrate and cantilever tip. This phenomenon was first demonstrated with titin [66] and later with the extra-cellular matrix protein tenascin [72, 73], which contains repeats of the fibronectin III (FN-III) domain (Fig. 5.7). The unfolding of each of the FN-III domains can be described accurately using the WLC model [72]. AFM measurements offer an opportunity to understand the characteristics that underlie the mechanical properties of proteins. Such force-extension curves, however, illustrates an important drawback in the use of native protein fragments for the study of mechanical properties. When pulling a heterogeneous multi-domain protein (*e.g.*, a sequence of mixed domains and/or structures), one usually cannot know which unfolding peak corresponds to which domain. The elastic properties of specific domains are therefore difficult, or impossible even, to identify. The AFM technique is also able to measure refolding of protein domains, as an extended protein can be relaxed by returning the substrate to its original position [69] (as depicted in Fig. 5.6). Subsequent re-extension of the protein demonstrates a recovery of folded domains that is dependent on the time interval between consecutive extensions [72].

Predictably, applying sufficiently high mechanical load by AFM causes proteins to unfold in subsequent steps, where each force peak in an F-D curve reflects an unfolding event. However, these experiments are far from *equilibrium* because the rapid increase in distance between the molecular anchor (on the substrate) and cantilever typically prevents rebinding or refolding of the protein. If we consider the folded protein structure as stochastic, then the probability that a domain will unfold is presumably dependent on the applied force, the extension required to break the bonds, and the rate at which the domain unfolds with no applied force. In a first approximation, unfolding of each structural segment can be considered as a two state process energetic process. A structural segment resides either in a low-energy conformation ($E_1$) representing the folded state or in a high-energy conformation ($E_2$), the unfolded state. An energy barrier has to be overcome when switching between both states ($E_b = E_2 - E_1$). Such a two state unfolding energy landscape can be characterized by the distance from the native to the transition state, $x_b$, and by the energy barrier, or the free energy of activation, at which a structural segment spontaneously unfolds in the absence of force. When axial force ($f$) is applied to a folded domain the protein will unravel. This transition can be described by an Arrhenius relation proposed by Bell in 1978, modified to account for force [75, 76]:

$$k = \omega_0 \exp\left(-\frac{E_b - f \cdot x_b}{k_B T}\right) \tag{5.10}$$

where $f$ is the applied force, $x_b$ is the distance over which the unfolding event occurs, $\omega_0$ is the rate constant (or diffuse relaxation frequency) in the absence of an applied force, and again, $k_B$ is Boltzmann's constant and $T$ is the absolute temperature. Thus, the rate at which protein unfolding occurs increases exponentially with the applied force. When $f \rightarrow 0$ this equation is similar to that describing the dissociation of non-covalent proposed by Bell [77] and describes the offrate, $k_0$. An externally applied force effectively lowers the unfolding energy barrier [75, 76]. Furthermore, the force required for unfolding depends on the rate at which force is applied, the so-called loading rate, $r_f = df/dt$. In general, unfolding of proteins at higher pulling velocities requires higher forces. Evans and Ritchie further showed [75, 78] that the most probable unfolding force, $f_r$, is a function of $\ln(r_f)$:

$$f_r = \frac{k_B T}{x_b} \ln\left(\frac{r_f x_b}{k_b T \cdot k_0}\right) \tag{5.11}$$

Here, we see that an increase in loading rate increases strength ($r_f \uparrow, f_r \uparrow$). This is a major issue in reconciling results of molecular simulation (limited to short timescales and thus high loading rates) with experimental results, further discussed in Chap. 7: Mechanical Characterization in Molecular Simulation. Force spectroscopy can be used to determine the most *probable* unfolding force through variation in different pulling velocities, *i.e.* loading rates. Through a spectrum of velocities (across an order of magnitude or more), fitting the force-rate behavior can yield both $x_b$, $k_0$ (Eq. 5.11), and ultimately $E_b$ (Eq. 5.10). These parameters can characterize the energy barrier underlying the unfolding reaction. It is noted that a two-state energy landscape is a gross simplification—macromolecular complexes are stabilized by a huge number of different interactions. Consequently, the energy landscape is rough with energy barriers of different heights and widths [79–82], and such approaches merely homogenize the prominent energy barriers along the reaction coordinate, which is given by the stressing force [76, 83].

Extensions of these AFM capabilities include the culture of living cells on the AFM tip to study cell-cell and cell-substrate adhesion [84, 85] and chemical functionalization of the AFM tip to facilitate chemomechanical interaction with cell surface receptors [86], specific to the cantilever material and molecule of interest. A broad spectrum of biological processes requires controlled cell adhesion, including embryonic development, assembly of tissues and the nervous system, cellular communication, inflammation and wound healing, tumor metastasis, cell culturing, and viral and bacterial infection. By promoting adhesion between an AFM cantilever tip and the sample surface, pN- to nN-scale tensile forces on specific surface features. Also referred to as single-cell force spectroscopy (SCFS), not only can overall cell adhesion be studied, but also the properties of single adhesion-receptor–ligand interactions [87]. Probing cell adhesion, the resulting SCFS force-displacement curves can capture a multitude of events and mechanisms, as depicted in Fig. 5.8.

SCFS is not without limitations. The fact that SCFS uses a cell as a probe can also complicate certain aspects of single-molecule measurements. In contrast to the rigid tip of the AFM that is used in single molecule assays, the applied forces in SCFS cause the cell to stretch and deform. In addition, because the mechanical response of cells to deformation is not necessarily linear, dynamic SCFS measurements are not as straightforward as single molecule measurements [89]. Perhaps more crucial is the multitude of possible specific and unspecific cell-surface interactions, extraneous to the receptor and ligand of interest. If progress can be made to convert the important qualitative findings and force displacement data obtained *via* AFM into quantitative measurements of useful mechanical parameters, this approach can be used to probe the role and mechanical properties of these cytoskeletal components—data upon which many analytical and computational modeling efforts rely.

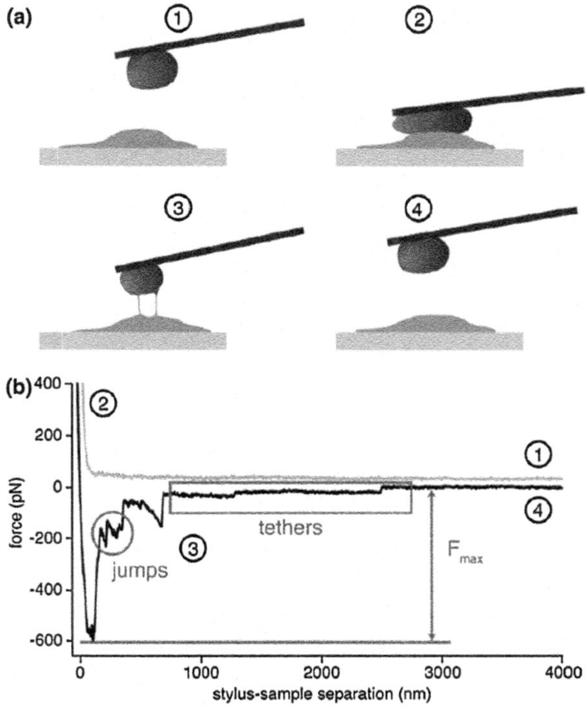


**Fig. 5.8** Principle of single-cell force spectroscopy. A living cell is attached to a functionalized cantilever (**1**) and lowered toward the substrate, commonly another cell (**2**). After reaching a certain contact force, the cell is kept in contact with the substrate for a predefined time to adhere before the cantilever is retracted at from the substrate (**3**) until substrate and cantilever-bound cell are fully separated (**4**). The resulting force-displacement data recorded during approach and retraction of the cantilever-bound cell depict unbinding events. As cell adhesion is a complex phenomenon involving a large number of molecules, the curves show an intricate spectrum of detachment events, classified as jumps and tethers (from attached molecules) and can be assigned in the retraction. Moreover, the maximum attachment force, $F_{max}$, can be determined. Adapted from Bippes and Muller [88], used with permission, copyright © 2011 IOP Publishing

AFM has been exploited as a research tool by the biophysics community because this technique affords Ångstrom-scale positioning accuracy, the ability to both image and mechanically manipulate a single biological structure with better than nm/pN resolution, and the potential to track biological processes in near physiological environments over time. Depending on the situation, forces that are measured in AFM include mechanical contact force, van der Waals forces, capillary forces, chemical bonding, electrostatic forces, magnetic forces, solvation forces, etc. It is important to note that, due to the contact-based nature of AFM, its use in cell mechanics is restricted to cells that can adhere tightly to a substrate over the course of an experiment. Multipurpose AFM assays are achieved through modification of the cantilever (probe) in order to attain the desired function [90]. Applications of AFM in various forms range from basic imaging, nanomechanics, viscoelastic characterization [91, 92], and electromechanical (*e.g.*, piezoelectric) coupling [93], to name a few.

As discussed in Chap. 3: The Challenges of Biological Materials, recently it was discovered that the soft substrates with physiological-relevant stiffness could promote (or sometimes inhibit) a number of significant biological transformations in stem cell differentiation. For example, the stiffness of a tissue scaffold can determine the cellular proliferation—human mesenchymal stem cells could choose their lineage specification according to the varied matrix elasticity [94]. AFM is truly one of the ideal instruments to measure the mechanical stiffness of cell culture scaffolds [95], and potentially monitor the changes of some essential (and as-yet unidentified) mechanical phenotypes (*i.e.*, cell body and nuclear stiffness, surface adhesion, membrane topography, etc.) in real time. To understand the mechanism of cell-substrate mechanical interaction, the precise measurement of substrate mechanical properties is essential.

The ability to function under relatively small forces (essential for soft biological materials) as well as its extension to operate under liquid environment (compatible with physiological condition) renders AFM as a powerful and indispensable tool in bioscience. The sensing and detecting the biomolecular interactions at high resolution may find future applications in the early diagnosis of diseases such as cancers. Moreover, AFM can provide imaging resolutions (on the scale of nanometers) beyond that of the diffraction-limited resolution of the optical microscopes (approximately 200 nm), resulting in an indispensable *skeleton key*—fit for all jobs—in the biological tool-box. However, by design, AFM is still restricted in to nanoscale resolutions—bridging hierarchical scales requires other suitable tools.

## 5.6 Microscale Approaches: Cells and Tissues

Again, with a cornucopia of potential methods for probing biological material systems at the cellular and tissue scales, discussion is inherently limited to a select few cases. Some successful approaches include magnetic twisting cytometry [96–98], particle tracking and substrate deformation methods [99–101], and arrays of vertical microcantilevers [102–104]. Like optical tweezers and AFM, such methods apply either a direct load (twisting cytometry applies a rotational torque, for example) to induce a mechanical response, or provide a means to measure strain/deformation and indirectly determine mechanical forces (such as particle tracking within a cell or substrate). Vertical microcantilevers, for example, can be thought of passively employing many cantilever "probes" on the same cell (as opposed to a single, AFM tip). Since the stiffness of the cantilever is known (*via* beam theory approximations) the forces on the cell can be determined through image post-processing of the deflections of each individual cantilever. The net forces at the cell-substrate interface—the traction applied by the cell—can then be determined (see Fig. 5.9). Here, we focus on two common assays: (1) micropipette aspiration, as it is applicable to living cells, and; (2) nanoindentation of biological tissues, encroaching the realm of bulk homogeneous material behavior and continuum formulations.

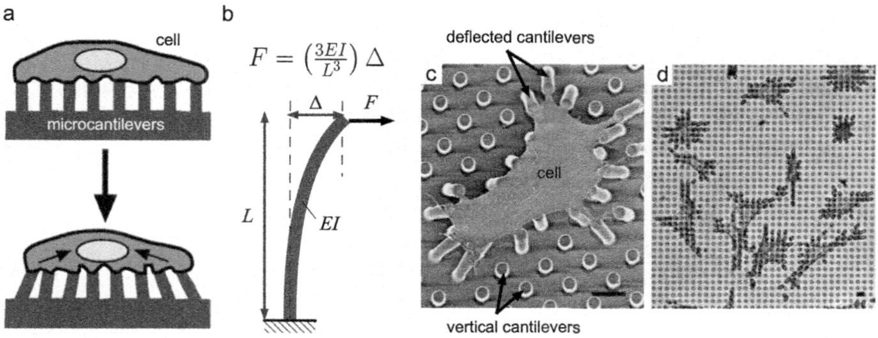


**Fig. 5.9** Principle of vertical microcantilever arrays. (**a**) With the appropriate surface density of vertical posts positioned on a substrate, a cell should spread across multiple posts as depicted. (**b**) Under the proper geometric constraints of post height and width (*e.g.*, constant $EI$, known $L$), cells exerting traction forces would deflect the posts as microcantilevers according to classic beam theory. (**c**) Scanning electron micrograph (SEM) and (**d**) Phase-contrast micrograph of representative smooth muscle cells attached to microcantilever array. The tips of the posts were functionalized with fibronectin to promote adhesion. Micrographs depict deflected cantilevers under traction force of the cell (scale bars indicate 10 μm). Adapted from [102], used with permission, copyright © 2003 National Academy of Sciences

The micropipette aspiration technique (MA) has been used widely to study time-dependent deformation of living, individual cells subjected to extracellular pressure [105]. Here, the cell is drawn into a glass tube (*e.g.*, pipette), the inner diameter of which is a chosen fraction of the nominal diameter of the cell, *via* stepwise application of aspiration pressure $\Delta P$ (*i.e.*, suction). The force $F_{MA}$ on a static cell in a micropipette is simply the suction pressure $P$ times the cross sectional area of the pipette, or $P\pi R^2$. However, if the cell is moving (traversing the pipette) due to the stretching of the microvillus, the force on the cell will be less than that exerted on the static cell. Its value is adjusted to account for the relative velocity of the cell [106]:

$$F_{MA} = \Delta P\pi R^2\left(1 - \frac{u_p}{u_a}\right) \tag{5.12}$$

where $R$ is the micropipette radius, and $u_p$ and $u_f$ are the velocities of the cell in the presence and absence of applied aspiration pressure $\Delta P$, respectively. Applied aspiration pressure ranges from 0.1–1000 Pa, with resolution of ≈0.1 Pa [105]. This pressure is maintained over a specified duration, and the attendant extension of the cell into the pipette is monitored *via* optical microscopy. The micropipettes and glass walls that define the fluid cell of the experiment are coated to inhibit cell adhesion. Thus, MA enables *real-time* correlation of pressure and whole-cell deformation.

Through application of a chosen viscoelastic model for the cell membrane, MA-induced deformation can be used to calculate elastic modulus $E$, apparent viscosity $\eta$ for the cell membrane and time constants of deformation and/or relaxation $\tau$. Selection of an appropriate viscoelastic model, however, is nontrivial. One must "smartly choose" an idealized representation. For example, when cells that behave similar to solids are aspirated into a micropipette, obviously they do not flow into the pipette when the aspiration length ($L_p$) exceeds the pipette radius ($R$). The aspiration length increases linearly with the suction pressure regardless of the value for $L_p$ as shown for chondrocytes [107] and endothelial cells [108], suggesting elastic-like behavior. In both cases the cells are spherical in shape, simplifying any geometric assumptions. Thus, a continuum description is formulated, assuming an infinite, homogeneous half-space drawn into a micropipette, resulting in:

$$\Delta P = \frac{2\pi}{3}\phi\left(\frac{L_p}{R}\right)E \tag{5.13}$$

where $E$ is the Young's modulus for the homogeneous solid, $\Delta P$ is again the suction pressure and $\phi$ is a term that depends weakly on the ratio of the thickness of the pipette wall to the radius of the pipette. From such formulations, the Young's modulus can be predicted (approximately 650 Pa for chondrocytes [107], for example). Of course, this presumes the cells behave as an infinite homogeneous half-space—prudence is necessary in such modeling judgements. For example, the above formulation neglects any time-dependent effects. Such experiments have been successfully extended to measure the viscoelastic behavior of the cells that flow and deform in narrow channels during physiological function, including erythrocytes (red blood cells), and granulocytes and neutrophils (two types of white blood cells) through the vasculature [105, 109].

More complex formulations are required to account for such time-dependent effects and properties such as viscosity ($\eta$) and relaxation time ($\tau$). A recent investigation [110] observed two distinct regimes during aspiration of biological "viscoelastic drops": a fast elastic deformation followed by a viscous flow (see Fig. 5.10). Cellular aggregates in solution round up to form "drops" in order to minimize their surface energy, similar to oil drops in water. A general model based on this viscoelastic behavior was formulated to deduce the surface tension ($\gamma$), viscosity ($\eta$), and Young's modulus ($E$). To combine the elastic and viscous regimes, a modified Maxwell model (depicted in Fig. 5.10(h)) was used, where the total displacement $L_p(t)$ is given by:

$$L_p(t) = \frac{F}{k_1}\left(1 - \frac{k_2}{k_1+k_2}e^{-t/\tau_c}\right) + \frac{F}{\xi_t}t \tag{5.14}$$

where $k_1 = \pi R_p E$ and $\xi_t = 3\pi^2\eta R_p$. The first term characterizes the elastic regime with $\tau_c = \xi_c(k_1+k_2)/(k_1k_2)$ being the raising time of the elastic deformation, $\delta$, and the second term characterizes the flow at constant velocity. The tissue relaxation time separating the elastic and viscous regimes is given by $\tau = \xi_t/k_1 = 3\pi\eta/E$. Using the values for $\Delta P$, the surface tension, $\gamma$, is derived from the Laplace law.

Such studies allow the determination of mechanical properties through "interpretation" of idealized theoretical models—it is apparent that the cellular aggregates are not composed of springs and dashpots of the Maxwell model. How the surface tension and the viscoelastic properties of an aggregate depend on the properties of the subunits and on their interconnection remains an open question [110]. However, through the balance of elasticity and viscous behavior, an increase of the

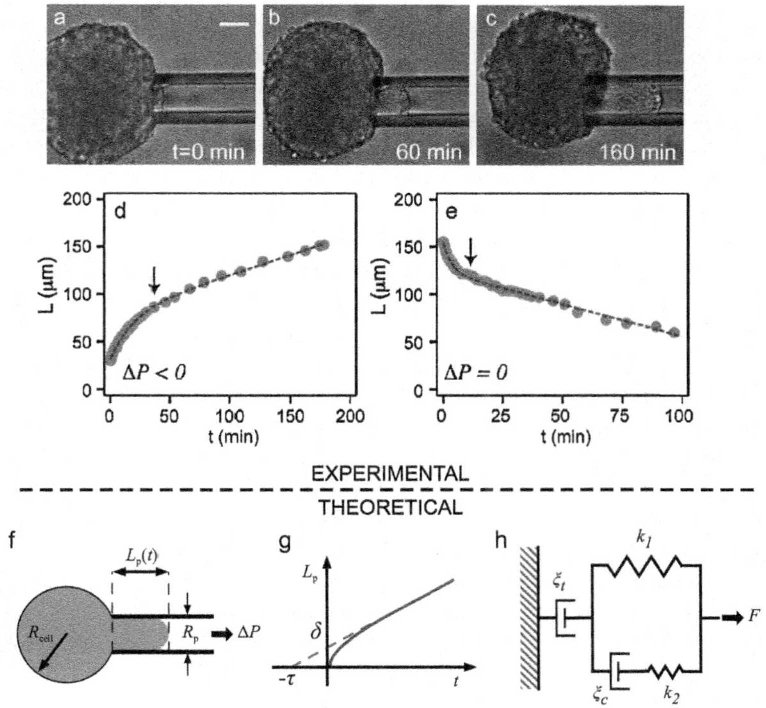


**Fig. 5.10** Micropipette aspiration of spherical cellular aggregates—using idealized theory to explain experimental phenomena. Experimental: (**a**) to (**c**) Aspiration of an aggregate, scale bar is 50 μm. (**d**) Aspiration and (**e**) retraction cycles for an aggregate at constant pressure ($\Delta P_{aspiration} = 1180$ Pa). *Arrows* indicate the transitions from elastic to viscous regimes. *Dotted lines* are fits to the experimental curves using the viscoelastic model. Theoretical: (**f**) Schematic presentation of an aspirated drop. (**g**) Creep curve showing a fast elastic deformation, $\delta$, followed by a viscous flow. (**h**) Modified Maxwell model. The initial elastic deformation is accounted by $k_1$, related to the elasticity of the aggregate, $k_2$ accounts for the initial jump in $L_p(t)$, and $\xi_c$ is a local friction coefficient, related to the raising time of the elastic deformation. The dashpot, denoted by $\xi_t$, represents the viscous dissipation of the flowing tissue. The model successfully decouples each contribution, whereas the relationship in the cellular aggregate is more complex. Adapted from [110], used with permission, copyright © 2010 American Physical Society

surface tension with the applied force was shown, interpreted as an effect of cellular mechanosensing. The micropipette aspiration technique can possibly relate the surface tension of aggregates to the cell-cell adhesion energy [111]. Combined with confocal microscopy, tissue relaxation under stress can be studied at microscopic level by probing the cellular rearrangements inside an aspirated aggregate.

In addition, micropipette aspiration has been applied to discern the viscoelastic contribution of the cortex, or outermost region of the cytoplasm that is rich in the cytoskeletal protein actin [112]. This experimental approach has also been applied to cell types that are present in load-bearing tissues such as cartilage. From these examples, it is clear that micropipette aspiration is a useful approach for cell types that undergo large, general deformation that contributes critically to cell and/or tissue function. Compared to more conventional methods, the micropipette aspiration technique is easy to set up and can be applied to *in vivo* examination of biological systems, such as living tissue or drug treated tumors. Although the applied stress state is relatively complex and based largely on fluid mechanics, continuum approximations have commonly been used to extract the mechanical and functional characteristics of the cell deformed by aspiration. Analytical models of viscoelasticity, though useful for constructing hypotheses that can be tested *via* MA, are restricted by the many assumptions required for mathematical tractability. Beyond closed-form analytical descriptions, computational models of MA cell deformation include finite element modeling (FEM) [113, 114] and boundary integral modeling (BIM) [115]. Increasingly, these models incorporate biphasic, or layered, structural assumptions that delineate the contributions of the cortex and of the cytoplasm and other multilayered biomaterials [116, 117]. As the predictions of these models depend on the choice of constitutive model, including the value of mechanical properties and time constants, the chief use of these simulations is to fit experimental data to a given constitutive model and thereby obtain estimates for elastic and viscous properties. Clearly, the choice of constitutive model dramatically affects the value of fitted parameters, *i.e.*, the calculated viscoelastic characteristics, but provide an integrated perspective of the relevant behaviors and parameters.

Finally, we review a technique typically reserved for tissues and other such "bulk" biomaterial samples: nanoindentation. Nanoindentation is a variety of indentation hardness tests applied to small volumes. While the indentation area can range from a few square nanometers to micrometers and larger, here, we maintain the "nano-" prefix for consistency. Moreover, the term "nanoindentation" has also been associated with compressive of bending load applications to biological materials (such as collagen fibrils [118], amyloids [119], and viruses [120]). Here, we focus on the classical connotation, wherein a material is literally *indented* until a local impression is formed on the surface of a sufficiently large sample.

The mechanistic principle is simple: during the course of the instrumented indentation process, a record of the depth of penetration is made, and then the area of the indent is determined using the known geometry of the indentation tip (see Fig. 5.11). Load and displacement are monitored during the loading and unloading, and properties such as hardness and reduced modulus are calculated from the unloading curves using well-established equations based on elastic contact theory [121]. This technique, traditionally performed *via* dedicated indenters can now be reliably achieved using AFM instrumentation, enabling mechanical property determination at the nanoscale using the high spatial and force resolutions of the AFM. With a development history linked to material hardness, nanoindentation has been particularly successful in the characterization of the mechanical properties of biomineralized materials ranging from bone and teeth [15, 122–125], to deep sea sponges and diatoms [126–128], to nacre, protective shells and fish scales [129–134], and even bird's beaks [135, 136] and human hair [137, 138]. Moreover, NI can be efficiently implemented to investigate variations in mechanical properties with changes in tissue organization (such as the variation in bone hydration) or composition in mineralized and soft tissues (such as the inorganic layers of nacre), and

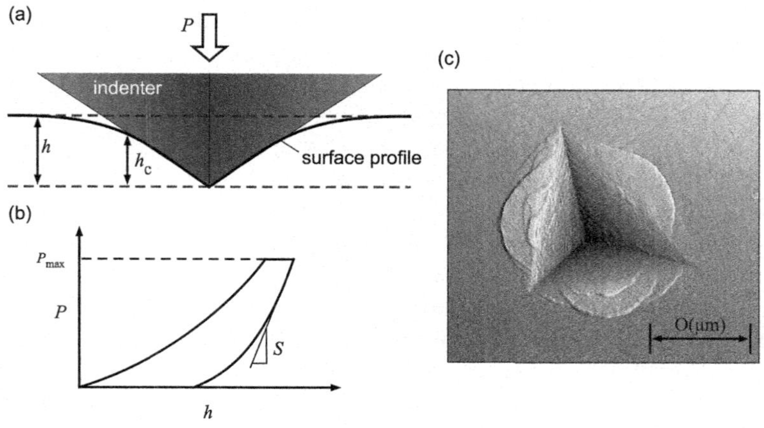


**Fig. 5.11** Schematic of nanoindentation (NI). (**a**) Illustration of the loading process showing parameters characterizing the contact geometry. A probe tip of known geometry (here, a Berkovich tip) is pressed into a sample surface with load, $P$, and the surface profile is deformed to a controlled depth, $h$. Note that the indenter contact depth, $h_c$, is not necessarily equal to the indentation depth—corrections are made depending on probe tip and surface type. (**b**) Typical force-indentation curve ($P - h$) showing elastic loading (to $P_{max}$), hold, and unloading regimes. The reduced modulus, $E_{eff}$, can be determined from the initial relaxation stiffness, $S$, *e.g.*, $dP/dh$ at initial unloading (see Eq. 5.15). (**c**) Example AFM image of the residual indent left by a Berkovitch tip during a nanoindentation experiment (on a Zr-Cu-Al metallic glass), representative of typical indent size on the order or micrometers (AFM image from Wikipedia, released to public domain)

map mechanical properties spatially in complex biomaterials. With a typical working force range of 1 μN to 500 mN and displacement range of 1 nm to 20 μm, this technique bridges the gap between atomic force microscopy (AFM) and macroscale mechanical testing.

The measurement of the elastic modulus ($E$) follows from its relationship to contact area and the measured unloading stiffness through the relation [121]:

$$S = \frac{dP}{dh} = \beta \frac{2}{\sqrt{\pi}} E_{eff} \sqrt{A(h)} \tag{5.15}$$

where $S$ is the initial *unloading* stiffness (the slope of the unloading curve $dP/dh$, evaluated at the maximum load), and $A(h)$ is the projected contact area between the indenter tip and the sample at maximum load (at a depth $h$). $E_{eff}$ is the effective elastic modulus defined by:

$$E_{eff} = \left[ \frac{1 - \nu^2}{E} + \frac{1 - \nu_{ind}^2}{E_{ind}} \right]^{-1} \tag{5.16}$$

The effective elastic modulus takes into account the fact that elastic displacements occur in both the specimen, with Young's modulus $E$ and Poisson's ratio $\nu$, and the indenter, with elastic constants $E_{ind}$ and $\nu_{ind}$. Note that Eq. 5.15 is a very general relation that applies to any axisymmetric indenter [139, 140]. It is not limited to a specific simple geometry, even though it is often associated with flat punch indentation. Although originally derived for elastic contact only [139], it has subsequently been shown to apply equally well to elastic–plastic contact [140] and that small perturbations from pure axisymmetry geometry do not effect it either [141]. One critical assumption, however, is that, to derive the elastic modulus of the sample directly from NI, the Poisson ratio of the sample is required.

For example, nanoindentation has been used to quantify the elastic properties of bone at the lamellar level [122], and compare these properties among osteonal, interstitial and trabecular microstructures from the *same* human femur. It was concluded that the nanostructure of bone tissue must differ substantially among lamellar types, anatomical sites and individuals and suggests that tissue heterogeneity is of potential importance in bone fragility and adaptation [122]. While the material is constant (bone), the materiome fundamentally changes (attributed to variations in porosity, mineralization, and collagen fiber orientation). *In vivo* and *in situ* analysis of native structures under physiological conditions as well as the rapid analysis of molecular species under a variety of experimental treatments are made possible *via* nanoindentation.

As a more exotic example, we consider a study of the exoskeleton the beetle: *Pachnoda marginata* [142, 143]. The bodies of insects such as beetles are completely covered by cuticle, which is a multi-functional interface between the animal and the environment [144, 145]. It serves primarily as an exoskeleton that gives the body its shape and stability. Like most biological materials, cuticle is a fiber composite [144]. The fibers mainly consist of chitin [146], and the matrix is formed by proteins, and found in other species such as lobsters (as mentioned in Chap. 2: The Materiome). Chitin, similar to collagen, is a natural polymer composed of 300 nm-long and 3 nm-thick nanofibrils [144]. The arthropod cuticle—such as that found in a beetle—has a multilayer structure typically consisting of three main layers: epicuticle, exocuticle and endocuticle [142]. The mechanical properties of the insect cuticle, like the properties of human bone, are adapted to the function of particular body parts.

A recent study by Arzt *et al.* [142] focused on the head articulation system of the beetle *Pachnoda marginata* (Fig. 5.12(a)). The surfaces in this system at this particular location must operate in contact and must be resistant to wear and friction and at the same time have to provide high mobility of the joint. Detailed information about the surface and internal structure of the cuticle was obtained, and mechanical properties (hardness, $H$, and elasticity modulus, $E$) of its superficial cuticle layers were measured using nanoindentation. The surface morphology and ultrastructure of the cuticle were studied by means of scanning electron microscopy (SEM). The samples were mechanically tested in the fresh, dry and chemically treated conditions, in order to identify the influence of desiccation (dry versus fresh conditions) and removal of an outer wax layer (dry versus chemically treated conditions). The question was simple: *Do the structural features and mechanical properties of the multilayer cuticle result in friction reducing and wear-resistant properties*?

Indentation tests on different parts of insect cuticle displayed a considerable difference in the mechanical behavior between fresh (hydrated) materials, with a modulus of $E = 1.5$ GPa and hardness of $H = 0.1$ GPa and dehydrated materials, $E = 7.5$ and $H = 0.50$ GPa, in agreement with previous results [143, 147, 148].

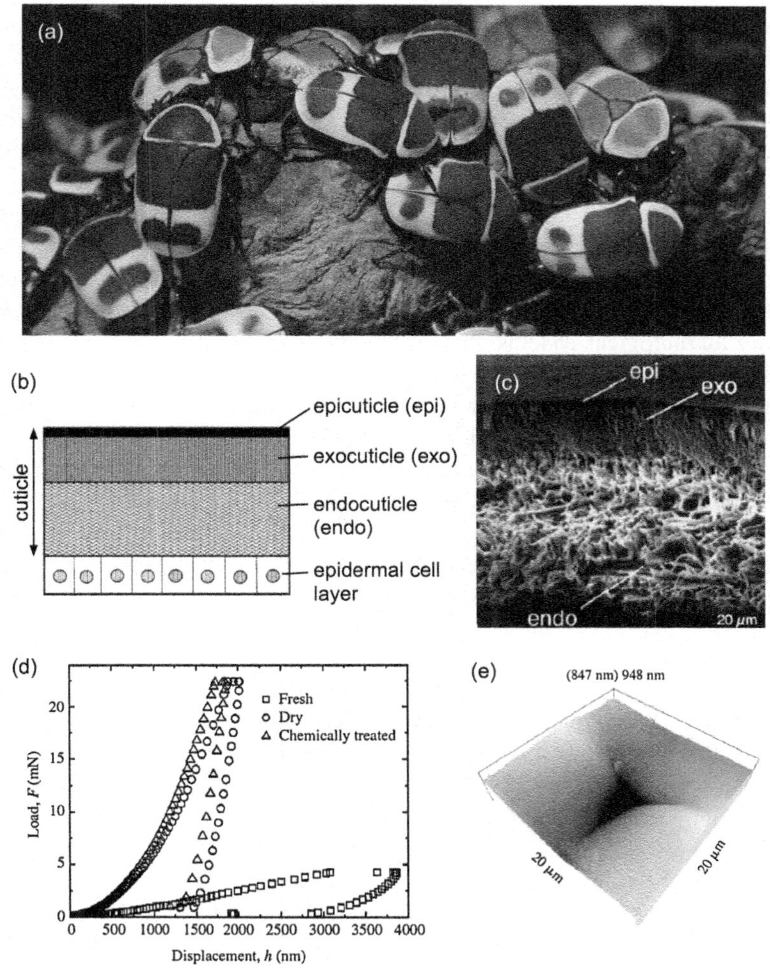


**Fig. 5.12** Nanoindentation of insect cuticles. (**a**) Photo of the beetles, *Pachnoda marginata* (Photo by G. Hume; permission under the Creative Commons Attribution-Share Alike license). (**b**) Diagram of the multilayered structure of the insect cuticle. The shaded areas represent the single cuticle layers: non-chitinous epicuticle; procuticle consisting of exocuticle and endocuticle with chitin fibers usually oriented parallel to the surface. An epidermal cell layer underlies the endocuticle. (**c**) Cross fracture of the cuticle showing the epicuticle (epi), exocuticle (exo) and endocuticle (endo). Fibres of the outer part of the exocuticle are oriented perpendicular to the surface but are parallel in the deeper layers of the exocuticle and in the endocuticle. (**d**) Typical nanoindentation load–displacement curves for one sample in the fresh (hydrated), dry and chemically treated conditions. (**e**) AFM image of the cuticle surface after the indentation test (shading depicting depth from 0 to 800 nm). Adapted from [142], used with permission, copyright © 2006 The Company of Biologists Ltd.

After drying, removal of cuticle lipids caused only small additional changes in the indentation results (a slight increase in hardness and stiffness). This supports that the mechanical behavior of the dry material is governed mainly by proteins and chitin which could not be removed by the lipid extraction—a structural reinforcement of the cuticle. Combing the structural and geometric information with the mechanical characterization through nanoindentation can elucidate new design approaches to lightweight and flexible yet strong and tough armors using Nature as a guide [132, 133, 145, 149, 150]. Or at least explain the "*crunch*" when you accidently step on a beetle.

Further, although the elastic theories of Hertz and Sneddon are often applied in the analysis of these data, both of these classic approaches are applicable to normal contact and time-independent, linear-elastic half-spaces. These assumptions are not valid for all biological systems, such as the deformation of living cells adhered to a substrate, effectively acting as (layered) thin films. There is significant effort underway within the biophysics community to address these limitations, and increased interaction between the mechanical and materials engineering communities who have made significant progress in the analysis and interpretation of contact-based experiments may lead to more rapid maturation of this important tool.

Another deviation from elastic theories resides in the inherent viscoelasticity of biomaterials. The most readily observed effect of viscoelasticity on indentation is creep, or a sinking of the tip into the sample under a constant load. The solution is a simple holding of peak load for a short interval prior to relaxation. In short, it was found that when loading is followed by unloading without a hold at peak load, displacement increases slightly in the initial portion of the unloading phase, because the creep rate of the material is initially higher than the imposed unloading rate [151]. This phenomenon results in a negative and changing slope in the initial unloading region, making it impossible to use Eq. 5.15 to measure modulus. To eliminate this effect a hold period at peak load allows the material to approach equilibrium prior to unloading [151]. This same approach has been used in many indentation studies of biomaterials to allow the creep rate to dissipate prior to unloading [14]. Even widespread and accepted methods can be subject to refinement.

Nanoindentation has emerged as a critical tool for the study of biological systems in their natural state, further contributing to both biomaterial design and pharmacological research, used to probe biomolecules and biological nanostructures from single proteins to complex assemblies [14, 152].

## 5.7 Summary

One criticism of the study and characterization of biological systems such as those described is that, despite the ability to show that applied mechanical stress alters structure and function—regardless of scale—the heterogeneity among responses is largely ignored. Simply put, while we may determine the mechanical properties of a bone specimen, that does not necessarily mean it is representative of my bones or your bones. Biology changes the game. Furthermore, the response of a single material or tissue to mechanical signals (*i.e.*, mechanisms of cellular deformation and mechanotransduction) cannot be decoupled easily from the response of a biological system. For instance, at the microscale, *in vitro* approaches such as elastic substrate deformation have been specifically adapted by researchers in an attempt to impose static and cyclic deformation representative of *in vivo* conditions—the testing apparatus approximates the biological interface. Thus the methods and objectives of such substrate manipulation studies are peculiar to the cell type of interest [153, 154] an intrinsically nontransferable to other material systems.

In essence, there can be a cascade of multi-scale mechanical responses, from the scale of tissues (scale: mm), to a cell population (10s μm), to individual cells (μm), to membranes (10s nm), to individual biomolecules (nm) and their constitutive proteins (Å). How can we make consistent connections?

We note the similarity between the governing equations for AFM (Eq. 5.3) and optical tweezers (Eq. 5.1), the applied load due to steered molecular dynamics (SMD; described in Chap. 7: Mechanical Characterization in Molecular Simulation), and Hooke's Law: $F = k \cdot \Delta$. Indeed, as mentioned, one of the intrinsic advantages of mechanical characterization lies in the applicability of concepts such as force, stiffness, and displacement. While the "stiffness" of an AFM cantilever (governed by classical Euler-Bernoulli beam theory) may differ intrinsically from the laser and dielectric dependent "stiffness" of optical tweezers, both simply represent a known (and deterministic) relation between applied load (force, $F$) and resulting deformation (displacement, $\Delta$). One need not be an expert in induced radiation pressures and gradients to interpret the F-D results of an OT investigation—the behavior becomes an abstraction of the experimental method. Undergraduates introduced to the finite element method (FEM) for the first time are sometimes asked to view the elements as "springy-legos"—*e.g.*, basic building blocks with known stiffness. This simplistic view is enough to understand the basic mechanics behind FEM, yet is obviously not sufficient to encompass the power and potential applications of finite element approaches. Being said, we have yet to define the basic "springy-legos" of biological materials, where the spring stiffness, $k$, is a complex function (dependent on local physiochemical environment, loading and growth history, functionality, etc.). While it is difficult to design "multiscale" experimental approaches, using current tools, from optical tweezers to nanoindentation, can elucidate the mechanical behaviors (the elusive "springs") necessary to assemble a cohesive and deterministic multiscale mechanistic model of the *biomateriome*—from nano to macro.

In this chapter, we have outlined an impressive array of experimental tools that can be utilized to further our understanding of whether and how the mechanical environment and mechanical behavior of a material affects its biological function. Despite the apparent simplicity of the force-displacement relationships that govern these techniques, the sophistication of our experimental tools far exceeds our current analytical and computational understanding. Rigorous interpretations of how force, stress and energy can be imposed and measured *via* these approaches will enable more accurate assessment of current results and also more lucid comparison of results among the different experimental tools across scales. Certainly, the coupled chemical interactions and the inherent time dependence of molecular and cellular processes present challenging opportunities for further research.

Further, in order to maximize the engineering potential of these capabilities, it is essential to recognize how biological systems are intrinsically different from synthetic material systems, as well as which outstanding questions in developmental and pathological biology can be addressed *via* quantification of mechanical responses and the materiome. The application of external mechanical stimuli can induce biochemical reactions that can generate additional mechanical forces. In contrast with most material systems, the mechanical behavior of *living* materials and systems cannot be characterized simply in terms of fixed "properties", as the structure is a dynamic system that adapts to its local mechanochemical environment. A mechanistic understanding of the relationships among extracellular environment and intracellular structure and function, however, requires meaningful quantification of these closely coupled fields, from experimental and computational approaches alike.

## 5.8 Suggested Readings

- Y. Dufrene (ed.), *Life at the Nanoscale: Atomic Force Microscopy of Live Cells* (Pan Stanford Publishing, 2011)
- A.-S. Duwez, N. Willet (eds.), *Molecular Manipulation with Atomic Force Microscopy* (CRC Press, 2011)
- J. Nagatomi (ed.), *Mechanobiology Handbook* (Taylor & Francis, 2010)
- M.L. Oyen (ed.), *Handbook of Nanoindentation With Biological Applications* (Taylor & Francis, 2010)
- M.J. Padgett, J. Molloy, D. McGloin (eds.), *Optical Tweezers Methods and Applications* (CRC Press, 2010)
- P.C. Braga, D. Ricci (eds.), *Atomic Force Microscopy in Biomedical Research* (Springer, 2008)
- F. Yang, J.C.M. Li (eds.), *Micro and Nano Mechanical Testing of Materials and Devices* (Springer, 2008)
- A. Noy (ed.), *Handbook of Molecular Force Spectroscopy* (Springer, 2008)
- A.C. Fischer-Cripps, *Nanoindentation* (Springer, 2004)
- Y.C. Fung, *Biomechanics: Mechanical Properties of Living Tissues* (Springer, 1993)

## References

1. J.C. Kendrew, G. Bodo, H.M. Dintzis, R.G. Parrish, H. Wyckoff, D.C. Phillips, A three-dimensional model of the myoglobin molecule obtained by x-ray analysis. Nature **181**(4610), 662–666 (1958)
2. S. Perumal, O. Antipova, J.P.R.O. Orgel, Collagen fibril architecture, domain organization, and triple-helical conformation govern its proteolysis. Proc. Natl. Acad. Sci. USA **105**(8), 2824–2829 (2008)

3. M. Dao, C.T. Lim, S. Suresh, Mechanics of the human red blood cell deformed by optical tweezers. J. Mech. Phys. Solids **53**(2), 493–494 (2005)
4. J. Li, M. Dao, C.T. Lim, S. Suresh, Spectrin-level modeling of the cytoskeleton and optical tweezers stretching of the erythrocyte. Biophys. J. **88**(5), 3707–3719 (2005)
5. K.J. Van Vliet, G. Bao, S. Suresh, The biomechanics toolbox: experimental approaches for living cells and biomolecules. Acta Mater. **51**(19), 5881–5905 (2003)
6. Y.C. Fung, *Biomechanics: Mechanical Properties of Living Tissues* (Springer, Berlin, 1993)
7. M.S.Z. Kellermayer, C. Bustamante, Folding-unfolding transitions in single titin molecules characterized with laser tweezers (vol. 276, p. 1112, 1997). Science **277**(5329), 1117 (1997)
8. A. Borgia, P.M. Williams, J. Clarke, Single-molecule studies of protein folding. Annu. Rev. Biochem. **77**, 101–125 (2008)
9. D.J. Muller, Y.F. Dufrene, Atomic force microscopy as a multifunctional molecular toolbox in nanobiotechnology. Nat. Nanotechnol. **3**(5), 261–269 (2008)
10. C.B. Prater, H.J. Butt, P.K. Hansma, Atomic force microscopy. Nature **345**(6278), 839–840 (1990)
11. B.L. Smith, T.E. Schaffer, M. Viani, J.B. Thompson, N.A. Frederick, J. Kindt, A. Belcher, G.D. Stucky, D.E. Morse, P.K. Hansma, Molecular mechanistic origin of the toughness of natural adhesives, fibres and composites. Nature **399**, 761–763 (1999)
12. S.E. Cross, J. Kreth, L. Zhu, R. Sullivan, W.Y. Shi, F.X. Qi, J.K. Gimzewski, Nanomechanical properties of glucans and associated cell-surface adhesion of streptococcus mutans probed by atomic force microscopy under in situ conditions. Microbiology **153**, 3124–3132 (2007)
13. Y.L. Sun, Z.P. Luo, A. Fertala, K.N. An, Stretching type ii collagen with optical tweezers. J. Biomech. **37**(11), 1665–1669 (2004)
14. D.M. Ebenstein, L.A. Pruitt, Nanoindentation of biological materials. Nano Today **1**(3), 26–33 (2006)
15. K. Tai, F.J. Ulm, C. Ortiz, Nanogranular origins of the strength of bone. Nano Lett. **6**(11), 2520–2525 (2006)
16. C.T. Lim, E.H. Zhou, A. Li, S.R.K. Vedula, H.X. Fu, Experimental techniques for single cell and single molecule biomechanics. Materials Sci. Eng. C Biomim. Supramol. Syst. **26**(8), 1278–1288 (2006)
17. K.A. Addae-Mensah, J.P. Wikswo, Measurement techniques for cellular biomechanics in vitro. Exp. Biol. Med. **233**(7), 792–809 (2008)
18. D.J. Gardiner, P.R. Graves, H.J. Bowley, *Practical Raman Spectroscopy* (Springer, Berlin, 1989)
19. K. Wuthrich, Protein-structure determination in solution by nmr-spectroscopy. J. Biol. Chem. **265**(36), 22059–22062 (1990)
20. S.B. Shuker, P.J. Hajduk, R.P. Meadows, S.W. Fesik, Discovering high-affinity ligands for proteins: sar by nmr. Science **274**(5292), 1531–1534 (1996)
21. D.M. Rissin, C.W. Kan, T.G. Campbell, S.C. Howes, D.R. Fournier, L. Song, T. Piech, P.P. Patel, L. Chang, A.J. Rivnak, E.P. Ferrell, J.D. Randall, G.K. Provuncher, D.R. Walt, D.C. Duffy, Single-molecule enzyme-linked immunosorbent assay detects serum proteins at sub-femtomolar concentrations. Nat. Biotechnol. **28**(6), 595–625 (2010)
22. H.J. Lee, A.W. Wark, R.M. Corn, Microarray methods for protein biomarker detection. Analyst **133**(8), 975–983 (2008)
23. M.D. Wang, H. Yin, R. Landick, J. Gelles, S.M. Block, Stretching DNA with optical tweezers. Biophys. J. **72**(3), 1335–1346 (1997)
24. M. Streichfuss, F. Erbs, K. Uhrig, R. Kurre, A.E.M. Clemen, C.H.J. Bohm, T. Haraszti, J.P. Spatz, Measuring forces between two single actin filaments during bundle formation. Nano Lett. **11**(9), 3676–3680 (2011)
25. A. Ashkin, J.M. Dziedzic, Optical trapping and manipulation of viruses and bacteria. Science **235**(4795), 1517–1520 (1987)
26. J.R. Moffitt, Y.R. Chemla, D. Izhaky, C. Bustamante, Differential detection of dual traps improves the spatial resolution of optical tweezers. Proc. Natl. Acad. Sci. USA **103**(24), 9006–9011 (2006)

27. M.C. Williams, Optical tweezers: measuring piconewton forces, in *Single Molecule Techniques*, ed. by P. Schwille (Biophysics Textbook Online, 2002)
28. C. Cecconi, E.A. Shank, C. Bustamante, S. Marqusee, Direct observation of the three-state folding of a single protein molecule. Science **309**(5743), 2057–2060 (2005)
29. A.D. Mehta, M. Rief, J.A. Spudich, D.A. Smith, R.M. Simmons, Single-molecule biomechanics with optical methods. Science **283**(5408), 1689–1695 (1999)
30. T.M. Raschke, J. Kho, S. Marqusee, Confirmation of the hierarchical folding of rnase h: a protein engineering study. Nat. Struct. Biol. **6**(9), 825–831 (1999)
31. R.L. Baldwin, G.D. Rose, Is protein folding hierarchic? ii. folding intermediates and transition states. Trends Biochem. Sci. **24**(2), 77–83 (1999)
32. K. Svoboda, S.M. Block, Force and velocity measured for single kinesin molecules. Cell **77**(5), 773–784 (1994)
33. C. Veigel, C.F. Schmidt, Moving into the cell: single-molecule studies of molecular motors in complex environments. Nat. Rev. Mol. Cell Biol. **12**(3), 163–176 (2011)
34. M. Bathe, C. Heussinger, M.M.A.E. Claessens, A.R. Bausch, E. Frey, Cytoskeletal bundle mechanics. Biophys. J. **94**(8), 2955–2964 (2008)
35. A.M. Goldyn, B.A. Rioja, J.P. Spatz, C. Ballestrem, R. Kemkemer, Force-induced cell polarisation is linked to rhoa-driven microtubule-independent focal-adhesion sliding. J. Cell Sci. **122**(20), 3644–3651 (2009)
36. T. Kuhne, R. Lipowsky, J. Kierfeld, Zipping mechanism for force generation by growing filament bundles. Epl **86**(6) (2009)
37. R. Nambiar, R.E. McConnell, M.J. Tyska, Myosin motor function: the ins and outs of actin-based membrane protrusions. Cell. Mol. Life Sci. **67**(8), 1239–1254 (2010)
38. D.G. Grier, A revolution in optical manipulation. Nature **424**(6950), 810–816 (2003)
39. S.B. Smith, L. Finzi, C. Bustamante, Direct mechanical measurements of the elasticity of single DNA-molecules by using magnetic beads. Science **258**(5085), 1122–1126 (1992)
40. T.R. Strick, J.F. Allemand, D. Bensimon, A. Bensimon, V. Croquette, The elasticity of a single supercoiled DNA molecule. Science **271**(5257), 1835–1837 (1996)
41. C. Gosse, V. Croquette, Magnetic tweezers: micromanipulation and force measurement at the molecular level. Biophys. J. **82**(6), 3314–3329 (2002)
42. A.R. Bausch, F. Ziemann, A.A. Boulbitch, K. Jacobson, E. Sackmann, Local measurements of viscoelastic parameters of adherent cell surfaces by magnetic bead microrheometry. Biophys. J. **75**(4), 2038–2049 (1998)
43. A.R. Bausch, W. Moller, E. Sackmann, Measurement of local viscoelasticity and forces in living cells by magnetic tweezers. Biophys. J. **76**(1), 573–579 (1999)
44. G. Binnig, C.F. Quate, C. Gerber, Atomic force microscope. Phys. Rev. Lett. **56**(9), 930–933 (1986)
45. J. Howard, *Mechanics of Motor Proteins and the Cytoskeleton* (Sinauer, Sunderland, 2001)
46. T.R. Albrecht, S. Akamine, T.E. Carver, C.F. Quate, Microfabrication of cantilever styli for the atomic force microscope. J. Vac. Sci. Technol., A, Vac. Surf. Films **8**(4), 3386–3396 (1990)
47. N.A. Burnham, X. Chen, C.S. Hodges, G.A. Matei, E.J. Thoreson, C.J. Roberts, M.C. Davies, S.J.B. Tendler, Comparison of calibration methods for atomic-force microscopy cantilevers. Nanotechnology **14**(1), 1–6 (2003)
48. J.P. Cleveland, S. Manne, D. Bocek, P.K. Hansma, A nondestructive method for determining the spring constant of cantilevers for scanning force microscopy. Rev. Sci. Instrum. **64**(2), 403–405 (1993)
49. J.E. Sader, J.W.M. Chon, P. Mulvaney, Calibration of rectangular atomic force microscope cantilevers. Rev. Sci. Instrum. **70**(10), 3967–3969 (1999)
50. J.W.M. Chon, P. Mulvaney, J.E. Sader, Experimental validation of theoretical models for the frequency response of atomic force microscope cantilever beams immersed in fluids. J. Appl. Phys. **87**(8), 3978–3988 (2000)
51. J.L. Hutter, J. Bechhoefer, Calibration of atomic-force microscope tips. Rev. Sci. Instrum. **64**(7), 1868–1873 (1993)

52. H.J. Butt, M. Jaschke, Calculation of thermal noise in atomic-force microscopy. Nanotechnology **6**(1), 1–7 (1995)
53. C. Rotsch, M. Radmacher, Drug-induced changes of cytoskeletal structure and mechanics in fibroblasts: an atomic force microscopy study. Biophys. J. **78**(1), 520–535 (2000)
54. M. Radmacher, Measuring the elastic properties of living cells by the atomic force microscope. Atom. Force Microsc. Cell Biol. **68**, 67–90 (2002)
55. G. Charras, P. Lehenkari, M. Horton, Biotechnological applications of atomic force microscopy. Atom. Force Microsc. Cell Biol. **68**, 171–191 (2002)
56. A. Alessandrini, P. Facci, Afm: a versatile tool in biophysics. Meas. Sci. Technol. **16**(6), 65–92 (2005)
57. B. Chen, M. Gao, J.M. Zuo, S. Qu, B. Liu, Y. Huang, Binding energy of parallel carbon nanotubes. Appl. Phys. Lett. **83**(17), 3570–3571 (2003)
58. A.E. Cohen, L. Mahadevan, Kinks, rings, and rackets in filamentous structures. Proc. Natl. Acad. Sci. USA **100**(21), 12141–12146 (2003)
59. J. Adamcik, J.M. Jung, J. Flakowski, P. De Los Rios, G. Dietler, R. Mezzenga, Understanding amyloid aggregation by statistical analysis of atomic force microscopy images. Nat. Nanotechnol. **5**(6), 423–428 (2010)
60. L.M.C. Sagis, C. Veerman, E. van der Linden, Mesoscopic properties of semiflexible amyloid fibrils. Langmuir **20**(3), 924–927 (2004)
61. A.K. Chamberlain, C.E. MacPhee, J. Zurdo, L.A. Morozova-Roche, H.A.O. Hill, C.M. Dobson, J.J. Davis, Ultrastructural organization of amyloid fibrils by atomic force microscopy. Biophys. J. **79**(6), 3282–3293 (2000)
62. R. Khurana, C. Ionescu-Zanetti, M. Pope, J. Li, L. Nielson, M. Ramirez-Alvarado, L. Regan, A.L. Fink, S.A. Carter, A general model for amyloid fibril assembly based on morphological studies using atomic force microscopy. Biophys. J. **85**(2), 1135–1144 (2003)
63. G. Witz, K. Rechendorff, J. Adamcik, G. Dietler, Conformation of circular DNA in two dimensions. Phys. Rev. Lett. **101**(14) (2008)
64. G.S. Manning, Correlation of polymer persistence length with Euler buckling fluctuations. Phys. Rev. A **34**(5), 4467–4468 (1986)
65. P. Hinterdorfer, Y.F. Dufrene, Detection and localization of single molecular recognition events using atomic force microscopy. Nat. Methods **3**(5), 347–355 (2006)
66. M. Rief, M. Gautel, F. Oesterhelt, J.M. Fernandez, H.E. Gaub, Reversible unfolding of individual titin immunoglobulin domains by afm. Science **276**(5315), 1109–1112 (1997)
67. O.H. Willemsen, M.M.E. Snel, A. Cambi, J. Greve, B.G. De Grooth, C.G. Figdor, Biomolecular interactions measured by atomic force microscopy. Biophys. J. **79**(6), 3267–3281 (2000)
68. B.T. Marshall, M. Long, J.W. Piper, T. Yago, R.P. McEver, C. Zhu, Direct observation of catch bonds involving cell-adhesion molecules. Nature **423**(6936), 190–193 (2003)
69. J.M. Fernandez, H.B. Li, Force-clamp spectroscopy monitors the folding trajectory of a single protein. Science **303**(5664), 1674–1678 (2004)
70. P.J. Flory, *Statistical Mechanics of Chain Molecules*, repr. edn. (Hanser, Munich, 1989, distributed in the USA by Oxford University Press, New York)
71. C. Bustamante, J.F. Marko, E.D. Siggia, S. Smith, Entropic elasticity of lambda-phage DNA. Science **265**(5178), 1599–1600 (1994)
72. A.F. Oberhauser, P.E. Marszalek, H.P. Erickson, J.M. Fernandez, The molecular elasticity of the extracellular matrix protein tenascin. Nature **393**(6681), 181–185 (1998)
73. M. Rief, M. Gautel, A. Schemmel, H.E. Gaub, The mechanical stability of immunoglobulin and fibronectin iii domains in the muscle protein titin measured by atomic force microscopy. Biophys. J. **75**(6), 3008–3014 (1998)
74. T.E. Fisher, A.F. Oberhauser, M. Carrion-Vazquez, P.E. Marszalek, J.M. Fernandez, The study of protein mechanics with the atomic force microscope. Trends Biochem. Sci. **24**(10), 379–384 (1999)
75. E. Evans, K. Ritchie, Dynamic strength of molecular adhesion bonds. Biophys. J. **72**, 1541–1555 (1997)

76. E. Evans, Probing the relation between force—lifetime—and chemistry in single molecular bonds. Annu. Rev. Biophys. Biomol. Struct. **30**, 105–128 (2001)
77. G.I. Bell, Models for the specific adhesion of cells to cells. Science **200**(4342), 618–627 (1978)
78. E. Evans, Energy landscapes of biomolecular adhesion and receptor anchoring at interfaces explored with dynamic force spectroscopy. Faraday Discuss. **111**, 1–16 (1998)
79. K.A. Dill, S. Bromberg, K.Z. Yue, K.M. Fiebig, D.P. Yee, P.D. Thomas, H.S. Chan, Principles of protein-folding—a perspective from simple exact models. Protein Sci. **4**(4), 561–602 (1995)
80. K.A. Dill, H.S. Chan, From Levinthal to pathways to funnels. Nat. Struct. Biol. **4**(1), 10–19 (1997)
81. K.A. Dill, Polymer principles and protein folding. Protein Sci. **8**(6), 1166–1180 (1999)
82. J.N. Onuchic, P.G. Wolynes, Z. Lutheyschulten, N.D. Socci, Toward an outline of the topography of a realistic protein-folding funnel. Proc. Natl. Acad. Sci. USA **92**(8), 3626–3630 (1995)
83. R. Merkel, P. Nassoy, A. Leung, K. Ritchie, E. Evans, Energy landscapes of receptor-ligand bonds explored with dynamic force spectroscopy. Nature **397**(6714), 50–53 (1999)
84. M. Benoit, D. Gabriel, G. Gerisch, H.E. Gaub, Discrete interactions in cell adhesion measured by single-molecule force spectroscopy. Nat. Cell Biol. **2**(6), 313–317 (2000)
85. M. Benoit, Cell adhesion measured by force spectroscopy on living cells. Atom. Force Microsc. Cell Biol. **68**, 91–114 (2002)
86. P. Hinterdorfer, Molecular recognition studies using the atomic force microscope. Atom. Force Microsc. Cell Biol. **68**, 115–139 (2002)
87. J. Helenius, C.P. Heisenberg, H.E. Gaub, D.J. Muller, Single-cell force spectroscopy. J. Cell Sci. **121**(11), 1785–1791 (2008)
88. C.A. Bippes, D.J. Muller, High-resolution atomic force microscopy and spectroscopy of native membrane proteins. Rep. Progr. Phys. **74**(8) (2011)
89. E.A. Evans, D.A. Calderwood, Forces and bond dynamics in cell adhesion. Science **316**(5828), 1148–1153 (2007)
90. A. Noy, D.V. Vezenov, C.M. Lieber, Chemical force microscopy. Annu. Rev. Mater. Sci. **27**, 381–421 (1997)
91. M. Radmacher, M. Fritz, C.M. Kacher, J.P. Cleveland, P.K. Hansma, Measuring the viscoelastic properties of human platelets with the atomic force microscope. Biophys. J. **70**(1), 556–567 (1996)
92. R.E. Mahaffy, S. Park, E. Gerde, J. Kas, C.K. Shih, Quantitative analysis of the viscoelastic properties of thin regions of fibroblasts using atomic force microscopy. Biophys. J. **86**(3), 1777–1793 (2004)
93. M. Minary-Jolandan, M.F. Yu, Uncovering nanoscale electromechanical heterogeneity in the subfibrillar structure of collagen fibrils responsible for the piezoelectricity of bone. ACS Nano **3**(7), 1859–1863 (2009)
94. A.J. Engler, S. Sen, H.L. Sweeney, D.E. Discher, Matrix elasticity directs stem cell lineage specification. Cell **126**(4), 677–689 (2006)
95. A.J. Engler, F. Rehfeldt, S. Sen, D.E. Discher, Microtissue elasticity: measurements by atomic force microscopy and its influence on cell differentiation. Cell Mech. **83**, 521 (2007)
96. D.E. Ingber, D. Prusty, Z.Q. Sun, H. Betensky, N. Wang, Cell shape, cytoskeletal mechanics, and cell cycle control in angiogenesis. J. Biomech. **28**(12), 1471–1484 (1995)
97. M. Puig-de-Morales, M. Grabulosa, J. Alcaraz, J. Mullol, G.N. Maksym, J.J. Fredberg, D. Navajas, Measurement of cell microrheology by magnetic twisting cytometry with frequency domain demodulation. J. Appl. Physiol. **91**(3), 1152–1159 (2001)
98. J.X. Chen, B. Fabry, E.L. Schiffrin, N. Wang, Twisting integrin receptors increases endothelin-1 gene expression in endothelial cells. Am. J. Physiol., Cell Physiol. **280**(6), 1475–1484 (2001)
99. R.J. Pelham, Y.L. Wang, Cell locomotion and focal adhesions are regulated by substrate flexibility. Proc. Natl. Acad. Sci. USA **94**(25), 13661–13665 (1997)

100. R.J. Pelham, Y.L. Wang, High resolution detection of mechanical forces exerted by locomoting fibroblasts on the substrate. Mol. Biol. Cell **10**(4), 935–945 (1999)
101. K.A. Beningo, Y.L. Wang, Flexible substrata for the detection of cellular traction forces. Trends Cell Biol. **12**(2), 79–84 (2002)
102. J.L. Tan, J. Tien, D.M. Pirone, D.S. Gray, K. Bhadriraju, C.S. Chen, Cells lying on a bed of microneedles: an approach to isolate mechanical force. Proc. Natl. Acad. Sci. USA **100**(4), 1484–1489 (2003)
103. O. du Roure, A. Saez, A. Buguin, R.H. Austin, P. Chavrier, P. Silberzan, B. Ladoux, Force mapping in epithelial cell migration. Proc. Natl. Acad. Sci. USA **102**(7), 2390–2395 (2005)
104. T. Braun, M.K. Ghatkesar, N. Backmann, W. Grange, P. Boulanger, L. Letellier, H.P. Lang, A. Bietsch, C. Gerber, M. Hegner, Quantitative time-resolved measurement of membrane protein-ligand interactions using microcantilever array sensors. Nat. Nanotechnol. **4**(3), 179–185 (2009)
105. R.M. Hochmuth, Micropipette aspiration of living cells. J. Biomech. **33**(1), 15–22 (2000)
106. J.Y. Shao, R.M. Hochmuth, Micropipette suction for measuring piconewton forces of adhesion and tether formation from neutrophil membranes. Biophys. J. **71**(5), 2892–2901 (1996)
107. W.R. Jones, H.P. Ting-Beall, G.M. Lee, S.S. Kelley, R.M. Hochmuth, F. Guilak, Alterations in the Young's modulus and volumetric properties of chondrocytes isolated from normal and osteoarthritic human cartilage. J. Biomech. **32**(2), 119–127 (1999)
108. D.P. Theret, M.J. Levesque, M. Sato, R.M. Nerem, L.T. Wheeler, The application of a homogeneous half-space model in the analysis of endothelial-cell micropipette measurements. J. Biomech. Eng. Trans. ASME **110**(3), 190–199 (1988)
109. R.M. Hochmuth, H.P. Tingbeall, B.B. Beaty, D. Needham, R. Transontay, Viscosity of passive human neutrophils undergoing small deformations. Biophys. J. **64**(5), 1596–1601 (1993)
110. K. Guevorkian, M.J. Colbert, M. Durth, S. Dufour, F. Brochard-Wyart, Aspiration of biological viscoelastic drops. Phys. Rev. Lett. **104**(21) (2010)
111. Y.S. Chu, W.A. Thomas, O. Eder, F. Pincet, E. Perez, J.P. Thiery, S. Dufour, Force measurements in e-cadherin-mediated cell doublets reveal rapid adhesion strengthened by actin cytoskeleton remodeling through rac and cdc42. J. Cell Biol. **167**(6), 1183–1194 (2004)
112. A. Yeung, E. Evans, Cortical shell-liquid core model for passive flow of liquid-like spherical cells into micropipets. Biophys. J. **56**(1), 139–149 (1989)
113. F. Guilak, V.C. Mow, The mechanical environment of the chondrocyte: a biphasic finite element model of cell-matrix interactions in articular cartilage. J. Biomech. **33**(12), 1663–1673 (2000)
114. R.G. Zhao, K. Wyss, C.A. Simmons, Comparison of analytical and inverse finite element approaches to estimate cell viscoelastic properties by micropipette aspiration. J. Biomech. **42**(16), 2768–2773 (2009)
115. E. Kim, F. Guilak, M.A. Haider, An axisymmetric boundary element model for determination of articular cartilage pericellular matrix properties in situ via inverse analysis of chondron deformation. J. Biomech. Eng. Trans. ASME **132**(3) (2010)
116. J. Brugues, B. Maugis, J. Casademunt, P. Nassoy, F. Amblard, P. Sens, Dynamical organization of the cytoskeletal cortex probed by micropipette aspiration. Proc. Natl. Acad. Sci. USA **107**(35), 15415–15420 (2010)
117. R.G. Zhao, K.L. Sider, C.A. Simmons, Measurement of layer-specific mechanical properties in multilayered biomaterials by micropipette aspiration. Acta Biomater. **7**(3), 1220–1227 (2011)
118. A.J. Heim, W.G. Matthews, T.J. Koob, Determination of the elastic modulus of native collagen fibrils via radial indentation. Appl. Phys. Lett. **89**(18) (2006)
119. K. Sweers, K. van der Werf, M. Bennink, V. Subramaniam, Nanomechanical properties of alpha-synuclein amyloid fibrils: a comparative study by nanoindentation, harmonic force microscopy, and peakforce qnm. Nanoscale Res. Lett. **6** (2011)
120. N. Kol, Y. Shi, M. Tsvitov, D. Barlam, R.Z. Shneck, M.S. Kay, I. Rousso, A stiffness switch in human immunodeficiency virus. Biophys. J. **92**(5), 1777–1783 (2007)

121. W.C. Oliver, G.M. Pharr, Measurement of hardness and elastic modulus by instrumented indentation: advances in understanding and refinements to methodology. J. Mater. Res. **19**(1), 3–20 (2004)
122. P.K. Zysset, X.E. Guo, C.E. Hoffler, K.E. Moore, S.A. Goldstein, Elastic modulus and hardness of cortical and trabecular bone lamellae measured by nanoindentation in the human femur. J. Biomech. **32**(10), 1005–1012 (1999)
123. S. Habelitz, G.W. Marshall, M. Balooch, S.J. Marshall, Nanoindentation and storage of teeth. J. Biomech. **35**(7), 995–998 (2002)
124. K.S. Tai, H.J. Qi, C. Ortiz, Effect of mineral content on the nanoindentation properties and nanoscale deformation mechanisms of bovine tibial cortical bone. J. Mater. Sci., Mater. Med. **16**(10), 947–959 (2005)
125. M.L. Oyen, Poroelastic nanoindentation responses of hydrated bone. J. Mater. Res. **23**(5), 1307–1314 (2008)
126. M. Sarikaya, H. Fong, N. Sunderland, B.D. Flinn, G. Mayer, A. Mescher, E. Gaino, Biomimetic model of a sponge-spicular optical fiber—mechanical properties and structure. J. Mater. Res. **16**(5), 1420–1428 (2001)
127. A. Woesz, J.C. Weaver, M. Kazanci, Y. Dauphin, J. Aizenberg, D.E. Morse, P. Fratzl, Micromechanical properties of biological silica in skeletons of deep-sea sponges. J. Mater. Res. **21**(8), 2068–2078 (2006)
128. D. Losic, K. Short, J.G. Mitchell, R. Lal, N.H. Voelcker, Afm nanoindentations of diatom biosilica surfaces. Langmuir **23**(9), 5014–5021 (2007)
129. B.J.F. Bruet, H.J. Qi, M.C. Boyce, R. Panas, K. Tai, L. Frick, C. Ortiz, Nanoscale morphology and indentation of individual nacre tablets from the gastropod mollusc trochus niloticus. J. Mater. Res. **20**(9), 2400–2419 (2005)
130. F. Barthelat, C.M. Li, C. Comi, H.D. Espinosa, Mechanical properties of nacre constituents and their impact on mechanical performance. J. Mater. Res. **21**(8), 1977–1986 (2006)
131. A. Perez-Huerta, M. Cusack, W.Z. Zhu, J. England, J. Hughes, Material properties of brachiopod shell ultrastructure by nanoindentation. J. R. Soc. Interface **4**(12), 33–39 (2007)
132. B.J.F. Bruet, J.H. Song, M.C. Boyce, C. Ortiz, Materials design principles of ancient fish armour. Nat. Mater. **7**(9), 748–756 (2008)
133. H.M. Yao, M. Dao, T. Imholt, J.M. Huang, K. Wheeler, A. Bonilla, S. Suresh, C. Ortiz, Protection mechanisms of the iron-plated armor of a deep-sea hydrothermal vent gastropod. Proc. Natl. Acad. Sci. USA **107**(3), 987–992 (2010)
134. D. Zhu, C.F. Ortega, R. Motamedi, L. Szewciw, F. Vernerey, F. Barthelat, Structure and mechanical performance of a "modern" fish scale. Eng. Mater. **14**(4), B185–B194 (2012)
135. Y. Seki, M.S. Schneider, M.A. Meyers, Structure and mechanical behavior of a toucan beak. Acta Mater. **53**(20), 5281–5296 (2005)
136. Y. Seki, S.G. Bodde, M.A. Meyers, Toucan and hornbill beaks: a comparative study. Acta Biomater. **6**(2), 331–343 (2010)
137. G.H. Wei, B. Bhushan, P.M. Torgerson, Nanomechanical characterization of human hair using nanoindentation and sem. Ultramicroscopy **105**(1–4), 248–266 (2005)
138. G.H. Wei, B. Bhushan, Nanotribological and nanomechanical characterization of human hair using a nanoscratch technique. Ultramicroscopy **106**(8–9), 742–754 (2006)
139. G.M. Pharr, W.C. Oliver, F.R. Brotzen, On the generality of the relationship among contact stiffness, contact area, and elastic-modulus during indentation. J. Mater. Res. **7**(3), 613–617 (1992)
140. C.M. Cheng, Y.T. Cheng, On the initial unloading slope in indentation of elastic-plastic solids by an indenter with an axisymmetric smooth profile. Appl. Phys. Lett. **71**(18), 2623–2625 (1997)
141. H.J. Gao, T.W. Wu, A note on the elastic contact stiffness of a layered medium. J. Mater. Res. **8**(12), 3229–3232 (1993)
142. N. Barbakadze, S. Enders, S. Gorb, E. Arzt, Local mechanical properties of the head articulation cuticle in the beetle pachnoda marginata (coleoptera, scarabaeidae). J. Exp. Biol. **209**(4), 722–730 (2006)

143. S. Enders, N. Barbakadse, S.N. Gorb, E. Arzt, Exploring biological surfaces by nanoindentation. J. Mater. Res. **19**(3), 880–887 (2004)
144. J.F.V. Vincent, Arthropod cuticle: a natural composite shell system. Composites, Part A, Appl. Sci. Manuf. **33**(10), 1311–1315 (2002)
145. J.F.V. Vincent, U.G.K. Wegst, Design and mechanical properties of insect cuticle. Arthropod Struct. Devel. **33**(3), 187–199 (2004)
146. S.O. Andersen, M.G. Peter, P. Roepstorff, Cuticular sclerotization in insects. Comp. Biochem. Physiol., Part B Biochem. Mol. Biol. **113**(4), 689–705 (1996)
147. J.E. Hillerton, S.E. Reynolds, J.F.V. Vincent, On the indentation hardness of insect cuticle. J. Exp. Biol. **96**(Feb), 45 (1982)
148. E. Arzt, S. Enders, S. Gorb, Towards a micromechanical understanding of biological surface devices. Z. Met.kd. **93**(5), 345–351 (2002)
149. J.H. Song, S. Reichert, I. Kallai, D. Gazit, M. Wund, M.C. Boyce, C. Ortiz, Quantitative microstructural studies of the armor of the marine threespine stickleback (gasterosteus aculeatus). J. Struct. Biol. **171**(3), 318–331 (2010)
150. J.H. Song, C. Ortiz, M.C. Boyce, Threat-protection mechanics of an armored fish. J. Mech. Behav. Biomed. Mat. **4**(5), 699–712 (2011)
151. B.J. Briscoe, L. Fiori, E. Pelillo, Nano-indentation of polymeric surfaces. J. Phys. D, Appl. Phys. **31**(19), 2395–2405 (1998)
152. O. Franke, M. Goken, A.M. Hodge, The nanoindentation of soft tissue: current and developing approaches. J. Met. **60**(6), 49–53 (2008)
153. T.M. Freyman, I.V. Yannas, R. Yokoo, L.J. Gibson, Fibroblast contraction of a collagen-gag matrix. Biomaterials **22**(21), 2883–2891 (2001)
154. D.P. Pioletti, J. Muller, L.R. Rakotomanana, J. Corbeil, E. Wild, Effect of micromechanical stimulations on osteoblasts: development of a device simulating the mechanical situation at the bone-implant interface. J. Biomech. **36**(1), 131–135 (2003)

# Chapter 6
# Computational Approaches and Simulation

**Abstract** Computation and simulation provides a means to investigate complex materiomic systems with unparalleled control and accuracy. At the same time, a holistic description of a material system necessitates knowledge of the lowest possible scale—atomistic and molecular interactions. While quantum level resolution provides a means to understand atom-to-atom interactions, molecular interactions provides the foundation for deterministic (or predictable) mechanistic behavior. In recent years, molecular dynamics has developed into a powerful tool to investigate biological systems such as the stretching of proteins and other macromolecules. The advent of reactive molecular dynamics (wherein chemical bonds can be formed or ruptured) has extended the range of applications at the nanoscale. Being said, the limitations of full atomistic simulation (in terms of accessible time and length scales) has necessitated coarse-grain and other multiscale methods, in a bottom-up "fine-trains-coarse" paradigm. Not unlike the reduction of engineering analysis to critical components, such multiscale methods can be used to bridge each structural hierarchy, characterize performance and behavior, and successfully explore the entire materiome *via* simulation.

*Computers are useless. They can only give you answers.*

*Pablo Picasso (1881–1973)*

## 6.1 Introduction

Recently, an increasing number of researchers have grown to consider the computer as a tool to do science and engineering, including the design of new materials, new structures, or the development of new drugs, without ever synthesizing them. Predicting the properties and behavior of materials by computer simulation from a fundamental or *ab initio* perspective has long been a vision of computational materials scientists. As computational and processing power increases, modeling and simulation tools become indispensable to the investigation of material systems. The key to achieving this goal is utilizing hierarchies of paradigms and scales that seamlessly connect macroscopic systems to first principal quantum mechanics (QM).

S.W. Cranford, M.J. Buehler, *Biomateriomics*, Springer Series in Materials Science 165,
DOI 10.1007/978-94-007-1611-7_6, © Springer Science+Business Media Dordrecht 2012

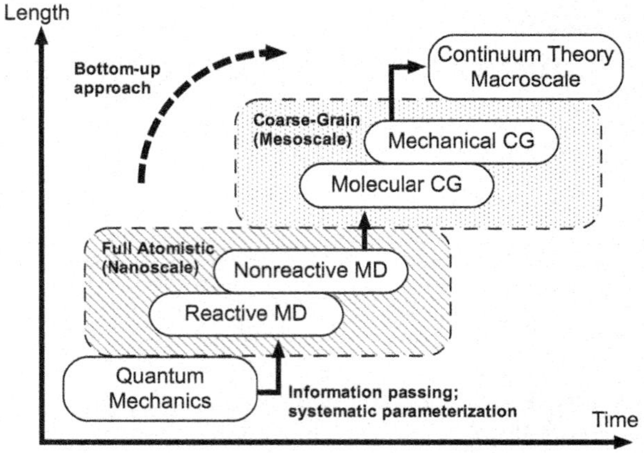


**Fig. 6.1** Hierarchy of characteristic/accessible time and length scales for computational materiomics. This chapter focuses on the transition between full atomistic molecular dynamics and mesoscopic coarse-grain approaches. There exist numerous approaches in achieving this connection, notably hierarchical (*i.e.*, information-passing) and concurrent (*i.e.*, simultaneous) methods

Particular progress has been made in relating molecular-scale chemistry to mesoscopic and macroscopic material properties essential to define the *materiome* (see Fig. 6.1). This chapter will review large-scale atomistic and molecular modeling methods commonly implemented to investigate the mechanical properties of natural and biological materials with nanostructured hierarchies. The intent is to present basic concepts of hierarchical multiscale modeling capable of providing a bottom-up description of chemically complex materials.[1]

Modeling and simulation have evolved into predictive tools that complement experimental analysis at comparable length- and time-scales. Whereas many questions regarding material systems can be studied with a single-scale approach at a single level of resolution, the use of theoretical and computational multiscale approaches enables critical progress in linking the chemical or molecular, and mesoscopic structures of these materials to macroscopic properties, across disparate scales, an integral aspect of materiomics. Multiscale simulation models for biological materials have become increasingly popular in recent years and have enabled the direct link between experiment and theoretical bottom-up descriptions of materials. Such methods are necessary when a single level of resolution is inadequate to probe complex interactions that precipitate between scales, and hierarchies are intimately interwoven. Such studies are crucial in understanding the impact of genetic mutations, structural flaws and defects, hierarchical structures, environmental variations (such as solvent or pH changes), and other chemical stimuli on the properties of materials.

In the study of these materials, the integrated use of a suite of computational and theoretical approaches is vital to cover all relevant material properties and scales, and includes first principles calculations, atomistic and molecular modeling,

---

[1] We note that the computational and simulation techniques discussed herein are by no means intended to be exhaustive, presented in full depth, or canonical. Focus is particularly given to molecular dynamics approaches and coarse-grain methodologies insofar as they are relevant to biological materials. The intent is to illustrate a multiscale paradigm necessary to a materiomic perspective, and not provide a robust technical guide or resource. Interested readers are directed to the suggested readings at the end of the chapter.

"coarse-grained" multiscale approaches as well as continuum theory based methods. Material properties of protein and biological systems are determined by processes and interactions on a wide range of length- and time-scales. As there is inherently mutual influence between scales ("scale-bridging" effects), it is not straightforward to provide quantitative information and understanding at a particular scale (whether at the ultimate functional level or some intermediate regime) without taking this properly into account. Be that as it may, in the field of atomistic-based multiscale stimulation, it is now possible to begin from the smallest scales (considering electrons and atoms), to reach all the way up to macroscopic scales of filaments, fibrils, fibers and entire tissues, by explicitly considering the characteristic structural features at multiple materiomic hierarchies. Both concurrent multiscale methods (*i.e.*, the use of methods of different accuracy in the same computational domain) and hierarchical multiscale methods (*i.e.*, the sequential, hierarchical use of methods through parameter passing) are applied. Such approaches are possible with the advent of first principles based multiscale simulation techniques (see, for instance, a review article for a broad introduction to this field [1]). The basic principle underlying these multiscale simulation methods is "*finer scales train coarser scales*"—a "bold" approach with enormous potential to change the way we analyze and design materials (Fig. 6.1).

**Multiscale Techniques**: Investigative methods, encompassing theoretical, experimental, and computational approaches, which probe material properties across a multitude of scales. Different computational hierarchies are combined and linked to simultaneously address phenomena of a given system at several levels of resolution and across several time- and length-scales. Multiscale techniques aim to establish cross-scale interactions and mechanisms that elucidate structure-property-process relations that supplement material characterization and properties at a single scale level.

Even though there are still major challenges ahead of us, this progress now provides one with many opportunities, transforming the materials science of biological and other materials as a discipline through increased integration of computational approaches in scientific research. Despite significant advances in the study of biological materials, the fundamental physics of many phenomena in biology continue to pose a substantial challenge in terms of modeling and simulation.

One strategy, which has been followed by many engineering disciplines, is to devise independent models, which can describe material behavior and aspects at a given scale. A collection of descriptions across scales can then provide a practical (and theoretically sound) description of the material. The link between scales is essentially given by the parameters characterizing each scale independently. A good example is the separation of quantum effects, dislocation theory, and continuum models of metals, for example. This approach is quite successful when it comes to the description of materials and utilized in many applications (consider the discussion of copper in Chap. 1: Introduction)—a quantum or atomistic perspective may be utilized for nanoelectronics, for example, whereas a continuum perspective for larger mechanical components. The intended *application scale* determines the *model resolution*.

While delineating the material properties (and thus suitability for a given application), such a piece-wise multiscale approach lends no insight into the *materiome*, no indication of how changes at one scale can be expressed at another. It is the contention of materiomics that complex hierarchical material systems, by definition, prohibit such "separation of scales"—they are intrinsically woven. A continuum model cannot be accurately developed without consideration of molecular phenomena, and molecular phenomena can be highly dependent on atomistic and quantum level effects. A complete description of the material system is required incorporating all scales—what we have labeled as the materiome. Thus, from a materiomic perspective, an integrative multiscale approach is not only desired but necessary, to not only describe material properties but to understand the structural organization, physical mechanisms throughout scales that contribute to material function.

## 6.2 Advantages of Modeling and Computation

There is an apparent disconnect between theoretical model building (*i.e.* closed-form solutions to problems) and experimental results. To validate any predictive theory (about the failure strength of a material, for example), one is inclined to make comparisons with experiments. However, analytical models typically contain unknown parameters, which have to be fitted to computational or empirical results. Indeed, many fields have retroactively uncovered the "truth" behind such parameters, and one can invoke the colloquial idiom "*hindsight is 20/20*". Modeling and computational simulation provides an alternative route. A well parameterized material model is as commensurable as a physical specimen. Moreover, unlike a physical sample, you cannot lose, break, or ruin a model (you can, however, forget how to formulate it!). Once created, a model, like an idea, cannot be destroyed.[2] The advantages of modeling and computational investigation are numerous, for example, through simulation you can:

- *Focus on specific material behaviors*: There can be a disconnect between experimental procedures and intended material response. Computational modeling allows a one-to-one correspondence between simulation and response, allowing idealized and isolated material testing.
- *Direct and precise control*: Simulated systems have the benefit that they are *designed* material samples. Essentially, the represent idealized, or "perfect" experimental conditions. There is complete control over boundary conditions (such

[2]This does not imply that all models are useful, merely the fact that models are more akin to a theory or piece of knowledge—abstract and nonphysical—than a tangible experimental specimen. At times "failed models" are most useful as they teach us what is missing but other models, like failed theories, are best forgotten.

as ideal fixity), specimen geometry (down to single atom placement and resolution) and experimental conditions (such as load magnitude applied to designated molecules or atoms, or applied strain rate maintaining a precise, constant, and fixed value).

- *Manipulate material system components and properties*: Simulation can explore potential "*What if?*" problems, unconstrained by realistic properties. This encompasses parametric studies (such as changes in pressure and temperature) as well as modification of the material itself (What if copper bonds were twice as strong? What if concrete was half as stiff? What if the moon was made of cheese?). Variations in true behavior can lead to better materials and optimized system responses, and explore the functionality of complex biomaterial systems.
- *Easily introduce system defects*: The failure of systems provides a vast amount of data, and are typically characterized by highly nonlinear and coupled phenomena that is difficult to physically monitor (such as crack propagation for fracture of a composite material). Simulation allows the introduction defects, with precise geometry and placement, allowing the systematic investigation of such nonlinear behaviors. For complex biological materials, such defects can are typically characterized by mutations (unnatural changes in protein sequence), which, unlike experimental samples, can be introduced by a simple change in computer code.
- *Quantify information/data inaccessible to experiment*: Even the common mechanical measurement of stress is dependent on measuring force and an associated cross-sectional area. Simulation allows the direct extraction of stress and strain, as well as more abstract quantities such as energy (potential energies, strain energies, free energies, etc.), local temperature, atom quantities (such bond energy, charge, etc.), and a host of statistical measures. If it can be calculated it theory, it can be "measured" by computational modeling.
- *Repeat and reproduce the results indefinitely*: The "sample supply" of a computational model is limitless. While experimental protocols may require economy in the design and undertaking of tests and measurement (especially if the tests are destructive), simulation allows endless test attempts (see Fig. 6.2). This can facilitate more creative tests, for unexpected responses; and is particularly feasible now with the broad availability of high-performance computing (*e.g.* massively parallelized computers).

With such advantages, if ideal models of materials and material systems were available, it appears we would no longer require experimental efforts. Biological tissues, drugs, high-performance materials, and nano-machines would all be designed *in silico* and merely need to be built-to-order. This, however, is obviously not the case. The primary caveat is that we currently do not have ideal models of materials and their interactions. For example, while there are sophisticated full atomistic models of protein interactions (such as the CHARMM force field, to be further discussed later within this chapter), such models are currently restricted to simple investigations, limited in terms of chemical reactivity, scale, and ultimately functional predictive power. The parameterization of such models requires either experimental

**Fig. 6.2** Limited versus unlimited material supply. One advantage of computer modeling and simulation is that many configurations of material systems can be stored and simulated with relative ease. Experimental samples need to be synthesized, and can be consumable or destroyed upon testing. Moreover, uncommon or complex materials (such as a rare mutations in proteins or unstable allotropes of molecules) may be difficult or uneconomical to procure. Material models, conversely, are inexhaustible and can easily be formulated for any imaginable chemistry. Note however, that current material models are not universal, and empirical methods are still required to parameterize molecular configurations (geometry and sequence, for example) and chemical interactions. Simply put, the computer drive on the right can contain >70,000 protein structures hosted at the Protein Data Bank (http://www.rcsb.org). Albeit, such a vast collection of structural data was obtained by X-ray crystallography or NMR spectroscopy… one sample at a time

data or *ab initio* information (*e.g.*, quantum mechanical behavior) to reflect variegated permutations of elements and chemical structures. As a shared knowledge base grows, computational models become more *independent* from such fittings, and converge to theoretical ideal. Only then will simulation provide a sufficient and adequate platform for pure *in silico* design.

Presently, simulation is unlimited in potential, but grounded in physical reality. Just because we can simulate it, doesn't mean it is worthwhile and does not mean it reflects physical realty. As a trivial example, one could (potentially) develop a alpha helical protein structure, and instead of weak hydrogen bonds stabilizing the helical convolutions as in Nature, introduce stronger covalent bonds (equivalent, say, to the carbon-backbone of the protein), and then investigate the mechanical behavior. These "*Franken*"-proteins would never be able to be synthesized, yet the potential advantages and pitfalls of such a material can be observed computationally. Experimental validation is crucial for such models and simulations as "simulated materials" can be considered useless unless they can be produced in a lab. Being said, the construction of some "theoretical materials" can lead to new potential discoveries. Balance is required between what is physically *possible* and what is *simulated*, and both efforts directed accordingly.

The ultimate goal of simulation is the development of accurate computer models, enabling the predictions about properties of materials built in ways that have never been made before. For engineers, there are models for how to make a car, or a building, for example, but for designing the basic structures of new materials, the technology today is really at an infant stage (but full of potential for the future!). As such models are developed, we can improve on Nature.

## 6.3 Necessity of Atomistic Investigation

Atomistic simulations have proved to be a unique and powerful way to investigate the complex behavior of phenomena such as dislocations, cracks, and grain boundary processes at a very fundamental level. Atomistic methods are often the core in modern materials modeling, but why? An atomistic perspective is a powerful tool in the understanding of matter and materials. Indeed, from an excerpt from the renowned physicist Richard Feynman's *The Feynman Lectures on Physics* (1964), he states:

> If, in some cataclysm, all scientific knowledge were to be destroyed, and only one sentence passed on to the next generation of creatures, what statement would contain the most information in the fewest words? I believe it is the atomic hypothesis that all things are made of atoms—little particles that move around in perpetual motion, attracting each other when they are a little distance apart, but repelling upon being squeezed into one another. In that one sentence you will see an enormous amount of information about the world, if just a little imagination and thinking are applied.

Materiomic characterization founded at the molecular scale is motivated by three fundamental factors:

1. Providing a common, universal basis for all materials;
2. Reflecting fundamental functional behaviors and properties (especially failure phenomena), and;
3. Capturing even the weakest contributions to materiomic functionality (which may be expressed at large scales).

It can indeed be argued that the common, universal basis is rooted in the quantum behavior of materials, providing a mathematical description of the wave–particle duality of matter and energy. This is undoubtedly the case. Atomistic simulation must develop along with such *ab initio* information, and, ideally, each molecular force and bond explained *via* subatomic interactions. However, the quantum realm is a tricky scale, full of uncertainty and probabilities. Definition of the materiome, however, is inherently multiscale, requiring the description of material interactions from the Ångstrom to the meter. If a common language is to be used for such a description (and seamlessly bridge scales), classical (or Newtonian) physics provides the fundamental basis and a common, mechanistic foothold, proven to be theoretically sound. While the seemingly exotic behavior of matter posited by quantum mechanics becomes more apparent when dealing with extremely tiny particles, the laws of classical Newtonian physics remain accurate in predicting the behavior of the vast majority of large objects, even on the order of the size of molecules and proteins. If we can consider full atomistic behavior the cornerstone in which to build our multiscale perspective, quantum mechanics provides the foundation of that stone, but they both support the same structure (see Fig. 6.3).

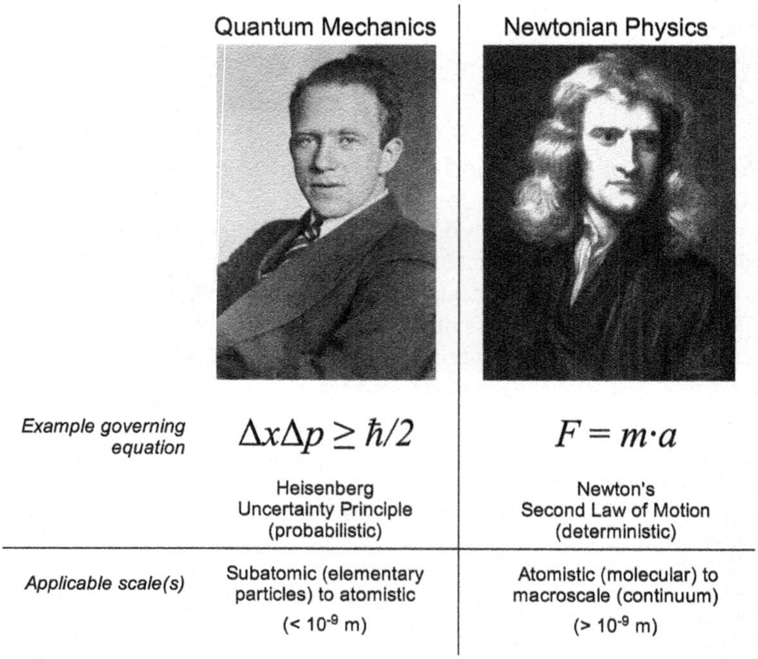


**Fig. 6.3** The scope of Quantum mechanics and Newtonian physics. The most fundamental theory of materials at our disposal is quantum mechanics, providing a mathematical description of the wave–particle duality of matter and energy. Unfortunately, the description, by definition, is *probabilistic* (*e.g.*, wave functions, quantum states, etc.). While necessary for the description of atomic and subatomic interactions (at nanometer scales and below), the description of a complex materiome is inherently multiscale. To allow consistent description across scales, classical or Newtonian physics is necessarily used, which is accurate from the molecular to macroscales and inherently *deterministic*. (*Image Left*: Werner Heisenberg, a German theoretical physicist who made foundational contributions to quantum mechanics and is best known for the uncertainty principle of quantum theory. *Image Right*: Portrait of Sir Isaac Newton, English physicist and mathematician who laid the foundation for most of classical mechanics; figures within public domain)

One of the strengths and a unique feature of atomistic methods is its very fundamental viewpoint of materials, or *materiomic*, phenomena. After all, there is only a finite number of elements and thus a finite (though vast) set of chemical interactions. From a modeling standpoint, the only physical law that is put into the full atomistic simulations is Newton's Second Law of Motion (*i.e.*, $F = ma$) and a definition of how atoms interact with each other (*e.g.*, an atomistic potential). Despite this simple basis, very complex phenomena can be simulated. Unlike many continuum mechanics approaches such as finite element methods (FEM), atomistic techniques require no *a priori* assumptions about the macroscale material description *e.g.*, elastic properties, linearity, isotropy, etc. Once the atomic interactions are chosen according to the specific bond properties and the chemical and structural makeup of the material (*e.g.*, materiomic information), the material behavior is determined by default, and mechanisms operating at multiple material scales are naturally captured (provided

**Table 6.1** Approximate strength of common chemical bonds in biology (note that there is also a dependence on the chemical microenvironment, such as surrounding charges, solvent, etc.)

| Bond type | Approximate bond strength (kcal/mol) |
|---|---|
| C≡C (Covalent) | 200 |
| C=C (Covalent) | 150 |
| C–C (Covalent) | 85 |
| S–S (Covalent) | 50 |
| $-COO^{-}\cdots{}^{+}H_3N-$ (Ionic) | 20 |
| $H_2O\cdots H_2O$ (H-bond) | 5 |
| $-CH_3\cdots H_3C-$ (London dispersion forces, part of the van der Waals forces) | <0.1 |

that a sufficiently large sample is simulated and proper averaging over the fluctuations is done).

We have seen that the material properties of biological materials are inherently linked to structures, processes, and interactions across a wide range of length and time scales. However, cross-scale interactions complicate a singular "snapshot" of material behavior at a single scale. For example, while a protein sequence can be known at the molecular level, it is very difficult to predict the function of that protein through knowledge of the amino acids. Quantitative information and understanding of the materiome must consider all hierarchies, which must include even the weakest interactions. Again, ultimately, this lies at the atomistic and molecular level, where system conformational and structural behavior can be greatly affected by relatively weak atomic interactions (such as London dispersion forces (part of the van der Waals forces) and H-bonding). To put the energy scale in perspective, we provide in Table 6.1 the typical strength of common chemical bonds in biology.

A typical covalent bond, *i.e.*, carbon-carbon, has an energy on the order of approximately 100 kcal/mol. Not only is the strength of the bond important, but also the stability. If no chemical reaction comes into play, such a carbon-carbon bond can be considered permanent. In contrast, for typical hydrogen bonds (H-bonds), the energy is on the order of approximately 5 kcal/mol (it is noted that this value is highly dependent on the donor/receptor properties of the bonded molecules as well as on the solvent environment). Further, they break and reform on a rather short time scale, and are a common example of reversible bonds. Nevertheless, H-bonding is an important (and almost universal) feature of biological materials, stabilizing protein structures and providing strength and robustness. Thus, an accurate representation of complex biological systems must incorporate molecular variations such as reversible weak bonding. Changes in H-bonding distribution, for example, can neither be directly captured by continuum methods nor coarse-grain molecular models.[3] By definition, the donor/receptor relation requires atomistic precision, and necessitates

[3]They can, however, be inferred by clever modeling.

full atomistic investigations. We next review the most common tool for atomistic exploration: *molecular dynamics*.

## 6.4 Overview of Molecular Dynamics

As discussed in Chap. 3: The Challenges of Biological Materials, a cursory inventory of biological materials (or, more specifically, protein-based materials) uncovers the basic building block at the atomistic scale: amino acids. It is through the combination of these components and at this scale that proteins are differentiated, and the resulting secondary, tertiary, and quaternary structures, *i.e.*, function, are determined *a priori*. This is one of the reasons we can root computation and simulation at the atomistic scale, and by-pass fundamental quantum level approaches.

As such, a powerful tool in the investigation of any materiome is molecular dynamics (MD). While continual computational advances have made MD approaches more accessible (and powerful) than ever before, the theoretical foundations can be traced to the mid-20th century. While full atomistic Monte Carlo methods can be attributed slightly earlier to the work of Metropolis, Ulam, Teller and co-workers [2, 3], the molecular dynamics method was first introduced by Alder and Wainwright in the late 1950's [4–6] to study the interactions of hard spheres (*i.e.*, hypothetical and ideal "atoms"). Even with such an "engineering approximation", many important insights concerning the behavior of simple liquids emerged from their studies. In 1964 the first molecular dynamics simulations using a realistic potential for liquid argon [7] were undertaken, followed shortly by simulation of liquid water in 1971 [8]. The first polypeptide and protein simulations—with a focus on macromolecular instead of bulk atomistic behavior—appeared in 1970's [9–12].

Molecular dynamics has evolved as a suitable tool for elucidating the atomistic mechanisms that control deformation at the nanoscale, and for relating this information to macroscopic material properties. Mechanistic behavior, function, and failure are directly linked to distinct atomistic mechanics and require atomistic and molecular level modeling as an indispensable tool for studying biological materials. Unlike a bulk metal (such as gold, for example), knowledge of a specific amino acid sequence that defines a specific protein material (such as DNA) begets new responsibility (nay, liability) on the researcher to both understand each amino acid interaction, and the resulting affect on structural and system-level behavior.

An understanding of molecular mechanisms, both theoretically and computationally, for different material systems is essential for a complete description of the materiome, and thus discussion of molecular dynamics is the central discourse of this chapter. However, it is strongly noted that, for the sake of brevity, a focus is given only to the essential theoretical basis of molecular simulation methods, including a limited selection of force fields. More thorough discussions can be found in the suggestion readings found at the end of the chapter.

**Molecular dynamics (MD)**: a computational methodology in which simulation of the physical movements (trajectories) of individual atoms is determined *via* Newtonian mechanics. Molecular dynamics simulation is frequently used in the study of proteins and biomolecules, in which dynamic, time-dependent phenomena is essential. A molecular dynamics simulation requires the definition of a potential function or force field—a description of the terms by which the atoms and molecules in the simulation will interact.

The sole objective of MD techniques is to accurately simulate the motion of a group of atoms, generally representing the fraction of a larger system, to observe a critical phenomenon of interest, and/or to get an estimate of the global system properties. This is similar to statistical mechanics, wherein the microscopic properties of individual atoms and molecules are related to the macroscopic bulk properties of a system *via* thermodynamic metrics. Rather than dealing with statistical ensembles, MD considers the evolution of a system over time, and the results of molecular dynamics simulations may be used to determine macroscopic thermodynamic properties of the system based on the *ergodic hypothesis*: the statistical ensemble averages are equal to time averages of the system. Being said, extraction of meaningful data through MD simulations requires proper averaging of the quantities that were sampled (to eliminate statistical "noise", so-to-speak), similar to the collective ensemble behavior of statistical mechanics. Like most science, a single result is not enough to provide confirmation of a hypothesis, but truth surfaces over multiple repeated trials.

The basic computational instrument behind MD is to calculate the dynamical trajectory (the spatial movement over time) of each atom in the material using atomic interaction potentials that describe attractive and repulsive forces in between pairs or larger groups of atoms. The interaction potentials are generally based on a mix of empirical data and first-principles (or *ab initio*) based information such as quantum mechanics calculations (see Fig. 6.4(a)). Solving each atom's equation of motion according to Newton's Law $F = ma$, enables the calculation of positions $\overline{r}_i(t)$, velocities $\overline{v}_i(t)$, and accelerations $\overline{a}_i(t)$ at each step in time, leading to atom trajectories that can reveal overall dynamics of the system as well as properties such as viscosity, bulk modulus, or fracture toughness. It can be considered an alternative approach to methods like Monte Carlo (repeated random sampling of atomistic systems), with the difference that MD actually provides full dynamical information (*i.e.*, material behavior "timelines") and deterministic trajectories.

Formally, the total energy ($E$) of the system is written as the sum of kinetic energy ($K$) and potential energy ($U$):

$$E = K + U \tag{6.1}$$

where the kinetic energy is given by

$$K = \frac{1}{2}\sum_{j=1}^{N} m_j \overline{v}_j^2 \tag{6.2}$$

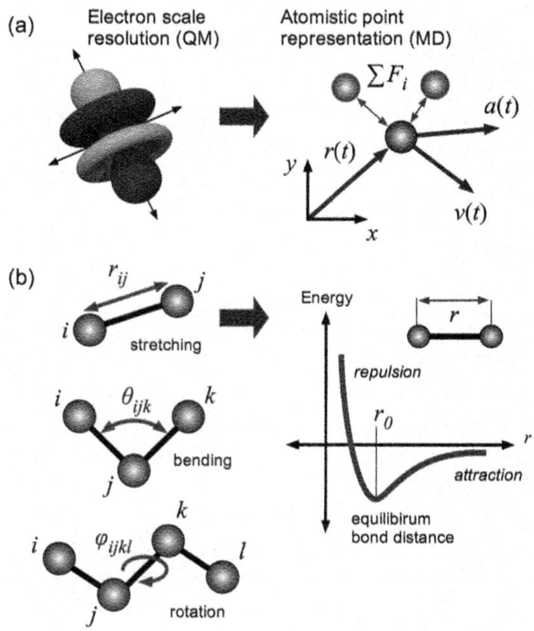


**Fig. 6.4** Basic approach of the molecular dynamics simulation method. (**a**) Atomistic structure (neutrons, electrons, and protons) replaced by a point representation in the molecular dynamics approach. (**b**) Illustration of the energy decomposition in classical molecular dynamics force fields, along with a representation of a simple potential function between pairs of atoms

and the potential energy is an effective function of the atomic coordinates, $\overline{r}_i(t)$, such that:

$$U = \sum_{j=1}^{N} U_j(\overline{r}_j) \tag{6.3}$$

with a properly defined potential energy surface, $U(\overline{r})$. Here, we emphasize the *effective* potential energy as a function of coordinates (or position), because more sophisticated atomistic force fields take into account not only position, but also environmental effects (such as atomic neighbors and local charge). In general, the potential, $U$ can be formulated as a function of multiple geometric constraints (summed for each atom), including bond stretching, $U(r_{ij})$, bending, $U(\theta_{ijk})$, and rotation or torsion, $U(\phi_{ijkl})$ (see Fig. 6.4(b)). Regardless of form, the forces are obtained from the potential energy surface (sometimes also called force field or potential) as $\overline{F} = -\nabla_{\overline{r}_j} U(\overline{r}_j)$. Simplification of atomic interactions by a closed-form potential significantly increases the computational efficiency of MD simulation. Thus, through Newton's Second Law, molecular dynamics is reduced:

$$m_j \frac{d^2\overline{r}_j}{dt^2} = -\nabla_{\overline{r}_j} U(\overline{r}_j) \tag{6.4}$$

For $j = 1 \ldots N$. The numerical problem to be solved is thus a system of coupled second-order nonlinear differential equations that can only be solved numerically given more than two particles (*i.e.*, the classical $n$-body problem). Typically, MD is based on updating schemes that yield new positions from the old positions, velocities, and the current accelerations of particles, as well as any imposed boundary conditions, forces, of constrains. For instance, in the commonly used Verlet scheme, this can be mathematically formulated as:

$$r_i(t_0 + \Delta t) = -r_i(t_0 - \Delta t) + 2r_i(t_0) + a_i(t_0)(\Delta t)^2 + O\left(\Delta t^4\right) \tag{6.5}$$

The forces and accelerations are related by $a_i = f_i/m$. The forces are obtained from the potential energy surface (or force field) as:

$$F = m_j \frac{d^2\bar{r}_j}{dt^2} = -\nabla_{\bar{r}_j} U(\bar{r}_j) \tag{6.6}$$

For $j = 1 \ldots N$. This technique—the time integration of particle positions based on applied potentials—cannot only be used for particles that are atoms, it can also be applied for particles that represent groups of atoms, such as in molecules or in coarse-grain models (to be discussed in Sect. 6.6).

Various fast numerical integration schemes are employed to solve the equations of motion and simulate a large ensemble of atoms representing a larger material volume; however, in particular for all-atom simulations, high-frequency vibrations of light atoms (such as hydrogen atoms) requires a time step in the order of femtoseconds ($10^{-15}$ seconds!) for accurate and numerically stable calculations. This limits the application of full-atomistic MD methods to nanometer size systems, at submicrosecond time scales. It is also noted, however, that simulations in excess of hundreds of nanoseconds typically run for weeks or months and large computer clusters with hundreds of processors.

## 6.5 Atomistic Force Fields: Conventional and Reactive

All-atom force fields are predominantly used in molecular dynamics simulations of materials at the nanoscale, as they generally are the most reliable yet computationally efficient way of studying dynamics of materials and molecules. Definition of atomic interactions by force fields is at the very crux of molecular dynamics method, as it defines the complete material behavior. The expressions for the potential energy landscape are used to determine the forces and accelerations used in a MD updating scheme (*i.e.*, Eqs. 6.5 and 6.6). Many flavors and types of force fields exist, from empirically derived formulations to those parameterized from detailed quantum calculations, implemented within a wide range of simulation programs which are continuously being refined and expanded. The GROMOS/GROMACS[4] [13] packages are also commonly used in all-atom molecular dynamics. The NAMD Scalable Molecular Dynamics [14] program[5] is a popular code that is capable of carrying out computations using CHARMM and other force fields, specifically focused

[4]http://www.gromacs.org/.

[5]http://www.ks.uiuc.edu/research/namd/.

on biomolecules. The LAMMPS ("Large-scale Atomic/Molecular Massively Parallel Simulator", originally developed by Steve Plimpton and coworkers at the Sandia National Laboratory[6]) molecular dynamics code is a highly flexible package that has potentials for soft materials (*e.g.*, biomolecules and polymers), solid-state materials (*e.g.*, metals, semiconductors), and coarse-grained or mesoscopic systems [15]. It can be used to model atoms or, more generically, as a parallel particle simulator at the atomistic, meso, or continuum scales.

Here, we briefly review some of the basic force field concepts and components, applied in material systems ranging from metals, polymers, and proteins. The methods developed within the realm of conventional force fields are widely applicable, and are the basis for mesoscopic descriptions of molecular behavior.

### *6.5.1 Conventional Force Fields*

Many force fields have been developed to describe crystalline systems, including metals and ceramics.These are typically separated into pair potentials and multi-body potentials. In pair potentials, the total potential energy of the system is expressed as the sum over all pairs of bonds. In multi-body potentials, the strength of the pair-wise interactions are modulated by the atomic environment (*i.e.*, proximity of neighbors, local charge, etc.). Examples for pair potentials include the Lennard-Jones potential [16] and the Morse function [17]. Popular multi-body potentials include the Tersoff potential [18], or the Stillinger-Weber potential [19] to model covalent interactions, or the EAM and MEAM potentials suitable for metals [20–22].

Popular force fields specific to organic systems and biomolecules include the DREIDING, AMBER, CHARMM force fields, widely used throughout the protein and biophysics community. Force fields of this type fall into a general category of *nonreactive* potentials. Simply put, a nonreactive potential is characterized by no chemical reactions, or permanent molecular connectivity and chemical state. Such force fields are based on harmonic and anharmonic terms describing covalent interactions, in addition to long-range contributions describing van der Waals (vdW) interactions, ionic (Coulombic) interactions, as well as hydrogen bonding. Since the bonds between atoms are modeled by harmonic springs or variations thereof, bonds between atoms can not be broken, and new bonds can not be formed, nor can equilibrium angles change and geometric configurations change. Further, the charges are fixed and can not change. Being said, the CHARMM and similar force fields provide a reasonable description of the structures and behavior of proteins. For example, a recent study implemented the CHARMM force field to investigate the transition from $\alpha$-helical to $\beta$-sheet conformations ($\alpha$-$\beta$ transition)—a universal deformation mechanism in alpha-helix rich protein materials such as wool, hair, hoof, and cellular proteins [23]. As this $\alpha$-$\beta$ transition does not involve any chemical reactions or bond breaking, the nonreactive force field can accurately capture

[6]http://lammps.sandia.gov/.

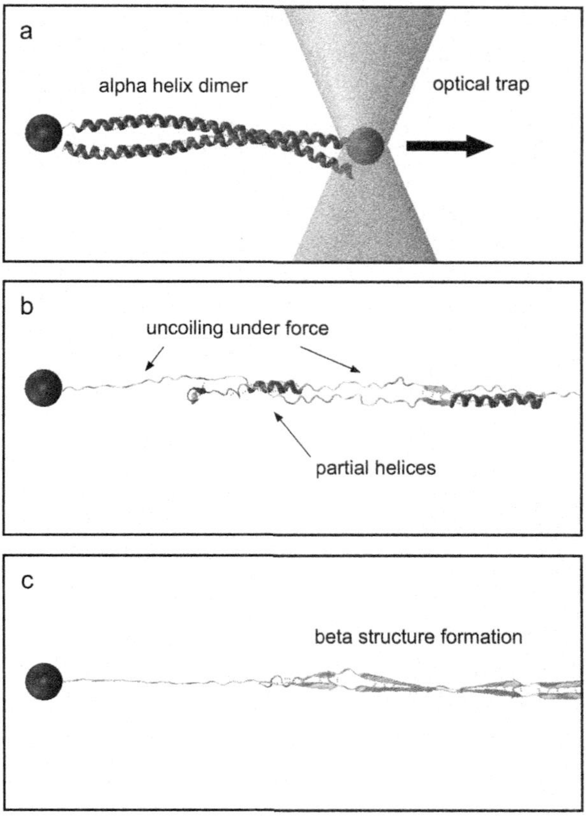


**Fig. 6.5** Implementing the nonreactive CHARMM potential to investigate fibrous proteins under mechanical load. Alpha helices and beta sheets are major secondary structural motifs that organize the three-dimensional structure of proteins—the transition between alpha-helices and beta sheets ($\alpha$-$\beta$ transition) is observed universally for a broad range of alpha-helix based protein filaments. The transition significantly thins filaments, but toughens and strengthens it at large deformation. The snapshots depict the structure of a coiled-coil alpha-helical dimer under mechanical load (representing the loading of experimental assays, such as optical tweezers). (**a**) Initial, alpha-helical structures. (**b**) Characteristic uncoiling and induced transverse alignment of the peptide sequence. (**c**) Formation of beta-sheet structure. Adapted from [23]

the change in protein structure. Moreover, as the transition is induced due to applied force, MD formulations are necessary to capture the dynamic evolution of the system (see Fig. 6.5).

**Nonreactive Potential**: Full atomistic molecular dynamics formulation which is characterized by permanent molecular connectivity (encompassing bonds and angles) and chemical state (such as ionic charge). Examples include the DREIDING, AMBER, CHARMM force fields. Such force fields are based on computationally efficient harmonic and anharmonic terms describing interactions. Parameters are specific to a known molecular structure, and not necessarily transferable depending on chemistry and molecular configuration. For example, a carbon atom within the backbone of a protein ($\alpha$-carbon) is parameterized differently than a carbon atom of a carbon nanotube.

A contributing factor to the accuracy of such nonreactive force fields is the relatively limited number of components to consider—*i.e.*, all combinations of amino acid interactions can be accounted for with relatively minimal effort. For the sake of brevity, the main aspects of the CHARMM force field will be discussed here; the basic concepts of the MD technique and force field formulations are common to all packages used.

The parameters in force fields are often determined from more accurate quantum chemical simulation models by using the concept of force field training. For example, parameters for the CHARMM force field have been meticulously optimized and revised over the years, taking into consideration a wide variety of input including *ab initio* results, *e.g.*, *via* Density Functional Theory (DFT), experimental crystal structures and geometries, as well as vibrational spectra [24]. Specifically, the CHARMM potential includes bonding and non-bonding interaction terms to describe short- and long-range forces between particles, where the contributions to bond stretching, bending, and rotation are individually expressed. For the three contributions in the figure shown (Fig. 6.4(b)), simple mathematical expressions are used. For example, both bond stretching can bond angles can be described by a simple harmonic spring functions, expressed for bond stretching as:

$$\phi(r_0) = \frac{1}{2} k_{bond}(r - r_0)^2 \tag{6.7}$$

In addition to simple harmonic parameters (*e.g.*, $k_{bond}$ representing bond *stiffness*, $r_0$ bond *length*), several other terms are included to model the chemical properties of proteins and nucleic acids correctly. In the CHARMM model, for example, the mathematical formulation for the empirical energy function that contains terms for both internal and external interactions has the form:

$$\begin{aligned} U_{CHARMM} &= U_{bonds} + U_{UB} + U_{angle} + U_{dihed.} + U_{improp.} \\ &\quad + U_{nonbond} + U_{Coulombic} \\ &= \sum_{bonds} K_b(b - b_0)^2 \\ &\quad + \sum_{UB} K_{UB}(S - S_0)^2 \\ &\quad + \sum_{angles} K_\theta(\theta - \theta_0)^2 \\ &\quad + \sum_{dihed.} K_\chi\left[1 + \cos(n\chi - \delta)\right] \\ &\quad + \sum_{improp.} K_{imp}(\phi - \phi_0)^2 \\ &\quad + \sum_{pairs} 4\varepsilon\left[\left(\frac{\sigma}{r_{ij}}\right)^{12} - \left(\frac{\sigma}{r_{ij}}\right)^{6}\right] \\ &\quad + \sum_{ions} \frac{q_i q_j}{\varepsilon_d r_{ij}} \end{aligned} \tag{6.8}$$

where $K_b$, $K_{UB}$, $K_\theta$, $K_\chi$, and $K_{imp}$ are the bond, Urey-Bradley, angle, dihedral angle, and improper angle force constants, respectively; $b$, $S$, $\theta$, $\chi$, and $\phi$ are the bond length, Urey-Bradley 1,3-distance, bond angle, dihedral angle, and improper torsion angle, respectively, with the subscript zero representing the equilibrium positions for the individual terms.

The Coulombic (*e.g.*, ionic) and Lennard-Jones 6:12 terms constitute the external or nonbonded (pair) interactions; $\varepsilon$ reflects the strength of the interaction, or the Lennard-Jones well depth, $\sigma$ is the distance at the Lennard-Jones minimum, $q_i$ is the partial atomic charge, $\varepsilon_d$ is the effective dielectric constant, and $r_{ij}$ is the distance between atoms $i$ and $j$. In the CHARMM force field, no additional terms are used for H-bonds, since the combination of charge and Lennard-Jones contributions were verified to be adequate for describing protein, solvent, and interface hydrogen bonding. In all-atom force fields, water molecules are generally also treated explicitly. Parameters of the force field generally are specified considering a specific water model (*e.g.*, TIP3P model for CHARMM [25, 26]).

The CHARMM force field belongs to a class of models with similar descriptions of the interatomic forces; where other models include the DREIDING force field, the UFF force field (Universal Force Field) [27], or the AMBER model [28, 29]. As discussed, in CHARMM and other classical force fields, bonded terms are modeled with harmonic springs or its variations, and therefore cannot be modified (*e.g.*, towards a different chemical state, such as from $sp^2$ to $sp^3$, or broken) once defined by the connectivity input obtain from the topology of the molecule. Further, the atomic charges are fixed and cannot change during a simulation. These simplifications improve the simulation speed drastically and are not a major issue for most simulations studying conformational changes of proteins under ambient physiological conditions. On the other hand, simulations in extreme conditions such as mechanical perturbations (*e.g.*, protein unfolding studies where the breaking of covalent bonds is involved) or modeling the properties under the exposure to harsh chemical environments may require reactive force fields that can take into account changes in fixed charges of the molecules, the formation and breaking of new bonds and variations in the bond order (*e.g.*, single versus double bond). We refer the interested reader to review articles for additional information, in particular regarding force field models [26, 29–32].

### 6.5.2 Reactive Force Fields

To fully describe complex chemomechanical interactions and processes—especially those encapsulating large deformation and failure—"static" nonreactive force fields are insufficient, and reactive force fields are necessary. Reactive force fields represent a major milestone in over-coming the limitations of classical nonreactive force fields, specifically their lack of the ability to describe rupture and formation of covalent bonds and their limitations in modeling chemical reactions. The covalent bond terms described using harmonic terms (see, for example, Eq. 6.7), do not provide an accurate description of the bond energetics at large bond stretch and during reformation of new bonds (in fact, the bonds become energetically "stronger" as they are stretched). For failure properties of materials which naturally involves large bond deformation and bond rupture mechanisms, this translates into the properties of molecules at large strain, a phenomenon also referred to as nonlinear elasticity or *hyperelasticity*. These effects can have profound impact for materials failure mechanisms, especially for materials subject to fracture under extreme mechanical conditions, such as brittle crystalline materials [33, 34].

Several flavors of reactive potentials have been proposed in recent years, such as ReaxFF [35, 36] and the AIREBO (Adaptive Intermolecular Reactive Empirical Bond Order) potential [37, 38]. Reactive potentials can overcome the limitations of empirical force fields and enable large-scale simulations of thousands of atoms with near quantum mechanics accuracy. The ReaxFF formulation of reactive potentials, originally developed for hydrocarbons [35], have now been extended to cover a wide range of materials, including metals, semiconductors, and organic chemistry in biological systems such as proteins [36, 39–44]. Once more, for brevity, we will focus discussion on only a general ReaxFF potential and note that this is not meant to lessen development and impact of other reactive potentials. There are other versions of the ReaxFF potential with slight modifications suitable for protein (*biomateriomic*) modeling, and the force field itself is constantly being refined and extended. The idiosyncrasies of such formulations are beyond the scope of the current discussion.

**Reactive Potential**: Full atomistic molecular dynamics formulation which is characterized individual element parameterization. Examples include ReaxFF and the AIREBO potentials. Reactive force fields take into account the local chemical environment to dynamically alter the atomistic and molecular interactions. Parameters are specific to elements only, and ideally are transferable to any chemistry and molecular configuration. However, the potential must be trained to account for all possible chemical reactions, and as such current reactive potentials are limited to few element types (the AIREBO potential, for example, is suitable for hydrocarbons only). Nevertheless, connectivity information such as bonds and angles are not required, as they are implicitly determined by the system geometry.

Reactive potentials are based on a more sophisticated formulation than most non-reactive potentials. To describe the details of bond stretching and breaking, a bond length-bond order relationship is employed to obtain smooth transition from non-bonded to single, double, and triple bonded systems, for example. All connectivity-dependent interactions (valence and torsion angles) are formulated to be bond-order (BO) dependent. This ensures that their energy contributions disappear upon bond dissociation (*i.e.*, breaking) so that no energy discontinuities appear during reactions. Similar to the CHARMM model, reactive potentials also feature non-bonded interactions such as shielded van der Waals and shielded Coulomb (ionic). Additionally, reactive formulations use a geometry-dependent charge calculation scheme

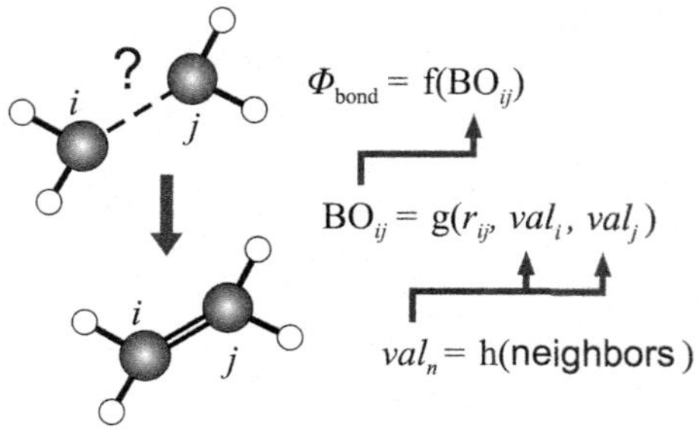


**Fig. 6.6** Simplified schematic of the hierarchical embedded bond-order functions within the ReaxFF potential formulation. During molecular dynamics computations, the appropriate bond force is determined for atom pairs, $ij$, as a function of bond order, $BO_{ij}$ rather than bond length, $r_{ij}$. Description of the energy as a function of bond order allows a more fundamental representation of the underlying chemistry. In contrast, conventional force fields such as CHARMM express energies as a function of distance only, and are constrained to nonreactive and defined connectivities. The bond order is dependent on the specific atom pairing, interatomic distance, and available valency of each atom. The valency, in turn, relies on the atom type and current neighbors (either bonded or nonbonded) that affect coordination. The embedded cascading functions can follow many permutations, dependent on the current state of the system, allowing for a much more robust chemical representation

that accounts for polarization effects and modeling of charge flow, assigning a partial charge to each atom at each integration step and thereby includes important quantum mechanical details about interatomic bonding. Due to the increased complexity, reactive potentials can be about twenty to thirty times more expensive (computationally) than conventional models. At the risk of sounding repetitious, a reactive potential can be thought of as a *hierarchical* force field where the functionality expressed between the atom-atom interactions is founded by embedded relations (*i.e.*, lower scales) for governing parameters as a function of geometry, bond-orders, and local environment (depicted schematically in Fig. 6.6). This is a critical breakthrough leading to a seamless multiscale bridge between QM and empirical force fields, illustrating a practical and fundamental theme of materiomic investigation.

To explicate a little further, in ReaxFF, the total energy of a system is expressed as the sum of different contributions that account for specific chemical interactions (similar to the CHARMM formulation as given by Eq. 6.8). The total energy is defined by:

$$E_{system} = E_{bond} + E_{vdW} + E_{Coulomb} + E_{angle} + E_{torsion} + E_{over} + E_{under} \tag{6.9}$$

The above formulation encompassing interactions ranging from simple two-body contributions ($E_{bond}$, $E_{vdW}$) to multi-body contributions due to local chemical environment ($E_{over}$, $E_{under}$). Deconstructing the term for bond energy as a summation of all bonds, we can write:

$$E_{bond} = \sum_{bonds} \phi_{bond} \tag{6.10}$$

where the bond potential can be expressed as:

$$\phi_{bond}(D_e, p_e, BO_{ij}) = -D_e BO_{ij} \cdot f(BO_{ij}) = -D_e BO_{ij} \exp\left[p_e\left(1 - BO_{ij}^{p_e}\right)\right] \tag{6.11}$$

where $D_e$ and $p_e$ are additional bond parameters, subject to the atom types. We note that, unlike the CHARMM formulation, the bond potential is a function of bond-order, $BO_{ij}$, as opposed to bond distance, $r_{ij}$, which is explicitly determined based on the current state of each atom. Typically, all connectivity-dependent interactions—such as bonds, angles, and torsions—are formulated to be bond-order dependent. The basic concept of bond orders is simple to explain where individual bond strength, formulation, or rupture is based on the local atomistic environment, taking advantage of some theoretical chemistry principles. For example:

$$BO_{ij} = \sum_j g(r_{ij}, r_{0,ij}, val_i, \alpha_j, \beta_j) \tag{6.12}$$

for $j = 1, 2, \ldots, N$ nearest neighbors. The terms $\alpha_j$ and $\beta_j$ are numerical fitting parameters that describe the dependence of the bond order on the bond geometry, adapted for each bond type, $ij$. The term $r_{ij}$ is the interatomic distance, $r_{0,ij}$ the equilibrium distance for a specific bond, and $val_i$ the atomic valency to account for over- and under-coordination (accounting for the fact that carbon, for instance, cannot have more than four bonds). Thus, in consideration of the atomistic environment of an individual atom, relevant bond strengths can be determined and varied during simulation. Similar functions are defined for all contributions. The embedded functions immediately lead to a relationship between bond length, binding energy, and coordination. This method has been very successful to describe the interatomic bonding in several materials, for example the C–C bonds in diamond, graphite, carbon nanotubes, and even hydrocarbons.

As an example of the formulation, we can look at the bond order/bond distance relationship for carbon-carbon interactions, including the contributions from sigma ($\sigma$), pi ($\pi$), and double-pi ($\pi\pi$) bonds. Each contribution is related to the interatomic distance ($r_{ij}$), *via* a smooth fitted function:

$$BO_{ij} = \exp\left[\alpha_\sigma \left(\frac{r_{ij}^{\sigma}}{r_0}\right)^{\beta_\sigma}\right] + \exp\left[\alpha_\pi \left(\frac{r_{ij}^{\pi}}{r_0}\right)^{\beta_\pi}\right] + \exp\left[\alpha_{\pi\pi} \left(\frac{r_{ij}^{\pi\pi}}{r_0}\right)^{\beta_{\pi\pi}}\right] \tag{6.13}$$

This function is depicted in Fig. 6.7. The expression above was a smooth fit for quantum-level calculations of the bond dissociation (*e.g.*, the $\alpha$ and $\beta$ parameters), ensuring a smooth transition in bond order [35, 41]. We note that here, the bond order is designated "uncorrected" as overcoodination and undercoordination terms are implemented to force systems toward the proper number of bonds (*e.g.*, valence terms as expressed in Eq. 6.12). Note that the contribution from the double-pi bond decays at a relatively short separation. The end result is a chemically accurate bond potential that is informed "on-the-fly", but requires many *ab initio* parameters (here, $BO_{ij}$ is dependent on *nine* fitted parameters, and only applicable to a single

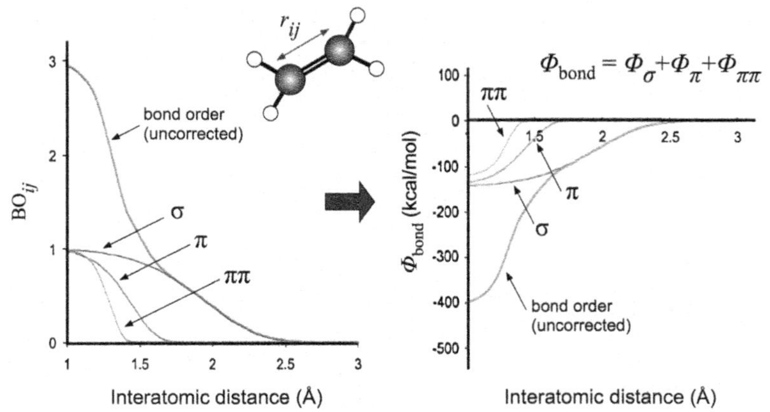


**Fig. 6.7** From bond order to bond potential. Carbon-carbon interactions in the ReaxFF potential are formulated *via* bond order/bond distance relationships, explicitly accounting for sigma ($\sigma$), pi ($\pi$), and double-pi ($\pi\pi$) bond contributions, determined by quantum-level investigations [35]. Bond order, $BO_{ij}$, is determined by interatomic distances (function given by Eq. 6.13), and treated as a variable within the bond potential function, $\phi_{bond}$ (see Eq. 6.14)

carbon-carbon interaction!). In implementation, the bond potential given in Eq. 6.11 explicitly accounts for the sigma, pi, and double-pi contributions, or:

$$\phi_{bond} = -D_e^\sigma BO_{ij}^\sigma \cdot f\left(BO_{ij}^\sigma\right) - D_e^\pi BO_{ij}^\pi \cdot f\left(BO_{ij}^\pi\right) - D_e^{\pi\pi} BO_{ij}^{\pi\pi} \cdot f\left(BO_{ij}^{\pi\pi}\right) \tag{6.14}$$

It is again stressed that a crucial aspect of the ReaxFF force field is that all parameters are derived from fitting against quantum mechanical data. This process is referred to as *force field training*. The basic concept is to require a close-as-possible agreement between the QM prediction and the force field results, for a wide range of properties and chemical or mechanical environments. Typically these properties include elastic properties and equation of state, surface energies, dissociation energies and landscapes, or the interaction energies of organic compounds with surfaces. This is a common theme to materiomics, where information from a lower hierarchical level—in this case DFT level calculations—is passed to a higher hierarchical level to describe larger-scale system properties and behavior.

Figure 6.8 shows the results of stretching and breaking a small protein molecule with covalent cross links under tensile loading [44]. The study shown here reveals the differences between a nonreactive CHARMM force field and the ReaxFF reactive force field. Figure 6.8 shows snapshots as the molecule is being stretched, modeled using the reactive model. As the molecule is being pulled, the covalent cross-linking disulfide bonds within the molecule break. These breaking points correspond to the peaks in the force-extension plot shown in Fig. 6.8, and the force drops significantly after each breaking point as the elastic energy stored in the protein is released. In the case of the CHARMM model, bond breaking cannot be described, and the force continues to rise once the covalent cross-link within the pro-

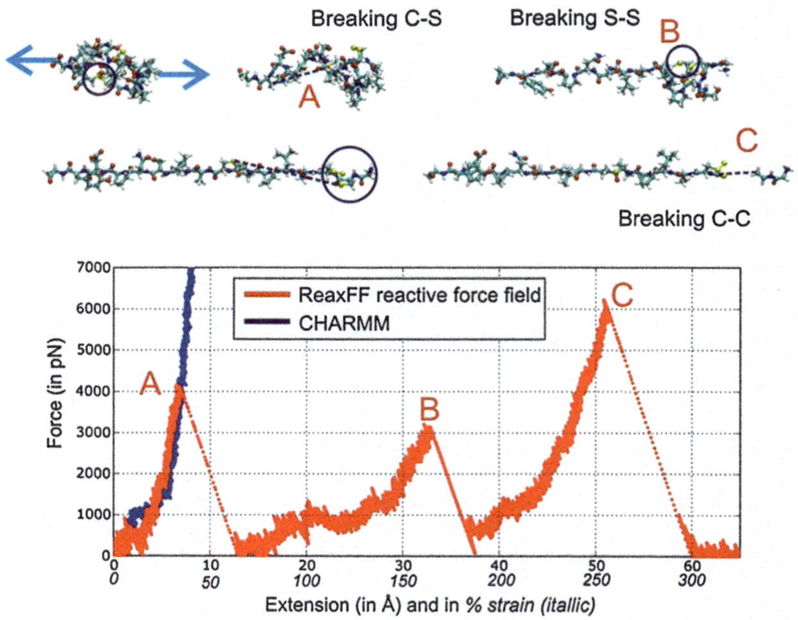


**Fig. 6.8** Reactive vs. nonreactive potentials. Breaking a single protein molecule by pulling at the ends of a small protein (conotoxin PnIB from *conus pennaceus*; PDB identification code 1AKG). The simulation shown here reveals the differences between a nonreactive (CHARMM) force field and the ReaxFF reactive force field. Snapshots depict the molecule as it is being stretched, modeled using the reactive model (ReaxFF). As the molecule is being pulled, the covalent cross-links (disulfide bonds) within the molecule break. These breaking points correspond to the peaks in the force-extension plot; and the force drops significantly after each breaking point as the elastic energy stored in the protein is released. In the case of the CHARMM model, bond breaking can not be described, and the force continues to rise once the covalent cross-link within the protein is being stretched. Modified from [44] and [32]

tein is being stretched. The results shown here clearly illustrate the significance of a reactive force field approach in modeling the failure of molecules, specifically when the breaking of covalent bonds is involved.

The utility of reactive potentials is obvious when failure (*e.g.*, bond breaking or dissociation) occurs, but reactive force fields can also reflect changes in chemical environments. For example, disulfide bonds are important chemical cross-links that control the elasticity of fibrous protein materials such as hair, feather, wool and gluten in breadmaking dough [45–48]. Indeed, as mentioned prior in Chap. 2: The Materiome, the disulfide bond bridges at the molecular level resulted in the popular "perm" hairstyle, thermally controlling (breaking and reforming) the disulfide-based molecular links in human hair. Disulfide bonds provide stability to folded structures of proteins, enabling specific mechanical functions such as molecular sensing, switching and signalling in biology [47, 49, 50]. The fracture strength of disulfide bonds, however, is decreased under the presence of reducing agents, due to a loss of cross-link stability controlled by the chemical microenvironment. A nonreactive potential would be forced to "assign" a single strength and stiffness to S–S bonds, regardless of environment. Recent simulations in explicit solvent and dithiothreitol (DTT) employing ReaxFF indicate potential tunability of the elasticity, rupture mechanism and strength of proteins, where factors such as accessibility, mechanical strain and local redox potential govern the dominating rupture mechanism and location [51].

By using the first principles based ReaxFF reactive force field, it was shown that the stability of the disulfide bond is strongly influenced by the redox potential of the chemical microenvironment, where variations in the concentration of reducing agents can trigger various fracture mechanisms in the protein [51]. A direct simulation of the fracture strength showed that the presence of a reducing agent drastically changes the strength of a protein disulfide bond, and controls the rupture location of a bond (Fig. 6.9)—the reducing agent decreases the energy barrier of disulfide bond rupture, an effect only possible *via* reactive modeling.

Similarly, experimental studies have also indicated a transition to elongated disulfide bonds upon introduction of a reducing agent to the system [52]. Key mechanistic insights can be gained from inspection of the deformation of such bonds under load—such as the elongation and softening of the S–S bond with respect to the competing candidate for rupture, the C–S bond, facilitating concentration of mechanical force on the weakened bond. Such effects are near impossible to directly observe experimentally.

Overall, the first principles based ReaxFF reactive force field, combined with mechanical perturbation and long-time scale structure and inventive free energy sampling protocols such as metadynamics [53, 54], provides a computational approach for studying the link between chemistry and mechanics of protein materials in various physical and chemical environmental conditions. It should be noted that the most recent formulations of the force field are generic versions for protein applications, and therefore may not capture all aspects of specific chemical reactions accurately. Future iterations will focus on fine-tuning of the force-field parameters to accurately describe a range of chemical reactions relevant to protein materials, and extending reactive capabilities to greater length and time-scales. Future computational studies of the structural dynamics of biomolecules subject to chemical cues will hopefully provide insight into the detailed mechanics of proteins and complex biological materials.

## 6.6 Limitations of Full Atomistic Simulation

The atomistic or molecular approach brings great promises. Yet, if atomistic methods are so powerful, one may ask why they have yet to become universal in material applications, biological and chemical analysis, and engineering design? The answer

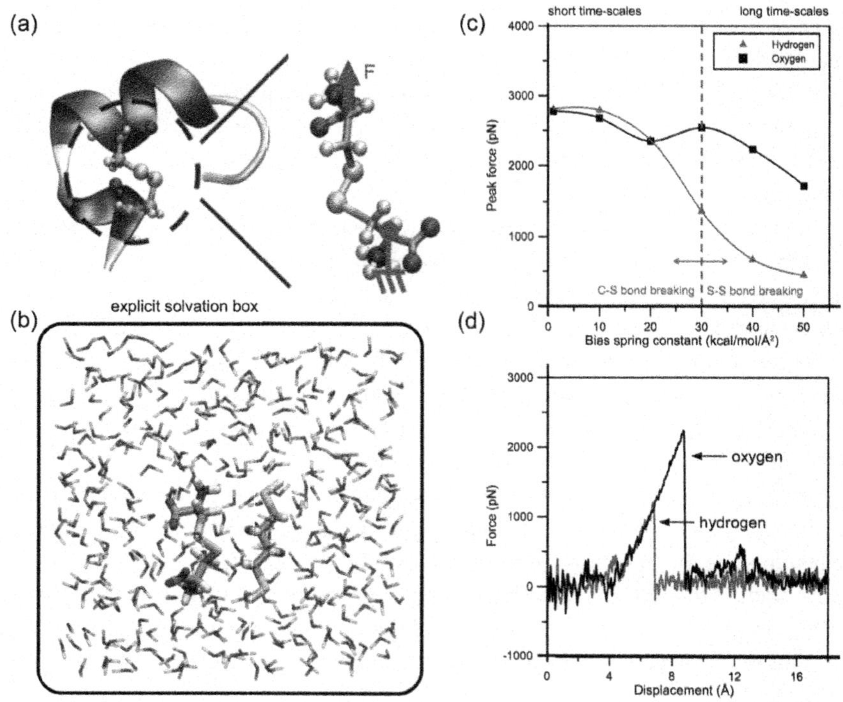


**Fig. 6.9** Simulating a reactive environment: quantifying disulfide bond strength in presence of reducing agent *via* ReaxFF. (**a**) The disulfide bond model is taken from a small protein (Protein Data Bank identification code 1AKG). Load is applied through steered molecular dynamics (SMD), where one of the amino acids is fixed and the other is pulled at the $\alpha$-carbon atom as depicted. To induce a reducing potential, hydrogen atoms (in the schematic shown as a hydrogen molecule) are tethered towards the S–S bond with a harmonic bias potential, thereby sampling the reaction at a shorter time scale. (**b**) Overall system solvated in an explicit water box (approximately 1,000 atoms) with periodic boundary conditions. (**c**) and (**d**) Changes of nanomechanical strength properties of disulfide bond under presence of a reducing and oxidizing agent. (**c**) Breaking force as a function of the bias potential spring constant, showing the breaking force is strongly dependant on the presence of a reducing or oxidizing environment. In the system with $O_2$, an oxidizing environment causes a small reduction in strength, but the C–S bond always breaks first. In stark contrast, the system with $H_2$ shows that the mechanism and force required to induce bond breaking is strongly affected in a reducing environment. As the bias potential constant increases (representing the behavior at longer time-scales) the peak forces drop to significantly smaller levels than in the oxidizing environment and the S–S breaks first rather than C–S bond. (**d**) Force-displacement curve showing a comparison of systems with $H_2$ and $O_2$ at a fixed bias potential. The result shows a much lower peak force due to the presence of $H_2$, very different from the $O_2$ case. Adapted from [51]

is, while powerful and fundamental in the description of atomistic interactions, current atomistic methods are inherently limited by three constraints:

1. Accessible timescale
2. Accessible length scales
3. Available potentials

The accessible timescale of atomistic simulations are limited by two factors: (1) the mathematical stability of the integration scheme; and (2) long-time scale phenomena at the molecular scale. The application of MD to long-time scale deformation and failure phenomena such as creep or fatigue, or dynamic processes such as protein folding, is particularly challenging. For example, even small and relatively "fast folding" proteins require timescales on the order of *micro*seconds to fold in isolated simulations [55], which requires significant computing effort. In recent years, this limitation has motivated efforts to enable atomistic and molecular level simulations of such long-time phenomena, extending potential simulation domains for protein-based materials, as well as polymers and crystalline materials [56–58]. Many of these applications are based on statistical models and include various levels of approximation for long-time scale mechanisms [59].

Complicating the matter, one of the key issues of MD is that a system may be trapped in a *local energy minimum*, and that the escape out of the local minimum is hindered due to the lack of accessible time scales effectively suppressing the exploration of the entire state space. For instance, unstable intermediate states may occur that physically persist for only a matter of microseconds, yet require substantial computational effort. In order to simulate certain phenomena such as protein folding, it is essential that the entire space of possible configurations can be explored. Methods such as metadynamics [56] applied to protein modeling overcome this limitation by enabling the system to escape out of local energy traps. Other approaches (such as the replica exchange method [60]) facilitate a more extensive exploration of the state space by running copies of the system at multiple temperatures (also called parallel tempering), effectively enabling us to simulate the long-term behavior of molecular assemblies. The autonomous basin climbing method is an algorithm that enables climbing out of potential minima, extending the metadynamics approach towards applications of modeling crystals and liquids and thereby facilitates the simulation of mechanisms such as creep in solids and viscosity effects in liquids at vast time scales.

A quick tally of the contributions of the CHARMM force field (Eq. 6.8) reveal seven different contributions for force, and thus seven different components to calculate, *per atom, per timestep*. If we consider the average interatomic distance is on the order of a few Ångstroms, there is an inherently limitation in the maximum number of atoms to efficiently simulate, and thus a corresponding limit in accessible length scale. For example, an all-atom model of a collagen microfibril—on the order of 300 nm in size—with full-length molecules including the explicit simulation of all water molecules with all chemical details results in a full-atomistic model contains of approximately 60,000 atoms [61], and requires about six hours of computational time per simulated nanosecond on thirty-two parallel processors (circa 2011). Since phenomena involving the molecular (and supramolecular) scale are often in the range of several nanoseconds or even microseconds, molecular dynamics simulations of the full-atomistic collagen fibril are at the limit of current computational capabilities. Recent advances in computational power now enable the simulation of billions of particles in massively parallelized MD simulations implemented on petaflop supercomputers, reaching dimensions on the order of micrometers [62].

One must also consider the large computational resources necessary to perform large-scale full atomistic simulations. This becomes more evident as the simulation sizes increase to systems with billions of atoms (yet still on length scales on the order of micrometers). Classical MD systems can be quite efficiently implemented on modern supercomputers using parallelized computing strategies. Such supercomputers can range from dozens to hundreds to thousands of processors. One of the largest simulations reported to date treated almost 320 billion particles—corresponding to a cubic piece of solid copper with an edge length of 1.56 μm, requiring a supercomputer with over 130,000 processors [63].

Aside from limitations with respect to the system size (*i.e.*, length scale) and the accessible time scale, MD has another important limitation related to the availability of interatomic potentials for a specific material. By design, once the atomic interactions are chosen, the complete material behavior is determined. These potential functions must be able to model the characteristic type of chemical bonding, which can be a limiting factor for the applicability of the MD method. Materials such as silicon, iron, and steel, or colloidal systems (*e.g.*, cement), as well as some polymers pose particular challenges with respect to the development of models for their chemical interactions and reactions. For modeling failure phenomena, this issue is particularly critical, as it involves bond breaking and bond rearrangements, which resembles a chemical reaction—think of bond breaking as the reverse reaction to bond formation in the synthesis of a molecule from its basic atomic constituents. Although in some cases is difficult to find the correct potential for a specific material, atomistic interactions can often be chosen such that generic properties common to a large class of materials, or those with common behaviors, are incorporated. This allows use to design "model materials" to study specific materials phenomena. Being said, potentials can be calculated very accurately for specific atomic interactions, using quantum mechanics methods such as the aforementioned Density Functional Theory (DFT). The development of better interatomic potentials—such as the aforementioned reactive force fields—combined with the increasing computational capability of systems approaching several million atoms—predicate major breakthroughs in the years to come.

## 6.7 Coarse-Graining and Other Multiscale Methods

Albeit providing a rather accurate description of macromolecules, all-atom modeling approaches have historically been prohibitively expensive when large systems and long simulation times must be considered. This led to the development of coarse-grained models and approaches [64]. By utilizing similar formulations and computational algorithms as full atomistic molecular dynamics, "coarse" model elements can be simulated employing simple bonded and nonbonded interactions that can be rapidly calculated in each time step. The integrated use of simulation methods with different computational expense and accuracy is referred to as multiscale modeling, where a systematic link is established between multiple scales. Although we focus on the coarse-graining of full atomistic systems, multiscale modeling in general encompasses any systematic passing of information from lower levels to coarser, larger scales. Coarse-grain methods can be broadly characterized as any model formulation (numerical or otherwise) that simplify *many* degrees of freedom to *few* degrees of freedom. The necessity for such methods arise when the properties of interest have causes, effects, and influences extending over multiple time and length scales—properties of many biological materials.

Such a reduction in degrees of freedom is not new—simplified systems for analysis can be found in many engineering disciplines. Complex electronic components are designed based on simplified models of circuits, with element behavior defined by such general properties as current, voltage, and resistance. Robust building structures are analyzed *via* notions of beam deflections and beam-column joint rotations, among other simplifying assumptions. In both cases, more detailed system representations are known and can be implemented (*e.g.*, implementation of temperature and material effects in a transistor, or a detailed frame analysis including stress concentrations in bolts or welds). It is apparent that such additions result in a more accurate representation of the modeled system, but also serve to increase the computational expense of analysis as well as introduce a more sophisticated theoretical framework (which subsequently requires a more detailed set of material and model parameters). Rarely, however, is the use of simplified and computationally efficient models justified by inaccessible time- and length-scales of the more detailed description. Such models are applied for analysis *in lieu* of a more detailed description because they provide an accurate representation of the system level behavior and response with confidence in the properties of the model components. Full atomistic simulations are used to probe and acquire the behavior of molecular components and parameterize coarse-grain potentials. These potentials, in turn, are implemented in coarse-grain representations that can facilitate the analysis of large-scale system level phenomenon. There is an inherent assumption that the coarse-grain theoretical framework provides an accurate representation of the atomistic behavior, just as application of elastic beam theory assumes the reliability of a steel girder within a structure.

It is stressed that even if full atomistic representations are computationally possible, a coarse-grain description can be still be suitable for systematic analysis of variable system configurations, which requires a large number of simulations. The is an alternative motivation for coarse-graining approach complementary to extension of accessible time- and length-scales, where the catalyst for coarse-grain potential development is not the extension of traditional molecular dynamics, but to provide an accurate and reliable method for system-level analysis and probe the mechanical response and structure-property relation for hierarchical systems. Furthermore, mesoscopic models provide simple and efficient modeling techniques for experimentalists, allowing a more direct comparison between simulation and a vast variety of experimental techniques, without requiring specialized molecular dynamics tools such as specialized computer clusters with complex software.

A developed coarse-grain model can only reflect the behavior included in the governing potentials and associated parameters, and consequently, the source of such parameters typically determines the accuracy and utility of the coarse-grain model. A complete theoretical foundation of the *materiome* requires synergistic multiscale transitions from atomic to mesoscale to macroscale descriptions. Hierarchical "handshaking" at each scale is crucial to predict structure-process-property relationships, to provide fundamental mechanistic understanding, and to enable predictive modeling and material optimization to guide synthetic design efforts. As suggested, a *finer-trains-coarser approach* is not limited to bridge atomistic to mesoscopic scales, but can also refer to hierarchical parameterization transcending any scales, such as mesoscopic to continuum levels, or beam-columns to buildings. Such a multiscale modeling paradigm establishes a fundamental link between component behavior and system-level function, providing a consistent theoretical approach to develop coarse-grain models for systems of various scales, constituent materials, and intended applications.

To illustrate, such multiscale model transitions can be compared to the parameterization of atomistic force fields by detailed quantum mechanical methods. In his paper, *The Art and Science of an Analytic Potential* [65], D.W. Brenner states:

> Analytical potential energy functions (sometimes referred to as empirical or classical potentials) are simplified mathematical expressions that attempt to model interatomic forces arising from the quantum mechanical interaction of electrons and nuclei. Their use is generally necessitated either by the desire to model systems with sizes and/or timescales that exceed available computing resources required for quantum calculations...

Brenner succinctly defines a coarse-grain model (simplified mathematical expressions) and their primary advantage (extension of accessible time- and length-scales). In development of MD potentials, it is possible to derive the force constant parameters used in full atomistic models from first principle quantum mechanical calculations in an inherent *fine-trains-coarse* approach [66].

**Fine-Trains-Coarse Method**: A bottom-up approach to multiscale model development where parameterization of material behavior at one level (*coarse*) is fitted from a more sophisticated and robust analysis at a smaller scale (*fine*) of higher resolution, such as fitting molecular force field parameters from quantum mechanical results. Allows efficient computation of subsequent scales with a logical basis in first principle theories. Most suitable if a clear separation of scales exists in the hierarchical structures of a material.

In effect, more sophisticated and/or computationally expensive techniques (*fine*), such as Density Functional Theory (DFT), are simplified in the parameters and functions of atomistic potentials (*coarse*). Here, the atomic structure (bosons, fermions, etc.) can be considered as the material building blocks, while bonded atoms (such as carbon-carbon bonds) can be thought of as the structural system. From this perspective, atomistic force fields can be considered as a thorough and systematic coarse-graining of quantum mechanics (Fig. 6.10(a) to (b)). The direct application of quantum mechanical results to large macromolecular systems (such as proteins) is not yet feasible because of the large number of atoms involved.

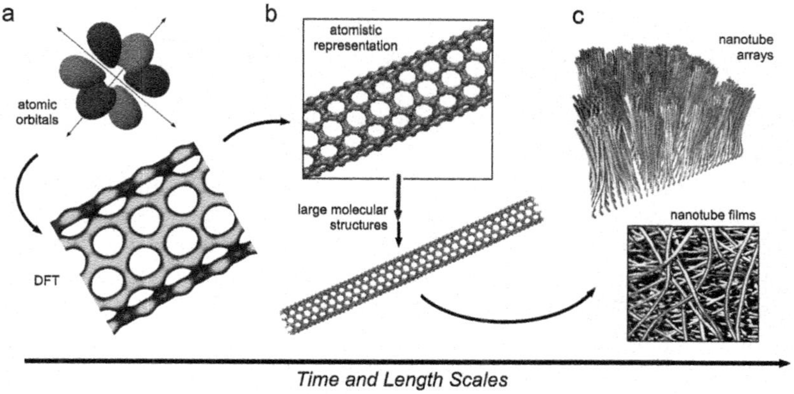


**Fig. 6.10** Jumping scales: quantum to atomistic, atomistic to mesoscopic. The rationale for coarse-graining approaches can be justified through analogous comparison between the parameterization of atomistic force fields by detailed quantum mechanical methods to the development of coarse-grain potentials by the results of full atomistic simulations. (**a**) Schematic of atomic orbitals (wave functions) leading to the electron charge isosurface of a carbon nanotube *via* Density Functional Theory (DFT), wherein the behavior of carbon atoms can be determined. (**b**) Full atomistic model of identical carbon nanotube using carbon-carbon interactions (modeled with ReaxFF atomistic potential) allowing the simulation of larger molecular structures. (**c**) Coarse-grain carbon nanotube systems, using potentials developed from atomistic results. Systems consisting of hundreds of nanotubes can be represented, reaching scales on the order of micrometers (and implicitly representing billions of atoms). Each transition from (**a**) to (**c**) involves an increase is accessible system size, length scale, and time scales, as well as a simplification of the governing theoretical models

As a logical extension, the results of full atomistic simulations can be implemented in the fitting of mesoscopic coarse-grain potentials (Fig. 6.10(b) to (c)). Each transition involves an increase in accessible system size, length scale, and time-scales, as well as a simplification of the governing theoretical models. As a consequence, details of interactions are lost, but the desired fundamental behavior is maintained. It is noted that this "desired fundamental behavior" is dependent on the system application, and can include mechanistic behaviors, thermal or electrical properties, molecular interactions, equilibrium configurations, etc. To illustrate, an atomistic representation of carbon bonding can include terms for atom separation, angle, and bond order (such as CHARMM [67] or AMBER [28] type force fields), but lacks any description of electron structure, band gaps, or transition states found in quantum mechanical approaches such as DFT. However, the atomistic model maintains the accurate behavior of the individual carbon atoms, and the neglecting the effects of electrons is deemed a necessary simplification. Likewise, the coarse-graining of a carbon nanotube integrates the effects of multiple carbon bonds into a single potential. The exact distribution of carbon interactions is lost, but the behavior of the molecular structure is maintained.

The coarse-graining procedures discussed hereafter focus on constituent materials that can be modeled by full atomistic techniques using classical molecular dynamics, while the system structure (and relevant behaviors) requires a fully informed coarse-grain representation. The atomistic model provides the underlying physics to the coarse-grain potentials, allowing investigation at the desired length scale.

The development of coarse-grain models allows the simulation of events on "physical" timescales and size scales, leading to a range of possible advances in nanoscale design and molecular engineering in a completely integrated bottom-up approach. A coarse-graining approach is intended to develop tools to investigate material properties and underlying mechanical behavior typically required for material design that systematically integrates characteristic chemical responses. However, it is emphasized that the intention is not to circumvent the need for full atomistic simulations—in contrast, coarse-graining *requires* accurate full atomistic representations to acquire the necessary potential parameters. The finer-trains-coarser procedure described here is an attempt to reconcile first principles derivations with hierarchical multiscale techniques by using atomistic theory with molecular dynamics simulations *in lieu* of empirical observations, in a unified and systematic approach.

## 6.8 Example Coarse-Grain Approaches

Atomistic force fields, which must encompass atom-atom bonds and interactions, can be developed in a general formulation due to the common molecular components. For example, the behavior of carbon-carbon bonds in a carbon nanotube is *similar* to the alpha-carbon backbone of a protein sequence in terms of strength and bond length (though not entirely equivalent). A thoroughly developed atomistic force field is capable of representing a vast assortment of molecular systems. Coarse-grained potentials, conversely, are usually developed to represent a particular system, and consequently feature unique idiosyncrasies in construction and behavior. Accordingly, differences in system complexity and goals of modeling lead to difficulties in developing a universal method for coarse-graining. Attempts to avoid this system-dependent approach (*i.e.*, generalized coarse-graining frameworks) can potentially result in a complex coarse-grain description to account for the multitude of molecular interactions a general description must encompass. The introduction of many potentials and parameters essentially mimic the function and form of full atomistic force fields, albeit at a coarse-grain scale.

In the last decade, various simple models have been used to describe the large-scale motions of complex molecular structures where more detailed classical phenomenological potentials [28, 67] involving all atoms cannot be used because of the restrictions on the amount of time that can be covered in computer simulations. The reader is referred to recent reviews for a more thorough discussion of techniques and applications [64, 68–70]. The intent here is to provide a brief summary/discussion of some of the approaches implemented for coarse-grain potentials as an (non-exhaustive) overview of methods.

Aggregated single-bead models are the most direct approach taken for studying macromolecules. The term single-bead derives from the idea of using single point

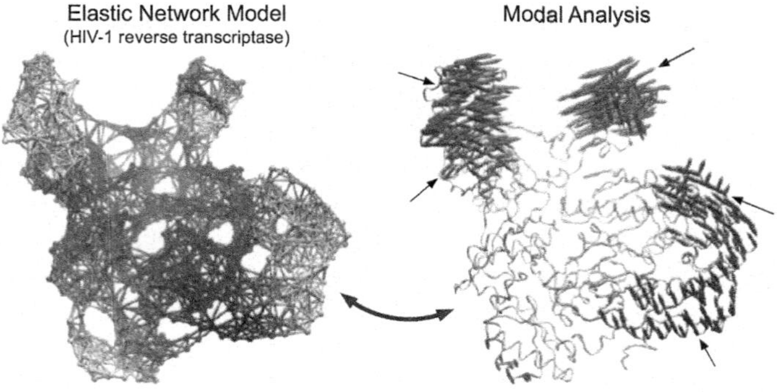


**Fig. 6.11** Predicting the "easy" modes *via* elastic network modeling (ENM). ENMs have been widely used in recent years for investigating the cooperative motions that biomolecular systems tend to undergo under equilibrium conditions. *Left*: Anisotropic network model (ANM) for the HIV-1 reverse transcriptase (RT). Node positions, identified by the coordinates of $\alpha$-carbons (along the protein backbone) known from experiments and each residue pair with $\alpha$-carbons located within a designated cutoff distance is connected by a spring of uniform stiffness. *Right*: Backbone structure. Modal analysis (eigenvalue decomposition) returns the softest—or most energetically favorable—motions of the molecular structure (directions indicated by arrows). Adapted from Bahar *et al.* [74], used with permission, copyright © 2011 The Protein Society

masses for describing a functional group, such as single amino acids, in a macromolecular protein structure. The elastic network model (ENM) [71], Gaussian Network Model [72], and Go-like models [73] are well-known examples that are based on such approximations. These models treat each amino acid as a single bead located at the $C_{alpha}$ position, with mass equal to the mass of the amino acid. Simple elastic network models have been used to study the properties of coarse-grained proteins and larger biomolecular complexes, focusing on the structural fluctuations about a prescribed equilibrium configuration, such as normal vibrational modes. In the simplest form, such coarse-grain models can be defined by a single potential for all bead types. For example, elastic network models can be thought of as a single pair potential between neighbor atoms with a constant stiffness. Although simple, such networks provide pertinent structural information for macromolecular conformations. Normal mode analysis of elastic network models can been advantageously applied to modeling protein flexibility associated with topographical changes, depicting close agreement between large-scale collective motions of proteins experimentally observed (see Fig. 6.11).

Elastic network potentials differ from the traditional bond potential (see Eq. 6.7) in that the spring does not explicitly represent a chemical bond—it encompasses all interactions between two network connections—thus a decrease in degrees of freedom. With the introduction of nonlinear spring behavior, complex model representations introducing directionality and anisotropy can be formulated [75]. As such, even a single potential description can become increasingly sophisticated as applications attempt to probe complex molecular deformations such as protein residue fluctuations [76] and equilibrium state transitions [77, 78]. In Go-like models, an additional Lennard-Jones-based term is included in the potential to describe short-range non-bonded interactions between atoms within a finite cutoff separation. Despite their simplicity, these models have been extremely successful in explaining thermal fluctuations of proteins and have also been implemented to model the unfolding problem to elucidate atomic-level details of deformation and rupture that complement experimental results [79–81]. A more recent direction is coupling of ENM models with a finite element-type framework for mechanistic studies of protein structures and assemblies [82]. Due their simplicity, single-bead models have several shortcomings. With classic ENM, only harmonic deviations from the initial configuration are possible. In the Go model, native interaction definitions lead to a minimally frustrated land-scape which is highly biased towards the input configuration of the molecule. Such models therefore cannot predict folding or unfolding intermediates and meta-stable states. The explicit treatment of protein-solvent interactions, non-native interactions and H-bonds is also not possible with single-bead models. It is now widely accepted that for protein unfolding studies, the results obtained using such models are only qualitative at best, although they may reveal important aspects of topology-dependent mechanical resistance [79–81] and can thus be used to improve our understanding of structure-property links.

Using more than one bead per amino acid provides a more sophisticated description of protein molecules. In the simplest case, the addition of another bead can be used to describe specific side-chain interactions in proteins [83]. Higher-level models, for instance, four- to six-bead descriptions, capture more details by explicit or united atom description for backbone carbon atoms, side chains, and carboxyl and amino groups of amino acids [84, 85]. Although multi-bead models have superior qualities compared to single-bead descriptions, dozens of additional energetic terms involving pseudo-bonds and other means to avoid complex dihedral or improper potentials that stabilize the conformation of the polypeptide chain have to be introduced for generic models. Even with the introduction of these terms, some of which are physically not intuitive, the models offer limited applicability, as the defined side chain interactions are only valid for simple residues such as glycine and alanine. More complex yet computationally efficient potentials that intrinsically take into account sequence specificity are extremely challenging to develop, thus making readily available all-atom descriptions and simulation packages more favorable for most applications.

Coarser-level multiscale modeling methods similar to those presented here have been reported more recently, applied to model biomolecular systems at larger time- and length scales. These models typically employ super-atom descriptions that treat clusters of amino acids as beads. In such models, the elasticity of the polypeptide chain is captured by simple harmonic or anharmonic (nonlinear) bond and angle terms. These methods are computationally quite efficient and capture shape-dependent mechanical phenomena in large biomolecular structures [86], and can also be applied to collagen fibrils in connective tissue [87] as well as mineralized composites such as nascent bone [88].

The subsequent step in coarse-grain model development is the combination of potentials to represent the inter- and intra-molecular interactions of individual macromolecules, representing the stretching, orientation, and deformation of simple biomolecules. Thus, for general applications, a more complex formulation is required—one that maintains a definitive structure of the macromolecule, as well as integrate the interactions between macromolecules. The development of the MARTINI force field [89–91] attempted to provide a general coarse-grain model that could be efficiently adapted for a multitude of biological systems by taking advantage of the fact that the majority of biological molecules (such as protein structures and lipids) are composed of the same categories of functional groups at the atomistic level. Such approaches were previously implemented with various degrees of complexity for application to specific systems, integrating multiple beads per functional group, or more complex parameter formulations (see [90] for a discussion and [92, 93] as examples). Essentially, by coarse-graining the functional groups, molecules with different architectures can be easily built and simulated. The aim of the MARTINI force field was to retain the chemical nature of the molecular components while defining as few bead types as possible. The applied philosophy was to avoid a focus on reproduction of structural details for a particular system, but rather aim for a broad range of applications without the need to re-parameterize the model each time. As a result, there is a slight trade-off between accuracy and applicability, with many different biological systems able to be modeled. Only four interaction types are considered: polar, nonpolar, apolar, and charged, and subtypes of these interactions are used to define hydrogen-bonding characteristics. The MARTINI force field can be represented by:

$$\begin{aligned} U_{MARTINI} &= U_{bonds} + U_{angles} + U_{nonbonded} \\ &= \sum \phi_{bond}(r) + \sum \phi_{angle}(\theta) + \sum \phi_{pair}(r) \end{aligned} \tag{6.15}$$

where harmonic potentials are implemented for the bonded and angle potentials, or:

$$\phi_{bond}(r) = \frac{1}{2}k_b(r - r_0)^2 \tag{6.16}$$

$$\phi_{angle}(\theta) = \frac{1}{2}k_\theta(\cos\theta - \cos\theta_0)^2 \tag{6.17}$$

It is not a coincidence that this formulation is similar to that of full atomistic force fields such as CHARMM (see Eq. 6.8) and ReaxFF (see Eq. 6.9). The harmonic potentials are parameterized by constants ($k_b$, $r_0$, $k_\theta$, and $\theta_0$) common to all bead-types. Such an approach increases the general applicability and versatility of the force field. Similar to full atomistic formulations, nonbonded interactions between interaction sites $i$ and $j$ are described by the simple Lennard-Jones 12:6 potential, arranged into distinct subsets of interaction groups, or *categorizations* of biomolecular structures and interactions.

The major advantage of the MARTINI force field is the simplicity and versatility of the parameterization—many proteins, lipids, and molecules can be built using the same set of coarse-grain building blocks, encapsulating the universality-diversity paradigm central to materiomics. However, these MARTINI-type formulations restrict the coarse-graining to distinct functional groups (*i.e.*, amino acids) and known secondary structures (such as alpha-helical conformations) to accurately maintain atomistic interactions while increasing computational efficiency. The obvious disadvantage of such approaches lies in the restriction of the coarse-graining scale. For example, on average, each bead of the MARTINI model has the volume of four water molecules. As a result, the time and length scales of MARTINI and MARTINI-type simulations are still limited, albeit much more efficient than full atomistic approaches. For appropriate systems such as biological membranes consisting of lipid bilayers and lipid-protein interactions—beyond the capacity of full atomistic representations—such approaches provide a powerful simulation framework (see [94, 95] for more thorough reviews and discussion of such methods).

To capture fundamental interactions, general coarse-grain potentials developed for application to multiple systems limited to few atom mappings (*i.e.*, 2-, 4-, and 6-bead models) and are beneficial and appropriate for systems where the interactions are still at the atomistic scale (*i.e.*, proteins, polymers, etc.). It is also apparent that different material systems can be characterized by mechanical properties at the atomistic to micro-scale, requiring a more general coarse-graining approach to model the intended structure-property behavior. A hierarchical multiscale system can benefit from a system-dependent approach to coarse-graining. Accurate system representation is maintained at a cost of coarse-grain potential versatility. By applying a system-dependent finer-trains-coarser approach, a variety of systems can be coarse-grained and modeled for different purposes, transcending different scales and functionalities. Such derived coarse-grain potentials are not meant to be universally applicable, but provide a means investigate a specific system under specific conditions.

It is emphasized that not all systems will benefit from a coarse-grain representation, and prudent consideration must be made regarding the system characterization and intent of the simulations. For some systems, complex behaviors may require full atomistic representation, or material inhomogeneities may not be able to be described by coarse-grain elements. Such systems may benefit and indeed require full atomistic representations. Typical motivations for a coarse-graining approach include:

1. Inaccessible time-scale for phenomenon or behavior *via* full atomistic representation.
2. Inaccessible length-scale for phenomenon or behavior *via* full atomistic representation.
3. Focus on global system properties and/or mechanical behavior rather than on molecular structure and/or chemical interactions.
4. Desire for a direct simplified analysis of simulation results and system behavior

As well as these motivating factors, some systems are more conducive to coarse-graining due to chemical composure and/or structural arrangement. Coarse-grain models are easily adapted for atomistically homogenous systems, consisting of repetitive structures, such as carbon nanotubes (with a uniform cylindrical nanostructure), or collagen microfibrils (with repeating tropocollagen units). By extension, atomistically heterogeneous materials, such as protein-based materials, are applicable to coarse-graining if they are mesoscopically homogenous, where the local effects of distinct amino acids (or other molecular inhomogeneities) produce similar global behaviors and deemed negligible. Other system properties to consider include the discretization or material units and/or mechanical behavior, with a logical correlation to coarse-grain elements, and any hierarchical structures in which the intended structure-property relation is to be investigated.

## 6.9 Bridging Hierarchies: Coarse-Grain Model Formulation

The aforementioned coarse-graining techniques—ENM models and MARTINI-type formulations, for example—have been successful in their own applications. Elastic network models reflect the atomistic structure, while MARTINI-type models accurately capture the interactions of functional groups for applications in assembly. However, such techniques focus on molecular geometry and interactions rather than reflect accurate mechanical properties and response. The materiomic investigation of the mechanical behavior of hierarchical systems requires a system-dependent parameterization of coarse-grain potentials, with a focus on maintaining both molecular interactions and molecular mechanics. Thus, we discuss a universal framework through a *finer-trains-coarser* multiscale paradigm, which effectively defines coarse-grain potentials *via* the *response* of full atomistic simulations, while introducing relevant mechanical properties at the mesoscopic scale. The approach is unique in its ability to transcend multiple scales enabling the investigation of a broad regime of systems from protein-based materials to synthetic composite structures, while emphasizing a theoretical foundation on full atomistic investigations and asserting energetic equivalence and consistent mechanical behavior between all levels of modeling. The result is a set of problem-specific coarse-grain representations with diverse applications. The ensuing discussion presents guiding principles behind model formulation.

It is apparent that a universal, stepwise procedure for coarse-graining a system is limited by the inherent simplifications and intent of the coarse-grain representation. A model developed to investigate arrays of carbon nanotubes will differ from a model developed for the simulation of protein unfolding, even though the coarse-grain elements and potentials are similar. The coarse-graining of a particular system can be characterized by trial-and-error and subjective omissions or inclusions of pertinent behaviors, and as such considered just as much an *art* as a *science*. Careful consideration must be taken for the intended application and purpose of the coarse-grain model. Indeed, such a judicious approach is the motivation for a finer-trains-coarser method. A deliberate focus on the atomistic behavior to characterize the coarse-grain model not only provides a complete "bottom-up" theoretical basis for a material system, but also assists in the formulation of the coarse-grain model by delineating relevant behavior and interactions. Nevertheless, although we contend that a step-by-step coarse-graining "recipe" is problematic, we can outline the "ingredients" in a general, coarse-graining framework:

1. Potentials required to define and characterize the mesoscale system.
2. Full atomistic "test suite" applied to determine required coarse-grain potential parameters.
3. Fitting of atomistic results to coarse-grain potential parameters *via* energy equivalence and consistent mechanical behavior.
4. Validation of coarse-grain model with full atomistic and/or empirical results.

We proceed to discuss each in detail.

### *6.9.1 Coarse-Grain Potential Type and Quantity*

The first step in the development of a coarse-grain model is to determine the necessary potentials to characterize the system. In general, we define the total energy of a coarse-grain system as the sum of these potentials:

$$U_{system} = U_{CG} = \sum \phi_{CG} \tag{6.18}$$

where $\phi_{CG}$ are the defined coarse-grain potentials. Although a simple statement, it is nontrivial in implementation. Indeed, the type of developed potentials can serve to both broaden and restrict the applications of the model. Specifically, the coarse-grain potentials must be complex enough to represent the molecule in the intended mesoscopic simulation. One must consider whether provisions are necessary for fracture, intermolecular interactions, plasticity, formation of secondary structures, or a host of other molecular behaviors depending on the intent of the investigation. It is again stressed that a coarse-grain model for a system is *not* universal. The goal is to utilize the fewest and simplest potentials as possible that represent the system structure(s), mechanical properties, and interactions. Such an approach can be facilitated by the utilization of harmonic spring potentials to reflect mechanical response, where:

$$\phi_i(\psi) = \frac{1}{2}k_i(\psi - \psi_0)^2 \tag{6.19}$$

Here, $k_i$ refers to a harmonic stiffness while $\psi$ typically refers to either an interatomic distance, $r$, or an angle, $\theta$, for stretching and bending deformations, respectively. It is noted that even this simple harmonic function is exploited for conventional (CHARMM) and reactive (ReaxFF) force fields, as well ENM and MARTINI-type coarse-grain formulations, and again surface here as a general coarse-graining approach. The harmonic spring potential results in a linear relationship between force and deformation, allowing efficient computation of interactions. Indeed, more complex nonlinear behavior can be approximated by combinations of linear functions (such as bilinear or trilinear formulations) to maintain computational efficiency *in lieu* of the introduction of more complex potentials.

A possible deficiency of harmonic potentials is the continuous increase in force with deformation. Systems subject to large deformations will subsequently deviate from true behavior, and thus, without provisions such as a potential cutoff, the use of the harmonic potential should be limited to small deformation. Selection of the harmonic spring potential can also be affected by the atomistic response by which it is parameterized. For example, a linear elastic strain response can sufficiently modeled by harmonic spring potentials, albeit limited to small deformation.

### 6.9.2 Full Atomistic Test Suite

An atomistic "test suite" is thus developed to obtain the necessary mechanical response and molecular interactions to be integrated into coarse-grain potentials. Typically, a relatively simple atomistic simulation is applied to isolate a single molecular behavior, mimicking simple material specimen tests. For example, uniaxial stretching can be applied to obtain the force-displacement or stress-strain response of a macromolecule, allowing the calculation of Young's modulus, and the parameterization of coarse-grain bond strength. Further, a three-point bending test can be utilized to determine the bending stiffness of a molecule, thereby allowing the parameterization of the coarse-grain bending potential. For molecular interactions, pairs of molecules are simulated to determine relative adhesion. In essence, a single atomistic investigation is implemented to characterize a single coarse-grain potential, ensuring accurate representation of each behavior. The number of required "tests" depends on the number of coarse-grain potentials (and associated parameters) implemented to describe the system.

### 6.9.3 Fitting Coarse-Grain Potentials

The fundamental principle underlying the parameterization of coarse-grain potentials is the assertion of energy conservation. Energy equivalence is imposed between potential energy functions and relevant atomistic energy results. The basis of the formulation of the energy functions, either through elastic strain energy, deformation energy, adhesion energy, or other techniques, essentially defines the accuracy and behavior of the representative coarse-grain system. Two such approaches of energy conservation warrant separate discussions: (1) the application of direct energy equivalence between atomistic and coarse-grain representations, and; (2) the assertion of energy conservation *via* consistent mechanical behavior.

*Direct Energy Equivalence*: Non-bonded interactions representing macromolecular adhesion, attraction, or repulsion innately have no mechanical property analogue for parameterization. For such energetic potentials as the Lennard-Jones 12:6 function, or Coulombic interactions, the coarse-grain equivalent is extracted directly from the energetic results of a full atomistic representation. Potential energy as a function of separation can be derived between individual macromolecules. The coarse-grain potential can then be expressed as:

$$\phi_{CG}(R) = \sum_{atoms} \omega_{ij}\phi_{ij}(r_{ij}) \tag{6.20}$$

where $\phi_{CG}$ is the coarse-grain potential, $R$ is the macromolecular separation, and the sum is a weighted combination of the full atomistic potentials, $\phi_{ij}$. This is essentially how coarse-grain potentials are developed for a small number of atoms per superatom (such as MARTINI-type force fields), where the superposition of multibody effects is apparent in the mapping. The relation becomes more complex as the number of incorporated/mapped atoms increase, required explicit atomistic simulation of the desired macromolecules. The above relation should not be thought of as a simple summation of potentials. The coarse-grain potential incorporates effects of the entire represented system (such as hydrophobic and other solvent effects, electrostatic interactions, hydrogen bonding between molecules, and/or entropic effects). We thereby introduce unknown weighting coefficients, $\omega_{ij}$ in the summation of Eq. 6.20. The complexity necessitates a full atomistic investigation to derive the effective coarse-grain potential parameters.

*Consistent Mechanical Behavior*: Bonded and angle potentials effectively represent molecular mechanical behavior. As there can be many contributions to even simple mechanical process (such as breaking of hydrogen bonds or solvent friction during molecular stretching or bending), we introduce conservation of energy for the relevant mechanical response (as opposed to the potential energy of direct energy equivalence) resulting in consistent mechanical behavior between full atomistic and coarse-grain representations. The approach is to apply energy equivalence between the coarse-grain potential and the observed strain or deformation energy of the full atomistic system:

$$\phi_{CG}(\psi) = U(\psi) \tag{6.21}$$

where $\phi_{CG}$ is again the coarse-grain potential, and $U(\psi)$ is the representative energy function for the mechanical response. It is an underlying assumption of the model that the response of the full atomistic system is accurately described by the energy function, $U(\psi)$. The choice of the strain or deformation energy function is dependent on the system to be coarse-grained, and assumptions of elastic/plastic behavior, material isotropy, system failure such as yielding or fracture, etc., affect the definition (and interpretation) of atomistic results. As an example, for an elastic-isotropic material, we can define strain energy as:

$$U(\varepsilon) = \frac{1}{2}\int_V [\sigma][\varepsilon]\,dV \tag{6.22}$$

To equate with $\phi_{CG}$, we must formulate $U$ as a function of $\psi$. For axial stretching, this requires the formulation of strain as a function of bond length, $\varepsilon(r)$, while for shearing, this requires the formulation of shear strain as a function of shearing angle, $\gamma(\theta)$. Individual parameterization of coarse-grain potentials *via* consistent mechanical behavior has an inherent limitation by the decoupling of mechanical behavior. To illustrate, as bond and angle coarse-grain potentials are independent under simulation, there is no coupling between stretching and bending of the molecule, and effects such as twisting or torque are completely neglected. However, under full atomistic representations, such coupling can serve to increase stresses and induce failure. Ultimately, the model is tailored to the desired application, and systems with obvious or significant coupling effects should be avoided.

### 6.9.4 Validation

Once the coarse-grain potentials are developed validation of the coarse-grain model is a necessary step to assure an accurate representation. The primary basis for validation is through a direct comparison to full atomistic simulations. The "test suites" applied were intended to result in a single behavior/response for each coarse-grain potential whereas the full coarse-grain representation can possibly involve the interaction and/or coupling of individual potentials. As such, the combination of two behaviors (bending and stretching for example) must be explored to justify the coarse-grain model. Relatively simple validation simulations can be designed, combining multiple system behaviors, and a one-to-one comparison made between full atomistic and coarse-grain results.

Secondary support and model validation is found in experimental data, and takes an important place in the development and application of any computational model. Experimental techniques such as nanoindentation, optical traps, or atomic or chemical force microscopy (AFM/CFM) can directly probe materials at the mesoscale and determine system level characteristics and mechanical properties. The development of coarse-grain models is intended to elucidate such system-level behavior and thus correlation with experimental results is essential. Reciprocally, an accurate coarse-grain model can serve to validate and support experimental results. It is noted that advances in experimental techniques are continually probing smaller scales, reaching atomistic precision and allowing single molecule investigations (such as optical tweezers methods, atomic force microscopy (AFM), etc.). Such data can serve to reinforce both the full atomistic representation and the developed coarse-grain model, thereby increasing confidence in both representations and supporting the finer-trains-coarser approach.

In the context of this discussion, coarse-grain models provide an efficient means to simulate and investigate systems in which the desired behavior, property or response is inherently at the mesoscale—those that are both inaccessible to full atomistic representations and inapplicable to continuum theory. Be that as it may, such approaches have been utilized for many years as a means for system analysis—albeit without the "coarse-grain" label. Examples can be found in structural dynamics (lumped-mass models), transportation planning (traffic flow analysis), and even economics (macroeconomics). Indeed, any system in which the aggregate behaviors of many small contributions are "homogenized" into effective or bulk properties system components can be considered "coarse-grained". To illustrate, we consider two extreme structures: (1) a three-story concrete building, and (2) an alpha-helical protein domains.

#### Macroscale Example: N-story Concrete Building

Comprehensive assessment of the risk of earthquake-induced structural collapse of a concrete building requires a robust analytical model that captures both nonlinear behavior and modes of failure. Ideally, the strength, stiffness, deformation capacity,

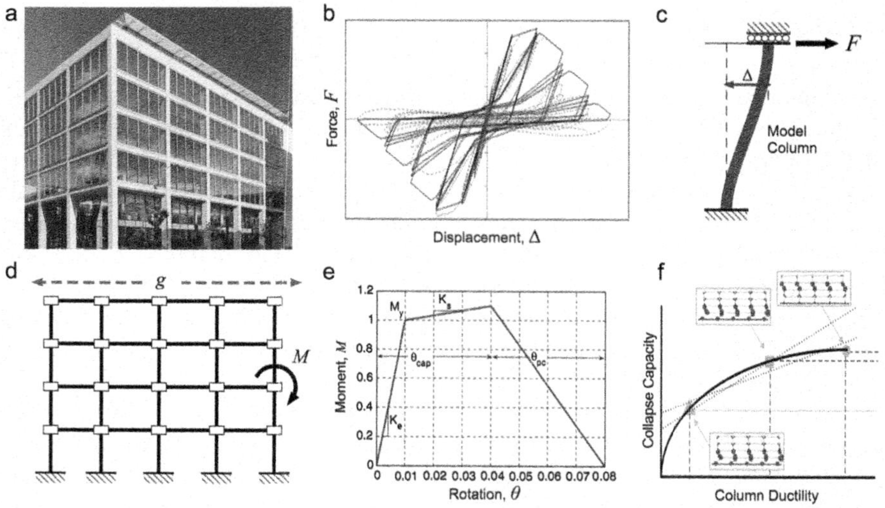


**Fig. 6.12** A "coarse-grain" approach for structural engineering by reducing the degrees of freedom within a building model to facilitate computationally intensive analysis. (**a**) A typical concrete-frame building representing all details (akin to a "full atomistic" molecular model). (**b**) Experimental results ("test suite") of primary system component. In this case, the force-displacement cyclical pushover response of a reinforced concrete column. (**c**) Schematic of ideal column model behavior. (**d**) Simplified representative model of building system, integrating the reduced and parameterized behavior of components, wherein applied seismic motions ($g$) are related to beam-column joint moments ($M$). (**e**) The reduced "coarse-grain" behavior of the model, the moment-rotation behavior of the beam-column joints derived from the backbone of the hysteresis in (**b**). (**f**) System-level analysis, determining collapse capacity as a function of the moment-rotation relation parameters (in this case, ductility). Adapted from [96]

and energy dissipation of all building characteristics—from beam-column joints to cubicle walls—should be accounted for. It should be obvious that accurately modeling the minutia of every nut and bolt is both computationally inaccessible and highly unnecessary. Useful models exploit simplified approximations of component behavior—in the case of concrete building, the collapse behavior can be accurately reflected in the post-peak softening response of the moment frame beams and columns.

Here, the main structural members (*e.g.*, beams and columns) are reduced to a single constitutive behavior derived from "fine-scale" results (loading of concrete column members). From the experimental results, an "ideal column model" can be formulated, and elements assembled into a representative building. This model can then be subject to complex loading scenarios (such as seismic accelerations) and the nonlinear response and collapse behavior can be assessed with great efficiency (see Fig. 6.12). Such simplifications in structural analysis are similar to molecular coarse-graining approaches, wherein complex behavior can be analyzed by through a reduction of system degrees of freedom, while maintaining relevant behavior and failure modes.

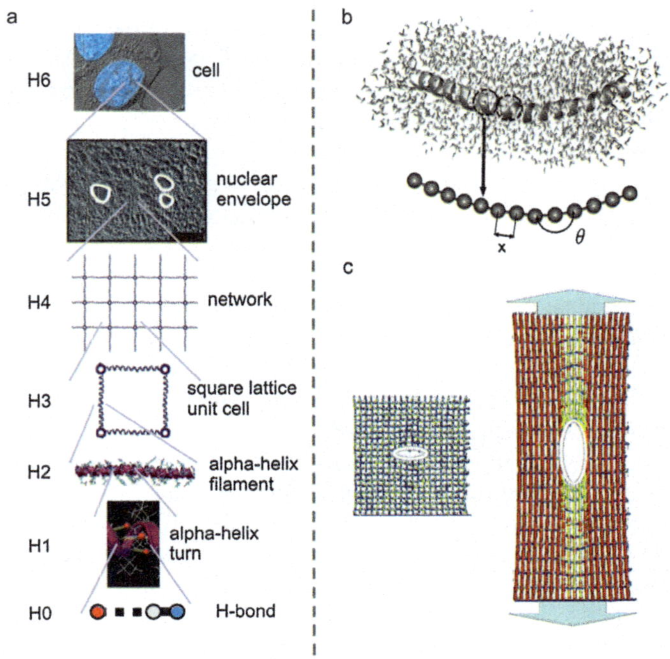


**Fig. 6.13** Coarse-graining approach exploring the hierarchical flaw tolerance mechanism in a cellular alpha-helix protein network. (**a**) Seven levels of hierarchies are considered, from intrabackbone hydrogen bond (H0), alpha-helical turns (H1), filaments of alpha-helices (H2), to the representative unit cell (H3) of protein networks (H4) that form the cell nucleus (H5) of eukaryotic cells (H6). The structure at each level is adapted to provide an optimal mechanical response and plays a key role in the overall mechanical behavior. Unfolding of alpha-helix turns (H1) proceeds *via* breaking of strong clusters of 3 to 4 H-bonds (H0). The lattice structure (H3) is the key to facilitate large strain gradients in the protein network, enabling gigantic strain gradients at virtually no energetic cost at the network level (H4). (**b**) Coarse-graining of the full atomistic alpha-helical protein within explicit solvent (water) to a fibrous chain model with reduced degrees of freedom and implicit solvent effects. (**c**) Protein network deformation implementing coarse-grain representation, illustrating the change of the crack orientation from a horizontal to a vertical one. The transformed vertical crack orientation features only marginal stresses at its corners. Adapted from [97, 98]

## Nanoscale Example: Alpha-Helical Protein Domains

Alpha-helical protein domains are the key constituents in a variety of biological materials, including cells, hair, hooves and wool. It has been determined both experimentally and *via* simulation that the mechanical response of biological materials is a combination of molecular unfolding or sliding, with a particular significance of rupture of reversible chemical bonds such as hydrogen bonds (H-bonds), covalent cross-links, or intermolecular entanglement. The dominance of specific mechanisms can emerge at different time- and length-scales and hierarchical arrangement/structure. As such, it is difficult to generalize fully atomistic results from nano to macro. Figure 6.13(a) displays an example hierarchical alpha-helical protein system.

A coarse-grain model can be developed to investigate the unfolding behavior of alpha-helical domains *via* data obtained from full atomistic molecular dynamics (Fig. 6.13(b)). The coarse-grain representation integrates parameters that define the energy landscape, including unfolding and refolding distances, and the location of folded and unfolded states. Although unfolding of short alpha-helical segments can be modeled using full atomistic techniques, a coarse-grain representation is required to full investigate the length dependence on mechanical response of a network of such proteins. Such an approach intends to extend a known atomistic behavior to larger systems *via* coarse-grain potentials.

Indeed, individual, isolated alpha-helices are rarely found in biology—alpha-helical based protein networks constitute the intermediate filament structure in the cell's cytoskeleton and the nuclear membrane. Using a network of coarse-grain bead-spring structures, large meshes reflecting an assembly of intermediate filaments can be constructed and probed for mechanical properties and behavior (Fig. 6.13(c)). It was found that the characteristic properties of alpha-helix based protein networks are due to the particular nanomechanical properties of the protein constituents, enabling the formation of large dissipative yield regions around structural flaws, effectively protecting the protein network against catastrophic failure [97]. The direct simulation of such large networks is only possible through the use of such multiscale coarse-grain models under discussion. The integration of coarse-grain and atomistic methods shed insight to the roles and mechanisms of individual levels of structural hierarchies [98], indicating multiscale coupling and unpredictable large-order effects. Table 6.2 describes the key mechanisms of the nuclear membrane, and how they contribute to the overall properties.

Coarse-grained techniques based on results from quantum mechanics or all-atom MD modeling approaches show great promise as they can run much faster than multi-atom descriptions for molecular building blocks such as proteins and are relatively easier to implement than multi-bead potentials that require complicated energy terms for achieving the correct molecular geometry. Since such high-level precision is not sought after in these coarser methods, simpler terms are generally used to achieve the global structure and dynamical information of the system.

In addition to methods that uniformly simplify a complex system by coarse graining, hybrid approaches that employ atomistic details at active regions of biomolecular systems or crystalline solids also show great promise [99]. The use of hybrid approaches in protein modeling has been pioneered early on through the use of so-called QM-MM methods where quantum mechanical descriptions are used at enzymatic sites, and non-reactive models are used elsewhere in the protein.

Overall, one can identify two fundamental viewpoints in multiscale modeling:

1. Hybrid approaches: Employing different levels of detail or fidelity in a single model simultaneously, requiring on-the-fly interaction between multiple scales.
2. Hierarchical approaches: Enabling sequential scale transitions by extraction or passing of key information *i.e.*, parameters, to higher scales based on more accurate simulations at smaller scales. In effect, scales are treated discretely.

Developing accurate models that can predict not only the overall structural behavior but also processes such as bond formation, self-assembly as well as molecular

**Table 6.2** Mechanisms across multiple hierarchies, which interact synergistically to create heightened system function. Adapted from [98]

| Hierarchy level | Description | Key mechanism(s) |
|---|---|---|
| H0 | Atomistic, chemical, intra-backbone H-bond | H-bonds form at moderate temperatures, drive formation of alpha-helices |
| H1 | Alpha-helical turn, H-bond cluster | Clusters of 3-4 H-bonds provide maximum mechanical strength at minimal material cost; coiled geometry of polypeptide provides significant entropic "hidden length" |
| H2 | Alpha-helix domain; secondary structure | Linear array of repeated alpha-helical turns enables large extensibility ($>150\,\%$ strain) and toughness *via* uncoiling deformation |
| H3 | Unit cell; intermediate filament network building block | Extension of alpha-helical properties in multiple directions |
| H4 | Protein network | Provide redundancy, flaw tolerance |
| H5 | Nuclear membrane | Structural integrity for cell, load bearing (external and internal) |
| H6 | Cytoskeleton; nuclear membrane | Biological function; aggregate effects of all mechanisms |

and macroscale failure of materials is an area of great interest for physics, materials science, and medical applications. Which of the two approaches *i.e.*, hybrid or hierarchical, is better suited depends strongly on the type of application and the specific properties that are simulated. Regardless, coarse-grain modeling can be described in terms of four guiding principles: (1) selection and characterization of governing potential types and quantity; (2) applied full atomistic "test suite" to reflect independent behaviors and determine relevant coarse-grain potential parameters; (3) fitting of atomistic (fine) results to necessary potential parameters (coarse), and; (4) validation of the developed model *via* comparison with full atomistic results (component-level) or correlation with empirical data (system-level).

In principle, coarse-graining and multiscale approaches are elegant solutions to model hierarchical systems through scale-by-scale dissection and analysis. Cascading mechanisms can be tracked, and multiscale coupling easily quantified. This problem in practice, however, is the uncertainty involved in the process (both within the presumed model and the physical system it represents). While impossible to eliminate altogether, there are two routes to mitigate such intrinsic uncertainties. The first is more extensive validation from "real-world" systems. In effect, if the coarse-grain model is reflective in actual behavior, the model can be deemed reliable. However, such "experimental analogues" are typically unattainable. Secondly, error quantification and stochastic variability must be developed and integrated with general coarse-graining approaches—similar to the introduction of stochastic variability in strength and behaviors for the analysis of macroscale system performance [100]. Homogenization of behavior (*e.g.*, effective stiffness and molecular interactions) can be useful if and only if the variability is minimal. Until such error propagation is well understood and quantified, prudence must be taken in the application (and interpretation) of coarse-grain and similar multiscale methods.

## 6.10 Summary

The applicability of continuum theories to biological (and other) materials remains subject of discussion. Certainly, at a rather macroscopic scale these methods can still be used. However, recent research indicates that the governing processes often occur at the nanoscale, where properties change rapidly and often irregularly across variations in space—thus, the basic assumptions of continuum theories are violated. This particularly applied to natural and biological materials, which often have structural features across many length scales, with smallest features at the nanoscale. The fact that these materials are used at dimensions comparable to their structural heterogeneity—for example, enzymes operate on substrates with sizes comparable to their diameter—complicates the situation and certainly questions the applicability of continuum-based methods.

Much like MD simulations, finite element analysis help researchers verify the analytical solutions and to visualize the experiments that are hard to see at nanoscale levels. Over the last decades, there has been a new realization that understanding atomistic or nanoscale behavior is required for understanding how materials function. This is partly because of the increasing trend to miniaturization as relevant length scales of materials approach several nanometers in modern technology. Once the dimensions of materials reach submicron length scales, the continuum description of materials is questionable and the full atomistic information becomes important to appropriately examine materials phenomena.

The understanding of how materials deform and break is often limited to phenomenological engineering theories that describe the macroscale materials behavior, neglecting the underlying atomistic microstructure. However, deformation and failure of materials is ultimately governed by quantum mechanics. These quantum mechanical effects that control chemical bonds are neglected in most existing theories. To include these effects, atomistic models are developed that are capable to simulate the motion of all atoms in the material, with systems comprising up to several billion particles, thus reaching macroscopic scales of material behavior that can be directly observed in experiment. The goal of such atomistic models is to understand the macroscopic response of materials, for example, under mechanical stimulation, based on their fundamental, atomistic ultrastructure.

Modern multiscale modeling techniques use a sequence of overlapping hierarchies encompassing various simulation tools to bridge the scales from nano to macro. Book-ending these techniques are quantum mechanics, encapsulating the basic constituents of matter (electrons, protons, etc.), and continuum theory, which, by definition, has no limiting scale. These modeling techniques allow a rigorous linking of material properties. While bridging scales ultimately results in loss of "mechanical resolution", this does not prevent the utility of such approaches. The reduction of quantum-level interactions to bond potentials in atomistic force fields (in essence simple harmonic springs) does not limit the application of molecular dynamics—the *necessary* behavior is maintained. This shares the same principle as coarse-graining a system of atoms into an effective "bond", perhaps representing a segment of alpha-helix. Being said, defining the necessary behavior (and thus interaction parameters) is nontrivial. Indeed, *two* variables for bond stiffness within the CHARMM force field ($r_0$, $K$) can easily extend to *nine* variable for the same chemical bond in ReaxFF (see Eq. 6.13), yet that does not devalue either formulation—the *intent* is simply different. There is inherent value in the abstraction of defining a chemical bond as a spring—effectively a mechanical description relating load and deformation—as certain patterns of interactions lead to characteristic functional units. Basic functions can manifest over and over—across scales—and appear universal between differing material systems. This is one of the motivations to characterize biological materials through category theory, abstract functional building blocks, and the *ologs* as discussed in Chap. 4: Universality-Diversity Paradigm: Music, Materiomics, and Category Theory. This fundamental viewpoint could revolutionize the engineering approach to use and create materials, by incorporating the atomistic to macroscopic scales into materials analysis and synthesis based on universal and cooperative functionalities.

Along such lines of thought, computer simulations have sometimes been referred to as "computational experiments". Designing smart computer experiments is the key to successful simulation. Computational modeling allows a bottom-up approach in understanding the *materiome* of a system, focusing on the finest, atomistic scales of detail governed by quantum mechanics as a *de facto* starting point, reaching up to large macroscopic scales described by continuum models, using hierarchical multiscale, or scale-bridging, modeling. As discussed, molecular simulation approaches to materiomic systems are determined by a wide range of scales and demand an equally wide range of simulation methods at various levels of resolution including a varying amount of degrees of freedom, from electrons to atoms to entire molecules to the "engineered" continuum. This approach is supplementary to many experimental efforts starting at the macroscopic level and reaching down to finer scales.

The interplay between material properties and structural arrangement of hierarchical systems provides unique and robust mechanical behavior that transcends multiple scales. Coarse-grain models provide an efficient means to simulate and investigate such systems in which the desired behavior, property or response is inherently at the mesoscale. The relatively few parameters and potentials that define a coarse-grain representation must accurately represent the full atomistic behavior of the structure in both mechanical response and intermolecular interactions. Concurrently, the diverse nature and possible components of hierarchical systems makes a general formulation difficult. We thus discussed a system-dependent finer-trains-coarser approach to coarse-grain model development, requiring the parameterization of coarse-grain potentials *via* full atomistic results in a consistent and systematic manner.

A well-defined coarse-grain representation simultaneously allows investigation of molecular structures and behavior at multiple levels of hierarchy while intrinsically maintaining full atomistic behavior. This simplification also allows a pragmatic approach to system analysis—similar to the simplification of macroscale systems—delineating global behavior from constituent element properties, thereby encompassing multiscale structure-property relations. Extension of accessible time- and length-scales can allow direct correlations with empirical investigations, providing a novel tool for experimental design and material characterization from the bottom-up and encompass the entire materiome.

## 6.11 Suggested Readings

On molecular dynamics techniques and applications:


- D. Frenkel, B. Smit, *Understanding Molecular Simulation, Second Edition: From Algorithms to Applications* (Academic Press, London, 2002)
- D.C. Rapaport, *The Art of Molecular Dynamics Simulation*, 2nd edn. (Cambridge University Press, Cambridge, 2004)
- J.M. Haile, *Molecular Dynamics Simulation: Elementary Methods* (Wiley, New York, 1997)
- A. Leach, *Molecular Modelling: Principles and Applications*, 2nd edn. (Prentice Hall, 2001)
- M.P. Allen, D.J. Tildesley, *Computer Simulation of Liquids* (Oxford University Press, Oxford, 1987)
- M.J. Buehler, *Atomistic Modeling of Materials Failure* (Springer, New York, 2008)


On multiscale methods and other coarse-grain approaches:


- P. Derosa, T. Cagin (eds.), *Multiscale Modeling: From Atoms to Devices* (CRC Press, 2010)
- M.O. Steinhauser, *Computational Multiscale Modeling of Fluids and Solids: Theory and Applications* (Springer, Berlin, 2008)
- J. Fish (ed.), *Multiscale Methods: Bridging the Scales in Science and Engineering* (Oxford University Press, New York, 2010)
- E.B. Tadmor, R.E. Miller, *Modeling Materials: Continuum, Atomistic and Multiscale Techniques* (Cambridge University Press, 2011)
- J. Fan *Multiscale Analysis of Deformation and Failure of Materials* (Wiley, 2011).
- S. Yip (ed.), *Handbook of Materials Modeling* (Springer, 2005)


On continuum methods and general material modeling:


- T.J.R. Hughes, *The Finite Element Method: Linear Static and Dynamic Finite Element Analysis* (Dover Publications, 2000)
- E.B. Tadmor, R.E. Miller, R.S. Elliott, *Continuum Mechanics and Thermodynamics: From Fundamental Concepts to Governing Equations* (Cambridge University Press, 2011)

## References

1. W. Goddard, A perspective of materials modeling, in *Handbook of Materials Modeling*, ed. by S. Yip (Springer, Berlin, 2006)
2. N. Metropolis, S. Ulam, The Monte Carlo method. J. Am. Stat. Assoc. **44**(247), 335–341 (1949)
3. N. Metropolis, A.W. Rosenbluth, M.N. Rosenbluth, A.H. Teller, E. Teller, Equation of state calculations by fast computing machines. J. Chem. Phys. **21**(6), 1087–1092 (1953)
4. B.J. Alder, T.E. Wainwright, Phase transition for a hard sphere system. J. Chem. Phys. **27**(5), 1208–1209 (1957)
5. B.J. Alder, T.E. Wainwright, Studies in molecular dynamics. 1. General method. J. Chem. Phys. **31**(2), 459–466 (1959)
6. B.J. Alder, T.E. Wainwright, Studies in molecular dynamics. 2. Behavior of a small number of elastic spheres. J. Chem. Phys. **33**(5), 1439–1451 (1960)
7. A. Rahman, Correlations in the motion of atoms in liquid argon. Phys. Rev. **136**, 405–411 (1964)
8. A. Rahman, Fh. Stilling, Molecular dynamics study of liquid water. J. Chem. Phys. **55**(7), 3336 (1971)
9. P.Y. Chou, G.D. Fasman, Prediction of protein conformation. Biochemistry **13**(2), 222–245 (1974)
10. M. Levitt, A. Warshel, Computer-simulation of protein folding. Nature **253**(5494), 694–698 (1975)
11. B.R. Gelin, M. Karplus, Sidechain torsional potentials and motion of amino-acids in proteins—bovine pancreatic trypsin-inhibitor. Proc. Natl. Acad. Sci. USA **72**(6), 2002–2006 (1975)
12. J.A. Mccammon, B.R. Gelin, M. Karplus, Dynamics of folded proteins. Nature **267**(5612), 585–590 (1977)
13. D. Van der Spoel, E. Lindahl, B. Hess, G. Groenhof, A.E. Mark, H.J.C. Berendsen, Gromacs: fast, flexible, and free. J. Comput. Chem. **26**(16), 1701–1718 (2005)
14. M.T. Nelson, W. Humphrey, A. Gursoy, A. Dalke, L.V. Kale, R.D. Skeel, K. Schulten, Namd: a parallel, object oriented molecular dynamics program. Int. J. Supercomput. Appl. High Perform. Comput. **10**(4), 251–268 (1996)
15. S. Plimpton, Fast parallel algorithms for short-range molecular-dynamics. J. Comput. Phys. **117**(1), 1–19 (1995)
16. M.P. Allen, D.J. Tildesley, *Computer Simulation of Liquids* (Oxford University Press, Oxford, 1987)
17. P.M. Morse, Diatomic molecules according to the wave mechanics. ii. vibrational levels. Phys. Rev. **34**(1), 57–64 (1929)
18. J. Tersoff, Empirical interatomic potential for carbon, with applications to amorphous-carbon. Phys. Rev. Lett. **61**(25), 2879–2882 (1988)
19. F.H. Stillinger, T.A. Weber, Computer-simulation of local order in condensed phases of silicon. Phys. Rev. B **31**(8), 5262–5271 (1985)
20. D.J. Oh, R.A. Johnson, Simple embedded atom method model for fcc and hcp metals. J. Mater. Res. **3**(3), 471–478 (1988)
21. J.E. Angelo, M.I. Baskes, Interfacial studies using the eam and meam. Interface Sci. **4**(1–2), 47–63 (1996)
22. M.S. Daw, M.I. Baskes, Embedded-atom method—derivation and application to impurities, surfaces, and other defects in metals. Phys. Rev. B **29**(12), 6443–6453 (1984)
23. Z. Qin, M.J. Buehler, Molecular dynamics simulation of the alpha-helix to beta-sheet transition in coiled protein filaments: evidence for a critical filament length scale. Phys. Rev. Lett. **104**(19) (2010)
24. A.D. MacKerell, D. Bashford, M. Bellott, R.L. Dunbrack, J.D. Evanseck, M.J. Field, S. Fischer, J. Gao, H. Guo, S. Ha, D. Joseph-McCarthy, L. Kuchnir, K. Kuczera, F.T.K. Lau, C. Mattos, S. Michnick, T. Ngo, D.T. Nguyen, B. Prodhom, W.E. Reiher, B. Roux, M. Schlenkrich,

J.C. Smith, R. Stote, J. Straub, M. Watanabe, J. Wirkiewicz-Kuczera, D. Yin, M. Karplus, All-atom empirical potential for molecular modeling and dynamics studies of proteins. J. Phys. Chem. B **102**(18), 3586–3616 (1998)

25. J.W. Ponder, D.A. Case, Force fields for protein simulations. Protein Simul. **66**, 27 (2003)
26. A.D. Mackerell, Empirical force fields for biological macromolecules: overview and issues. J. Comput. Chem. **25**(13), 1584–1604 (2004)
27. A.K. Rappe, C.J. Casewit, K.S. Colwell, W.A. Goddard, W.M. Skiff, Uff, a full periodic-table force-field for molecular mechanics and molecular-dynamics simulations. J. Am. Chem. Soc. **114**(25), 10024–10035 (1992)
28. D.A. Pearlman, D.A. Case, J.W. Caldwell, W.S. Ross, I. Cheatham, S. DeBolt, D. Ferguson, G. Seibel, P. Kollman, Amber, a package of computer programs for applying molecular mechanics, normal mode analysis, molecular dynamics and free energy calculations to simulate the structural and energetic properties of molecules. Comput. Phys. Commun. **91**(1), 1–41 (1995)
29. W. Wang, O. Donini, C.M. Reyes, P.A. Kollman, Biomolecular simulations: recent developments in force fields, simulations of enzyme catalysis, protein-ligand, protein-protein, and protein-nucleic acid noncovalent interactions. Annu. Rev. Biophys. Biomol. Struct. **30**, 211–243 (2001)
30. H.A. Scheraga, M. Khalili, A. Liwo, Protein-folding dynamics: overview of molecular simulation techniques. Annu. Rev. Biophys. Bioeng. **58**, 57–83 (2007)
31. A.A. Deniz, S. Mukhopadhyay, E.A. Lemke, Single-molecule biophysics: at the interface of biology, physics and chemistry. J. R. Soc. Interface **5**(18), 15–45 (2008)
32. M.J. Buehler, S. Keten, Colloquium: failure of molecules, bones, and the earth itself. Rev. Mod. Phys. **82**(2), 1459 (2010)
33. H.J. Gao, A theory of local limiting speed in dynamic fracture. J. Mech. Phys. Solids **44**(9), 1453–1474 (1996)
34. M.J. Buehler, F.F. Abraham, H.J. Gao, Hyperelasticity governs dynamic fracture at a critical length scale. Nature **426**(6963), 141–146 (2003)
35. A.C.T. van Duin, S. Dasgupta, F. Lorant, W.A. Goddard, Reaxff: a reactive force field for hydrocarbons. J. Phys. Chem. A **105**(41), 9396–9409 (2001)
36. A.C.T. van Duin, A. Strachan, S. Stewman, Q.S. Zhang, X. Xu, W.A. Goddard, Reaxff(sio) reactive force field for silicon and silicon oxide systems. J. Phys. Chem. A **107**(19), 3803–3811 (2003)
37. D.W. Brenner, O.A. Shenderova, J.A. Harrison, S.J. Stuart, B. Ni, S.B. Sinnott, A second-generation reactive empirical bond order (rebo) potential energy expression for hydrocarbons. J. Phys., Condens. Matter **14**(4), 783–802 (2002)
38. S.J. Stuart, A.B. Tutein, J.A. Harrison, A reactive potential for hydrocarbons with intermolecular interactions. J. Chem. Phys. **112**(14), 6472–6486 (2000)
39. A. Strachan, E.M. Kober, A.C.T. van Duin, J. Oxgaard, W.A. Goddard, Thermal decomposition of rdx from reactive molecular dynamics. J. Chem. Phys. **122**(5), (2005)
40. K. Chenoweth, S. Cheung, A.C.T. van Duin, W.A. Goddard, E.M. Kober, Simulations on the thermal decomposition of a poly(dimethylsiloxane) polymer using the reaxff reactive force field. J. Am. Chem. Soc. **127**(19), 7192–7202 (2005)
41. K.D. Nielson, A.C.T. van Duin, J. Oxgaard, W.Q. Deng, W.A. Goddard, Development of the reaxff reactive force field for describing transition metal catalyzed reactions, with application to the initial stages of the catalytic formation of carbon nanotubes. J. Phys. Chem. A **109**(3), 493–499 (2005)
42. S.S. Han, A.C.T. van Duin, W.A. Goddard, H.M. Lee, Optimization and application of lithium parameters for the reactive force field, reaxff. J. Phys. Chem. A **109**(20), 4575–4582 (2005)
43. S. Cheung, W.Q. Deng, A.C.T. van Duin, W.A. Goddard, Reaxff(mgh) reactive force field for magnesium hydride systems. J. Phys. Chem. A **109**(5), 851–859 (2005)
44. M.J. Buehler, Hierarchical chemo-nanomechanics of proteins: entropic elasticity, protein unfolding and molecular fracture. J. Mech. Mater. Struct. **2**(6), 1019–1057 (2007)

45. A.N. Parbhu, W.G. Bryson, R. Lal, Disulfide bonds in the outer layer of keratin fibers confer higher mechanical rigidity: correlative nano-indentation and elasticity measurement with an afm. Biochemistry **38**(36), 11755–11761 (1999)
46. H. Wang, D.A.D. Parry, L.N. Jones, W.W. Idler, L.N. Marekov, P.M. Steinert, In vitro assembly and structure of trichocyte keratin intermediate filaments: a novel role for stabilization by disulfide bonding. J. Cell Biol. **151**(7), 1459–1468 (2000)
47. O. Mayans, J. Wuerges, S. Canela, M. Gautel, M. Wilmanns, Structural evidence for a possible role of reversible disulphide bridge formation in the elasticity of the muscle protein titin. Structure **9**(4), 331–340 (2001)
48. N. Mucke, L. Kreplak, R. Kirmse, T. Wedig, H. Herrmann, U. Aebi, J. Langowski, Assessing the flexibility of intermediate filaments by atomic force microscopy. J. Mol. Biol. **335**(5), 1241–1250 (2004)
49. F. Aslund, J. Beckwith, Bridge over troubled waters: sensing stress by disulfide bond formation. Cell **96**(6), 751–753 (1999)
50. P.J. Hogg, Disulfide bonds as switches for protein function. Trends Biochem. Sci. **28**(4), 210–214 (2003)
51. S. Keten, C.-C. Chou, A.C.T. van Duin, M.J. Buehler, Tunable nanomechanics of protein disulfide bonds in redox microenvironments. J. Mech. Behav. Biomed. Mater. **5**(1), 32–40 (2012)
52. A.P. Wiita, S.R.K. Ainavarapu, H.H. Huang, J.M. Fernandez, Force-dependent chemical kinetics of disulfide bond reduction observed with single-molecule techniques. Proc. Natl. Acad. Sci. USA **103**(19), 7222–7227 (2006)
53. M. Bonomi, D. Branduardi, G. Bussi, C. Camilloni, D. Provasi, P. Raiteri, D. Donadio, F. Marinelli, F. Pietrucci, R.A. Broglia, M. Parrinello, Plumed: a portable plugin for free-energy calculations with molecular dynamics. Comput. Phys. Commun. **180**(10), 1961–1972 (2009)
54. M. Bonomi, M. Parrinello, Enhanced sampling in the well-tempered ensemble. Phys. Rev. Lett. **104**(19), 190601 (2010)
55. J. Kubelka, J. Hofrichter, W.A. Eaton, The protein folding 'speed limit'. Curr. Opin. Struct. Biol. **14**(1), 76–88 (2004)
56. A. Laio, M. Parrinello, Escaping free-energy minima. Proc. Natl. Acad. Sci. USA **99**(20), 12562–12566 (2002)
57. A.F. Voter, F. Montalenti, T.C. Germann, Extending the time scale in atomistic simulation of materials. Annu. Rev. Mater. Res. **32**, 321–346 (2002)
58. A. Kushima, X. Lin, J. Li, J. Eapen, J.C. Mauro, X.F. Qian, P. Diep, S. Yip, Computing the viscosity of supercooled liquids. J. Chem. Phys. **130**(22), (2009)
59. M.J. Alava, P.K.V.V. Nukalaz, S. Zapperi, Statistical models of fracture. Adv. Phys. **55**(3–4), 349–476 (2006)
60. Y. Sugita, Y. Okamoto, Replica-exchange molecular dynamics method for protein folding. Chem. Phys. Lett. **314**, 141–151 (1999)
61. A. Gautieri, S. Vesentini, A. Redaelli, M.J. Buehler, Hierarchical structure and nanomechanics of collagen microfibrils from the atomistic scale up. Nano Lett. **11**(2), 757–766 (2011)
62. K.Y. Sanbonmatsu, C.S. Tung, High performance computing in biology: multimillion atom simulations of nanoscale systems. J. Struct. Biol. **157**(3), 470–480 (2007)
63. K. Kadau, T.C. Germann, P.S. Lomdahl, Molecular dynamics comes of age: 320 billion atom simulation on bluegene/l. Int. J. Mod. Phys. C **17**(12), 1755–1761 (2006)
64. V. Tozzini, Coarse-grained models for proteins. Curr. Opin. Struct. Biol. **15**, 144–150 (2005)
65. D.W. Brenner, The art and science of an analytic potential. Phys. Status Solidi B **217**(1), 23–40 (2000)
66. R. Car, M. Parrinello, Unified approach for molecular dynamics and density-functionaly theory. Phys. Rev. Lett. **55**(22), 2471–2474 (1985)
67. B.R. Brooks, R.E. Bruccoleri, B.D. Olafson, D.J. States, S. Swaminathan, M. Karplus, Charmm: a program for macromolecular energy, minimization, and dynamics calculations. J. Comput. Chem. **4**(2), 187–217 (1983)

68. P. Sherwood, B.R. Brooks, M.S.P. Sansom, Multiscale methods for macromolecular simulations. Curr. Opin. Struct. Biol. **18**, 630–640 (2008)
69. F. Tama, I.C.L. Brooks, Symmetry, form, and shape: guiding principles for robustness in macromolecular machines. Annu. Rev. Biophys. Biomol. Struct. **35**, 115–133 (2006)
70. I. Bahar, A.J. Rader, Coarse-grain normal model analysis in structural biology. Curr. Opin. Struct. Biol. **15**, 586–592 (2005)
71. M.M. Tirion, Large amplitude elastic motions in proteins from a single-parameter, atomic analysis. Phys. Rev. Lett. **77**(9), 1905–1908 (1996)
72. T. Haliloglu, I. Bahar, B. Erman, Gaussian dynamics of folded proteins. Phys. Rev. Lett. **79**(16), 3090–3093 (1997)
73. S. Hayward, N. Go, Collective variable description of native protein dynamics. Annu. Rev. Biophys. Bioeng. **46**, 223–250 (1995)
74. L. Meireles, M. Gur, A. Bakan, I. Bahar, Pre-existing soft modes of motion uniquely defined by native contact topology facilitate ligand binding to proteins. Protein Sci. **20**(10), 1645–1658 (2011)
75. A.R. Atilgan, S.R. Durell, R.L. Jernigan, M.C. Demirel, O. Keskin, I. Bahar, Anisotropy of fluctuation dynamics of proteins with an elastic network model. Biophys. J. **80**, 505–515 (2001)
76. P. Doruker, R.L. Jernigan, I. Bahar, Dynamics of large proteins through hierarchical levels of coarse-grained structures. J. Comput. Chem. **23**(1), 119–127 (2002)
77. I. Navizet, R. Lavery, R.L. Jernigan, Myosin flexibility: structural domains and collective vibrations. Protein. Struct. Funct. Bioinform. **54**, 384–393 (2004)
78. W. Zheng, S. Doniach, A comparative study of motor-protein motions by using a simple elastic-network model. Proc. Natl. Acad. Sci. USA **100**(23), 13253–13258 (2003)
79. H. Dietz, M. Rief, Elastic bond network model for protein unfolding mechanics. Phys. Rev. Lett. **100**, 098101 (2008)
80. D.K. West, D.J. Brockwell, P.D. Olmsted, S.E. Radford, E. Paci, Mechanical resistance of proteins explained using simple molecular models. Biophys. J. **90**(1), 287–297 (2006)
81. J.I. Sulkowska, M. Cieplak, Mechanical stretching of proteins—a theoretical survey of the protein data bank. J. Phys., Condens. Matter **19**, 283201 (2007)
82. M. Bathe, A finite element framework for computation of protein normal modes and mechanical response. Protein. Struct. Funct. Bioinform. **70**(4), 1595–1609 (2007)
83. I. Bahar, R.L. Jernigan, Inter-residue potentials in globular proteins and the dominance of highly specific hydrophilic interactions at close separation. J. Mol. Biol. **266**(1), 195–214 (1997)
84. H.D. Nguyen, C.K. Hall, Molecular dynamics simulations of spontaneous fibril formation by random-coil peptides. Proc. Natl. Acad. Sci. USA **101**(46), 16180–16185 (2004)
85. H.D. Nguyen, C.K. Hall, Spontaneous fibril formation by polyalanines; discontinuous molecular dynamic simulations. J. Am. Chem. Soc. **128**(6), 1890–1901 (2006)
86. A. Arkhipov, P.L. Freddolino, K. Imada, K. Namba, K. Schulten, Coarse-grained molecular dynamics simulations of a rotating bacterial flagellum. Biophys. J. **91**, 4589–4597 (2006)
87. M.J. Buehler, Nature designs tough collagen: explaining the nanostructure of collagen fibrils. Proc. Natl. Acad. Sci. USA **103**(33), 12285–12290 (2006)
88. M.J. Buehler, Molecular nanomechanics of nascent bone: fibrillar toughening by mineralization. Nanotechnology **18**, 295102 (2007)
89. S.J. Marrink, H.J. Risselada, S. Yefimov, D.P. Tieleman, A.H. de Vries, The martini force filed: coarse grained model for biomolecular structures. J. Phys. Chem. B **111**, 7812–7824 (2007)
90. S.J. Marrink, A.H. de Vries, A.E. Mark, Coarse grained model for semiquantitative lipid simulations. J. Phys. Chem. B **108**, 750–760 (2004)
91. L. Monticelli, S.K. Kandasamy, X. Periole, R.G. Larson, D.P. Tieleman, S.J. Marrink, The martini coarse-grained force field: extension to proteins. J. Chem. Theory Comput. **4**, 819–834 (2008)

92. J.C. Shelley, M.Y. Shelley, R.C. Reeder, S. Bandyopadhyay, M.L. Klein, A coarse grain model for phospholipid simulations. J. Phys. Chem. B **105**, 4464–4470 (2001)
93. J.C. Shelley, M.Y. Shelley, R.C. Reeder, S. Bandyopadhyay, P.B. Moore, M.L. Klein, Simulations of phospholipids using a coarse-grain model. J. Phys. Chem. B **105**, 9785–9792 (2001)
94. S.O. Nielson, C.F. Lopez, G. Srinivas, M.L. Klein, Coarse grain models and the computer simulation of soft materials. J. Phys., Condens. Matter **16**, 481–512 (2004)
95. M. Venturoli, M.M. Sperotto, M. Kranenburg, B. Smit, Mesoscopic models of biological membranes. Phys. Rep. **437**, 1–54 (2006)
96. A.B. Liel, C.B. Haselton, G.G. Deierlein, J.W. Baker, Incorporating modeling uncertainties in the assessment of seismic collapse risk of buildings. Struct. Saf. **31**, 197–211 (2009)
97. T. Ackbarow, D. Sen, C. Thaulow, M.J. Buehler, Alpha-helical protein networks are self-protective and flaw-tolerant. PLoS ONE **4**(6), e6015 (2009)
98. Z. Qin, L. Kreplak, M.J. Buehler, Hierarchical structure controls nanomechanical properties of vimentin intermediate filaments. PLoS ONE **4**(10), e7294 (2009)
99. M. Neri, C. Anselmi, M. Cascella, A. Maritan, P. Carloni, Coarse-grained model of proteins incorporating atomistic detail of the active site. Phys. Rev. Lett. **95**(21), 218102 (2005)
100. G. Stefanou, M. Fragiadakis, Nonlinear dynamic analysis of frames with stochastic non-gaussian material properties. Eng. Struct. **31**(8), 1841–1850 (2009)

# Chapter 7
# Mechanical Characterization in Molecular Simulation

**Abstract** A standardized procedure for extracting mechanical properties from molecular simulation unfortunately does not exist. The challenge is to construct a suitable procedure and extract useful mechanical measures comparable to macroscale metrics (stiffness, strength, fracture toughness, etc.). As a result, there have been many methods developed to exploit the capabilities of atomistic simulation (such as steered molecular dynamics, direct calculation of virial stress and strain, free energy minimization, etc.), and various analytical tools and models to interpret such results (*e.g.*, the classical Bell model for rate dependence, worm-like chain models for entropic unfolding, etc.). Here, we look at a sample of approaches used for mechanical characterization in molecular simulation.

*You have your way. I have my way. As for the right way, the correct way, and the only way, it does not exist.*

*Friedrich Nietzsche (1844–1900)*

## 7.1 Introduction

Molecular simulation has developed as an indispensable tool for the investigation of material systems at the atomistic level, yet there is very few "standardized methods" for the determination of mechanical properties. In this chapter—to stress the atomistic basis and primary hierarchical scale of the materiome—we review various atomistic based modeling approaches applied to study the mechanical behavior of molecular systems, and common methods to calculate mechanical properties. This can be thought of as "steps" in characterizing the materiome—extracting the necessary mechanical information from the modeling methods discussed in Chap. 6: Computational Approaches and Simulation, to inform the categorical representation presented in Chap. 4: Universality-Diversity Paradigm: Music, Materiomics, and Category Theory.

Due to the accelerating development and unique perspectives of the fields that contribute to advances in simulation, there is a diverse array of interests, approaches and tools implemented with the primary goal of probing the mechanical properties of molecular systems. To this end, we review a selection of common analytical

S.W. Cranford, M.J. Buehler, *Biomateriomics*, Springer Series in Materials Science 165, 
DOI 10.1007/978-94-007-1611-7_7, 

models and computational approaches with the aim to delineate the benefits and pitfalls of differing methodologies. The intent is to demonstrate that no approach is universal, and prudent judgment is essentially when linking atomistic results to mesoscopic and larger continuum scales.

The computational approaches discussed herein are a useful means for studying material science phenomena and can be used to complement experimental studies and theory to provide a novel viewpoint in the analysis of complex systems. However, the knowledge attained by atomistic and molecular simulation only derives practical value when applied to physical problems. While simulation of non-physical situations (such as nonexistent protein mutations) can shed light into some problems, care most be taken to avoid a "slippery slope" of simulation. Otherwise, the result could be computational *alchemy*—proving the transmutation of lead into gold on a computer cluster does not make it so! Thus, while a necessary theoretical tool, the value of simulation results must be considered alongside physical observations and/or experimental results (discussed in Chap. 5: Experimental Approaches).

## 7.2 Objective: Up-scaling Molecular Mechanics

An integrated approach that uses experiment and predictive simulation concurrently has evolved into a successful research paradigm of material (or materiomic) characterization. Structural design and materials engineering is unified through the understanding of hierarchical features with atomic resolution, from nano to macro, through an effective "cataloguing" of behaviors on a scale-by-scale basis. Materiomics attempts to elucidate the cross-scale effects, wherein all scales must be considered collectively and emergent phenomena predictable, controlled, and harnessed with designed intent. There is an intimate coupling between disparate length scales, as the long-range mechanical effects can cause drive phenomena from nano to macro *via* higher organizing principles. Previous approaches to such complex problems attempted to subjugate small scale phenomena into "effective" properties. Here, we focus on mechanical properties at nano- and mesoscale, as the *mechanics* of molecular and macromolecular interactions provides a general basis for many relevant material properties.

Critical to multi-scale mechanical modeling and simulation is the development of definitions for mechanical properties that are *calculable* within an atomistic or molecular system. Behaviors and effective properties determined at one hierarchical scale should be described in the same "mechanical language" across scales in a consistent manner (one of the motivations for building a foundation through atomistic/molecular behaviors, rather than quantum mechanics—an "uncertain" foothold—as discussed in Chap. 6: Computational Approaches and Simulation). The most developed mechanical descriptors and metrics are found in the field of *continuum mechanics*—a branch of mechanics that deals with the analysis of the kinematics and the mechanical behavior of materials modeled as a continuous mass

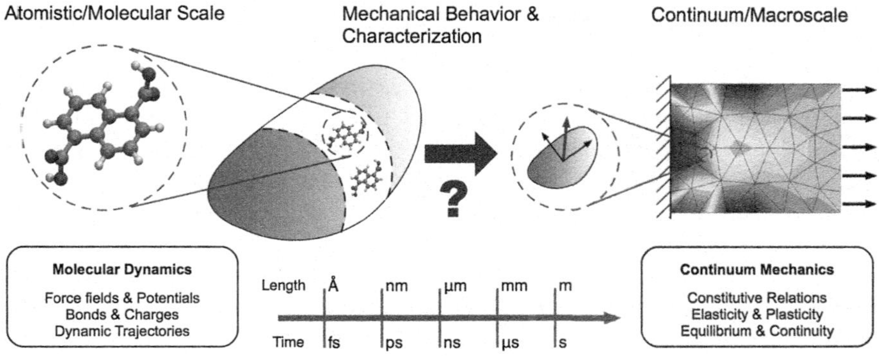


**Fig. 7.1** In search of a common language—the atomistic to continuum challenge. The fundamental behavior of materials is rooted in the atomistic and molecular scale, which is unavoidable discrete (noncontinuous) and defined by chemical bonds and ionic charges. Such systems are simulated by parameterized force fields and potentials, and allow characterization of mechanical behaviors. Large-scale applications and design necessitate a continuum perspective, defined by constitutive relations and fundamental mechanical formulations (*e.g.*, equilibrium, continuity). While atomistic behavior can be formulated in terms of continuum models, hierarchies and cross-scale interactions limit and universal scaling laws

rather than as discrete particles. By definition, a continuum perspective is *scale-free*, ignoring the fact that matter is made of atoms (not continuous). However, on length scales much greater than that of inter-atomic distances, such models are highly accurate and encompass fundamental physical laws such as the conservation of mass, the conservation of momentum, and the conservation of energy. Scaling atomistic and molecular behavior to the continuum scale (*i.e.*, scales wherein the hypothesis of continuous matter holds) would facilitate the atomistic design of complex materials (see Fig. 7.1). As expected, such an "atomistic to continuum" scaling is more difficult in practice. While atomistic behavior can be formulated in terms of continuum models (and has been successful for homogeneous materials, such as crystalline metals [1–4]), hierarchies and cross-scale interactions (*i.e.*, systems with a complex materiome) limit any universal scaling laws, wherein behavior "*A*" at the molecular scale can be associated with continuum model "*B*" at the macroscale. This is similar in to the concept of *material behavior scaling* as discussed in Chap. 2: The Materiome. Such a formulation is a practical metric—we have become accustomed to the concepts such as stress, stiffness, and viscosity, for example, and require common heuristics to characterize protein materials, metals, ceramics, and rubbers alike.

Examination of the state of a simulated material and the determination of its mechanical properties is accomplished by various means *via* the direct analysis and inspection of the behavior and response of a material system. Similar to experimental endeavors, this typically relies on the determination and analysis of data pertaining to force and stress states, displacement and strain gradients, changes in geometrical configurations, temperature, etc. Beyond the capabilities of physical assays, one of the primary benefits of computational approaches is the ability to directly calculate intangible (*qua* immeasurable) quantities such as *free energy* or *entropy*, an relate such metrics to fundamental theories (*i.e.*, laws of thermodynamics).

One is often interested in long length scale and long time scale phenomena in solids (*e.g.*, solid state phase transformations, fracture, friction, etc.), and such phenomena have been previously described by continuum theories. Microscopic simulations of complex systems, on the other hand, whether *via* molecular dynamics, lattice Boltzmann or Monte Carlo methods, deal with microscopic variables like the positions and velocities of constituent particles and together with detailed knowledge of interatomic potentials, hope to build up a description of the macro system from a knowledge of these "micro-variables". The fundamental principle underlying the mechanical characterization is the assertion of energy equivalence between mechanical response (classically formulated from a continuum perspective) and atomistic interactions (through potential energy functions). Simply put, to relate continuum theories of mechanical phenomena to typical atomic or molecular theories, methods of correlating changes in positions and energetics of the atomistic behavior with descriptions of deformation used in a continuum is required.

A continuum description typically must incorporate effects of the entire represented molecular system (such as hydrophobic and other solvent effects, electrostatic interactions, hydrogen bonding between molecules, and/or entropic effects), while asserting a homogeneous description of the material. As there can be many contributions to even simple mechanical process (such as breaking of hydrogen bonds or solvent friction during molecular stretching or bending), one could potentially introduce conservation of energy for the relevant mechanical response, and associate all factors to a single continuum response. The approach is to apply energy equivalence between the full atomistic system and continuum theory, or:

$$\sum \phi_{atomistic} \approx U_{continuum} \tag{7.1}$$

where $\phi$ is the per-atom potential effects, and $U$ is the representative energy function for the mechanical response. It is an underlying assumption of all approaches presented here that the response of the full atomistic system is *sufficiently* described by the continuum-based energy function, $U$. The choice of the strain or deformation energy function is dependent on the system to be investigated, and assumptions of elastic/plastic behavior, material isotropy, system failure such as yielding or fracture, etc., affect the definition (and interpretation) of atomistic results. The purpose of such a strong assumption is twofold: primarily to define material behavior using conventional mechanical property vernacular (stress, strain, Young's modulus, bending stiffness, fracture toughness, etc.) and allow comparisons between novel and well-established material systems, and secondarily to facilitate a consistent "bottom-up" description of the material, where properties at the nanoscale ultimately affect the behavior at the macroscale.

Mechanical characterization can be generally broken into two distinct approaches:

1. *Response based*: thought of as a simulated experiment, an indirect means to elicit a system response. Typically, a driving agent is applied to the system in the form

of applied forces, applied displacements, controlled variations in temperature or pressure, etc., and the resulting system evolution (such as deformation, incurred stresses, etc.) are recorded. These data from these computational "experiments" can then be fitted to various mechanical models, or phenomenological formulas.

2. *Energy based*: thought of as idealized material systems, designed to directly extract immeasurable quantities (such as the elastic strain of specific molecular bonds). For energy based methods, the simulated system is coerced into a multitude of configurations (such as stretching of compressive modes) and the energetics analyzed. Such results can then be used to potentially formulate new mechanical models, with insights gained from the system behavior.

Undoubtedly, many computational investigations can be considered a hybrid of the above approaches, applying a driving agent (such as a unfolding a $\alpha$-helix molecule) and tracking the energetics (such as energy variation due to breaking and reforming of H-bonds), for example. Like any physical assay, there a multiple issues concerned with the "design" of a successful computational experiment. A major influence is the effect of imposed boundary conditions. Molecular simulation can be greatly affected by constraints such as choice of ensemble (*e.g.*, microcanonical or canonical), periodicity, and/or imposed fixities—such conditions have physical analogues and are also considered in physical experimentation (the ensemble, for example, encompassing the environmental temperature or pressure). However, simulation inherently applies "computational conditions" that arise in the form of temperature damping parameters, minimization schemes, or long-range interaction solvers. A overlooked contribution can cause significant divergence between simulated and physical behavior, and even between similar simulation set-ups. Moreover, caution must be taken concerning the coupling of deformation modes. For example, at finite deformation, modes such as tension, bending and twisting can be coupled with one another, and thus the relation between applied loads and measured displacements may not be adequate to extract accurate mechanical properties. The validity of a presumed theoretical model must correspond to the design of the simulation. Being said, in spite of such concerns, molecular dynamics (MD) provides full dynamical information and deterministic trajectories, crucial to describe highly driven and coupled phenomena and failure.[1] Unlike many continuum mechanics approaches, atomistic techniques require no *a priori* assumption on the defect dynamics. Once the atomic interactions are chosen, the material behavior is determined completely.

The perspectives and overviews presented for the remainder of the chapter are intended to provide a broad synopsis only. Equations and formulations are presented to outline basic theories behind analytical techniques, but calculation details and specific numerical results are omitted. Further details regarding various methodologies can be found in the papers cited.

---

[1]MD provides an alternative approach to methods like Monte Carlo (MC), which provide certain advantages as well. However, this point will not be discussed further here as the simulation studies reviewed here are carried out with a MD approach.

## 7.3 Caveats and Missing Metrics

The ability to design synthetic materials at the same level of control Nature has perfected over biological materials at all length–scales has challenged scientists for decades. One of the outstanding issues is an inability to translate specific functionalities to quantifiable metrics. Nature does not care for quantifiable properties as long as the intended system function is achieved. Engineers, however, rely on mechanical properties to ensure adequate design of systems—choice of components may or may not meet certain requirements. This paradigm of "property selection" is superficial and merely a convenient (and efficient) means for successful, reliable and repeatable designs. For example, an (astute) undergraduate should be able to estimate (within an order of magnitude) the stiffness, failure strength, and ultimate extension of common engineering materials such as steel, concrete, aluminum, and even plastics. Such mechanical properties are critical to benchmark a materials performance, and indeed necessary for design. The properties dictate relevant *applications*, but, as previously discussed, unlike biological systems, such materials lack any intrinsic *function*, as discussed in Chap. 2: The Materiome.

Consider a physical material property such as Young's modulus ($E$)—describing a phenomenological behavior inherently linking the uniaxial stress to the strain of a material. A high Young's modulus indicates a small changes in strain with relatively large increases in stress, while a small Young's modulus indicates the opposite. One does not need to know anything about the material system, the structural arrangement, or the underlying physics to compare the representative stress-strain behavior. It has an underlying physical response.

Without defined properties necessary to *prescribe* design/engineering constraints, the exploitation of biological materials is more akin to *performance-based engineering* (PBE). Performance-based engineering is the design, evaluation and construction of engineered systems that meet, as economically as possible, the uncertain future demands that both owner/users and nature will put upon them. In effect, performance-based engineering attempts to balance possible hazards, damage, and system demands such that functionality can be maintained; rational trade-offs are made based on life-cycle considerations rather than structural considerations alone. Such systems are designed, implemented, and operationally supported to meet the non-functional performance requirements defined. In effect, the system *function* is considered, without any preconditioned requirements of components.

Currently, there is a lack of metrics to objectively measure properties such as "hierarchies" or "robustness", which are typically defined on a case-by-case basis. This need not be the case. Prior to Maxwell's equations (circa 1860's), it was difficult to reconcile electrodynamics, magnetism, and optics—what we now understand as electromagnetic waves. We do not assert that there is a set of differential equations that can quantitatively describe "hierarchies" and "robustness", but a universally accepted and objective definition is indeed possible (and necessary). From a similar perspective, there is no scale for structural arrangements—the carbon bonds composing a carbon nanotube resemble a macroscale truss structure, for example, but to they behave as one? Successful models have been formulated for nanotubes [5, 6], but then the question remains if such approaches are transferable to soft matter such as polymer melts and proteins? The application of self-similarity and fractal dimensions has been considered for polymer systems [7], introducing concepts such as scale invariance to molecular material systems. But do such concepts hold for rigid systems such a metals and ceramics? While shedding insight to particular systems, specific modeling approaches (while successful) must balance a reductionist approach to understanding with a holistic interpretation. A materiomic paradigm should not degrade into a "hierarchical theory of things" [8] with a theory for each of particular material system at each scale branching to a new set theories, but rather make the universal (functional) connections between disparate material systems.

## 7.4 Response Based Mechanical Characterization

One of the most logical approaches to characterize the mechanical properties and behavior of an atomistic system is to emulate macroscale material testing procedures and measurement, including tensile and compression tests, three-point bending tests, torsion, etc. For example, a small model system can be constructed as a computational "test specimen" with direct control over the loading and boundary conditions, from which the specimen can be subject to various testing procedures and mechanical properties directly determined. The dominance of specific mechanisms is controlled by geometrical parameters, the chemical nature of the molecular interactions, as well as the structural arrangement. To illustrate, for a specimen subject to tension, we can determine Young's modulus, $E$, indirectly *via* force-displacement results where:

$$E = \frac{\partial \sigma}{\partial \varepsilon} \approx \frac{\Delta \sigma}{\Delta \varepsilon} = \frac{L_0}{A_c} \frac{\Delta F}{\Delta r} \tag{7.2}$$

where $\sigma$ and $\varepsilon$ are the stress and strain, $F$ and $r$ are the force and displacement, and $A_c$ and $L_0$ are the cross-sectional area (assumed constant) and initial length. We note that the stress ($\sigma$) and the strain ($\varepsilon$) need not be calculated to determine the modulus $E$. Pure axial stretching would result in a one-to-one correspondence between results from $\sigma - \varepsilon$ and $F - r$ formulations, but a response based approach $(F - r)$ can capture the *effective* modulus—essentially defining the required load (force) to induce deformation. A comparison with $\sigma - \varepsilon$ derived values can indicate the coupling of mechanisms (such as unfolding or torsion, for example). For nonlinear behavior, an effective stiffness can be calculated for each regime independently, or fitted by a function, *e.g.*, $E(\varepsilon) = f(\varepsilon)$.

In a similar manner, the bending rigidity if a molecule can be determined *via* applied point loading to a molecule or molecular structure. Such an approach relies heavily on the assumption that bending behavior can be suitable approximated by conventional beam models (such as Euler-Bernoulli or Timoshenko beams). For example, if we consider a Euler-Bernoulli formulation, the bending stiffness, $EI$,

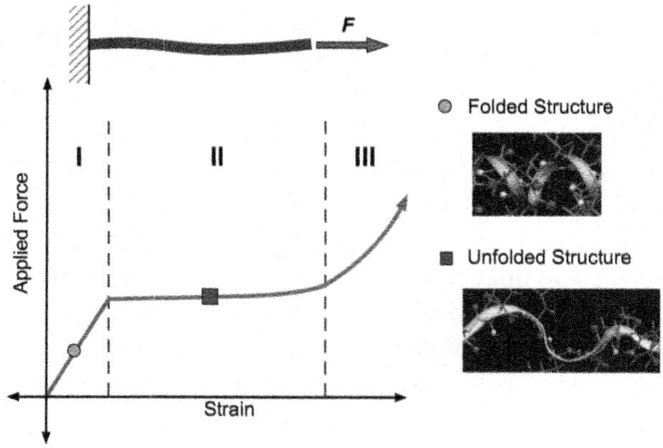


**Fig. 7.2** Qualitative tensile force-strain behavior of $\alpha$-helical protein (it is noted that *beta*-helical protein filaments (see *e.g.* Fig. 4.12) behave similarly). Where is the modulus? The first regime (**I**) consists of a steep increase in force until a critical strain induces a rupture of the H-bonds that stabilize the helical structure. The first regime is followed by the second regime (**II**) during which entropic unfolding of the alpha-helices occurs *via* successive rupture of all hydrogen bonds. In the third regime (**III**), a non-linear increase in strain by stretching the backbone is observed. For detailed analysis, see [11–13]. Traditional stiffness measures (such as Young's modulus) fail to capture the nonlinear effects, requiring more sophisticated formulations to transition to the macroscale (such as multi-parameter hyperelastic constitutive models)

can be related to the displacement measured by a three-point bending test:

$$EI = \frac{L_0^3}{48}\frac{\Delta F}{\Delta z} \tag{7.3}$$

where $L_0$ is the span of the "beam", $F$ the applied force (at the center of the span) and $z$ the displacement. Such models are based on the assumption of pure bending states while neglecting contributions from rotations and shear. Again, here the term $EI$ can be considered merely a label for "bending stiffness", and need not relate to the traditional interpretation of modulus, $E$, multiplied by the area moment of inertia, $I$. Even so, Euler-Bernoulli beam models are widely used to identify a set of uncorrelated parameters, including: torsional modulus, bending rigidity and Young's modulus, in both experimental and simulation studies [9, 10].

Such response-based approaches are advantageous when there is inherent difficulty in defining traditional elastic "properties" such as bending stiffness or Young's modulus. For example, in contrast with most material systems, protein-based materials cannot be described by a set of "fixed" properties, as the structure itself is a dynamic system that adapts to its local mechanochemical environment. Consider the unfolding of an $\alpha$-helix [11–13] one of the simplest protein structures, as depicted in Fig. 7.2.

It remains controversial whether the free energy landscape of such unfolding behavior of proteins consists of multiple, discrete transition states or the location of the transition state changes continuously with the pulling velocity [12]. This lack in

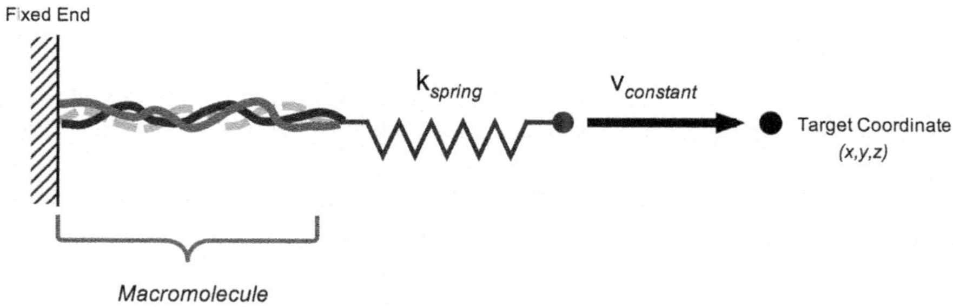


**Fig. 7.3** Schematic of steered molecular dynamics (SMD) simulations [17–20]. General constant velocity SMD approach where macromolecule is loaded by harmonic spring with defined stiffness, $k_{spring}$, and a fixed velocity, $v_{constant}$, towards a target coordinate $(x, y, z)$. The application of a constant load-rate in simulation (as opposed to constant load or constant displacement) approximates experimental techniques such as AFM and optical tweezer techniques where the load slowly increases from zero

understanding has prevented the development predictive strength models of protein materials, and complicates the application of continuum-based approaches.

### *7.4.1 Steered Molecular Dynamics*

Continuing advancement and emergence of experimental tools such as atomic force microscopy (AFM) [14] and optical tweezers [15] enable quantitative studies of deformation of individual molecules (both methods are discussed at length in Chap. 5: Experimental Approaches). Briefly, AFM consists of a micro-cantilever with a sharp tip at its end, which can be used to apply force. The mechanical signature of single molecules can be obtained by AFM, where the molecule (for example, an individual protein domain) is attached to a surface and manipulated by the tip that pulls the molecule at constant pulling speed. Likewise, optical tweezers apply forces on molecules that result from focused laser beam, a concept called optical trapping. The force can catch, hold and move dielectric particles ranging from nanometers to micrometers. Both methods essentially "pull" a molecule, and the force-displacement behavior is recorded, with possible resolutions of 0.1 pN and 1 Å currently accessible [16], also attainable by MD simulation. Such single-molecular level techniques provide extremely useful information to guide and validate atomistic-based multiscale computer simulation and facilitate the corroboration between molecular simulation and laboratory assays.

For example, steered molecular dynamics (SMD) simulations [17–20], in which external forces are used to explore the response and function of macromolecules, have become a powerful tool complementing and guiding *in vitro* single-molecule experiments (see Fig. 7.3). The simulated procedure mimics experimental protocols, allowing one-to-one comparisons of simulation and empirical data. Such simulations have increasingly been implemented to investigate complex biological phenomenon such as the unfolding behavior of protein materials [12, 13], ligand-receptor kinetics [21], and the unbinding trajectories of other biological and soft matter molecules [22, 23].

The SMD approach applies a moving spring force (see Fig. 7.3), such that the molecule can behave in a manner not captured by either force or displacement loading alone, allowing induced *conformational changes* in a system along a prescribed reaction vector. A driving force is applied to a selected atom, atom group, or molecule of magnitude:

$$F_{SMD} = k_{spring}(R - R_0) \tag{7.4}$$

where $k_{spring}$ is the spring constant, and $R_0$ is the distance from the end of spring from the tether point. A constant velocity is prescribed which monotonously increments or decrements the distance $R$ towards the tether point, or target coordinate. Alternatively, SMD can be used to apply a constant force, in which case the spring extension, $\Delta L$ is maintained throughout a simulation, where $F_{SMD} = k_{spring}\Delta L$. The application of SMD simulations corroborates experiments at the atomic scale.

## 7.5 Rate Dependence

As discussed in Chap. 6: Computational Approaches and Simulation, full atomistic MD simulation is inherently limited in both time and length scales. Due to computational limitations, overall time spans in such studies are limited to tens of nanoseconds, or, in very long simulation studies, to fractions of microseconds. As a consequence, simulation results of dynamically loading molecules have been carried out at large deformation rates, often exceeding several m/s (even approaching 10 or more m/s). Often, the disparity between the timescales accessible to experimental and simulation results differ by several orders of magnitude, it is difficult to formulate a direct link between simulation and experimental results. This is especially critical in molecular systems dominated by molecular unfolding or sliding, with a particular significance of rupture of weak chemical bonds such as hydrogen bonds, covalent cross-links or intermolecular entanglement, typified by biological protein materials (see Fig. 7.4).

To reconcile rate-dependent discrepancies, the Bell Model has been frequently utilized. The "classical" Bell model is a simple phenomenological model that describes the frequency of failure of reversible bonds [25]. The concept of reversibility thereby means that an individual bond can break under no force if one waits a sufficiently long time, and that it can likewise reform spontaneously. Such bonds may be associated with electrostatic, van der Waals (vdW), or weak H-bond interactions. The Bell expression for this dissociation, or off rate, $k$, is given by [25]:

$$k = \omega_0 \exp\left(-\frac{E_b - f \cdot x_b}{k_B T}\right) \tag{7.5}$$

where $\omega_0$ is the natural vibration frequency of the bond and $k_B T$ the thermal energy ($k_B$ being the Boltzmann constant, and $T$ the temperature). This conjecture was established long before single molecule experiments were performed. Later, it became

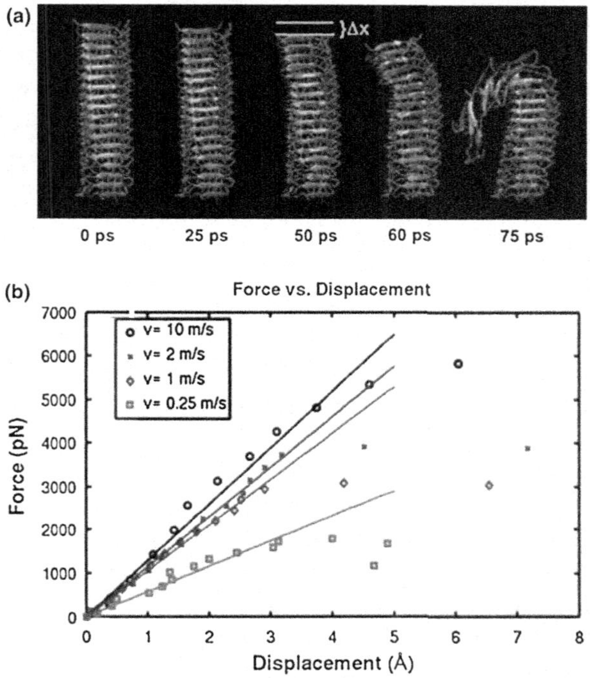


**Fig. 7.4** Snapshots of deformation and rate dependent force-extension plots of a $\beta$-solenoid structure (a triple helical fibrous protein). (**a**) Typical deformation profile of the triple-$\beta$-helix under compression (results from simulation with deformation speed $v = 25$ m/s). Significant compression of the molecule can be observed before failure, in particular at faster loading rates. Failure occurs when the structure buckles with respect to a random weak axis due to non-heterogeneous evolution of the topology under external force, H-bond rupture mediated by water molecules and thermal vibrations. (**b**) Force extension profiles for different loading rates are shown in *subplot*. The initial force extension behavior can be estimated by a linear regime followed by nonlinear plastic-like behavior toward failure. Adapted from [24]

very successful especially to describe forced unfolding of biological molecules. Indeed, the model enables one to characterize the bonds, their ruptures and their energy landscape profiles from the fitting with experimental results. The Bell model, despite its limitations, is one of the most well-known, simplest and most widely used strength models. Several recent papers have introduced advanced models for strength [26–31], but the phenomenological basis for rate dependence is sufficiently encompassed by the "classical" Bell model and the reader is referred to the cited references for more sophisticated formulations and applications.

A major limitation is that Bell's theory deals only with constant external force and does not explain loading rate dependence of strength—the critical variable to link simulation and empirical results. This limitation is important since, unlike simulation, a constant force cannot be applied in experiments due to the very short thermal impulse time compared with the measurement time [21, 32, 33]. Variations of pulling rates often induce a change in failure mechanism, characterized by an energy barrier, $E_b$, and a transition state distance, $x_b$. The analysis of the results of the pulling rate dependence of forces enables the interpretation of experimental and computational results through the determination of effective energy barriers and transition states, provides insight into microscopic failure mechanisms. One can demonstrate the logarithmic dependence of strength on the loading rate as follows [12, 21, 32, 33]:

$$f_r = \frac{k_B T}{x_b} \ln\left(\frac{r_f x_b}{k_B T \cdot k_0}\right) \tag{7.6}$$

where again $k_B$ is the Boltzmann constant, $T$ is absolute temperature, $x_b$ is the distance between the bound state and the energetic maximum, $r_f$ is the loading rate, and $k_0$ is the kinetic rate of binding dissociation at equilibrium (from Eq. 7.5 with $f = 0$).

The original Bell model does not provide a direct link between the atomistic chemical structure of a bond (or arrangements of bonds) and the resulting behavior. For instance, the effective energy barrier may represent a cluster of several weak bonds that break simultaneously, or a single, stronger bond. The model further does not capture the effect that the energy landscape itself may change under deformation of the protein. Other models have been reported that describes the strength behavior of proteins in dependence of the force rate, the rate of increase of force over time, as well as the transition between multiple transition states [26, 27, 32–35].

The Bell model and various extensions have been remarkably successful in fitting much of the data on forced unfolding of biological molecules. It is again noted that several other attempts have been made to extend and refine the simple "classical" Bell's model. For instance, it has been shown that rebinding can have a great impact on strength [33, 36, 37], as well as the energy landscape effect on the unfolding pathway of convoluted protein structures [12, 32]. In non-equilibrium pulling regimes, other models attempt also to explain non-logarithmic and probabilistic loading-rate dependence [26, 30]. Moreover, other extensions try to implement the influence of the transducer stiffness in order to explain the disparities in measured unbinding force among different methods [21]. As a last example of extension, we can mention the existence of models which take into account the energy landscape roughness of bonds [38, 39].

A dependence on rate introduces a challenge in correlating experimental results (timescales on the order of seconds and minutes) with computational results (timescales on the order of nano- and microseconds). The non-overlapping regimes can paint different pictures of the same molecular system, illustrated schematically by Fig. 7.5, that are difficult to reconcile. In addition to the affect of rate-dependence, molecular systems frequently depict length-dependencies [40–42], as well as variation in mechanical behavior due to prescribed boundary conditions [43, 44]. Such issues warrant prudent consideration for the interpretation of experimental results and the comparison with theoretical and numerical models.

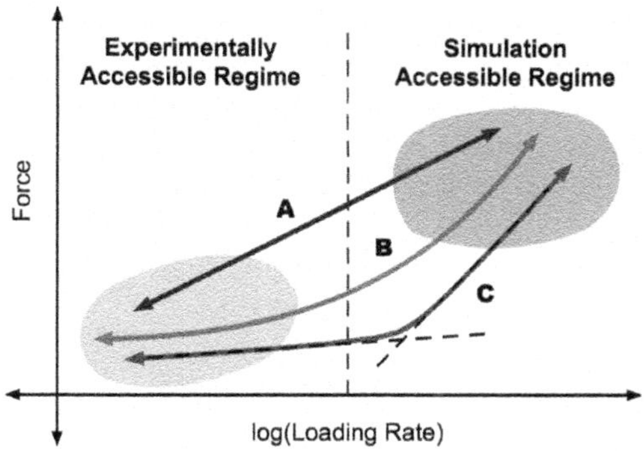


**Fig. 7.5** The fallacy of loading rate. Experimental results and computational results can paint a different picture of the same molecular system. If the system scales well with rate dependence (*e.g.*, no change in rupture mechanism), the formulation expressed by Eq. 7.6 can easily relate the loading rate to molecular strength (*e.g.*, energy barrier), as depicted by the linear relation in **A**. Extrapolating experimental or computationally data will result in the same parameters. Similarly, **B** depicts a case where there is a changing mechanism which is constant across scales. Again, experimental and computational results can be correlated (although Eq. 7.6 will fail). The problem arises if a transition state exists at some loading rate (**C**). Trends observed in the experimental regime (low loading rates) may completely differ from those observed in the computational regime (high loading rates)

## 7.6 Worm-Like Chain Models

Complementary to such rate dependent studies, a major success is the use of the Worm-Like-Chain (WLC) model to describe entropic elasticity [45, 46]. The use of Bell-type strength models in conjunction with Worm-Like-Chain models resulted in quantitative agreement between theory and experiment for tropocollagen molecules, for example [15, 47]. The results of empirical WLC fits were also previously discussed in Chap. 5: Experimental Approaches. In a simplified picture, the elastic stiffness of convoluted biomolecule (*e.g.*, DNA or a protein) may be parameterized by its contour length under zero tension, $l_c$, a persistence length, $l_p$ (and optionally, an elastic modulus, $K_0$). Persistence length measures the tendency of a uniform, flexible polymer to point in the same direction. Entropic compliance results when the length of a polymer is much greater than this persistence length, owing to the numerous configurations that a polymer may adopt. The elastic modulus, in contrast, measures the polymer's intrinsic resistance to longitudinal strain and reflects enthalpic contributions. The two parameters are generally correlated in worm-like polymers: the longer the persistence length, the greater the stretch modulus. Table 7.1 summarizes theoretical models for which analytical expressions have been derived; these treat the polymer backbone either as a worm-like chain (WLC) or a freely jointed chain (FJC), and incorporate entropic and/or enthalpic contributions. To illustrate, one form of the resulting force-extension relationship can be expressed as [45, 48]:

$$F(x) = \frac{k_B T}{l_p}\left[\frac{1}{4}\left(1 - \frac{x}{l_c}\right)^{-2} - \frac{1}{4} + \frac{x}{l_c}\right] \tag{7.7}$$

**Table 7.1** Entropic elasticity models. Adapted from [49]

| Model | Formula | Comments |
|---|---|---|
| Marko-Siggia [45, 48] | $F(x)=\frac{k_BT}{l_p}\left[\frac{1}{4}\left(1-\frac{x}{l_c}\right)^{-2}-\frac{1}{4}+\frac{x}{l_c}\right]$ | Derived from entropic theory; exact when salt concentration is high; applicable when $F \ll (1/4)(k_BTK_0^2/l_p)^{1/3}$ |
| Odijk [50] | $x(F)=l_c\left[1-\frac{1}{2}\left(\frac{k_BT}{Fl_p}\right)^{1/2}+\frac{F}{K_0}\right]$ | Derived from entropic/enthalpic theory; applicable for high-force regime, or $\lvert x-l_c\rvert/l_c \ll 1$ |
| Wang [49] | $F(x)=\frac{k_BT}{l_p}\left[\frac{1}{4}\left(1-\frac{x}{l_c}+\frac{F}{K_0}\right)^{-2}-\frac{1}{4}+\frac{x}{l_c}-\frac{F}{K_0}\right]$ | Derived from entropic/enthalpic theory; modified Marko-Siggia to account for enthalpic stretching; force range limitations similar to Marko-Siggia |

expressing the relationship between force ($F$) and extension ($x$) of a protein using its persistence length ($l_p$) and its contour length ($l_c$). Again, $k_B$ is the Boltzmann constant and $T$ is the absolute temperature. Other WLC formulations are given in Table 7.1.

Note that the various expressions have limited ranges of applicability, depending on a variety of underlying theoretical assumptions. Moreover, under large forces and displacements, the Worm-Like-Chain model fails and rupture of H-bonds within protein domains dominates [46]—*e.g.* a transition from unfolding-like behavior to unfolding with bond rupture. This behavior has been captured in quantitative models that combine the Worm-Like-Chain model with the Bell model [44, 51], in good comparison with experimental results, suggesting that such formulations can accurately reflect the interplay of strength and elasticity.

Nevertheless, despite the diversity of constructed force-displacement relationships, such response-based methods are transparent in their intent: apply a force or displacement and determine the mechanical characteristics (*e.g.*, $l_p$, $l_c$, $K_0$). Additionally, they provide a direct analogue to experimental approaches where force and displacement measurements provide the primary data [16]. Advances in experimental methods now enable one to probe time- and length-scales that are also directly accessible to large-scale atomistic based simulation, the combination of experiment and simulation might lead to a particularly fruitful interaction (such as those discussed in Chap. 5: Experimental Approaches). However, ambiguities and uncertainties may still arise in the consolidation and interpretation of the mechanical behavior into consistent material properties. Rigorous interpretations of how force, stress, and energy influence a material system are subject constant development and modification, and caution must be taken when directly comparing experimental and simulation results.

One of the advantageous of simulation, however, is the ability to extract data that is inaccessible to physical measurement and experimental techniques, such as the direct quantification of the potential energy of a material system, single molecule, or even single atom. We thus proceed to discuss methods that fully take advantage of explicit definition of a molecular system (*i.e.*, the potentials used and geometry prescribed), and begin with an approach that provides a direct analogue to continuum theory: *virial stress*.

## 7.7 Virial Stress

Mechanical stress is inherently a continuum concept—a measure of the internal forces acting within a deformable body. Many undergraduates struggle with the concept of tractions balanced by internal forces and the applicability of the divergence theorem, and find comfort in the more pedestrian definition of stress as force over area ($\sigma = F/A$). That does not help a molecular interpretation when there is neither a continuous "deformable body" nor an "area", but a set of discrete atoms (bonded or not). All the engineering materials we use (and subject to stress by any definition of the term) are indeed made up of a collection of such atoms. To reconcile this fact, we fundamentally define material stress as the change in free energy density as a function of material strain, or:

$$\tau_{ij} = \frac{\partial \psi}{\partial \varepsilon_{ij}} \tag{7.8}$$

where $\tau_{ij}$ is the component of stress induced by a strain, $\varepsilon_{ij}$, and the free energy density, $\psi$ is defined as energy per unit volume (the indices $i$ and $j$ refer to the basis vectors of a coordinate system). The formulation of stress in terms of energy landscape is convenient, as it is also a common description of atomistic systems in terms of potential energy functions (or force fields, as discussed in Chap. 6: Computational Approaches and Simulation), where:

$$\psi = \frac{1}{\Omega} \sum_{a \in \Omega} \phi_a(\chi) \tag{7.9}$$

where $\Omega$ is the considered system volume, $a$ the atoms within that volume, and $\phi$ the potential energy function dependent on some state variable, $\chi$. The *virial stress* is commonly used to find the macroscopic (continuum) stress in molecular dynamics computations [52]. The components of the macroscopic stress tensor, $\tau_{ij}$, in a macroscopically small, but microscopically large, volume $\Omega$ is typically taken to be:

$$\tau_{ij} = \frac{1}{\Omega} \sum_{a \in \Omega} \left[ -m^{(a)} v_i^{(a)} v_j^{(a)} + \frac{1}{2} \sum_{b \in \Omega} \left[ \left( r_i^{(a)} - r_i^{(b)} \right) F_j^{(ab)} \right] \right] \tag{7.10}$$

where $m_{(a)}$ is the mass of atom '$a$', $v_i^{(a)}$ and $v_i^{(a)}$ are the velocities in the $i$th and $j$th vector component basis, $r_i^{(a)} - r_i^{(b)}$ denotes the distance between atom '$a$' and atom '$b$' along the $i$th vector component, while $F_j^{(ab)}$ is the force on atom '$a$' exerted by atom '$b$' along the $j$th vector component, and $\Omega$ is the considered volume. To reduce random fluctuations, in addition to averaging over the representative spatial volume, $\Omega$, it is recommended to average further over small time interval around the desired time of the stress.

Although the expression developed is a time and spatial averaged quantity, it is often inappropriately used to obtain a point-wise, local stress, which contradict stress equilibrium of a system (see Zimmerman *et al.* [53] for a detailed discussion of virial stress). The intent here is not to delve into the accuracy and idiosyncrasies of the virial stress approach (such as the inclusion of the kinetic term, $m^{(a)} v_i^{(a)} v_j^{(a)}$ [54]), but merely to outline its potential application.

One primary advantage of the viral stress approach is that it can shed insight to the coupling of deformation modes and stress distributions in a molecular system, by simultaneously and directly evaluating all components of stress. For example, a model system can be subject to tensile force *via* SMD simulation as previously discussed. By calculating the virial stress, the force-extension data from the applied spring load can be directly compared to the virial stress and strain along the axis of extension. If there is limited coupling (for example, no shear modes under pure tension), the results of the virial stress-strain and the applied load-displacement should agree both qualitatively and quantitatively. Disagreement between the axial stress and "force over area" calculations can indicate either the dominance or coupling with a second-order mode of deformation, such as shearing or twist. Moreover, while per atom virial stresses may be inappropriate (in terms of a equivalence to the continuum interpretation of stress), they can be used to map a virial stress field, and indicate potential localizations of high force and failure events. This can shed great insight into heterogeneous material systems (such as a protein sequence) to indicate any potential "weak links".

Virial stress combined with explicitly applied loads and displacements are powerful tools in the determination of the mechanical properties of complex molecular systems. Ultimately, they rely on fundamental stress-strain relationships formulated by continuum theory, and effectively "translate" simulation results into representative mechanical characterization. However, some systems require a more fundamental interpretation of mechanical behavior when the system is unsuitable for applied loads or prescribed boundary conditions. For such systems, we rely *energy methods*.

## 7.8 Energy Methods

Here, we broadly categorize "energy methods" as approaches that assert mechanical conservation of energy, *i.e.*, the work done by external forces equals the energy stored in a deformed structure or molecular system. Since the total energy is constant, energy calculated by atomistic potentials is considered equivalent to energy determined by the imposed continuum theory (for example, linear elasticity), or:

$$\Psi(\chi) = \int_{\Omega} U(\chi)\, d\Omega = \sum_{\Omega} \phi(\chi) \tag{7.11}$$

where $\phi$ represents the atomistic potentials, and $U$ is the representative energy function for an arbitrary mechanical response, $\chi$, and, once more, $\Omega$ is the volume under consideration. For example, we again consider the uniaxial tensile behavior of a model system, where $\chi = \varepsilon_{uniaxial}$. The elastic strain energy density can be formulated as:

$$U(\chi) = U(\varepsilon) = \frac{1}{2}\sigma\varepsilon \tag{7.12}$$

Assuming linear elasticity, we find:

$$\sigma = E\varepsilon \rightarrow U(\varepsilon) = \frac{1}{2}E\varepsilon^2 \tag{7.13}$$

For the total volume,

$$\Psi(\varepsilon) = \frac{1}{2}E\varepsilon^2\Omega = \frac{1}{2}AL_0E\varepsilon^2 = \frac{1}{2}\frac{AE}{L_0}\Delta L^2 \tag{7.14}$$

where volume is defined as $\Omega = AL_0$ and strain is defined as $\varepsilon = \Delta L/L_0$. From Eq. 7.14 it is evident:

$$\Psi(\Delta L) = \alpha \cdot \Delta L^2 \tag{7.15}$$

where $\alpha$ ($\simeq AE/(2L_0)$) can then be a fitted parameter from extracted energy-displacement ($\Psi - \Delta L$) data. Note that, for comparisons across molecular systems, the definitions of $E$ or $A$ are somewhat arbitrary, representing an intrinsic stiffness of the molecule ($EA = K_0$). Similar formulations can be made regarding bending rigidity ($EI$) and shear modulus ($G$), for example. From this approach, one can consider the total system energy to derive mechanical characteristics. It should be noted that any explicitly determined properties (such as Young's modulus in the above example) are limited to the assumptions of the formulation (here, linear elasticity), but the introduced parameter, $\alpha$, is independent of such assumptions, allowing for quantitative comparisons across systems.

Moreover, due to the piece-wise formulation of the MD potentials (*e.g.*, bond, angle, dihedral, etc.), it is possible to decouple the contributions during a particular mechanical response, such as comparing the bond-stretching contribution to the weak van der Waals interactions during the stretching of a coiled-coil $\alpha$-helix dimer [12], for example. A *global* energy perspective (*e.g.*, encompassing all energetic contributions simultaneously) is the main difference from other energetic methods, which may consider the behavior as the summation of atom by atom contributions, such as the basis of the Cauchy-Born approach.

### *7.8.1 Similarity to Cauchy-Born Approaches*

Energetic elasticity is characterized by the stretching of atomic bonds, metallic, covalent, or ionic bonds, that lead to a change in potential energy in the material volume. Many techniques have been proposed to link micro-structural deformation with macroscopic concepts of elasticity. A prominent example is the Cauchy-Born (CB) rule [55], which relates the microscopic deformation fields with macroscopic stresses and strains. The CB rule, or different implementations, is the basis for several multi-scale modeling algorithms, including the quasi-continuum method [1, 56, 57], or the virtual internal bond method [58, 59]. A limitation of these methods is that they are largely limited to zero temperature (molecular mechanics/statics), and that the technique is only applicable far from defects or large stress or strain gradients (homogeneous bulk material properties). Recently, the CB approach has been extended to capture entropic effects as well, overcoming some of the limitations of the classical CB rule, and making such multi-scale techniques applicable to finite temperature studies [57].

In the simplest form, the Cauchy-Born rule [55] relates macroscopic deformation of crystals to changes in lattice vectors that allows direct link of potential to macroscopic continuum elasticity. The central assumption of the Cauchy-Born rule is to express the energy of an atomic system as a function of an applied strain tensor, $\varepsilon_{ij}$, and it is assumed that the continuum fields can immediately be *mapped* to the atomic scale. In general, the Cauchy-Born rule can be expressed as:

$$\mathbf{r} = \mathbf{F} \cdot \mathbf{R} \tag{7.16}$$

where $\mathbf{r}$ is the deformed lattice vector, $\mathbf{F}$ is an arbitrary (but admissible) deformation gradient, and $\mathbf{R}$ represents an undeformed lattice vector of a known (*a priori*) molecular structure. Cauchy assumed that macroscopic and atomic movement are the same—thus, a macroscopic deformation gradient can be transformed and scaled to the atomistic deformed crystal lattice vectors. Such deformed lattices are subject to stress equilibrium conditions, governed by the associated atomistic potentials, from which elastic properties can be derived [60]. Consider an arbitrary deformation gradient, $\mathbf{F}$, such that the Green's strain tensor is defined as:

$$\mathbf{E} = \frac{1}{2}(\mathbf{F}^T \cdot \mathbf{F} - \mathbf{I}) \tag{7.17}$$

or:

$$\varepsilon_{ij} = \frac{1}{2}\left(\frac{\partial u_i}{\partial X_j} + \frac{\partial u_j}{\partial X_i} + \frac{\partial u_k}{\partial X_i}\frac{\partial u_k}{\partial X_j}\right) \tag{7.18}$$

The goal is to express the strain energy density (SED) of an atomic unit cell as a function of the macroscopic strain field. It follows that first derivatives of the SED with respect to strain yields the stress, while the second derivatives yield the elastic constants. In crystal lattices with pair-wise interactions, the sum over all bonds in a unit cell yields the SED. With $\varepsilon_{ij}$ known, we can express the potential energy $\Psi$ for an atomistic representative volume element (RVE) as a function of applied strain:

$$\Psi(\varepsilon_{ij}) = \int_{\Omega} U(\varepsilon_{ij})\, d\Omega = \sum_{\Omega} \phi(\varepsilon_{ij}) \tag{7.19}$$

We note the similarity with Eq. 7.11, with the parameter $\chi = \varepsilon_{ij}$. We assume the imposed macroscopic deformation gradient on atomistic volume element corresponds to the macroscopic stress. Then using the continuum mechanics analogue:

$$\sigma_{ij} = \frac{\partial \Psi(\varepsilon_{ij})}{\partial \varepsilon_{ij}} \tag{7.20}$$

and

$$C_{ijkl} = \frac{\partial^2 \Psi(\varepsilon_{ij})}{\partial \varepsilon_{ij} \varepsilon_{kl}} \tag{7.21}$$

Typically, this calculation is simplified by truncated energy functions *via* a Taylor expansion of the associated atomistic potentials, $\phi(\varepsilon_{ij})$. The Cauchy-Born approach is limited to elastic behavior, and typically cannot describe phenomena classified as plasticity effects (such as dislocations), and valid only far away from defects in periodic lattice (*i.e.*, limited to homogeneous deformation of perfect lattice). This is because in such systems, atomic displacements within the presumed unit cell do not necessarily correspond to the continuum displacement field. Other systems where this is relevant are amorphous materials where the Cauchy-Born rule can only be applied in a statistical sense.

Herein lies the difference between a general energy approach and the Cauchy-Born rule. For the Cauchy-Born rule to be applicable, it as assumed the deformation gradient is equivalent for both the macroscale (continuum) and nanoscale (atomistic). This powerful assumption allows continuum formulations (*i.e.*, stress as a change in free energy, Eq. 7.20; elastic constants from stress as a change in strain, Eq. 7.21) at the molecular scale (free energy defined as summation of atomistic potentials). General energy methods, however, presume the overall system behavior is compatible with continuum energy methods—there can be coupling of modes that is intrinsically captured. The deformation of individual atoms need not be known, only the deformation of the molecular system. Mechanical properties are then derived from the applicable elastic energy formulations.

### 7.8.2 Fixed Configuration System Minimization

Inherently, energy methods are tied to methods of system minimization, such as conjugate gradient or steepest descent algorithms. As such, it is equally suitable for both molecular dynamic (MD) and Monte Carlo (MC) simulations. Indeed, the molecular trajectory from state to state need not be necessary, just the minimized conformation at the desired configuration (*e.g.*, with applied strain, curvature, torsion, etc.). In other words, one need not simulate the stretching of a protein molecule under axial force, but rather apply a set of fixed extensions, and use a minimization algorithm to extract the state. Such approaches are sometimes called molecular statics, effectively implementing "static" energy minimization methods to study the potential energy surfaces of mechanically constrained systems. This method eliminates any kinetic of temperature effects (which may not be of importance), and can resolve some time-scale and/or load-rate limitations. For some systems, this approach can be advantageous as the system can be prudently constructed for the desired result.

For example, the isotropic bending stiffness of mono-layer graphene is intrinsically difficult to determine as a direct consequence of its planar structure. Under load-driven dynamic trajectories, desired curvature is difficult to attain by either applied forces or displacement. Similarly, stress and strain approaches to determine

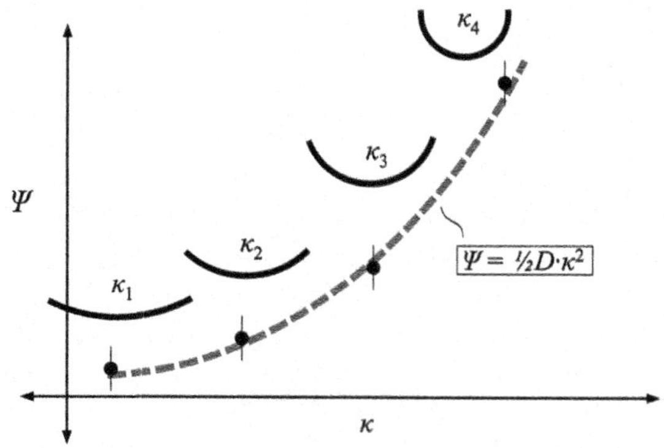


**Fig. 7.6** Schematic of a molecular statics approach *via* fixed configuration system minimization. Here, curvature ($\kappa$) is imposed to a molecular system that would otherwise be difficult to dynamically load and/or bend (such as monolayer graphene). Four independent simulations ($n = 4$) are undertaken to minimize the system energy. Difference in energy between systems is only incurred due to a difference in bending strain energy, which then can be fitted to the energy-curvature formulation (Eq. 7.22). At no point was the model explicitly "bent" from curvature to curvature. Minimization removes any dependence on time, temperature, or other kinetics

bending stiffness are inadequate due to the 1D nature of a bent graphene sheet. However, isotropic bending modulus can be easily determined for mono- and multi-layer graphene *via ab initio* energy methods [61]. To calculate the bending modulus of graphene, a rectangular sheet can be initially bent into a section of a cylinder with constant radius of curvature. Subsequently, the edges of the bent sheet are kept fixed and the bulk of the sheet is allowed to relax through system energy minimization. The bending modulus can be calculated by fitting the energy-curvature data to the following expression [62]:

$$\Psi(\kappa) = \frac{1}{2} D \cdot \kappa^2 \tag{7.22}$$

here $\Psi$ is the system strain energy per unit basal plane area, $D$ is the bending modulus, and $\kappa$ is the curvature. While similar to dynamic energy methods, the key difference is that fixed configuration system minimization approaches (*i.e.*, molecular statics) select and impose discrete states of deformation (or load). This is illustrated in Fig. 7.6.

The governing formula, such as that given by Eq. 7.22, is fitted to the set of points, determined from $n$ different minimization simulations (*e.g.* $\kappa_i$, for $i = 1$ to $n$). For response based, the governing formula, such as that giving by Eq. 7.15, is fitted to the trajectory of the dependent variable, *e.g.* $\Delta L(t)$. While molecular statics approaches are typically more stable, they can not reveal any dynamic phenomena that may occur (*e.g.*, energy barriers) between selected states, and frequently require a large set of data points, $n$.

## 7.9 Alternative Approaches

The approaches discussed up to this point encompass the most common paradigm of molecular characterization by imposing well-known mechanical formulations on system behavior. However, critical issues including the anisotropy of the structure, the size-dependence of the mechanical parameters, and the coupling between different deformation modes are typically neglected. Moreover, many molecular systems are hierarchically layered, entangled, or helical structures, inherently coupling behavior under force or strain load investigations. As such, there is no set of characterization approaches that can be universally applied. In this section, we outline a selection of "alternative approaches" that attempt to reduce the complexity of molecular behavior to characterize mechanical properties, including:

1. Wave Front Tracking
2. Normal Mode Analysis
3. Fluctuation Techniques

What these approaches have in common is that they all can be directly applied to complex systems (*e.g.*, heterogeneous, multi-phase, disordered, etc.) without any explicit consideration of molecular structure. The trade-off is that the discussed methods typically return "bulk" system properties (such as stiffness) that are difficult to break into specific coupled molecular mechanisms, about equilibrium conditions (*e.g.*, small perturbations similar to macroscale "elastic" analysis). Being said, they can successfully characterize system-level behavior and are advantageous when investigating complex geometries and systems with weak interactions (*e.g.*, H-bond networks, folded geometries, solvent effects, etc.).

### *7.9.1 Wave Front Tracking*

A novel method to characterize the inherent stiffness of a molecular system is by application and propagation of a compressive shock wave, or *wave-front tracking* (WFT), which provides a direct approach to measure the elastic modulus of a complex molecular structure. The approach is simple: at the beginning of a simulation, one end of a representative model is fixed, while the opposite end is subject to a displacement-based compressive pulse load or direct displacement (restricted in amplitude to maintain linear elastic behavior), initiating a self-propagating disturbance. The resulting stress wave propagation is tracked subsequently after removing the pulse load and keeping both ends fixed (schematically depicted in Fig. 7.7).

By tracking the compressive wave front position (*i.e.*, the boundary that separates the deformed and undeformed regions) as a function of time, the speed of the propagating wave, $c$, can be determined. This approach, complementary to the widely used SMD method and atom force microscopy (AFM) measurements, can provide detailed structural and dynamical information of mechanical properties and energy transport from a molecular perspective. Importantly, it rules out ambiguous

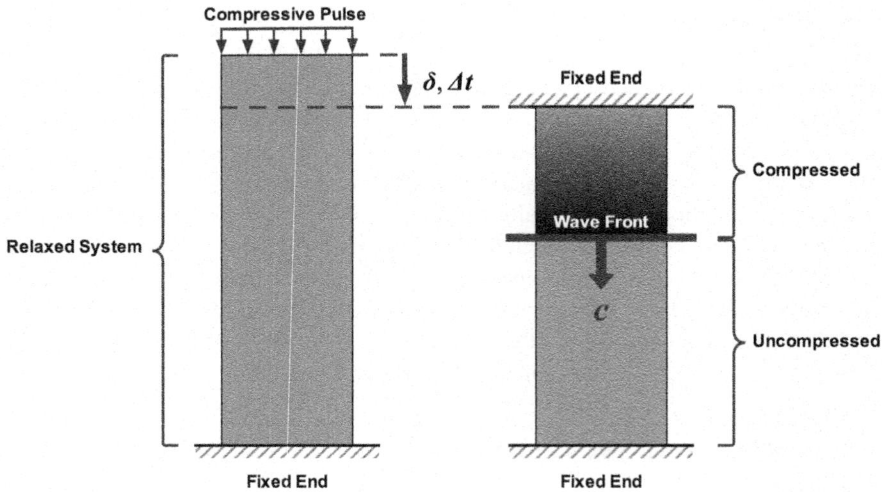


**Fig. 7.7** Schematic of the loading condition used in the wave-front tracking (WFT) approach. In the WFT approach, a displacement-based or force-based compressive pulse load with an amplitude of $\delta$ and impulse time, $\Delta t$ (*e.g.*, a known impulse velocity, $v = \delta/\Delta t$) is applied to one end of the material system at the beginning of the simulation (relaxed system). A small value of $\delta$ is required to maintain small deformation (*e.g.*, perturbations about equilibrium). The stress wave front, moving with speed $c$, is tracked subsequently after removing the impulse and keeping both ends fixed. Adapted from [63]

issues such as loading rate dependence and deformation localization at the loading boundaries [43, 44] both limitations of the conventional SMD approach and related methods as previously described. Once nucleated, the wave front travels at the speed of a longitudinal elastic wave, $c$. The elastic properties can be extracted based on:

$$c = \sqrt{\frac{E}{\rho}} \tag{7.23}$$

where $E$ is the Young's modulus and $\rho$ is the mass density of the structure. It is noted that $\rho$ is held constant, effectively assuming a homogeneous material system.

To illustrate, a recent molecular simulation investigation focused on the mechanical energy transfer processes in fibrous $\beta$-sheet-rich proteins that consist of highly ordered H-bond networks [63]. As was demonstrated in recent experimental studies [64, 65], the particular nanostructure found in these materials can be crucial in determining their macroscale properties. Yet, fundamental issues related to the mechanical properties of key protein constituents such as $\beta$-sheet-rich building blocks remain unknown, in particular their response to extreme mechanical loading in the context of mechanical energy transfer and dissipation. By investigating four model proteins including two morphologies of amyloids, beta solenoids, and silk $\beta$-sheet nanocrystals, it was found find that all $\beta$-sheet-rich protein fibrils provide outstanding small-deformation elastic moduli, where the silk nanocrystal reaches the highest value of $\approx$40 GPa. However, their capacities to dissipate mechanical energy differ significantly and are controlled strongly by the underlying molecular structure of

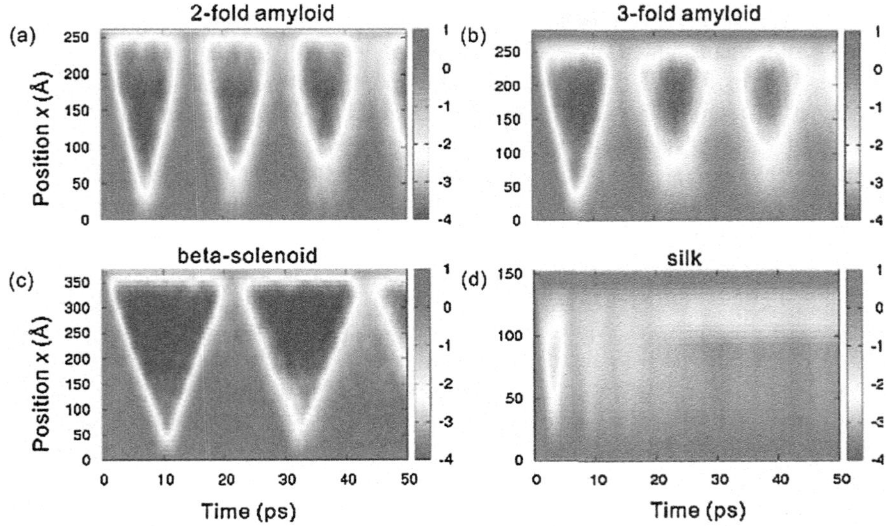


**Fig. 7.8** Energy dissipation of protein materials *via* wave front tracking (WFT) method. Plots depict axial displacement of constituting protofibrils (cross $\beta$-sheet structural arrangements) obtained from applied compressive wave. In comparison with amyloid fibrils (**a**) and (**b**), the compressive wave in the beta-solenoid protein (**c**) features less energy dissipation due to the molecular backbone along the axial direction. In the silk nanocrystal (**d**), transverse flexibility introduces severe damping and the wave propagation decays rather quickly. Stress wave characteristics such as propagation at a specific constant sound speed, reflection at the boundary, and energy dissipation are observed. A sharp and straight stress wave-front is indicative of linear-elastic response. From [63]

H-bond network. Notably, silk $\beta$-sheet nanocrystals feature a ten times higher energy damping coefficient than others, owing to flexible intra-strand motions in the transverse directions. Such dissipation properties was only elucidated by the WFT tracking approach. The results demonstrate a unique feature of silk nanocrystals, their capacity to simultaneously provide extreme stiffness and energy dissipation capacity (see Fig. 7.8).

Another important finding derived from the data shown in Fig. 7.8 is that mechanical energy emitted from compressive wave begins to dissipate within the time scale of hundreds of picoseconds. The main dissipation sources are viscous damping, converting stress wave energy into heat, and energy loss when the stress wave is reflected at the end, where the stress is multiplied. The first mechanism characterizes intrinsic material properties and is determined by the hierarchical structure of H-bonds in protein fibrils. The second mechanism is defined by the boundary condition rather than an intrinsic material property, which is confirmed by the blunt wave fronts. The rate of energy dissipation depends on interactions that couple an effective continuum stress wave and the thermal vibration of individual atoms.

The WFT approach can inherently account for the effects of the mechanochemical environment (such as hydrophobic effects in solvation) that may not arise in measures of molecular stress or energy. Innately, the method is restricted to directionally dependent stiffness, and requires a system with sufficient structural stability to withstand the compressive shock. In order to observe a distinct wave propagation feature, the displacement-based loading is applied with a rather large speed (*i.e.*, on the order of 100 m/s). In the discussed $\beta$-sheet structures, when the loading rate exceeded sound speeds, the simulated fibers started to fail (destruction of H-bond network). To avoid an inelastic response of protein materials, stress wave speeds were limited in a range of from 10 to 1000 m/s without altering the material properties extracted from the simulation. Crystalline or covalently bonded materials (*i.e.*, metals or carbon nanotubes) could be subject to a wider range of speeds.

As the wave front ultimately propagates towards the span of the model system, continued simulation can give insight towards the mechanical energy dissipation capacity of the material. Thus, WFT provides a novel means of probing the relationship between stiffness and toughness (in terms of energy dissipation and impact resistance) inaccessible by methods previously presented. Additionally, due to the relatively high propagation speeds, the time required (both computational time and represented physical time) for WFT simulations are typically less than those of similar force or strain loaded applications *via* response-based or energy-based methods previously described. Although limited in scope, the WFT method is presented here to further illustrate the application of continuum-based theory and relations (*i.e.*, longitudinal elastic wave propagation) to discrete molecular-scale investigations.

### 7.9.2 Normal Mode Analysis

If we could characterize a system solely based on structure, it would eliminate the need for any loading or deformation altogether. Normal mode analysis provides such an approach, only requiring a system's geometric and chemical conformation as an input. In continuum dynamics, the vibration of structures can be described by the equations of motion through eigenmode analysis, for example. Similarly, the structural stability and mechanical properties of complex molecular systems can be determined by analyzing the associated normal modes. For example, for a linear rod, the resistance to torsional loading along the axis can be quantified as torsional or shear modulus $G$. The equation of motion is written as:

$$\rho I \frac{\partial^2 \varphi}{\partial t^2} = GI \frac{\partial^2 \varphi}{\partial z^2} \tag{7.24}$$

where $\rho$ is the material density and $I$ is the moment of inertia, and $\varphi$ representing the torsional angle (along the $z$-axis at time $t$). With the necessary boundary conditions, the angular frequency $\omega$ for $n$-th order torsional mode can then be formulated as:

$$\omega = \left(\frac{n\pi}{L}\right)\sqrt{\frac{G}{\rho}} \tag{7.25}$$

where $L$ is the total molecular length along the $z$-axis, length and $n = 1, 2, 3$ is an integer representing the mode orders. Similarly, the vibrational equation for transverse bending motion of the Euler-Bernoulli beam model can be written as:

$$\rho A \frac{\partial^2 u}{\partial t^2} = -EI \frac{\partial^2 u}{\partial z^2} \tag{7.26}$$

where $A$ is the cross-sectional area of the molecule, $EI$ is the bending rigidity, and $u$ is the transverse displacement (along the $z$-axis at time $t$). Subsequently, *via* appropriate boundary conditions, the frequency for $n$-th order bending mode can be formulated as:

$$\omega = \left(\frac{\alpha_n}{L}\right)^2 \sqrt{\frac{EI}{\rho A}} \tag{7.27}$$

where $\alpha_n$ are numerical factors obtained by solving for the eigenvalues [66]. It should be noted that the expression given in Eq. 7.27 does not take into account any rotational and shear contribution. To include such effects, more sophisticated models such as Timoshenko beam theory could be used [66]. For the axial stretching and compression modes, the longitudinal equation of motion is given:

$$\rho A \frac{\partial^2 u}{\partial t^2} = EA \frac{\partial^2 u}{\partial z^2} \tag{7.28}$$

where $u$ is the displacement along molecular axis direction and $E$ is the Young's modulus (considered the effective mechanical stiffness). Again, by using appropriate boundary conditions, the solution for $n$-th order stretching mode can be formulated as:

$$\omega = \left(\frac{n\pi}{L}\right) \sqrt{\frac{E}{\rho}} \tag{7.29}$$

For each of the mechanical properties described above, the associated frequency ($\omega$) is required and correlated with a mechanical response. The low-frequency modes are related to rigid-body motions and continuum-like deformations, such as twisting, bending, and stretching. To construct a representative model for the determination of vibration frequencies, elastic network models (ENMs) provide an efficient means [63, 67–70], a coarse-graining technique introduced in Chap. 6: Computational Approaches and Simulation. In an elastic network model, all atoms are represented by sites connected through linear Hookean springs with energy:

$$\phi(r_{ij}) = k_{ENM}(r_{ij} - r_0)^2 \tag{7.30}$$

where the spring has length of its initial value as the reference distance $r_0$ and the hierarchy of interatomic interactions is described by either a constant $k_{ENM}$ or by a function $k_{ENM} = f(r_{ij})$ decaying with distance $r_{ij}$, which is tunable by the constant $r_0$. Such a formulation of $k_{ENM}$ can represent weak interactions of atoms at a distance (such as van der Waals interactions) and provide a more complex description of the system [68, 69]. See Fig. 7.9.

The elastic network model we evaluate the frequencies and mode shapes by diagonalizing the second derivative of the potential energy (or Hessian matrix), defined by:

$$H_{ij} = \frac{\partial E}{\partial x_i \partial x_j} \tag{7.31}$$

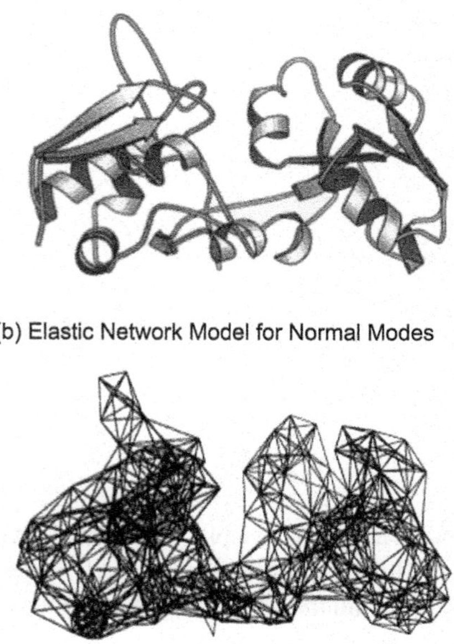


**Fig. 7.9** Extracting normal modes from a molecular "structure". (**a**) Schematic of complex protein structure (lysine-arginine-ornithine (LAO) binding protein). (**b**) Simplified (coarse-grain) structural representation *via* elastic network model (ENM) where backbone $\alpha$-carbons are connected by harmonic springs (Eq. 7.30). From Tama and Sanejouand [71], used with permission, copyright © 2001 Oxford University Press

where

$$E = \sum_{ENM\ springs} \phi(r_{ij}) \tag{7.32}$$

where $E$ is the total potential energy of system, and $x_i$, $x_j$ are perturbation of atomic position on atom $i$ and $j$. The eigenvalues of mass-weighted Hessian matrix, $\mathbf{H}'$, give the frequencies of corresponding modes, defined by:

$$\mathbf{H}' = \mathbf{M}^{-1/2}\mathbf{H}\mathbf{M}^{-1/2} \tag{7.33}$$

where $\mathbf{M}$ is a diagonal matrix of atomic masses. The corresponding eigenvectors of represent the mode shapes required to correlate to a mechanical response. Six rigid body modes are indicated by zero frequencies (including three translation and three rotation modes). Most of the next-higher modes (modes are ranked by eigenvalues from low to high) with low frequencies are collective modes, typically corresponding to elastic deformations such as twisting, bending and stretching. In the elastic network model, the Hessian matrices are often very sparse after introducing a cut-off range for the interaction. Thus diagonalizing these matrices requires much less computational effort and memory storage in comparison with full atomistic force field models.

Most importantly, the normal-mode-based approach makes it possible to analyze the intrinsic mechanical properties of a system without the need for applied loads, boundary conditions and removes the constraint of accessible timescales in direct molecular dynamics simulations. In contrast to applying force or strain loads, the normal mode decomposition of macromolecular motions provides a specific frequency and stiffness for each mode, and thus offers direct information on the elastic constants of an equivalent continuum model. Provided that the collective motion of the molecular system corresponds directly (or closely) to the deformation modes assumed in a continuum model, the parameters extracted from a normal mode analysis are be related to the elastic constants in continuum models [70–72]. However, normal-mode analysis is also limited by the simplicity of the representative elastic network model. That is, if full-atomistic, conventional force-field methods were to be used in the formulation of the Hessian, the resulting analysis would quickly become computationally intractable. Further, the predicted mechanical properties are restricted to elastic response, with no indication of potential failure mechanisms. While beneficial, the results of normal mode analysis can be complementary to properties extracted by other methods—*e.g.*, predicting critical modes of deformation to probe *via* full atomistic means [73].

### 7.9.3 Fluctuation Techniques

As illustrated, various methods have been developed for the numerical calculation of elastic constants of atomistic systems for which the interatomic forces are described by an appropriate interatomic potential. One approach is broadly classified as *fluctuation techniques*—exploiting thermal averages of fluctuations of the stress or the strain—from which elastic properties may be obtained involving the calculation of correlation functions [74, 75]. Simply put, the thermally induced atomic displacement fluctuations scale with the elastic stiffness of the material system, where deviations in atomistic displacement along a direction are proportional to the material stiffness in that direction [76]. Fluctuation methods, though requiring longer runs for accumulating statistically significant data, are often preferred because the entire matrix of elastic constants can be evaluated in a single run, whereas in response- or energy-based methods, every elastic constant requires the application of an appropriate strain (or stress). Moreover, mapping out a direct stress-strain curve can be challenging, especially for soft systems such as biological materials that are easily perturbed. The elastic constants can be calculated from the thermal strain fluctuations *via* the Parrinello-Rahman [76] fluctuation formula:

$$C_{ijkl} = \frac{k_B T}{\langle \Omega \rangle} \langle \varepsilon_{ij} \varepsilon_{kl} \rangle^{-1} \tag{7.34}$$

where $C_{ijkl}$ are the standard elastic constants, $\varepsilon_{ij}$ are the components of the strain tensor (the indices, $i$, $j$, $k$, $l$ run from 1 to 3), $\langle \Omega \rangle$ is the average volume of the system, $k_B$ is the Boltzmann constant, $T$ is the temperature, and the brackets, $\langle \rangle$ denote the ensemble average [77]. However, to avoid a dependence on the definition of *strain*, it is desired that the formulation rely on *displacement* fluctuations only. By formulation of the Helmholtz free energy, combined with the discrete Fourier transform, a relation between elastic constants and the displacement correlation function can be derived, where [78, 79]:

$$\langle u_i(\mathbf{k}) \cdot u_k(-\mathbf{k}) \rangle = \frac{k_B T}{\Omega} (C_{ijkl} k_j k_l)^{-1} \tag{7.35}$$

where $u_i$ and $u_k$ are the displacements along a vector, $\mathbf{k}$, where $\langle\rangle$ denotes thermal averaging, and again $k_B$ is the Boltzmann constant, $T$ temperature, $\Omega$ system volume, and $C_{ijkl}$ are the elastic constants. In general, such fluctuation techniques and the elastic constant calculations are amenable to both molecular-dynamics (MD) and Monte Carlo (MC) simulations, allowing application to a diverse range of molecular systems and conditions. Benefits of the approach include:

- Elastic constant calculated from geometric data only (displacement correlation function).
- Independent of molecular stress or strain measures, thus void of the approximations contained within such measures.
- Does not require the calculation of derivatives of the interatomic potential (thus inherently compatible with complex, multi-body potentials).
- Can be extended to spatially inhomogeneous systems to obtain local elastic constants, $C_{ijkl}(\mathbf{r})$.
- Elastic constants can be defined for any geometric direction.

A fluctuation approach can be advantageous for nonhomogenous and weakly bonded structures that are (a) difficult to apply consistent forces (*e.g.*, soft matter, fluids, hydrogels) or (b) difficult to determine local strain and stress measures (heterogeneous structures). Further, the correlation function inherently accounts for long-range effects such as charged environments and solvation effects. For example, such a technique was recently applied to determine the out-of-plane behavior of variable hydrated hydrogen bond networks between graphite oxide sheets [80]—*via* atomic displacement correlation functions, the determined mechanical stiffness incorporates the network of weakly bonded (and randomly distributed) water molecules between graphene oxide layers under equilibrium conditions.

## 7.10 Summary

A "well behaved" system should result in the same mechanical description regardless of the approach utilized. Ultimately, the analysis and approach is tailored to the desired phenomena, property, or behavior. In this chapter, we have outlined an diverse array of analytical tools that can be utilized to further our understanding of whether and how the mechanical environment and mechanical behavior of a biological materials may affect its biological function. In order to maximize the engineering potential of these capabilities, it is essential to recognize how biological systems are often intrinsically different from synthetic material systems, as well as which outstanding questions in developmental and pathological biology can be addressed *via* quantification of mechanical responses. The design and constraints of a simulated "experiment" defines the associated utility of the defined mechanical property—*e.g.* determining the elastic modulus in tension for a molecule typically does not hold for the same molecule in compression.

While this chapter focused on atomistic and molecular behavior, full atomistic description of mechanistic behavior alone does not provide for the determination of macroscopic behavior, since the higher scale defect interactions collectively operate to drive large-scale actions. It is now achievable to reach all the way up to larger mesoscopic and macroscopic scales of biological filaments, fibrils, fibers and entire tissues, by explicitly considering the characteristic structural features at multiple hierarchical levels, and the understanding and characterization of *all* scales (and across scales) is necessary to unlock the materiome. Even though there are still major challenges ahead of us, this progress is amazing and provides one with a large number of opportunities, transforming materials science as a discipline through increased integration of computational approaches in scientific research. Discoveries in the area of material properties of biological materials will extend our ability to carry out structural engineering, as used for buildings or bridges today, to the ultimate scale—nanoscale, and may be a vital component of the realization of nanotechnology through the merger of material and structure. The merits of biological materials, polymers, membranes and other nanostructures could not be utilized sufficiently unless the signal or motion at atomistic scale can be represented though macroscopic functions.

## 7.11 Suggested Readings

- E.B. Tadmor, R.E. Miller, *Modeling Materials: Continuum, Atomistic and Multiscale Techniques* (Cambridge University Press, 2011)
- E.B. Tadmor, R.E. Miller, R.S. Elliott, *Continuum Mechanics and Thermodynamics: From Fundamental Concepts to Governing Equations* (Cambridge University Press, 2011)
- W.K. Liu, E.G. Karpov, H.S. Park, *Nano Mechanics and Materials: Theory, Multiscale Methods and Applications* (Wiley, 2006)
- M.J. Buehler, *Atomistic Modeling of Materials Failure* (Springer, New York, 2008)

## References


1. E.B. Tadmor, M. Ortiz, R. Phillips, Quasicontinuum analysis of defects in solids. Philos. Mag., A, Phys. Condens. Matter, Struct. Defects Mech. Prop. **73**(6), 1529–1563 (1996)
2. W.A. Curtin, R.E. Miller, Atomistic/continuum coupling in computational materials science. Model. Simul. Mater. Sci. Eng. **11**(3), R33–R68 (2003)
3. W.K. Liu, E.G. Karpov, S. Zhang, H.S. Park, An introduction to computational nanomechanics and materials. Comput. Methods Appl. Mech. Eng. **193**(17–20), 1529–1578 (2004)
4. H.S. Park, P.A. Klein, Surface Cauchy-Born analysis of surface stress effects on metallic nanowires. Phys. Rev. B **75**(8) (2007)
5. B. Liu, H. Jiang, Y. Huang, S. Qu, M.F. Yu, K.C. Hwang, Atomic-scale finite element method in multiscale computation with applications to carbon nanotubes. Phys. Rev. B **72**(3) (2005)
6. G.I. Giannopoulos, P.A. Kakavas, N.K. Anifantis, Evaluation of the effective mechanical properties of single walled carbon nanotubes using a spring based finite element approach. Comput. Mater. Sci. **41**(4), 561–569 (2008)

7. T.A. Witten, Polymer solutions: a geometric introduction. Rev. Mod. Phys. **70**(4), 1531–1544 (1998)
8. R.B. Laughlin, D. Pines, The theory of everything. Proc. Natl. Acad. Sci. USA **97**(1), 28–31 (2000)
9. J.F. Smith, T.P.J. Knowles, C.M. Dobson, C.E. MacPhee, M.E. Welland, Characterization of the nanoscale properties of individual amyloid fibrils. Proc. Natl. Acad. Sci. USA **103**(43), 15806–15811 (2006)
10. T.P. Knowles, A.W. Fitzpatrick, S. Meehan, H.R. Mott, M. Vendruscolo, C.M. Dobson, M.E. Welland, Role of intermolecular forces in defining material properties of protein nanofibrils. Science **318**(5858), 1900–1903 (2007)
11. T. Ackbarow, M.J. Buehler, Superelasticity, energy dissipation and strain hardening of vimentin coiled-coil intermediate filaments: atomistic and continuum studies. J. Mater. Sci. **42**(21), 8771–8787 (2007)
12. T. Ackbarow, X. Chen, S. Keten, M.J. Buehler, Hierarchies, multiple energy barriers, and robustness govern the fracture mechanics of alpha-helical and beta-sheet protein domains. Proc. Natl. Acad. Sci. USA **104**(42), 16410–16415 (2007)
13. T. Ackbarow, S. Keten, M.J. Buehler, A multi-timescale strength model of alpha helical protein domains. J. Phys., Condens. Matter **21**, 035111 (2009)
14. C.B. Prater, H.J. Butt, P.K. Hansma, Atomic force microscopy. Nature **345**(6278), 839–840 (1990)
15. Y.L. Sun, Z.P. Luo, A. Fertala, K.N. An, Stretching type ii collagen with optical tweezers. J. Biomech. **37**(11), 1665–1669 (2004)
16. K.J. Van Vliet, G. Bao, S. Suresh, The biomechanics toolbox: experimental approaches for living cells and biomolecules. Acta Mater. **51**(19), 5881–5905 (2003)
17. K. Schulten, Manipulating proteins by steered molecular dynamics. J. Mol. Graph. Model. **16**(4-6), 289 (1998)
18. H. Lu, B. Isralewitz, A. Krammer, V. Vogel, K. Schulten, Unfolding of titin immunoglobulin domains by steered molecular dynamics simulation. Biophys. J. **75**(2), 662–671 (1998)
19. S. Izrailev, S. Stepaniants, K. Schulten, Applications of steered molecular dynamics to protein-ligand/membrane binding. Biophys. J. **74**(2), 177 (1998)
20. B. Isralewitz, M. Gao, K. Schulten, Steered molecular dynamics and mechanical functions of proteins. Curr. Opin. Struct. Biol. **11**(2), 224–230 (2001)
21. E.B. Walton, S. Lee, K.J. Van Vliet, Extending bell's model: how force transducer stiffness alters measures unbinding forces and kinetics of molecular complexes. Biophys. J. **94**, 2621–2630 (2008)
22. M. Sotomayer, K. Schulten, Single-molecule experiments in vitro and in silico. Science **316**, 1144–1148 (2007)
23. S.W. Cranford, C. Ortiz, M.J. Buehler, Mechanomutable properties of a paa/pah polyelectrolyte complex: rate dependence and ionization effects on tunable adhesion strength. Soft Matter **6**(17), 4175–4188 (2010)
24. S. Keten, J.F.R. Alvarado, S. Muftu, M.J. Buehler, Nanomechanical characterization of the triple beta-helix domain in the cell puncture needle of bacteriophage t4 virus. Cell. Mol. Bioeng. **2**(1), 66–74 (2009)
25. G.I. Bell, Models for the specific adhesion of cells to cells. Science **200**(4342), 618–627 (1978)
26. G. Hummer, A. Szabo, Kinetics from nonequilibrium single-molecule pulling experiments. Biophys. J. **85**(1), 5–15 (2003)
27. O.K. Dudko, G. Hummer, A. Szabo, Intrinsic rates and activation free energies from single-molecule pulling experiments. Phys. Rev. Lett. **96**(10) (2006)
28. H.J. Lin, H.Y. Chen, Y.J. Sheng, H.K. Tsao, Bell's expression and the generalized garg form for forced dissociation of a biomolecular complex. Phys. Rev. Lett. **98**(8) (2007)
29. R.W. Friddle, Unified model of dynamic forced barrier crossing in single molecules. Phys. Rev. Lett. **100**(13) (2008)

30. O.K. Dudko, G. Hummer, A. Szabo, Theory, analysis, and interpretation of single-molecule force spectroscopy experiments. Proc. Natl. Acad. Sci. USA **105**(41), 15755–15760 (2008)
31. O.K. Dudko, Single-molecule mechanics: new insights from the escape-over-a-barrier problem. Proc. Natl. Acad. Sci. USA **106**(22), 8795–8796 (2009)
32. E. Evans, K. Ritchie, Dynamic strength of molecular adhesion bonds. Biophys. J. **72**, 1541–1555 (1997)
33. E. Evans, Probing the relation between force, lifetime, and chemistry in single molecular bonds. Annu. Rev. Biophys. Biomol. Struct. **30**(1), 105–128 (2001)
34. R. Merkel, P. Nassoy, A. Leung, K. Ritchie, E. Evans, Energy landscapes of receptor-ligand bonds explored with dynamic force spectroscopy. Nature **397**(6714), 50–53 (1999)
35. G. Hummer, A. Szabo, Free energy reconstruction from nonequilibrium single-molecule pulling experiments. Proc. Natl. Acad. Sci. USA **98**(7), 3658–3661 (2001)
36. U. Seifert, Rupture of multiple parallel molecular bonds under dynamic loading. Phys. Rev. Lett. **84**(12), 2750–2753 (2000)
37. U. Seifert, Dynamic strength of adhesion molecules: role of rebinding and self-consistent rates. Europhys. Lett. **58**(5), 792–798 (2002)
38. R. Zwanzig, Diffusion in a rough potential. Proc. Natl. Acad. Sci. USA **85**(7), 2029–2090 (1988)
39. C. Hyeon, D. Thirumalai, Measuring the energy landscape roughness and the transition state location of biomolecules using single molecule mechanical unfolding experiments. J. Phys., Condens. Matter **19**(11), 113101 (2007)
40. S. Keten, Z.P. Xu, B. Ihle, M.J. Buehler, Nanoconfinement controls stiffness, strength and mechanical toughness of beta-sheet crystals in silk. Nat. Mater. **9**(4), 359–367 (2010)
41. Z.P. Xu, M.J. Buehler, Mechanical energy transfer and dissipation in fibrous beta-sheet-rich proteins. Phys. Rev. E **81**(6) (2010)
42. R. Paparcone, M.J. Buehler, Failure of a beta(1–40) amyloid fibrils under tensile loading. Biomaterials **32**(13), 3367–3374 (2011)
43. D.J. Brockwell, E. Paci, R.C. Zinober, G.S. Beddard, P.D. Olmsted, D.A. Smith, R.N. Perham, S.E. Radford, Pulling geometry defines the mechanical resistance of a beta-sheet protein. Nat. Struct. Biol. **10**(10), 731–737 (2003)
44. S. Keten, M.J. Buehler, Geometric confinement governs the rupture strength of h-bond assemblies at a critical length scale. Nano Lett. **8**(2), 743–748 (2008)
45. C. Bustamante, J.F. Marko, E.D. Siggia, S. Smith, Entropic elasticity of lambda-phage DNA. Science **265**(5178), 1599–1600 (1994)
46. P.E. Marszalek, H. Lu, H.B. Li, M. Carrion-Vazquez, A.F. Oberhauser, K. Schulten, J.M. Fernandez, Mechanical unfolding intermediates in titin modules. Nature **402**(6757), 100–103 (1999)
47. M.J. Buehler, S.Y. Wong, Entropic elasticity controls nanomechanics of single tropocollagen molecules. Biophys. J. **93**(1), 37–43 (2007)
48. J.F. Marko, E.D. Siggia, Stretching DNA. Macromolecules **28**(26), 8759–8770 (1995)
49. M.D. Wang, H. Yin, R. Landick, J. Gelles, S.M. Block, Stretching DNA with optical tweezers. Biophys. J. **72**(3), 1335–1346 (1997)
50. T. Odijk, Stiff chains and filaments under tension. Macromolecules **28**(20), 7016–7018 (1995)
51. M. Rief, J.M. Fernandez, H.E. Gaub, Elastically coupled two-level systems as a model for biopolymer extensibility. Phys. Rev. Lett. **81**(21), 4764–4767 (1998)
52. D.H. Tsai, Virial theorem and stress calculation in molecular-dynamics. J. Chem. Phys. **70**(3), 1375–1382 (1979)
53. J.A. Zimmerman, E.B. Webb, J.J. Hoyt, R.E. Jones, P.A. Klein, D.J. Bammann, Calculation of stress in atomistic simulation. Model. Simul. Mater. Sci. Eng. **12**(4), 319–332 (2004)
54. M. Zhou, The virial stress is not a measure of mechanical stress. Model. Numer. Simul. Mater. Behav. Evol. **731**, 59–70 (2002)
55. M. Born, K. Huang, *Dynamical Theory of Crystal Lattices* (Oxford University Press, London, 1954)

56. J. Knap, M. Ortiz, An analysis of the quasicontinuum method. J. Mech. Phys. Solids **49**(9), 1899–1923 (2001)
57. L.M. Dupuy, E.B. Tadmor, R.E. Miller, R. Phillips, Finite-temperature quasicontinuum: molecular dynamics without all the atoms. Phys. Rev. Lett. **95**(6) (2005)
58. H.J. Gao, P. Klein, Numerical simulation of crack growth in an isotropic solid with randomized internal cohesive bonds. J. Mech. Phys. Solids **46**(2), 187–218 (1998)
59. H.J. Gao, B.H. Ji, Modeling fracture in nanomaterials via a virtual internal bond method. Eng. Fract. Mech. **70**(14), 1777–1791 (2003)
60. J.L. Ericksen, On the Cauchy-Born rule. Math. Mech. Solids **13**(3-4), 199–220 (2008)
61. S. Cranford, D. Sen, M.J. Buehler, Meso-origami: folding multilayer graphene sheets. Appl. Phys. Lett. **95**, 123121 (2009)
62. L.D. Landau, E.M. Lifshitz, *Theory of Elasticity*, Vol. 7 (Butterworth–Heinemann, Oxford, 1995)
63. Z. Xu, R. Paparcone, M.J. Buehler, Alzheimer's a(1–40) amyloid fibrils feature size-dependent mechanical properties. Biophys. J. **98**, 2053–2062 (2010)
64. N. Du, J. Narayanan, L.A. Li, M.L.M. Lim, D.Q. Li, X.Y. Liu, Design of superior spider silk: from nanostructure to mechanical properties. Biophys. J. **91**(12), 4528–4535 (2006)
65. S.M. Lee, E. Pippel, U. Gosele, C. Dresbach, Y. Qin, C.V. Chandran, T. Brauniger, G. Hause, M. Knez, Greatly increased toughness of infiltrated spider silk. Science **324**(5926), 488–492 (2009)
66. W.J. Weaver, S.P. Timoshenko, D.H. Young, *Vibration Problems in Engineering* (Wiley-Interscience, Hoboken, 2001)
67. M.M. Tirion, Large amplitude elastic motions in proteins from a single-parameter, atomic analysis. Phys. Rev. Lett. **77**(9), 1905–1908 (1996)
68. K. Hinsen, Analysis of domain motions by approximate normal mode calculations. Protein. Struct. Funct. Genet. **33**, 417–429 (1998)
69. K. Hinsen, A. Thomas, M.J. Field, Analysis of domain motions in large proteins. Protein. Struct. Funct. Genet. **34**, 369–382 (1999)
70. F. Tama, Normal mode analysis with simplified models to investigate the global dynamics of biological systems. Prot. Peptide Letters **10**(2), 119–132 (2003)
71. F. Tama, Y.H. Sanejouand, Conformational change of proteins arising from normal mode calculations. Protein Eng. **14**(1), 1–6 (2001)
72. M. Delarue, Y.H. Sanejouand, Simplified normal mode analysis of conformational transitions in DNA-dependent polymerases: the elastic network model. J. Mol. Biol. **320**(5), 1011–1024 (2002)
73. S.E. Dobbins, V.I. Lesk, M.J.E. Sternberg, Insights into protein flexibility: the relationship between normal modes and conformational change upon protein-protein docking. Proc. Natl. Acad. Sci. USA **105**(30), 10390–10395 (2008)
74. L.R. Pratt, Fluctuation method for calculation of elastic-constants of solids. J. Chem. Phys. **87**(2), 1245–1247 (1987)
75. S. Sengupta, P. Nielaba, M. Rao, K. Binder, Elastic constants from microscopic strain fluctuations. Phys. Rev. E **61**(2), 1072–1080 (2000)
76. M. Parrinello, A. Rahman, Strain fluctuations and elastic-constants. J. Chem. Phys. **76**(5), 2662–2666 (1982)
77. A.A. Gusev, M.M. Zehnder, U.W. Suter, Fluctuation formula for elastic constants. Phys. Rev. B **54**(1), 1–4 (1996)
78. L.D. Landau, E.M. Lifshitz, L.P. Pitaevskii, *Statistical Physics* (Pergamon Press, Oxford, 1980)
79. M.T. Meyers, J.M. Rickman, T.J. Delph, The calculation of elastic constants from displacement fluctuations. J. Appl. Phys. **98**(6), (2005)
80. N.V. Medhekar, A. Ramasubramaniam, R.S. Ruoff, V.B. Shenoy, Hydrogen bond networks in graphene oxide composite paper: structure and mechanical properties. ACS Nano **4**(4), 2300–2306 (2010)

# Part III
# Applied Biomateriomics

# Chapter 8
# Unlocking Nature: Case Studies

**Abstract** Defining the materiome consists of linking material properties and function across multiple scales, from nano to macro. The key to "unlocking" Nature is not the replication of biological materials, but full understanding of the mechanisms and material interactions that result in system-level functionality. Here, we look at two biological materials in detail: silk and bone. Silk—a high-performance polymer-like fiber—can be thought of as two phases synergetically acting together. The H-bonded $\beta$-sheet nanocrystals provide strength and toughness, while disordered semi-amorphous domains imbue extensibility.The resulting hyperelastic stiffening behavior is critical to the flaw tolerance and robustness of a spider's web. Bone, in contrast, is a composite material made of relatively stiff mineral (hydroxyapatite) and compliant protein (tropocollagen) components. Bone achieves strength and toughness through a variety of mechanisms, including nanoconfinement, fibrillar sliding, and crack bridging. The general insights gained from the investigation of such systems can be applied a vast array of technological applications.

*As the poet said, 'Only God can make a tree', probably because it's so hard to figure out how to get the bark on.*

*Woody Allen, American Film director, actor, screenwriter*

## 8.1 Introduction

What constitutes a *materiomic* investigation? From a general perspective, *any* study that elucidates *any* material behavior or property at *any* scale is, by definition, investigating the materiome. A complete materiomic investigation, however, generally goes further and attempts to link structure to behavior, bridge multiple scales, and translate mechanisms to function. As we have seen, the complexity and multiscale structure of biological materials require a balance of computational and experimental approaches. To illustrate, this chapter focuses on two well-researched material systems from Nature, from nano to macro: silk and bone. Purely synthetic systems are discussed further in Chap. 10: Synthesis and Design.

The exceptional mechanical properties of both silk and bone have been extensively studied experimentally and in computational investigations, with new insights

S.W. Cranford, M.J. Buehler, *Biomateriomics*, Springer Series in Materials Science 165, 
DOI 10.1007/978-94-007-1611-7_8, 

being published regularly. A series of recent studies has been conducted to examine structure-function relationships across different length scales in silk, ranging from atomistic models of protein constituents to the spider web architecture, motivated by attempts to replicate the high performance in synthetic fibers. Similarly, bone has been studied with more immediate appeal, to fundamentally understand our own biological workings, and potentially advance medical applications such as regenerative medicine and tissue engineering.

The analysis of material properties across multiple scales is a crucial issue in understanding biological materials, as their structure changes with each hierarchical level (and thus length-scale), and therefore most material properties are strongly dependent on the scale of observation. Multi-scale experimental and simulation analyses are the key to improve our systematic understanding of how structure and properties are linked. Typically this is achieved from a *bottom-up* approach, linking more sophisticated lower length-scale parameters, which form the building blocks of the system at that level, to coarser, larger length-scales. Purely "bottom-up" approaches, however, are incomplete if they lack the interpretation of large-scale behavior to small-scale phenomena, an iterative "nano-to-macro-to-nano" paradigm. Full stratification of different levels of hierarchy using such analysis develops a powerful feedback loop where the bottom-up modeling approach catalyzes the insights we gain at each layer of the material ladder, with the possibility of controlling properties at multiple scales simultaneously, and to examine their effect on ultimate system function.

We note that the cases discussed in this chapter emphasize the *mechanistic* function of the particular material systems. Undoubtedly, mechanical performance provides a means to characterize behaviors and properties analogous to engineering practices—we may want to mimic silk to attain a high-performance synthetic fiber, for example, and not a web-like meshwork to capture flies. Ultimately, the understanding of the materiome must encompass all functionalities, including physiological purposes such as growth, chemical transport, etc., but a materiomic framework lies in a deterministic/mechanistic approach.

## 8.2 Spider Silk: From the Molecules to the Web

Spider silk as a biological material—from its exemplary mechanical properties to its multifunction usage in cocoons and webs—has been continually intriguing materials scientists, biologists, and engineers alike [1–9]. Since its development as an engineered material in China thousands of years ago, silk has been used throughout history not only as luxury fabrics but also in various technological applications including parachutes, medical sutures and, more recently, tissue regeneration [10, 11]. More generally, spider silk is researched intensively to better understand the interplay between molecular structure and performance, in part hoping to exploit silks' desirable properties in *de novo* bio-inspired materials [3, 12–17]. Silk is an exemplary natural material because its superior properties stem intrinsically from the cooperativity of hierarchically-organized components across scales (see Fig. 8.1).

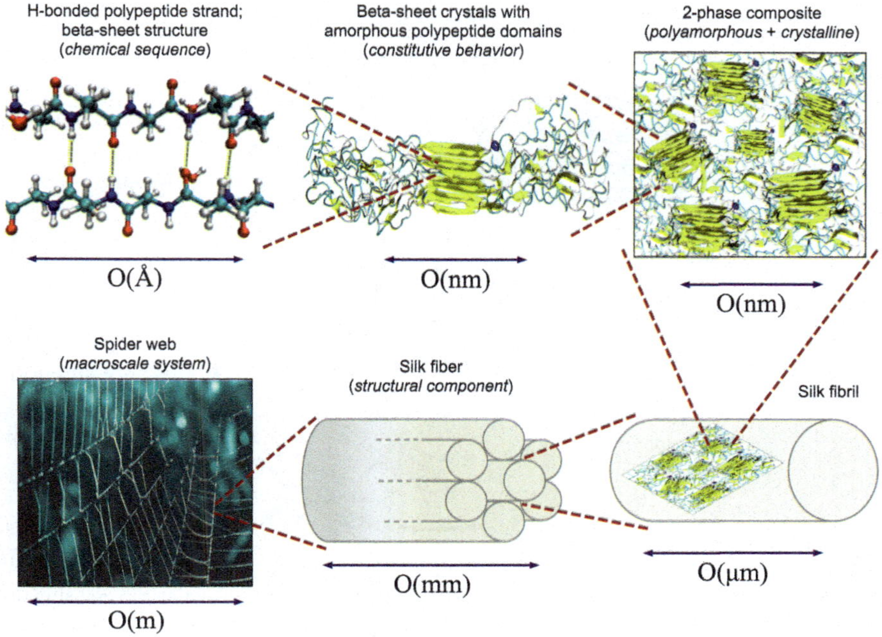


**Fig. 8.1** Hierarchical spider silk structure that ranges from nano (Ångstrom) to macro (meters). The image displays key structural features of silk, including the chemical sequence found at the level of polypeptide hydrogen bonded $\beta$-strands, the secondary structure $\beta$-sheet nanocrystals embedded in a softer semi-amorphous phase (affecting constitutive behavior), bulk assembly of poly-crystalline components which assemble into silk fibrils (*e.g.*, a two-phase composite), fibrils bundled into fibers or threads (bundle intended as schematic only; the number of fibrils in a bundle is larger than shown here) acting as structural components, and finally the functional web-structure itself

Our knowledge of silk molecular biology and material structure is growing rapidly and it may provide a rational basis for the design of structural materials through genetic engineering. A key to the understanding of silk is a complete, multiscale description of the materiome, *e.g.*, how the specific protein sequence of dragline silk is related to its function within a spider web (see Fig. 8.2). Rather than plagiarize Nature, we need to move beyond total mimicry and enhance our understanding of the underlying physical phenomena (as discussed in Chap. 1: Introduction)—the functional relations that make silk a successful material. The more general our understanding, the more powerful the concept, thereby increasing potential applications. We do not study silk with the hopes of building man-made webs or cocoons—we study silk to learn how to balance material structure and function from the protein sequence to a final functional system. The potential gain is the ability to make materials with similar function (and performance) but from different (and improved) building blocks.

An inherent dichotomy exists between the properties at the nanoscale (weak H-bonding) and its function at the macroscale (a strong and robust web). This con-

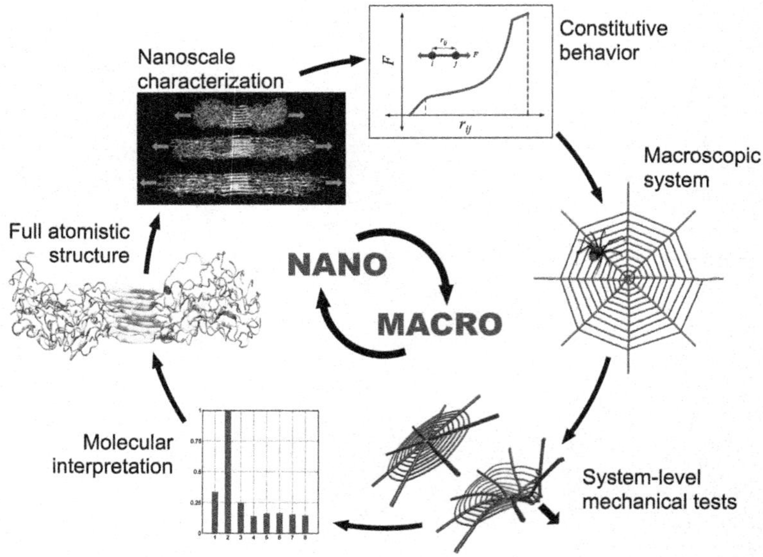


**Fig. 8.2** Holistic *nano-to-macro-to-nano* approach discussed herein, where an integrated multi-scale analysis is applied throughout the materiome, incorporating relevant scales from nano to macro in the description of system-level behavior and function. A simple material model of silk derived from prior molecular investigations can be used to investigate a complete web structure. The goal is to capture known molecular behavior at the largest scale, allowing interpretation of macroscale functionality in terms of molecular mechanisms, in a reciprocal manner. Such an approach was successfully implemented in a multi-scale study [18]

sequently results in a modeling *impasse*—the effect of such hierarchical configurations must be represented, but is limited by the length-scales accessible to simulations. Specifically, unlocking which molecular mechanisms of silk play a critical role in the natural function of an entire web is crucial to decipher the design principles that can be exploited in emerging nanotechnologies, using Nature's progress as a proven guide. Only recently has an integrated multiscale understanding of web mechanics been implemented in the investigation of web performance (as illustrated by Fig. 8.2).

Previous attempts to produce synthetic fibres that matched the performance of natural silks ultimately failed in their endeavors to "copy Nature" [17, 19, 20]. But that need not be the case if we fully understand the material we propose to emulate—the *materiome* of silk—from nano to macro. The consideration of structure (*i.e.*, web or cocoon) and material (*i.e.*, silk) separately is only part of the story: the material properties govern the structure and *vice versa*, creating heightened functionality through synergistic interactions. We thus begin our examination at the atomistic scale, and the defining molecular characteristics that differentiates "silk" from other protein-based materials.

### 8.2.1 Nanoconfinement and β-Sheet Crystals

Silk's lowest level of hierarchy is rooted in its primary protein structure, defined by a sequence of amino acid residues, which is responsible for subsequent folding mechanisms that leads to the defining molecular structure. Recent investigations revealed that one of the primary "building blocks" of silk are antiparallel $\beta$-sheet crystals at the nanoscale, consisting of highly conserved poly-Gly-Ala and poly-Ala repeats found in both commercial and spider silk [21]. These crystals provide stiff orderly cross-linking domains embedded in a semi-amorphous matrix that consists predominantly of less orderly $\beta$-structures, $3_1$-helices and $\beta$-turns [22–24]. The $\beta$-sheet nanocrystals, bonded by means of assemblies of hydrogen bonds, have dimensions of a few nanometers and constitute roughly 10 % to 15 % of the silk volume, whereas with less orderly extended structures the $\beta$-sheet content can exceed 50 % for spider and silkworm silks [7, 25]. When silk fibres are exposed to stretch, $\beta$-sheet nanocrystals reinforce the partially extended and oriented macromolecular chains by forming interlocking regions that transfer the load between chains under lateral loading [21], similar to their function in other mechanical proteins [23, 26]. Thereby, $\beta$-sheet nanocrystals provide cohesion between the long polypeptide strands, enabling the amorphous domains to stretch significantly. Eventually, fracture of $\beta$-sheet nanocrystals occurs at large deformation and large loads. It is apparent that these crystals are critical the mechanical function of silk, serving both as a stiffening (load transfer) and cross-linking (assembly) agent. But how they function in this role is less clear and has only recently been investigated.

A series of full atomistic nanomechanical computational experiments (using methods described by Chap. 6: Computational Approaches and Simulation) enables the characterization of the strain distribution within individual nanocrystals. To examine the key mechanical parameters of the silk $\beta$-sheet nanocrystals, a series of simple pull-out simulations was recently undertaken [27]. The choice of a pull-out scheme was motivated by the key loading condition of nanocrystals—namely, the lateral loading *via* the pulling of individual (and assumed independent) peptide chains as the molecular assembly is subject to tension, for example. Under load, the resultant beam-like bending behavior suggests that the mechanical response may be accurately represented by classical continuum beam theory. Indeed, the H-bonds can be thought of in terms of axial and shear load contributions to the "cross-section" of the beam, *i.e.* a single $\beta$-strand unit. From this perspective, it became clear that the $\beta$-sheet nanocrystals behaved in a similar fashion to *deep beams*, where shear deformations are critical. The relative importance of shear contributions in the deformation of a $\beta$-sheet nanocrystal of size $L$ is given by the shear contribution ratio:

$$s(L) = \frac{3D_B}{L^2 D_T} = \frac{1}{4}\frac{Eh^2}{GL^2} \sim \frac{h}{L} \tag{8.1}$$

where the terms $D_B = EI$ and $D_T = GA_s$ denote the effective bending and shear rigidities, respectively. In this notation, $E$ and $G$ stand for elastic and shear moduli, respectively, and $A_s$ is the assumed cross-sectional area. Again, caution should be taken as taking these terms as their strict definitions—the "shear resistance" is different from that of a continuum description. For a rectangular cross-section, the moment of inertia $I = bh^3/12$ and $A_s = bh$, where $b$ is the base length (related here to the number of $\beta$-sheets) and $h$ is the height of the cross-section (related here to the length of a $\beta$-strand). The shear contribution ratio $s$ demonstrates that as the size of the nanocrystal gets smaller, the loading scenario becomes predominantly shear ($s > 1$), whereas for large $\beta$-sheet nanocrystals pure bending controls deformation ($s < 1$). But does this affect the performance of silk? Is there a preferred mode of deformation? Does silk prefer bending or shear?

The measured size of the nanocrystals [7] indicate that silk has an inherent fondness for shear, but why? The change in deformation mode has important repercussions on its ultimate fracture behavior of the crystal beyond that of load capacity. In a bending mode, the H-bonds are competing across the $\beta$-strand (where some are subject to compression, some are in tension), whereas in shear, the H-bonds are cooperatively resisting the shear load. Larger $\beta$-sheet nanocrystals are inherently softer due to bending and fail due to crack-like flaws arising from the tension side of the bending profile (not unlike tension cracks forming in reinforced concrete members). This crack opening, unfortunately, is exasperated by water molecules competing for H-bonds [28, 29] and facilitates rupture, leading to catastrophic breakdown of the crystal. Smaller shear-dominated crystals, on the other hand, feature a self-healing ability until complete rupture occurs (H-bonds can reform during stick-slip, significantly increasing the total dissipated energy), indirectly protecting H-bonds from adverse exposure to surrounding water. Shear is therefore a more robust mode of deformation (see Fig. 8.3).

Moreover, as shown in earlier studies (and previously discussed in Chap. 4: Universality-Diversity Paradigm: Music, Materiomics, and Category Theory), homogeneous strain distribution across H-bonds facilitates cooperativity of the bonds, maximizing the contribution of each. An important implication of this model involves the length of individual $\beta$-strands. The effectiveness of each $\beta$-strand to carry load is limited regardless of deformation mode—although small groups of H-bonds can indeed work cooperatively under homogeneous shear, this mechanism breaks down beyond a critical number of H-bonds (approx. 3 to 4 H-bonds). Thus, while a shear mode of deformation is preferred, there is a characteristic size ($h$ in Eq. 8.1) where the cooperativity is maximum.

The existence of cooperativity depends quite strongly on the size of the crystals and breaks down once $\beta$-sheet nanocrystals exceed a critical size (see Fig. 8.4). In terms of a beam analogy, cooperativity of H-bonds limits the beam depth ($h$) while a preference for shear limits the beam length ($L$). Smaller $\beta$-strand nanocrystals provide a greater stiffness and fracture resistance, as they are predominantly loaded in uniform shear, which leads to cooperative rupture of H-bonds and stick-slip energy dissipation mechanisms. Through nanoconfinement, H-bonded $\beta$-sheet nanocrystals can reach great strength and toughness seemingly beyond the limitations of their relatively weak building blocks. Being said, silk is not solely composed of such nanocrystals—it must be understood as a two-phase material.

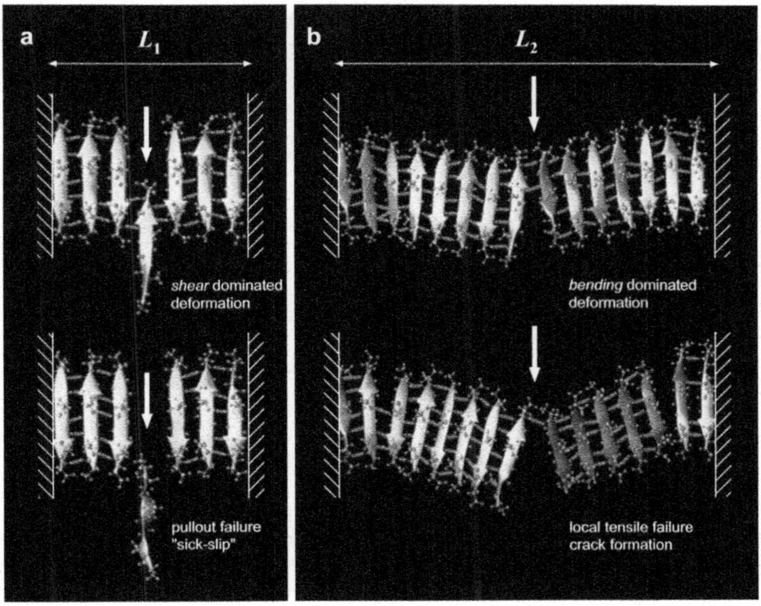


**Fig. 8.3** Length-dependent deformation mechanism and failure mode of $\beta$-sheet nanocrystals. (**a**) Small ($L_1 = 2.83$ nm) crystals undergo shear-dominated deformation (characterized by homogeneous bond strain), and failure is incurred by pull-out of the loaded $\beta$-strand, facilitating a "stick-slip" mechanisms of energy dissipation as H-bonds rupture and reform. (**b**) Larger $\beta$-sheet nanocrystals ($L_2 = 6.56$ nm) deform primarily through bending. Longer structures fail by significant bending, which leads to a crack-like flaw formation owing to non-uniform tensile deformation of hydrogen bonds. The crack-formation is further exasperated by surrounding solvent (water molecules; not shown for clarity). Silk exploits the self-healing capacity of weaker chemical building blocks, responsible for the load-withstanding capacity of smaller $\beta$-sheet nanocrystals, turning a weakness into strength by invoking highly dissipative deformation mechanisms. An optimal size for enhanced mechanical properties is found between 2–4 nm crystal size. Adapted from [27]

### *8.2.2 A Two-Phase Protein Composite*

Experimental studies, focused on developing a mapping between the repeating sequence units of spider silk and the basic structural building blocks of fibrils [21, 30–32], have confirmed that the nanocrystalline structure of silk proteins (and associated nanoconfinement) has a direct influence on the stiffness, toughness and failure strength of silk [22, 23, 33]. Being said, there are few molecular-level analysis of the complete nanostructure of silk assemblies. Indeed, the $\beta$-sheet crystalline structures arise from a small portion of the protein sequence (~6 alanine residues; about one-fifth the total sequence). One limiting factor from a computational perspective is the relatively large size of the molecular "building block" of silk and required solvent.

Recently, atomic-level structures of major ampullate silk proteins from the *Nephila clavipes* spider dragline silk sequence were obtained using *replica ex-*

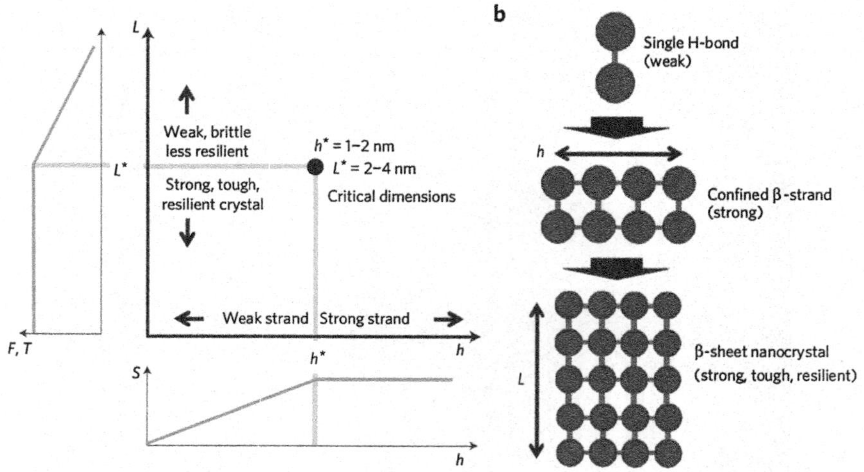


**Fig. 8.4** Effects of the geometric dimensions of spider silk nanocrystals. The formation of confined $\beta$-sheet nanocrystals with critical strand length ($h$) and critical nanocrystal size ($L$) provides maximum strength, toughness and stiffness. (**a**) Schematic phase diagrams to show the interplay of the parameters $h$ and $L$ in defining the properties of nanocrystals: $S$: strength of a $\beta$-strand as a function of strand length $h$, $F$: strength of nanocrystal as a function of crystal size $L$, $T$: toughness of nanocrystal as a function of crystal size $L$. Although increasing the number of hydrogen bonds in a $\beta$-strand increases its mechanical stability for small numbers of hydrogen bonds, the effect does not continue for strands that contain more than roughly four hydrogen bonds, resulting in a plateau of the strength $S$ for lengths in excess of $h^*$. (**b**) Schematic illustration of how hierarchical structure formation in the strand length $h$ and nanocrystal size $L$ lead to the formation of high--performance $\beta$-sheet nanocrystals that combine strength, toughness and resilience despite being composed of weak building blocks (*i.e.*, H-bonds). From Keten and Buehler *et al.* [27], copyright © 2010 Nature Publishing Group

*change molecular dynamics* (REMD), enabling robust nanomechanical analysis [24, 34, 35]. The challenges of reaching native (*e.g.*, equilibrium) structures within the time scales accessible to conventional molecular dynamics simulation (one of the limitations discussed in Chap. 6: Computational Approaches and Simulation) require enhanced sampling methods such as REMD [36]. Along with other protein structure prediction approaches [37, 38], REMD is considered to be an effective tool for investigating folding and aggregation of proteins (see Chap. 3: The Challenges of Biological Materials), as it reduces the likelihood of kinetic trapping at non-native states through temperature assisted conformational sampling [39–42]. The overall approach is summarized in Fig. 8.5. In effect, the REMD method attempts to computationally "spin" the silk molecules, acting as a computational surrogate for the spider's major ampullate glands and spinneret. Indeed, the simulations started from an extended conformation, which is very relevant in the processing of silk where elongational flow in the spinning duct leads to stretching and alignment of monomers in the concentrated dope, as shown in recent experimental work [43] and depicted in Fig. 8.6. At high-temperature replicas, weak interactions between the strands can be broken; that is, the system 'melts' and reforms in a wide range of

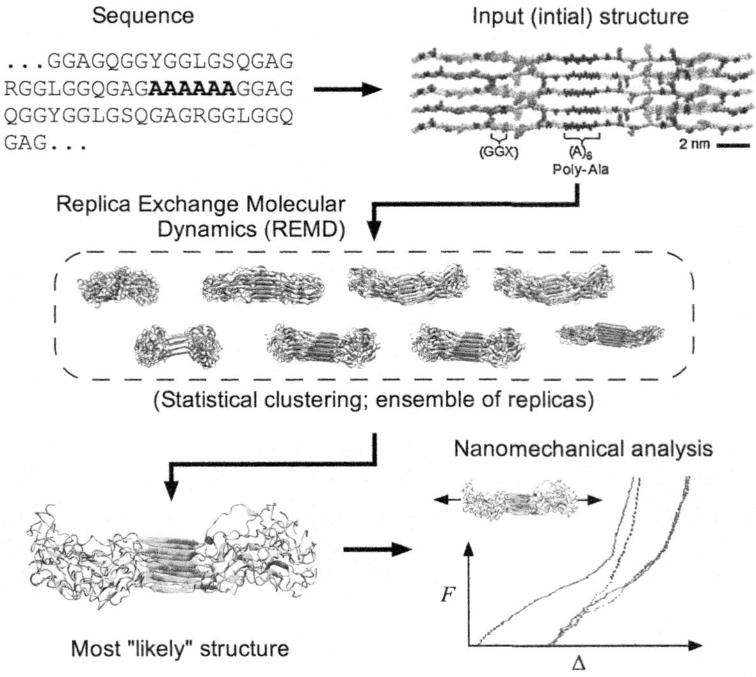


**Fig. 8.5** Approach for protein structure identification and validation *via* replica exchange molecular dynamics (REMD). The building blocks of silk proteins are a series of monomer units, consisting of a poly-alanine region (indicated by bold **A**'s) between two glycine rich domains (repeated up to 100 times in each chain), resembling a unit cell of silk. Replica exchange simulations [24] are performed on a lattice of aligned strands (*e.g.*, initial geometry) of the *N. clavipes* MaSp1 peptide sequence. Statistical clustering of the 300 K replica timeline determines the most probable native structures (most "likely" structure or representative building block). The secondary structure is then refined by equilibration with explicit solvent. Analysis of the secondary structure and dihedral angles determines the cases of poly-Ala length that result in the most defined $\beta$-sheet nanocrystals. These structure predictions can then be validated against experiment [22]. The resultant structures then allow for nanomechanical characterization. Once the structure of nanocrystals is optimized, further modifications to the sequence may be explored to optimize other features using this general approach

structures that can be investigated in detail at lower temperatures. A key assumption for using an initial orderly lattice arrangement is that edge-to-edge aggregation of strands is the driving force for formation of mono-crystals, rather than local folding of the backbone onto itself through short turn structures (*e.g.*, stable alignment over folding).

Structural predictions can be validated against experimental data based on dihedral angle pair calculations presented in Ramachandran plots, $\alpha$-carbon atomic distances, as well as secondary structure content (see Fig. 8.7). The structural analysis confirmed that poly-Ala regions in silk predominantly form distinct and orderly $\beta$-sheet crystal domains, while disorderly regions are formed by glycine-rich repeats that consist of $3_1$-helix-type structures and $\beta$-turns. This "semi-amorphous" domain

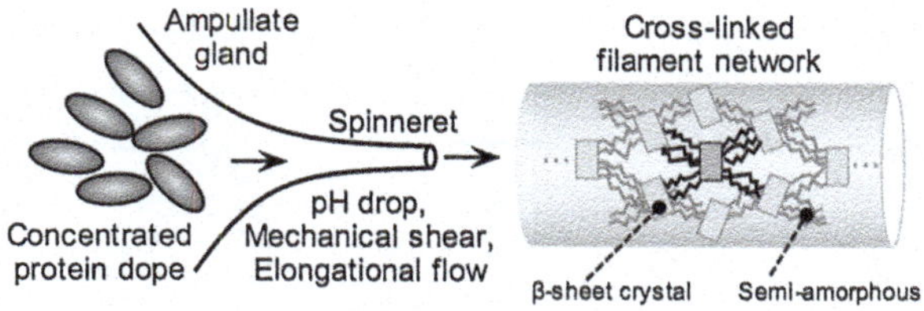


**Fig. 8.6** Schematic of the spinning process of spider silk in the spinneret, where elongational flow, resulting mechanical shear combined with a pH drop leads to the assembly of the protein monomers into a cross-linked filament network. Schematic from [35] and based on [43]

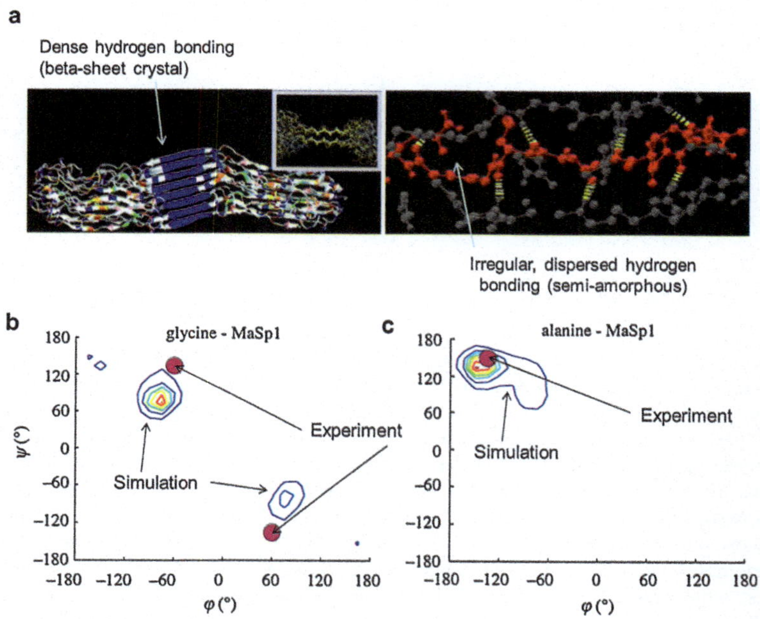


**Fig. 8.7** Results from silk ultrastructure identification and validation *via* replica exchange molecular dynamics (REMD). The images in (**a**) show detailed views of the hydrogen bonding structure within different domains. Dense hydrogen bonding is found within the $\beta$-sheet nanocrystals. More dispersed hydrogen bonding is found in amorphous regions, leading to a more irregular shape. Panels (**b**) and (**c**) show examples of structure validation, where Ramachandran maps are used to compare the dihedral angles of glycine and alanine residues (experimental and computational results shown). The variables $\varphi$ and $\Psi$ refer to the angles around N-$C_\alpha$ and $C_\alpha$-C bonds, respectively. Purple circles represent experimental values while colored maps refer to distributions of angles derived from the model, both of which show angle values that match rather well. Figure modified from prior work [24]

is critical the mechanical performance of silk, providing extensibility through hidden length formation, and influence the unique thermomechanical properties of spider silk [21, 44, 45].

Mechanical shearing simulations on selected structures illustrate that the nanoscale behavior of silk protein assemblies is controlled by the distinctly different secondary structure content and H-bonding in *both* the crystalline and semi-amorphous regions. For loading, half of the chains are randomly selected to be pulled in one direction while the other half is pulled in the opposite direction, to impose the characteristic lateral loading of the crystals as experienced in the native silk structure. Force-extension plots are based on the forces applied, and the measured distance between the center of mass of pulled atoms (a response based approach; see Chap. 7: Mechanical Characterization in Molecular Simulation).

We note that the REMD approach discussed here, providing a means of full atomistic resolution, is only one method complementary to other computational and experimental approaches. It is anticipated that novel statistical mechanics approaches [46], empirical methods, such as X-ray diffraction and scattering [47, 48], solid-state nuclear magnetic resonance (*e.g.*, ssNMR) [32, 49–51] and Raman spectroscopy [23, 25, 52], combined with pioneering multiscale atomistic-modeling methods such as those based on Density Functional Theory (DFT) [29] or molecular dynamics [27], will provide more insight into the atomic resolution structure for complex materials such as spider silk.

The natural variability in mechanical behavior makes it difficult to define "dragline silk" by a simple set parameters. Some view the use of molecular mechanics and dynamics as largely inappropriate for modeling the highly nonlinear viscoelastic properties of silks [17], limited to calculating a linear elastic modulus of a generic silk polymer within the variability range of experimental observations [16]. In a similar vein, micro-mechanics based on continuum tools such as finite element methods are limited in delineating structural mechanisms if incorrect nonlinear viscoelastic parameters or morphologies are used within the model [53]. Indeed, simpler polymer models are able predict a modulus, yield stress, and failure initiation guides based on generic observations on synthetic polymer parameters and morphologies such as cohesive energy density of molecular interactions, and such structure-property rules can be applied very successfully to native silks [16, 54, 55]. Such models have been extended to encapsulate the storage of mechanical energy by ordered ("crystal") domains and the dissipation of energy by disordered ("amorphous") domains [16, 56]. Successful in mapping nonlinear silk property profiles to an envelope of "representative" curves [17, 56], such methods reciprocally complement nanomechanical studies.

The prime advantage of atomistic modeling lies in the fact that structural transformations and failure mechanisms can be explicitly *observed* and/or *discovered* from the trajectories obtained from stretching simulations of the spider silk assemblies. For example, the amorphous domains stretch significantly with applied force, and a transition from turn to $\beta$-sheet structures can be quantified. A key observation is that failure of the system happens by sliding of $\beta$-strands with respect to each other, which can occur only upon breaking of the H-bonds in the crystalline domain, an observation only able to be fully captured by full atomistic modeling. Close investigation unveils a number of generic interactions between silk properties and molecular structures, that can be generalized as "representative behavior". For

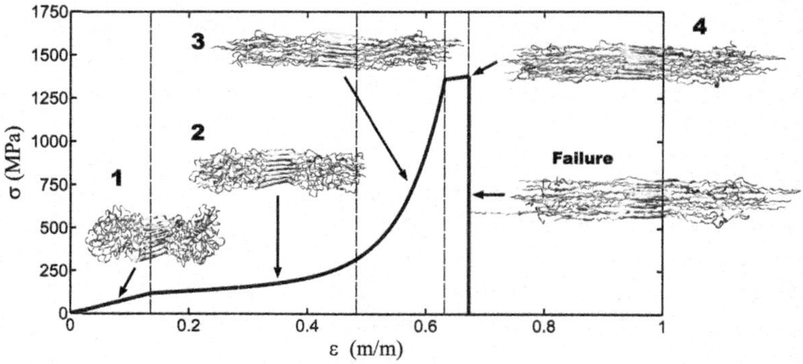


**Fig. 8.8** Derived constitutive behavior parameterized from full atomistic simulations of dragline spider silk and validated against experimentally measured behaviors [7, 24, 27, 57]. *Insets* depict the molecular nanostructure under deformation, which accounts for the two-phase behavior of $\beta$-sheet nanocrystals and semi-amorphous protein domains. The behavior can be separated into four distinct regimes: (**1**) Initial loading in which stress is transferred *via* the poly-amorphous polypeptides, followed by a defined yield point. (**2**) Entropic unfolding where "hidden length" is achieved *via* sequential rupturing of $3_1$-helices and $\beta$-turns. (**3**) Stiffening due to alignment of chains and $\beta$-sheet nanocrystals sustaining increases in applied stress. (**4**) Stick-slip mechanism of the $\beta$-sheet nanocrystal is triggered for a relatively small regime, and governs further deformation until ultimate rupture

example, the force-extension profiles of the modeled silk sequences depict a distinct characteristic curve that can be described by four regimes (see Fig. 8.8):

1. Initial rigidity until a yield point; the semi-amorphous domains homogeneously stretch and bear load until the disordered H-bonded structure fails;
2. The entropic unfolding regime where the associated $3_1$-helices and $\beta$-turns begin to rupture resulting in a high extensibility (hidden length) under relatively low stress;
3. Severe stiffening as aligned amorphous strands begin to form addition H-bonds, and load directly transferred to the $\beta$-sheet crystalline regions;
4. Ultimate rupture as H-bonds break in the crystalline regions, which triggers the sliding of $\beta$-sheet strands (stick-slip mechanism).

Moreover, the transition between and relative moduli (stiffness) of these regimes depend on the secondary structure content. For example, for molecules with high turn ratio and low $\beta$-sheet content, we observe an initial stiff regime, followed by a softer regime, followed by a very stiff regime leading to failure. On the other hand, systems with very high $\beta$-sheet content show a monotonically stiffening force-extension response. This suggests that the characteristic yielding behavior at the molecular level of the hierarchy of silk is controlled by the ratio of turn to $\beta$-sheet structures, where a higher "turn ratio" leads to the emergence of this phenomenon. This is an alternative means of achieving high initial stiffness in comparison with $\beta$-sheet content alone, which exhibits high initial stiffness as well, but provides much less extensibility. Moreover, wider secondary structure distribution exhibits a larger deviation in the initial end-to-end length, and increasing the turn content and inter-chain hydrogen bonding increases the initial stiffness and extensibility of the assembly, thereby improving the toughness as well. The failure strengths are of the same magnitude, regardless of structure content. The force-extension curves of molecular silk also show minor deviations from the inextensible chain models commonly used for polymer materials (*e.g.*, WLC models; see Chap. 7: Mechanical Characterization in Molecular Simulation). For example, the initial stiffness at low forces until yield, due to the rearrangement and rupture of H-bonds in the amorphous domains.

From a reductionist perspective, the material behavior of silk is achieved through the combination of distinct secondary structure units in silk protein, where the interplay of the *two* distinct constituent domains characterizes the nanoscale deformation mechanisms of the emerging nanocomposite structure (see Fig. 8.9). Here, we can effectively "divide and conquer", with mechanical characterization of each domain independently.[1] The benefit of such a procedure—essentially deconstructing silk into two distinct "building blocks"—is that each component can then be manipulated and potentially designed separately. Moreover, we can associate the composite behavior with the governing mechanism(s). The simplest approach to such a model is consideration of each phase as a multilinear spring [57]. The equation for two springs (*e.g.*, $k = g(\Delta)$) can be formulated as:

$$k_{\text{silk}}(\Delta_{\text{silk}}) = \left( \frac{1}{k_{\text{crys}}(\Delta_{\text{crys}})} + \frac{1}{k_{\text{amorph}}(\Delta_{\text{amorph}})} \right)^{-1} \tag{8.2}$$

where $k_{\text{crys}}$ represents the nonlinear stiffness of the crystalline phase, $k_{\text{amorph}}$ the semi-amorphous domain, and the deformation of each represented by $\Delta$ (note that for a given load, $\Delta_{\text{crys}} \neq \Delta_{\text{amorph}}$). This simple "two-phase" structure of silk, with associated nonlinear force-displacement behaviors, is sufficient to describe the four regimes characteristic of dragline silk (Fig. 8.9). More importantly, the separation of phases can quantify the contributions and molecular mechanisms of each component—the crystalline or semi-amorphous domains—to performance metrics of silk such as strength, extensibility, and toughness.

The constitutive sigmoidal behavior of the force-displacement response agrees qualitatively with the macroscale response of spider silk. Considering that the initial end-to-end length of the systems studied ranged from 8 nm to 14 nm and failure occurred at around 25 nm or more, the failure strains observed here seem to be much larger than that of experimental studies that show failure strains around 40 % [7]. The source of this discrepancy may be several factors in the idealized simulations that differ from physical experiments, including:

1. The results for strain depend on sequence; the idealized structure only consists of two amorphous domains and one crystal domain, hence the system is likely to be more extensible.

---

[1] Seemingly opposed to a "holistic" perspective stressed throughout this book, such a division is employed to separate a *presumed* composite behavior—similar to the separation of an ideal Kelvin–Voigt material into separate (but coupled) viscous and elastic components.

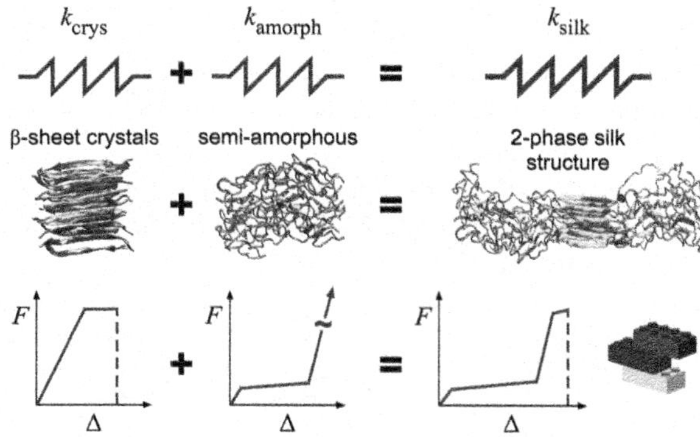


**Fig. 8.9** Making new building blocks from small building blocks: silk as a simple serial spring model. The constitutive behavior of two elements ($\beta$-sheet nanocrystals and semi-amorphous domains) can be represented by simple spring relations to generate a model of silk. The mechanical behavior of $\beta$-sheet nanocrystals under lateral loading depicts linear stiffness until the onset of the characteristic stick-slip phenomenon due to repeated breaking and reformation of H-bonds (a type of elastic-plastic yielding behavior). The mechanical behavior of the glycine rich semi-amorphous domain consists of three regimes (i. homogeneous stretching of the protein structure; ii. onset of yielding and unraveling *via* the breaking of H-bonds; iii. stretching of protein backbone). Note the serial combination of these regimes results in the 4-regime composite behavior depicted in Fig. 8.8. Such a separation of behavior allows efficient exploration of potential behaviors of silk, with parametric investigation possible through variations of crystal size, strength, "hidden length", backbone stiffness, etc. Adapted from [57]

2. In the bulk morphology of silk, there will be heterogeneity and defects, and orientation of the forces with respect to chains will not be perfect. Hence, the extensibility of the chains will be further reduced.
3. The REMD simulations correspond to hydrated systems with precise control on prestretching (*e.g.*, initial extended configuration). Prestretching of the strands during silk synthesis may contribute to the lower failure strain of the actual systems.

The failure strength of the molecular assembly shows less variation between sequences; rather, it is observed to depend more on the morphology of the crystal and the amorphous structures. More analysis and upscaling of our results will have to be done to make a rigorous link between nanoscale results and macroscale experiments.

### *8.2.3 Fibrils to Fibers*

Growing interest to better understand the nanomechanics of silk has prompted these and other recent studies at the atomistic and molecular scale, and may lend insight to the development of synthetic polymeric materials (such as cross-linked polyethylene with regular crystalline structures [58], for example). However, from the spider's perspective, he (or she) neither cares about the molecular structure nor the

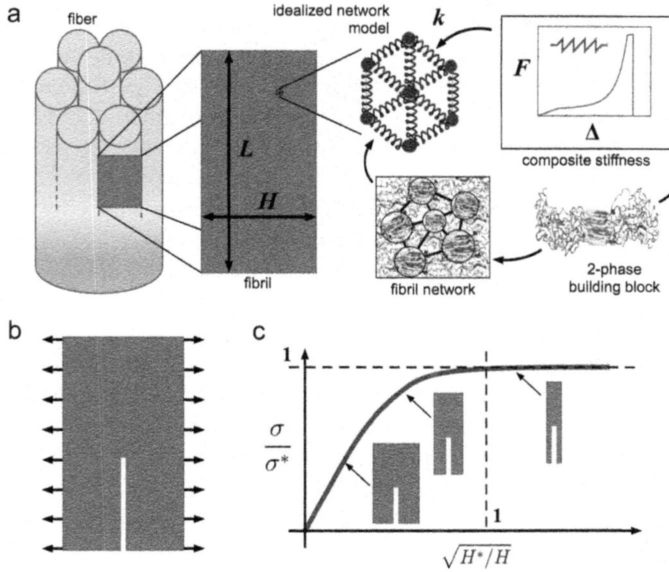


**Fig. 8.10** Exploration of fibril behavior *via* nanoscale mechanics. (**a**) A model of the silk fibril is reduced to a two-dimensional triangulated network model, with similar length-scales to that of the molecular structure (spacing corresponds to the distance between $\beta$-sheet nanocrystals). An ideal triangulation is motivated by the distribution of $\beta$-sheet nanocrystals amongst an amorphous matrix (*e.g.*, fibril network). This representative model can then be used to explore larger-scale silk fibril with dimensions $L \times H$ of several micrometers, and interpreted *via* nanomechanical parameters. (**b**) Example loading conditions used in investigation, reflecting classical fracture mechanics (*e.g.*, tensile mode I fracture with sharp crack) (**c**) Loading with varied aspect ratio (length $L$ is kept constant while the fibril width $H$ is varied) unveils a critical length scale, $H^*$ in which the fracture strength ($\sigma$) approaches the theoretical "defect free" strength, $\sigma^*$. Adapted from [59]

nanomechanics of dragline silk. Thus, we progress to the next level of hierarchy—the formation of silk and *fibers* (*i.e.*, threads) by the bundling (*i.e.*, spinning) of silk *fibrils*. The characteristic behavior derived from molecular protein studies provides the key to understanding behavior and function at both the fibril and spider web scales [18, 59].

The three dimensional arrangement of the secondary structures (*e.g.*, crystalline plus semi-amorphous domains) form a higher-order assembly, where the $\beta$-sheet crystals serve as cross-links embedded in a amorphous meshwork of less orderly structures. Protein chains combine into silk fibrils which bundle together into the fibers forming threads in the web of a spider (as depicted in Fig. 8.1). At this larger scale, experimental studies have shown that silk fibers contain many defects that act as stress concentrators, including cavities, surfaces, and tears [60–62]. These defects feature sizes that reach several hundred nanometers—clearly beyond the range of full atomistic simulation—and are crucial in the consideration of mechanical properties, as they serve as seeds for material failure through localized deformation (in fracture mechanics, for example, defects are known to lead to local stress concentrations).

To model the fibril, a coarse-grain spring-model of the protein network was recently developed [59], depicted in Fig. 8.10. This coarse-grain mesoscale model incorporated the material behavior at the protein scale derived from atomistic simulation into a spring network, reaching length scales of micrometers while relaying information about the underlying molecular mechanisms. To characterize the mechanical properties of individual fibrils with various types of defects and under different geometrical constraints, various loading scenarios were considered (motivated by traditional fracture mechanics; see Fig. 8.11). A series of *in silico* tensile tests were conducted for varying fibril sizes ($H$), where the dimensions were selected to resemble the typical size found in natural silk fibers with a diameter $H$ on the order of micrometers and a length that is several times larger than the width ($L \gg H$). Each fibril model contains a defect (*i.e.*, crack) running through fifty percent of the respective length, to account for localized stress concentrations that occur as a result of defects in naturally occurring fibrils. Failure stress and strain were recorded for varying loading conditions and different fibril sizes, and compared with experimental data for three types of silk fibers. As the width $H$ is reduced, the fiber's mechanical properties undergo a drastic change and reach successively higher failure stresses and strains. A critical length scale exists, $H^* = 50 \pm 30$ nm, at which the molecular properties of the protein network are most effectively scaled up to the micrometer scale and as seen in experimental stress-strain graphs typical of silk (*e.g.*, defective fibril approaches the strength of a perfect fibril). Indeed, the length scale of $H^*$ is quite similar to experimentally measured fibril diameter lengths on the order of 20 to 150 nm [7, 63, 64].

The underlying mechanism responsible for this remarkable behavior is a homogeneity of stress distributions at fibril sizes approaching the critical length scale [59]. Hence, as the fibril size approaches $H^*$, the majority of the material contributes to resisting failure, exhibiting failure stresses corresponding to the protein unfolding regimes. In the analysis of deformation fields at the crack tip in fibrils of different sizes, it was found an increased delocalization of stress at the crack tip as fibrils reach the critical fibril size $H^*$, resulting in completely homogeneous deformation where the entire sample reaches uniform stress concentrations before failure, pointing to a cohesive material mechanism to improve performance. As $H \to H^*$, hundreds of thousands of repeat units (and thus, protein domains) interact holistically in defining the mechanical properties of the fibril through a completely percolated network of unfolding protein domains. This also explains why the predictions from molecular simulations of the two-phase "building blocks" agreed well with experimental testing of macroscopic silk fibers. The synergy of multiple protein components within the fibril takes advantage of the effective strength and extensibility of each individual protein unit, yielding superior toughness moduli of 190–200 MJ/m$^3$ [59], observed for fibrils of critical length scales, favorably comparable to the range of experimental findings of 100–350 MJ/m$^3$ across different species [8]. Interestingly, a similar strength-enhancing nanoconfinement mechanism operates at the $\beta$-sheet nanocrystal scale. The similarity of the basic concepts behind these mechanisms is apparent, and reflects the universality of structure-function relationships in hierarchical biomaterials more generally. A final question remains: how is such exemplary behavior beneficial to a web as a functional material system?

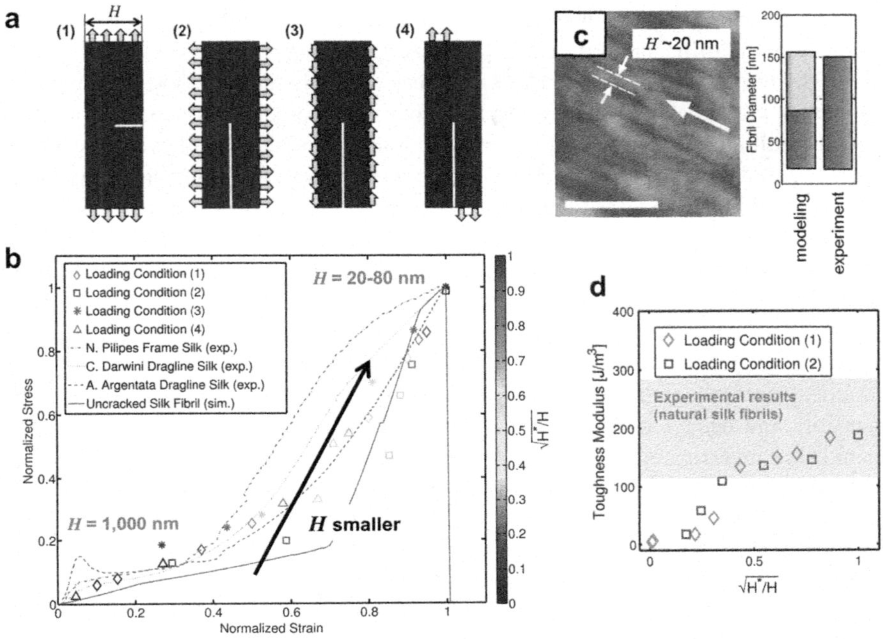


**Fig. 8.11** Analysis of mechanical properties of silk fibrils and dependence of the failure strain and failure stress on the fibril diameter $H$, with comparison to experimental results. (**a**) Four loading conditions considered for a two-dimensional network representative of silk protein fibrils, to include shear and tensile loading under varied fibril diameter $H$. A crack is introduced in each loading mode to represent naturally occurring flaws in the material. Fibril length $L$ is kept constant while the fibril width $H$ is varied. (**b**) Dependence of failure stress and strain on fibril size for the four loading conditions in panel, as size varies from ~1,000 nm to ~50 nm. Experimental data of three different silks is included for comparison. *Solid line* corresponds to loading profile of an uncracked fibril. Optimized material behavior is reached as fibril size approaches $H^* \approx 50 \pm 30$ nm. (**c**) Experimental image of a silk fibril (for *B. mori* silk) indicating a diameter of 20 nm (scale bar 150 nm). Reprinted with permission from Du *et al.* [64], copyright © 2011. (**d**) Modulus of toughness considered for varying fibril sizes under loading conditions (**1**) and (**2**). Both approach natural silk fibril values 100–350 MJ/m$^3$ (indicated in *shaded block* in the graph) as the fibril size approached the critical size $H^*$. Figure adapted from [59] and [65]

## 8.2.4 Web Performance

The spider web, although a common and recognized biological structure, is an evolutionary product with a myriad of functions, including the capturing of prey [66]. Architecturally, a spider web is constructed from two primary silk elements—radial threads composed of dragline silk and spiral threads composed of viscid (or sticky) capture silk [2]. Up to this point we have been primarily concerned with dragline silk, due to its exemplary and well-researched mechanical properties (compared to other silk types). From a structural perspective, radial threads constitute the supporting framework for the web, whereas spiral threads provide functionality (*e.g.*, capturing prey) and act as a medium of load transfer (akin to girders and joists in a building, for example). In spiders, dragline and capture silks are produced independently and differ in their detailed molecular structure (dragline threads are produced by the major ampullate glands, whereas viscid threads are flagelliform glands [67, 68]). Complicating matters, even though most spiders produce some form of dragline silk, the specific material properties vary among different evolutionary lineages of spiders [8, 15, 66, 69, 70]. Even among orb web weaving spiders (in which dragline silk serves a similar purpose) the material properties of dragline silk vary by more than 100 %, and across all spiders toughness varies over twenty-fold in species examined to date [8]. Is it even possible to deduce any *general* material-structure relationships between silk and a web?

Structural engineers exploit the fact that the performance of a building can resolved considering the key moment frames. Like a macroscale structure, the mechanical strength, stability, and deformation of the web is ultimately based on the capacity and behavior of the support elements—the dragline radial threads. A recent study investigated the distinct nonlinear behavior of these critical threads, and the associated effects on an entire web [18], advantageously drawing from the discussed nanomechanics of dragline silk [24, 27]. Moreover, the constitutive behavior of "derived" dragline silk, reflected in Fig. 8.8, can directly account for the combined behavior of $\beta$-sheet nanocrystals and semi-amorphous protein domains within the system-level performance of a web. The benefit of such a simple model is that each regime of the stress-strain behavior can be directly linked to phenomena at the molecular scale—completing the nano-to-macro cycle as depicted in Fig. 8.2. Moreover, *de facto* independence from empirical data is maintained—the molecular behavior considered for web model parameterization was the result of computational work only.[2]

As a functional structure for a spider, it is rare to see a perfectly intact spider web—*in situ* debris, loading, or changes in anchoring conditions—can easily lead to loss of silk threads. Yet, given the effort required by the spider for rebuilding a web, the web must still function well for prey capture in spite of minor damage. Indeed it has been shown that the introduction of flaws (missing web sections and/or broken threads) has only a local effect and that a spider web is still capable of functioning effectively as a net for catching prey [71]—removal of elements and even a large portion of threads away from the load have very little impact on the failure mechanism of the web as a whole. This suggests a remarkable damage tolerance of webs [5, 72]. However, the missing piece of understanding is where this damage tolerance is rooted and, specifically, how it relates to the unique molecular structure of silk.

Resistance to load, and subsequent damage, is a highly local phenomena within a web. The unique localization of damage can be directly explicated through the behavior of silk protein constituents on the molecular level, which express the underlying mechanism behind this remarkable structural property. Loading a web by a single radial thread, the vast majority of threads in the web exhibit relatively small

[2]Being said, the maximum stress level, on the order of 1–2 GPa, is in quantitative agreement with results from experimental studies [7].

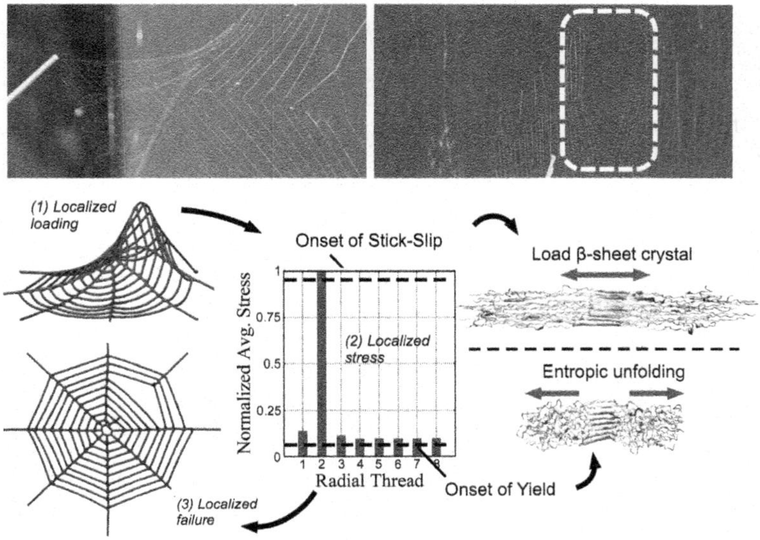


**Fig. 8.12** Local loading results in localized stress and failure in a spider web. Pulling a radial thread of a spider web (photos, *top*) results in relatively large web deformation, as the stiffer and stronger dragline radial threads deform the more compliant viscid silk of the spiral threads. Upon failure, only the loaded radial thread ruptures. The web is still intact and functional beyond the loaded area, indicating a robust structure. A model web (*bottom*) subject to local loading (**1**) indicates that stress is localized to the radial thread being pulled (**2**). Due to the yielding and entropic unfolding regime of the silks molecular structure, adjacent do not build up significant stress beyond the yield value (yet do contribute in energy dissipation; a toughening mechanism). Only $\beta$-crystals within the loaded thread undergo failure (**3**), in a manner similar to the crystals depicted in Fig. 8.3(*left*). Adapted from [18]

stress states, equivalent to the yielding regimes of the semi-amorphous regions. Large stress and thus deformation is highly localized on the radial thread where load is applied. Notably, for a radial thread subject to load, the semi-amorphous domains are almost fully unfolded (enhanced due to the fibrillar microstructure) and $\beta$-sheet nanocrystals begin to stretch at the particular segment of the thread to which force is applied. One thread is loaded, one thread breaks. Damage does not spread beyond the area of "targeted" load (see Fig. 8.12).

Computational evidence indicates that such localized failure is a universal characteristic of spider webs [18], which is directly related to the molecular structure. But is such behavior is unique to silk material or rather the unique web-architecture? It appears that, in opposition to popular thinking, the remarkable strength and extensibility of spider silk are not the primary factors, but rather it is the constitutive material behavior—the unique relation between an applied load and extension—that is essential to the function of a web. A systematic comparison with drastically different material behaviors illustrate the effect of constitutive relationship [18]. The analysis of the deformation mechanisms of a web indicates that while all radial elements contribute partially to the resistance of the applied load, the sudden yield and

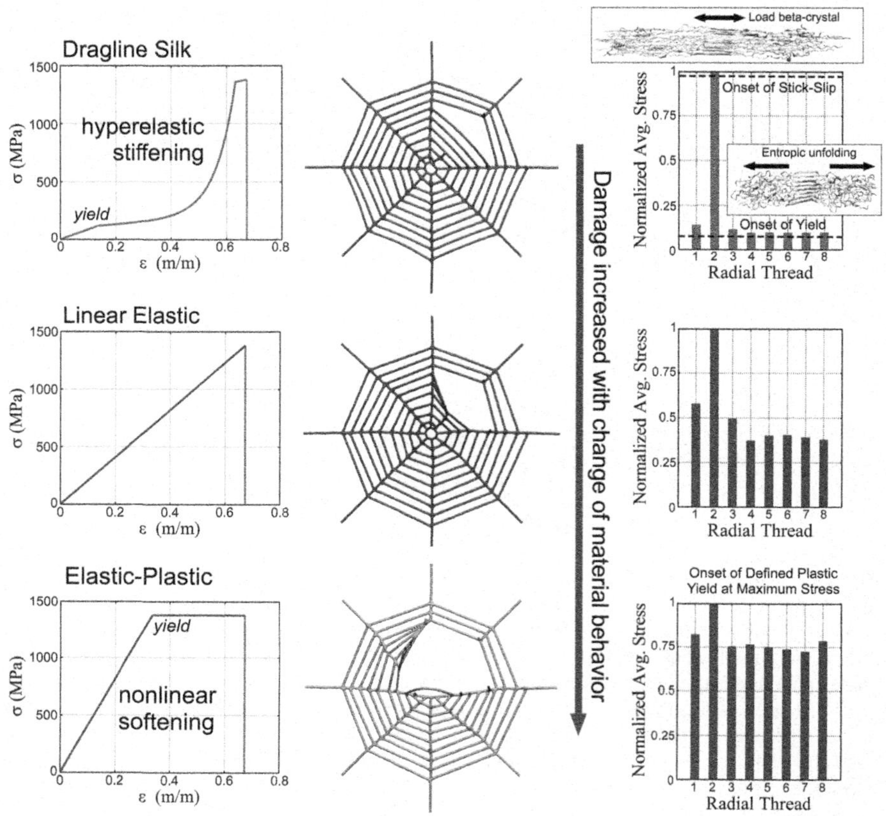


**Fig. 8.13** Is the secret the material (silk) or the structure (web)? Constructing webs of vastly different idealized material behaviors (atomistically derived dragline silk, linear elastic, and elastic-perfectly-plastic) with equivalent ultimate stress and strain demonstrates the advantage of silk as a material. Comparison of failure between the three models (*center panels*) confirms localized stresses and minimized damage for the natural hyperelastic stiffening silk behavior. A basis for macroscale-to-molecular interpretation lies in the analysis of the average stress of individual silk threads (*bar plots*; *right*) to determine distributions of molecular deformation states in the web. When load is applied locally to a radial thread, other radial threads not subject to applied force reach a maximum stress corresponding to the onset of yielding. The elastic-perfectly-plastic behavior (nonlinear softening) leads to an almost homogeneous distribution of stress in the web radials. Silk threads act as sacrificial structural elements—the easily fail to protect the web from incurring significant damage. Adapted from [18]

softening of the material introduces a weaker load path, and the loaded radial continues to resist the majority of force. The subsequent stiffening behavior facilitates the transfer of load to a localized, single radial element until ultimate failure occurs. Introduced linear-elastic and elastic-plastic models result in a more catastrophic, brittle-like failure with significantly increased damage zones (see Fig. 8.13). For the natural silk behavior, the radial threads effectively become the sole sacrificial elements, while the majority of the web-structure remains intact and functional. The

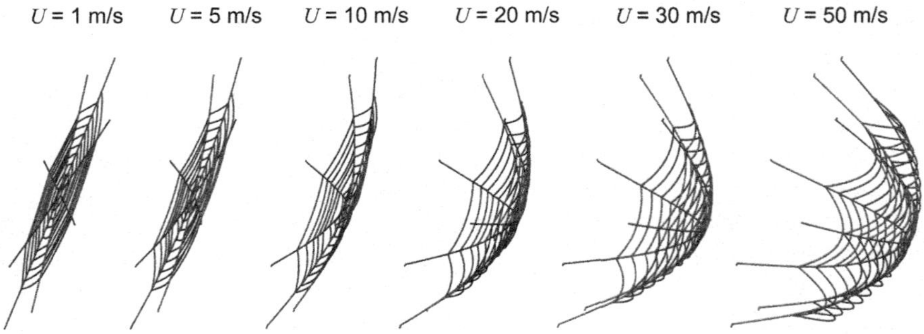


**Fig. 8.14** Web response under increasing wind speeds ($U = 1$ to 50 m/s). Little deformation occurs until wind speeds increase beyond ~5 m/s. The homogenous distribution of the drag force results in all radial threads undergoing simultaneous yield, and large subsequent deflections until ultimate failure of the entire structure. The sacrificial behavior is negated when all radials are subject to load. Thus, the initial stiffness up to yield, imbued by the semi-amorphous molecular regime, is necessary to beget structural integrity to a web under normal "operational" wind speeds. Adapted from [18]

discrete nonlinear architecture works as a mechanical "fuse" against the damage propagation. Under a critical stress, individual elements fail and thereby prevent further load propagation throughout the structure. This is a principle similar as used in electrical fuses, which fail under a critical electrical current to prevent the overall system failure by sacrificing a single element.

The small decrease in ultimate strength in natural silk is superseded by greatly enhanced structural robustness, allowing a spider to repair rather than rebuild, should a failure event occur. This suggests that more deformable, yet weaker, web construction results in an overall more efficient web in its natural environment. Unlike engineered structures, which are typically designed for ultimate loads, the forces a web could be naturally subject to are great relative to the web strength, *e.g.* a falling branch, a passing animal or a gust of wind. With the impossible task of designing for ultimate load negated, deformation capacity and structural robustness are likely more critical for the species' survival, properties that the tacit behavior of dragline silk provide as a construction material for webs. Such mechanistic evidence has significant evolutionary implications [15, 66], ensuring that debris or prey too large for the web to support will not cause irreparable damage, costing the spider energy to rebuild. Furthermore, the regulation of damage through control of material behavior has implications beyond webs to any structural design which would benefit from contained damage.

This sacrificial role of silk is apparent when single threads are loaded individually—the spider would rather a local failure than a system-wide collapse (*e.g.*, mitigated losses). But what about loads that effect the entire web structure? To understand the response of a web under global, distributed, loads, one can investigate the effect of *wind*, a common environmental condition (see Fig. 8.14). A recent study subjected the same web model discussed to drag forces due to wind speeds up to 70 m/s (a threshold at which the model fails). The system-level deflection curves are indicative of the mechanical behavior of radial threads, which ultimately transfer load to the web's anchoring points. While the spiral threads undergo increased deflection and capture more of the wind load due to their larger exposed length, they are effectively pinned to the much stiffer dragline radial threads, which limit web deflection. For wind speeds $<10$ m/s there is little difference between silk, linear elastic, or elastic-perfectly-plastic thread behaviors, and deflections are limited. This is because the structural rigidity of the web is attributed to the *initial* stiffness of the dragline silk prior to yielding—pure hyperelastic stiffening (with a nominal initial stiffness) would induce high levels of deformation under relatively low global loading. This can be demonstrated under higher wind loads, where the softening behavior of dragline silk at moderate deformation results in significant web deflection, greater than in the cases of linear elastic and the elastic-perfectly-plastic behavior. The yield load in the radial threads defines the regime of "operational wind speeds" for the spider (wind speeds exceeding $\sim$5 m/s for the model discussed).

This wind load investigation demonstrates that the initial stiffness of dragline silk provides structural integrity under functional, or operational, conditions (expected typical wind speeds). These findings suggest that whereas the nonlinear stiffening response of dragline silk is crucial to reduce damage of localized loading, it is disadvantageous to global (distributed) loading scenarios. Yielding caused by the high wind loading results in web displacements that could cause large areas of the sticky catching-spiral to impact surrounding environment (such as vegetation), which would result in the large-scale destruction of a web. However, under moderate wind loading, the linear regime of the dragline silk dominates behavior, and silk performs as well as the other material behaviors. The wind load cases illustrate that the initial stiffness of the dragline provide structural integrity under such "global" loading conditions. If this loading becomes "extreme" there is no benefit, and the silk yields. Presumably, what is considered "extreme" or "normal" environmental is dependent on the locale of the spider, and the approach applied in this case study can systematically link variation in the mechanical properties between silks of different species to such environmental loading conditions. Interspecies variation of this yield point may predict under what wind conditions a web is subjected. For example, the species *Caerostris darwini* builds some of the largest orb webs known among spiders and constructs longer bridge lines than other giant orb-weavers, spanning across even large rivers [8], and likely subject to relatively higher wind loads. The initial yielding of *C. Darwini* dragline silk occurs at a stress on the order of $\sim$300 MPa [8], maintaining structural rigidity at loads far surpassing the atomistically derived dragline silk model (yield at approximately 120 MPa). Yet, post-yield, the dragline silk of *C. Darwini* still exhibits the characteristic nonlinear stiffening.

The nanomechanics of dragline silk have been successfully connected to the functional performance of a web, through key mechanisms at critical length scales (summarized in Table 8.1). Note that this is a purely mechanistic perspective. While providing a description of the materiome from "nano-to-macro", it does not reflect the *complete* materiome because it neglects such mechanical effects as viscoelasticity, biological requirements to produce dragline silk (*e.g.*, metabolic cost), function

**Table 8.1** Summary of structural hierarchy of spider silk spanning length scales from Ångstroms to centimeters, along with key mechanisms that define its mechanical function. Adapted from [59, 65]

| Structure | Hierarchical level (scale) | Description | Key mechanism(s) |
|---|---|---|---|
| Amino acid sequence | Protein building block (Å) | Poly-alanine regions responsible for stiff cross-linking domains; glycine rich regions define structures within amorphous domains. | Basic building block that dictates secondary protein structure. |
| $\beta$-strand | Elementary protein structure; (Å to nm) | Aligned (parallel or anti-parallel) H-bonded protein segments; poly-alanine regions in silk. | Critical $\beta$-strand length for heightened stability of H-bond clusters; cooperativity of H-bonds maximized at a critical scale of 3 to 4 H-bonds. |
| $\beta$-sheet nanocrystals | Secondary protein structure (nm) | Defines local conformations (*e.g.*, $\alpha$ helices, $\beta$-turns); $\beta$-sheet crystals in silk, clusters of $\beta$-strands, have a size of 2 to 4 nm. | Nanoconfinement optimizes cooperative H-bonding, resulting in shear deformation shown to result in least catastrophic failure, through stick-slip that provides the basis for larger-scale toughness. |
| Two-phase protein composite | Tertiary protein structure (~10 to 20 nm) | Three-dimensional arrangement of interacting secondary structures; bulk material. | The cooperative interplay and load-balance of the protein composite domains achieves a unique hyperelastic material behavior. |
| Fibrils | Emergence of structural form (~100 nm) | Spun two-phase protein strands into continuous interwoven chains. | Homogeneous deformation state reached if fibril dimension confined to critical width, $H^* = 50 \pm 30$ nm; enables homogeneous load distribution, ensuring optimal strength across entire fibril length. |
| Fibers, threads | Structural component (~1 to 2 μm) | Bundles of fibrils. | Bundling of several geometrically confined fibrils into a fiber enhances material properties through reaching homogeneous deformation state (flaw tolerance). |
| Web | Functional system, material structure (cm to m) | Spider web geometry (*e.g.*, spiral) and material composition determine the deformation mechanisms of the web, responsible for optimized prey capture mechanisms. | Localization of failure due to a combined yielding-stiffening behavior at deformation; sacrificial elements imbue flaw tolerance and robustness. |

and performance of silk in nonstructural roles (*e.g.*, signalling prey capture through structural vibration), etc. Such omissions do not, however, devalue such mechanistic descriptions. Looking forward, the entirety of a bottom-up design paradigm illustrated here—from protein sequence to web structure—where function can be manipulated through alteration of structure on different hierarchical scales, presents opportunities in design of enhanced and novel synthetic, biomimetic and biological materials that provide a wide array of material function at low weight. The implications can be explored either from a *structural* or *material* perspective.

### *8.2.5 A Route to Structural Robustness*

From a structural perspective, can we generalize behavior of webs for non-silk materials and non-web architectures? A recent study successfully invoked the principles of fracture mechanics in the behavior of a spider's web [18]. It is well-known in linear elastic fracture mechanics (LEFM) that the size of the damaged zone in front of a crack, the so called *damage* or *plastic zone*, becomes larger for materials that can sustain larger plasticity. We can define the damage zone by the spatial dimension, $r$, which is proportional to is the material fracture toughness ($K_c$) and the material ultimate strength ($\sigma_c$). This gives us an idea about the scaling of the extension of the damaged zone in the case of linear elasticity, *i.e.* assuming the stress $\sigma$ directly proportional to the strain, $\varepsilon$. In order to generalize the concept to any constitutive law, particularly the hyperelastic stiffening of spider silk, we can consider a general nonlinear law, in the form of a simple power law, *i.e.* $\sigma \sim \varepsilon^{\kappa}$. The power of the stress-singularity at a crack tip (*e.g.*, defect) will accordingly be modified from the classical value by $\alpha = \kappa/(\kappa + 1)$, where $\alpha$ is defined as the hyperelastic or nonlinearity parameter [73]. Thus, the singularity changes similarly to what occurs at the tip of an edge cut [74], and consequently from its theory [75] we can directly predict the following fracture stresses ($\sigma_f$) for *nonlinear* materials, as a function of that related to *linear* elasticity:

$$\frac{\sigma_f^{(\alpha)}}{\sigma_c} = \left(\frac{\sigma_f^{(1/2)}}{\sigma_c}\right)^{2\alpha} \tag{8.3}$$

where $\sigma_f^{(1/2)}$ reflects the fracture strength if the material was linear elastic. More importantly, Eq. 8.3 is general and not related to a particular spider web or silk-like material. It shows that the damaged zone extension in the structure changes as a function of the material constitutive law. Accordingly, the fraction of damaged zone in the structure can be defined as:

$$\phi^{\alpha} = \left(\frac{r^{(1/2)}}{r^{(0)}}\right)^{2\alpha} = \left(\phi^{(1/2)}\right)^{(2\alpha)} = S^{2\alpha} \tag{8.4}$$

Equation 8.4 explicitly shows that the larger stiffening effect, the more localized the damaged zone. Note that $\phi^{(1/2)}$ represents the structural response for the classical case of linear elasticity and thus represents a structural parameter ($S$). From

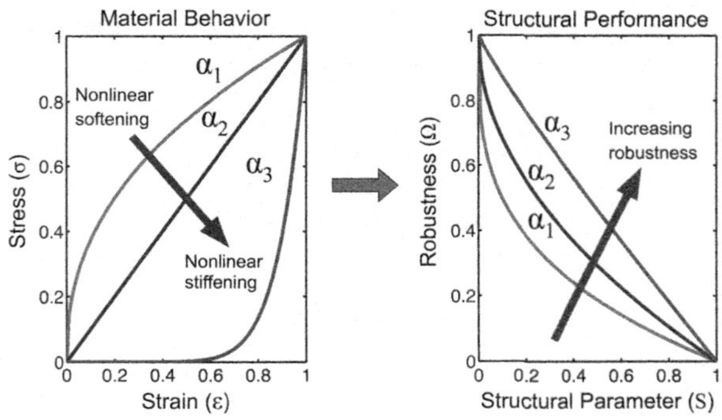


**Fig. 8.15** Effects of nonlinear stress-strain behavior of silk threads on web robustness *via* fracture mechanics analysis. *Left*: Plots of considered stress-strain curves (material behavior) demonstrating the transition from elastic-plastic (nonlinear softening) to nonlinear stiffening behaviors by the nonlinear parameter $\alpha$ ($\alpha_1 = 0.3$, $\alpha_2 = 0.5$, $\alpha_3 = 0.9$). *Right*: Structural robustness, $\Omega$, versus system-dependent constant, $S$, for the same nonlinear parameters. This shows that universally the robustness increases with an increase in $\alpha$ ($\alpha_1 < \alpha_2 < \alpha_3$). In general the larger nonlinear stiffening the larger the structural robustness, and thus, less damage. Adapted from [18]

the given arguments, $S$ is a structural property, dependent on the specific material properties (such as fracture toughness, $K_c$), system geometry (*i.e.* crack width, $a$, or element length), and applied loading conditions, representing the relative damage associated with linear elastic behavior (*i.e.*, from the original LEFM assumption). The "structural robustness", $\Omega$, is defined as the fraction of surviving material in the structure, that is, directly as:

$$\Omega^{(\alpha)} = 1 - \phi^{(\alpha)} = 1 - S^{2\alpha} \tag{8.5}$$

Consequently, the relative size of the damage zone is a function of the material constitutive law, where the parameter $\Omega$ is defined as the fraction of surviving material in the structure. Equation 8.5 is depicted in Fig. 8.15, where we can recognize that in general, fixing the other variables, the larger the material hyperelastic or nonlinearity parameter the larger the structural robustness.

We find that based on this theory, akin to the finding our simulations and experiments in webs, the key to reach minimal structural damage is that the material undergoes a severe increase in stiffness under applied stress—the size of the damaged zone in the proximity to a defect increases for materials that tend to feature a softening behavior as realized in the elastic-plastic model [73]. Conversely, the realization of a stiffening material behavior (*e.g.*, silk-like) results in a decrease of the damage zone. This directly shows that the introduction of nonlinear behavior dictates the relative size of the damage zone [75]. In spider webs we see that the cooperative action of a stiffening structural member (the radial silk thread under load) and yielding (or softening) of ancillary members result in a localization of elastic resistance. Concurrently, adjacent radials reveal partly unfolded semi-amorphous regions, or a partially stiffened state—immediately after failure, load is redistributed between these two adjacent threads, keeping the rest of the web intact and functional. Indeed, sections of the web removed from loading undergo limited deformation and strain. This property of the web architecture and silk's unique nonlinear behavior combine to form a structure designed to tolerate damage by inducing localized failure of redundant structural elements. In sum, for spider webs, when small loads are applied to the web, the entire web structure deforms and contributes to mechanical resistance. However, once loading exceeds the critical load necessary for molecular unfolding, localization of deformation occurs, and thus very few web elements fail at the ultimate breaking point.

Such investigations show that independent from the specific material behavior of silk, a more pronounced the material stiffening with strain, the larger the structural robustness as failure is increasingly localized, thus resulting in failure of a minimal number of elements in a discrete mesh-like structure. As discussed, this phenomenon is duly exemplified in webs, where the combined yielding-stiffening nonlinear material behavior provides an ideal medium for localized damage and thus structural robustness. This behavior can be deemed a structural liability if local failure (such as cracking) can lead to catastrophic collapse, irreparable damage, mechanical instabilities, or degradation of system function. The existence of a web-like mesh structure, however, is critical as the failure of few elements does not lead to the catastrophic breakdown of the material. For spider webs such localized failure is preferred, as structural elements effectively dissipate energy from the applied load and can be sacrificed (and replaced) without comprising the structural integrity and functionality of the entire system. This suggests the use as a practical design principle where the same "localized damage" can be expressed in synthetic systems by implementing a constitutive behavior inspired by silk, or, more generally, a deliberate increase in nonlinear material behavior to realize a significant stiffening as strain increases. Thus, the simple relation of structural failure mechanisms to nonlinear material behavior revealed here can be easily applied in the design of structural components in which minimal damage is desired. For example it motivates the use of sacrificial elements introduced in buildings and bridges for seismic and terrorist protection, or generally for the design of highly robust structures and armors.

A key interpretation is that superior performance does not need to rely on the strength of elementary building blocks of the material, but rather, on the integration of material and structure (or the web materiome). The superior functionality of spider silk to the overall performance of webs is based on the unique yielding-stiffening behavior rather than the relatively high values of ultimate strength and strain. Silk's inherent material nonlinearity is responsible for its advantage over linear-elastic or elastic-plastic materials such as steel or concrete, making it the preferred building material for robust web structures. Spider webs illustrate the fundamental issue of how the functionality of biological construction materials—in this case, spider silk—extends beyond traditional concepts of strength and strain by utilizing specific deformation mechanisms as loading is increased. As demonstrated by the theoretical analysis, these findings generally hold for other materials and nanostructures in which specific behavior can dictate functionality beyond limiting parameters, such as ultimate strength. The approach presented here can be adopted by scientists working on a diverse range of interdisciplinary fields related to polymers, biological materials, and other synthetic systems, where the inclusion of nonlinear material behavior in the design process can impede structural loss, minimize damage, localize failure, and increase system robustness and integrity. Such a holistic and multi-scale optimization of function through molecular structure and mechanical behavior—a materiomic approach—can lead to the development of further fundamental discoveries.

### *8.2.6 Beyond the Web: Synthetic Silks*

From a material perspective—can we improve upon Nature's silk? There are many routes to modify silk. Studies have successfully shown that variations in synthesis conditions lead to silk-based peptide fibrils or protein aggregates with rather different structural and mechanical properties. For example, the hydration level and solvent conditions (*e.g.*, ion content and pH) play a large role in the structure and mechanical properties of silk proteins [43, 76] and even the transition from concentrated dope to final silk in the spinning duct. Variations in crystallinity and alignment within the silk thread, for example due to the reeling speed of the collected sample, have also been mapped to macroscopic mechanical properties [7, 77]. Thus, we can potentially "tune" *natural* silk to meet specific *unnatural* requirements if desired.

Through a purely synthetic approach, a variety of silk sequence motifs could be theoretically designed, expressed, spun into fibres and then tested for desirable mechanical properties—ranging from simple manipulation of wild-type silk sequences [34] to addition of nonbiological functional groups (so-called *chimeric* silks, [78–80]), through both *in silico* and *in vitro* analysis. For example, a recent study expressed a recombinant protein of the spider *Nephila clavipes* in metabolically engineered *Escherichia coli*, from which it was spun into a fiber displaying mechanical properties comparable to those of the native silk [81]. In addition to illustrating a successful protocol to produce synthetic silks, it was also found that the recombinant proteins of lower molecular weight versions yielded inferior fiber properties. The size of a recombinant dragline silk protein is hypothesized to be a key factor in controlling the mechanical properties of the spun fiber. From an evolutionary point of view, spiders have evolved natural dragline silk proteins having very high molecular weights [82]. Given the fact that dragline silk proteins contain repeating motifs [21]—resulting in repeating semi-amorphous and crystalline domains—larger proteins possess more repeating units than the smaller ones, allowing increased inter- and intra-chain interactions. Moreover, smaller proteins result in more chain end defects in fibers, reducing mechanical properties. It was shown that larger molecular weights indeed increase mechanical performance (*e.g.*, stiffness and yield stress), providing insight into the evolution of silk protein size and suggesting why the previous spinning of lower molecular weight proteins did not successfully recapitulate the properties of native fibers [81].

As Nature demonstrates, silk is not limited to structural or prey capturing elements in webs, so we should not limit potential applications to super-fibers or mesh-like structures. Recently, regenerated silk solutions have been used to form a variety of biomaterials, such as gels, sponges and films, for medical applications [9, 11, 83].

Silks in these structures may not necessarily express (or require) the localized failure as seen in aerial webs. Silkworm silk fibres, for example, are comparable in strength and toughness to spider silk [6] but the tensile behavior of silkworm silk, although exhibiting a similar small strain yielding, is characterized by strain softening at large deformation [6, 13]. Such silk is assembled in film-like structures for cocoons [84] and as such, under load, optimal performance may require the distribution of stress over a larger material area, as opposed to localized failure. Even local rupture or breakage of the cocoon may be critical to the life cycle of the silkworm, whereas, in contrast, a spider can sacrifice portions of the web and rebuild.

Such analysis projects the possibility of new design paradigms in areas of application ranging from high performance fibers and multifunctional materials produced from artificially spun silk or derived materials [9, 43], to novel material platforms for tissue engineering or even structural applications in the defense or aerospace industry [11]. Despite decades of research in silk production and mechanical properties, the mass production of silk and biomimetic materials remains a challenge, particularly because of silk's unique features that can be achieved only by the controlled self-assembly of the macromolecular constructs with molecular precision at the nanoscale [17].

## 8.3 Rethinking 'Brittle' Materials: The Cases of Bone and Nacre

As previously discussed in Chap. 3: The Challenges of Biological Materials, many biological materials are composed solely of polymer-like polypeptide chains (such as silk), and typically considered a class of *soft matter*. What happens when there is a biological need for a stiffer material system such as bone or shell? One simple solution is to integrate minerals and proteins together. To attain strength, toughness, and rigidity, Nature implements a *composite* material. Using materials available in the environment that typically exhibit poor macro-scale mechanical properties (brittle biological minerals and compliant proteins), Nature has achieved orders-of-magnitude increases in strength and toughness that, in many cases, occurs in a nonadditive manner that goes beyond the simple composite rule of mixture formulations [85].

Throughout this text, the protein collagen has been used as an example protein material, and is one of the primary components of bone. But the tropocollagen molecule alone is not sufficient to understand its widespread success. Collagen, the most abundant protein on Earth, is a fibrous structural protein with superior material properties, and is the main mechanical component of connective tissue. In the form of elongated fibrils, is mostly found in fibrous tissues such as tendon, ligament and skin, and is also abundant in cornea, cartilage, bone, and blood vessels. Collagen is found in a number of tissues exposed to severe tensile or compressive loading, including tendon, bone, teeth, cartilage and the cornea [86–91]. However, the function (and performance) of collagen materials such as tendon and bone is quite different.

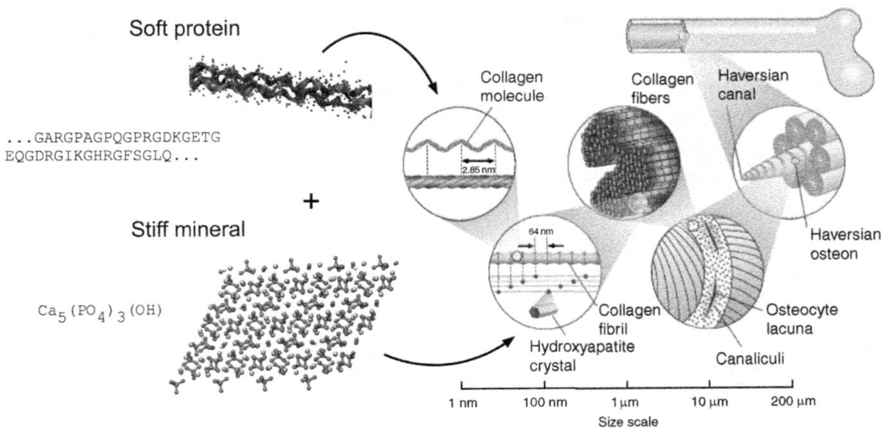


**Fig. 8.16** Hierarchical structure of bone, nano (Ångstrom) to macro (meters). The schematic displays key structural hierarchies of bone—primarily a composite combining a soft, protein material (tropocollagen molecules) with a hard, mineral phase (hydroxyapatite). While the chemical sequence and structure are well defined, the properties of bone arise across multiple scales, including the secondary structure of the protein (*e.g.*, triple helix), the size of the mineral platelets embedded in the protein matrix, assembly of mineralized collagen fibrils which assemble into collagen fibers, fibers within lamellar osteons (facilitating nutrient transport *via* Haversian canals) acting as structural components, and finally the functional bone itself. Adapted from Ritchie *et al.* [92]. Reprinted from Materials Science and Engineering: C, *Role of microstructure in the aging-related deterioration of the toughness of human cortical bone*, Vol. 26, 1251–1260, copyright © 2006, with permission from Elsevier

Tendon is a soft (yet tough) connecting material, serving a functional role connecting muscle and bone, whereas bone is a strong material that provides a kind of structure of scaffold for the entire body. Bone is one of the most widely studied protein materials found in Nature, a material composed out of assemblies of tropocollagen molecules, water, and mineral nanoparticles made of carbonated hydroxyapatite, representing an extremely tough, yet lightweight, adaptive and multi-functional material. Bone has evolved to provide structural support to organisms, and therefore, its mechanical properties are of great physiological relevance.

## 8.3.1 Bone as a Biocomposite

Bone is often stereotyped as simply a protective and supportive framework for the body; though it does perform this function, it is actually a very dynamic organ that is constantly remodeling and changing shape to adapt to the daily forces placed upon it. Like all natural materials, its mechanical properties are determined by its structure [91, 93, 94], which in turn is motivated by its (primarily mechanical) function [95]. Bone can be considered a *biocomposite*—a soft protein material combined with a hard mineral phase in a hierarchical structure (see Fig. 8.16). Structurally, the basic building blocks of such bone are an organic matrix (90 % type-I collagen, 10 % amorphous) and a mineral phase of calcium phosphate based hydroxyapatite, although its composition and structure are not invariant; they vary with factors such as skeletal site, age, sex, physiological function and mechanical loading, making bone a very heterogeneous structure [96].

While a diversity of structural motifs exist between bone tissues at different microstructural length-scales [89, 97], the most common in bone is the lamellar unit [98], with adjacent lamellae some 3 to 7 μm thick [99]. Several studies have suggested the existence of a "ply"-type structure, where the fibrils rotate with respect to their own axis across the width of the lamella [97, 98]. Indeed, novel measurement methods combining synchrotron X-ray texture measurements and scanning of thin sections of bone [100, 101] indicate that each lamellae consists of a series of fibril layers oriented at different angles (similar to the variation of orientation angles in engineered fiber-reinforced composites).

At the nanoscale level of hierarchy, each individual fiber is composed of an array of mineralized collagen fibrils. The mineralized collagen fibril, which is roughly 100 nm in diameter and ~5 to 10 μm or more in length, is the basic building block of the bone matrix. The fibrils consist of an assembly of 300 nm long and 1.5 nm thick collagen molecules, which are deposited by the osteoblasts (bone forming cells) into the extracellular space and then self-assemble into fibrils. The collagen fibrils are filled and coated by tiny crystals of hydroxyapatite that assemble in the gap between collagen fibrils, and the fibrils become mineralized as the bone tissue grows and matures. These crystals are mainly flat and are mostly arranged parallel to each other and to the long axis of the collagen fibrils in the bone composite, in a regularly repeating, staggered arrangement [102, 103]. The mineral hydroxyapatite crystals are relatively small with a thickness of 1.5 to 4.5 nm [104–106] with critical implications for its fracture behavior.

At the lowest dimension level, the collagen fibrils are formed by staggered arrays of tropocollagen molecules, each built from three polypeptides arranged in a 300 nm long triple-helical "nanorope" geometry and stabilized by hydrogen bonding between residues [107, 108]. Every third residue in each of these molecules is a glycine amino acid, and about one fourth of the tropocollagen molecule consists of proline and hydroxyproline [109]. Notably, even minute structural changes in the collagen molecule's architecture can have severe consequences for the macroscale tissue behavior, as is evident in genetic bone diseases, leading to mechanically compromised tissues (*e.g.*, brittle bone disease *osteogenesis imperfecta*, discussed further in Chap. 9: Pathological Materiomics).

From a mechanical perspective, the protein matrix absorbs and distributes shocks, but its softness is buttressed by hard, embedded minerals. The result is a composite material that can be thousands of times less susceptible to fracture than the pure mineral, and orders of magnitude stronger than pure protein [110]. The combination of protein and mineral is more than an efficient use of building materials (in a kind of brick and mortar arrangement). Indeed, the combination of each phase is a kind of "material symbiosis"—improving the performance of each component. The protein—the active biological component—benefits from the mineral by attaining strength and toughness otherwise unachievable. The benefit to the mineral is not so clear. Fracture of a pure mineral initiates at flaws, and in the physical world, flaws are unavoidable (entropic effects are sufficient to form impurities). In the case of biocomposites, flexible proteins will sneak into these mineral matrix. This initiates larger defects (as the protein pushes its way around the mineral) which, at first glance, may indicate a weakening mechanism. However, the protein serves two beneficial roles—it provides a glue to bind the mineral phase together and it effectively partitions the mineral matrix into smaller regions than would otherwise naturally occur. These nanoscale mineral sizes improve system-level behavior. The smaller the nanoparticles, the smaller the volume, so the strain energy stored in the mineral is too small to break and cracks will not propagate. Such flaws will not matter (*e.g.*, the composite system is *flaw tolerant* [110]).

### *8.3.2 Linking Collagen and Aging*

Like the case study of silk, we can make a direct link from nanoscale properties (in this case, the cross-linking of tropocollagen molecules) to macroscale performance (here, the process of aging). As for the case of silk, this nano to macro link is presented schematically in Fig. 8.17. It is known that excessive bone remodeling and other aging-related changes to the musculoskeletal system inevitably increase susceptibility to bone fracture [111]. In the case of the elderly, such changes are critical as the consequent fractures can lead to significant mortality. While a number of physiological variables, such as loading regime, incidence of traumatic falls, prior fractures, etc., are involved, it is apparent that the bone tissue itself deteriorates with age [111–113].

Fracture toughness measurements on 34 to 99 year old human cortical bone show some 40 % reduction in the crack-initiation toughness and almost complete elimination of the crack-growth toughness with increasing age [115], yet the reasons for this are not totally clear [92, 115]. Traditional thinking on the deterioration of bone with age has focused on bone quantity as a predictor of such fracture risk, where quantity is described by the bone mass or bone-mineral density, defined as the amount of bone mineral per unit cross-sectional area. For example, the elevation in bone remodeling activity, concurrent with menopause in aging women, can lead to osteoporosis, a condition of low bone mass associated with an increased risk of fracture. However, although bone mass can explain some of the fracture risk, there is now mounting evidence that low density alone is not the sole factor responsible for the aging-induced fracture risk. A landmark investigation by Hui *et al.* revealed a roughly ten-fold increase in fracture risk with aging that was *independent* of bone-mineral density [111]. This result and the concurrent realization that bone mass alone cannot explain the therapeutic benefits of anti-resorptive agents in treating osteoporosis [113], has re-emphasized the necessity for understanding the factors that control bone quality. Could a possible explanation lie at the nanoscale?

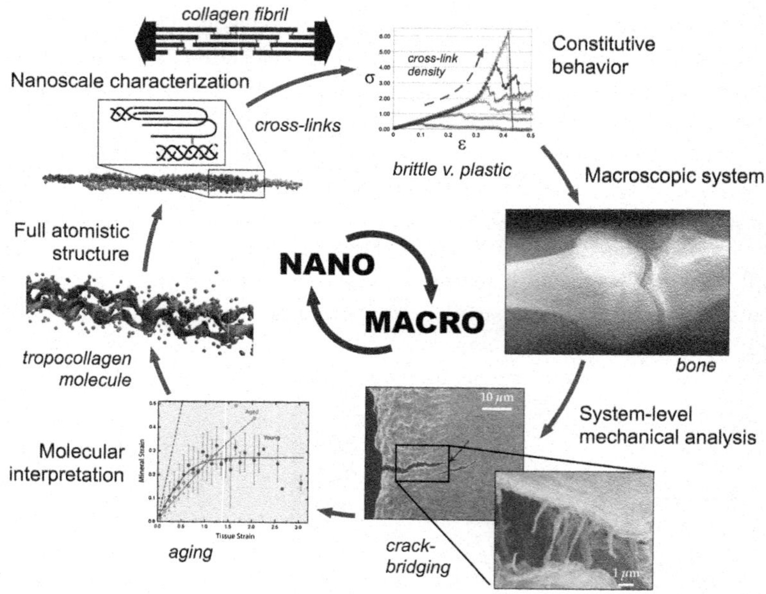


**Fig. 8.17** Holistic *nano-to-macro-to-nano* approach used here for bone, where, again, an integrated multiscale analysis is applied throughout the materiome, incorporating relevant scales from nano to macro in the description of system-level behavior and function—in this case, degradation of bone performance with aging. A model of cross-linked collagen derived from prior molecular investigations can be used to investigate the material properties of bone. Experimental studies indicated an increase in cross-links with age, as well as decreased toughness of bone. The goal is to capture known molecular behavior at the largest scale, allowing interpretation of macroscale functionality in terms of molecular mechanisms, in a reciprocal manner. *Inset photos* of bone fracture from Nalla *et al.*, reprinted by permission from Macmillan Publishers Ltd: *Nature Materials* [90], copyright © 2003. Experimental plot (molecular interpretation) from Zimmermann *et al.* [114], used with permission, copyright © 2011

With aging, at the nanoscale there are distinct changes in the collagen environment specifically with regard to an increased degree of cross-linking, at the submicrometer scale the nature and properties of individual collagen fibrils deteriorate. In human cortical bone, cross-links occur in two forms: enzymatic and nonenzymatic (advanced glycation end products that form both intra- and interfibrillar links along the collagen backbone [116]). While the level of enzymatic cross-links stabilizes around 10 to 15 years of age [117], nonenzymatic cross-links can increase up to fivefold with age [118, 119], which has been correlated to reduced bone toughness and fracture resistance [120–122].

Using coarse-grain molecular dynamics approaches as described in Chap. 6: Computational Approaches and Simulation, a recent investigation quantified the effects on nonenzymatic cross-links on the mechanical performance of collagen fibrils [123, 124]. In order to develop a fundamental and quantitative understanding of collagen mechanics, theoretical models encompassing the mesoscopic scales between the atomistic and the macroscopic level, considering atomistic and chemical interac-

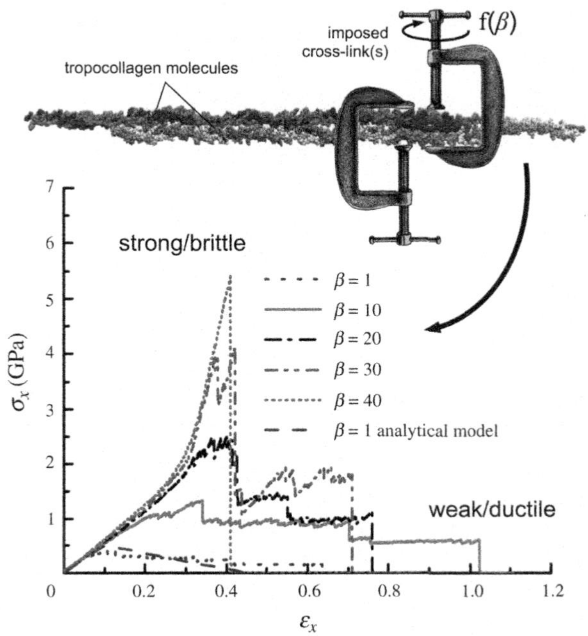


**Fig. 8.18** Cross-links and collagen fibril behavior. *Schematic*: Modeling allows the introduction of arbitrary cross-link strengths between two tropocollagen molecules in a fibril (indicated by the artificial "clamps"), *via* the introduction of the cross-link density parameter, $\beta$. *Plot*: Stress versus strain of a collagen fibril, for different cross-link densities (a function of the parameter, $\beta$). The results clearly show that larger cross-link densities lead to smaller yield strains, larger yield stresses as well as larger fracture stresses (strong but brittle). For larger cross-link densities, the second elastic regime (seen as much steeper, second slope) is activated. As the cross-link density increases, the collagen fibril shows a more 'brittle-like' deformation behavior. For values of $\beta > 25$, the deformation mechanisms is characterized by molecular fracture, and as a consequence, the maximum fracture stress of the collagen fibril does not increase with increasing cross-link densities. This cross-link density physically corresponds to the case when two cross-links per tropocollagen molecule are present. Below this value, the response is weaker and more ductile. Adapted from [123, 124]

tions during deformation are vital [125, 126]. Such a study highlights one advantage of *in silico* investigation—the cross-link densities and strengths were not limited to physical conditions, thereby probing extreme conditions and scenarios to illustrate system function (see Fig. 8.18). It was found that cross-link deficient collagen fibrils show a highly dissipative deformation behavior with large yield regimes, while increasing cross-link densities (such as those due to aging) lead to stronger fibrils, yet display an increasingly brittle (*e.g.*, non-plastic) deformation character [123]. The model predicted that collagen fibrils are capable of undergoing extremely large deformation without fracturing; how much of this is elastic or dissipative depends on the cross-link densities, and thus the *age* of the molecule. More recent, full atomistic studies enabled high resolution analysis of the molecular mechanics, cross-links, and deformation [127] between a pair of tropocollagen molecules. It was further

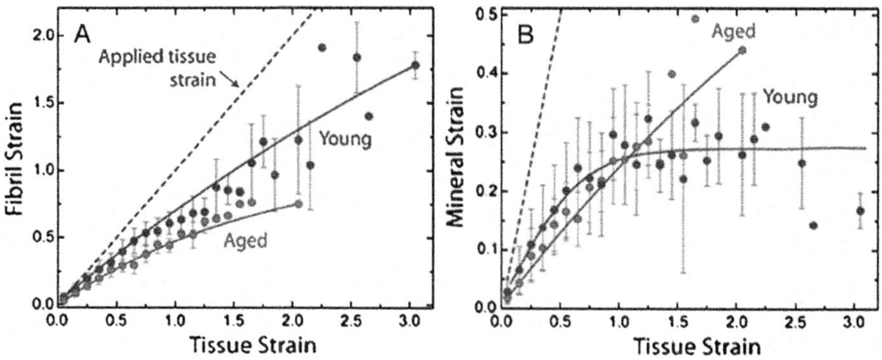


**Fig. 8.19** Young *v.* old bone. The results of experimental tensile testing of young and aged cortical bone samples in the longitudinal orientation are separated into the contributions of the fibril (A) and mineral (B). At a fixed tissue strain, the individual strain in the fibrils is approximately 25 % smaller in aged bone than young bone, while changes in the mineral strain are not statistically significant. The process of aging changes the deformation behavior of the mineralized collagen fibrils in part due to increased inter- and intrafibrillar cross-linking. The resulting elevated stiffness of the collagen negatively affects the bone's ability to plastically deform by fibrillar sliding (the protein molecules are tightly bound), which then must be accommodated at higher structural levels, by increased microcracking, for example. From Zimmermann et al. [114], used with permission, copyright © 2011

confirmed that the presence of a cross-link results in greater strength during deformation as complete intermolecular slip is prevented, and thereby particularly affects larger deformation levels [127].

A more recent experimental study found that age-related structural changes can significantly degrade the fracture resistance over multiple length scales [114]. Results from the small-angle X-ray scattering (SAXS) and wide-angle X-ray diffraction (WAXD) experiments are shown determined that, for a given tissue strain, the strain in the mineralized collagen fibrils is more than 25 % lower in aged bone than in young bone, implying that the fibrils have become stiffer with age (see Fig. 8.19). For the stiffness of the fibrils to change with age, structural changes must be occurring at the molecular level. Quantified nonenzymatic collagen cross-linking confirmed a higher level of cross-links in aged bone than in young bone [114]. The observations clearly indicate that these increased levels of nonenzymatic cross-links with age stiffen the collagen fibrils—they act to degrade the structural integrity of the fibrils by stiffening them to restrict fibrillar sliding (plasticity) and consequently degrade the ductility, strength, and intrinsic toughness of the bone. These recent experimental results are consistent with previous computational models of fibril deformation [123, 127, 128], which predict an inhibition of fibrillar sliding with cross-linking, and with experiments [120] showing a reduced toughness in bone glycated *in vitro*.

Taken as a whole, the evidence at submicrometer length scales strongly suggests that aging results in changes to the collagen environment, which specifically act to constrain the fibril's capacity to deform. The loss in plasticity at these nanometer length scales directly degrades the intrinsic toughness, thereby contributing to the increased fracture risk. As illustrated by both experimental and computational means, increased nonenzymatic collagen cross-linking suppresses plasticity at nanoscale dimensions. The link between increased cross-links and decreased bone toughness is, however, not immediately apparent. Simply put, the increased stiffness of the cross-linked collagen requires energy to be absorbed by "plastic" deformation at higher structural levels, which occurs by the process of microcracking. With progressive aging, bone is remodeled, resulting in an increased density of bone "cement lines" increases with age [92]. As these cement lines provide the prime sites for microcrack formation, the increased density gives rise to a higher microcrack density. This results in smaller ligament bridges spanning the crack, which in turn significantly degrades the crack-growth toughness. Such age-induced degradation in the potency of crack bridging in bone can has been directly measured, and is a major reason for the severe reduction in fracture resistance of bone with aging [90, 92].

The specific structure of intermolecular cross-links thereby plays a crucial role in defining the particular deformation paths. Aged collagen tends to show a high cross-link density, whereas young collagen features few cross-links. The larger the cross-link density, the lower the material's ability to dissipate energy without failure; at large cross-link densities collagen fibrils tend to involve molecular fracture and breaking of cross-links [123, 129] leading to increasingly brittle material behavior. This is certainly one factor associated with the increased fracture risk in older bone, but the marked deterioration in bone toughness involved the coupling with larger scale mechanisms—in this case crack bridging. What other toughening mechanisms exist within the material hierarchies of bone?

### *8.3.3 Multiple Mechanisms for Toughness*

Cross-linking of tropocollagen molecules provides one metric for the mechanical performance of bone, but it is not the sole factor. The mechanical properties of the constituents of bone largely control its strength and plasticity. To understand the toughness of bone and discern the roles of the observed toughening mechanisms, it is pertinent to note that the fracture resistance is a multiple-scale process with each level of structural hierarchy adapted to provide optimal toughness. Recent experimental and theoretical studies have provided new insight into the mechanisms underlying this behavior, most notably through the use of advanced instrumentation that permits the examination of bone properties at ever decreasing length scales, *e.g.*, transmission electron and X-ray microscopy, atomic force microscopy and Raman spectroscopy, as well as "bottom-up" multi-scale simulation modeling [96]. Bone derives its resistance to fracture (in terms of both the initiation and subsequent propagation of cracks) with a multitude of deformation and toughening mechanisms at many size-scales, ranging from the nanoscale structure of its protein molecules to its macroscopic physiological scale, summarized in Fig. 8.20 and Table 8.2.

The staggered arrangement of molecules into fibrils provides the basis for its ability to dissipate mechanical energy through molecular sliding rather than leading

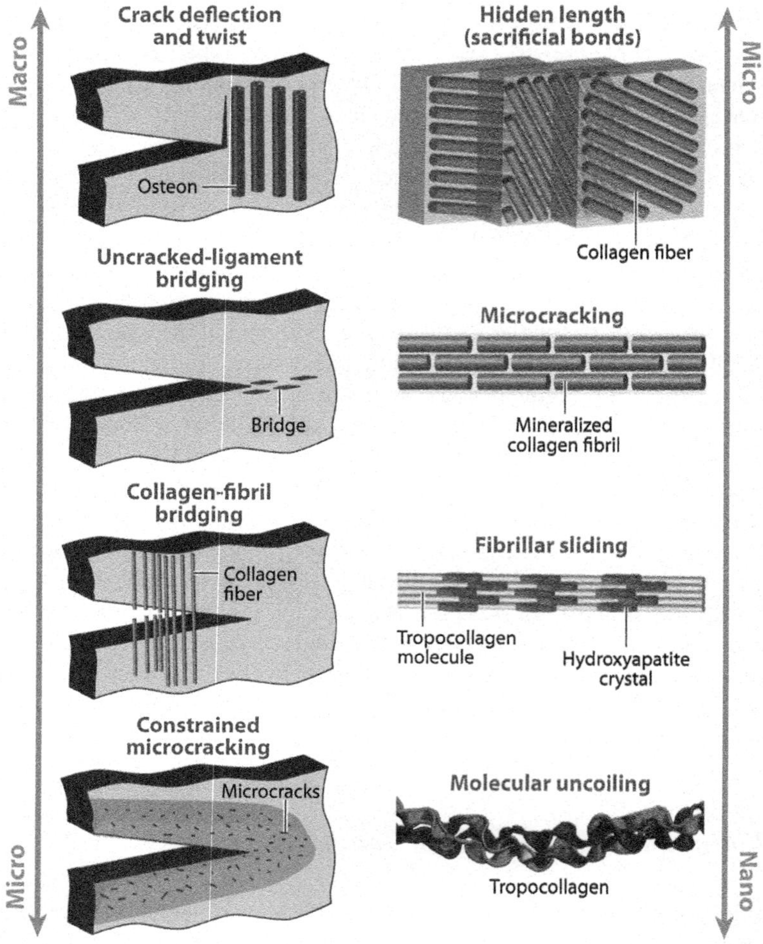


**Fig. 8.20** Multi-scale toughening mechanisms in bone. The toughness of bone results from a mutual competition between *extrinsic* or crack-tip shielding toughening mechanisms (*left*), which predominate at length scales at more than 1 μm, and *intrinsic* or plastic deformation toughening mechanisms (*right*), which are active at length scales at primarily less than 1 μm. Distinct toughening mechanisms occur at each level of hierarchy. Molecular uncoiling and intermolecular sliding of molecules are observed at the smallest level of tropocollagen molecules and mineralized collagen fibrils. Microcracking and fibrillar sliding are observed at the level of fibril arrays. At larger levels, the breaking of sacrificial bonds contributes to increasing the energy dissipation capacity of bone at the interface of fibril arrays, together with crack bridging by collagen fibrils. At the largest length scales in the 10 to 100 μm range, the primary sources of toughening are extrinsic and result from extensive crack deflection and crack bridging by uncracked ligaments, both mechanisms that are motivated by the occurrence of microcracking. From Launey *et al.* [96]

to catastrophic failure [125, 129–131]. In collagen fibrils, molecular stretching competes with intermolecular sliding and breaking of weak and strong bonds between tropocollagen molecules. The staggered architecture plays a key role in enhanc-

**Table 8.2** Summary of multiple mechanisms for toughness (energy dissipation) in bone under loading [96]

| Mechanism | Hierarchical level | Description |
|---|---|---|
| H-bond rupture | Protein building blocks; amino acid sequence | Rupture of inter- and intramolecular H-bonds of peptide strands composing tropocollagen molecules; reversible. |
| Tropocollagen unfolding | Molecular building blocks; multiple peptide strands | Entropic molecular stretching competes with intermolecular sliding and breaking of weak and strong bonds between tropocollagen molecules; enhanced by staggered architecture of tropocollagen; reversible. |
| Fibrillar sliding | Macromolecular; mineralized collagen fibrils; tropocollagen and hydroxyapatite | Continuous glide between tropocollagen molecules and mineral particles, initiated by slip at the hydroxyapatite-tropocollagen interface, enhanced by mineral presence. |
| Fiber sliding | Microscale; multiple collagen fibrils compose fibers embedded in extracellular matrix | Mineralized collagen fibrils are themselves twisted into larger collagen fibers, "glued" together by a thin layer of extra-fibrillar matrix, providing additional interfaces, friction, and hidden length for dissipation. |
| Crack deflection | Bone; transverse direction | Microcracks perpendicular to crack path blunt crack tip, deflect propagation, and result in tortuous fracture. |
| Crack bridging | Bone; longitudinal direction | Microcracks coaxial to crack path result in fracture coalescence, leading to the formation of uncracked regions along the crack length which 'bridge' the crack and carry load. |

ing the toughness of various collagen materials, exploited in both tendon (without mineral) or bone (biocomposite). Severe mechanical tensile loading of collagen is significant under physiological conditions in various related tissues, representing one of its key performance features, as shown by its ability to stretch up to 50 % tensile strain before breaking while reaching force levels of more than 10 nN (per molecule), or 10–20 GPa stress (obtained by normalization of the force by the cross-sectional area of the molecule) [125, 132, 133].

Whereas it has been shown that the breaking of H-bonds is a crucial mechanism of large and permanent deformation of protein molecules, fibrils and fibers, this mechanism is actually reversible when applied forces are removed [133]. The force required to break these bonds is highly dependent on the deformation rate (*i.e.*, time-scale), which provides a mechanism to dissipate energy under rapid loading. Indeed, such breaking of H-bonds can be regarded as a *viscoelastic* effect. This suggests that plastic deformation in bone must also involve larger-scale mechanisms such as intermolecular sliding, which would not be reversible unless remodeling of bone occurs.

At the next hierarchical level, sliding of mineralized collagen fibrils provides additional energy dissipation. The mineralized collagen fibrils are of particular significance for bone as they are the nanostructural building blocks. As the mineral phase has over an order of magnitude higher elastic modulus than the collagen, the presence of the hydroxyapatite phase is critical to the stiffness of bone [93]. Molecular modeling and experimental X-ray analysis suggest that continuous glide between tropocollagen molecules and between hydroxyapatite particles and tropocollagen molecules, initiated by slip at the hydroxyapatite-tropocollagen interface, enables a large regime of dissipative deformation once yielding begins, thus effectively increasing the resistance to fracture [129]. The larger stresses in the fibrils, preserved due to mineralization, can be maintained after slip initiation due to additional resistance to slip at the interface between the tropocollagen molecules and hydroxyapatite particles; this leads to a several fold increase in dissipation, exceeding the energy compared to fibrils without mineral phase. Thus, a key component of fibrillar sliding is the combined interaction of protein and mineral.

Sliding of multiple fibers provides additional toughness. The long (>5 to 10 μm) and thin (~100 nm) mineralized collagen fibrils are themselves twisted into larger collagen fibers, "glued" together by a thin layer (1 to 2 nm thick) of extra-fibrillar matrix [134, 135]. When the tissue is externally loaded in tension, the load is resolved into tensile deformation of the mineralized fibrils and shearing deformation in the extra-fibrillar matrix [130]. While no precise data on its mechanical behavior or its composition are available, it is likely that the extra-fibrillar matrix is comprised of non-collageneous proteins, such as osteopontin, and proteoglycans, such as decorin [96]. Indeed, single-molecule spectroscopy of fractured bone surfaces has confirmed that the extra-fibrillar matrix has properties similar to a glue-layer between the fibrils—specifically, it is relatively weak but ductile and deforms by the successive breaking of a series of "sacrificial bonds" [135, 136]. These results point towards a deformation mechanism where the matrix/fiber interface is disrupted beyond the yield point, and the matrix moves past the fibers forming and reforming the matrix/fiber bonds. An alternative explanation would be the disruption of bonds between the matrix and hydroxyapatite particles and a modification of the frictional stress between individual fibers [137].

The sub-micrometer deformation mechanisms described above contribute intrinsically to the fracture toughness of bone by forming "plastic zones" around crack-like defects, thereby protecting the integrity of the entire structure by allowing for localized failure through energy dissipation. However, at micro- to macroscale dimensions, the toughness of cortical bone is associated with a very different kind of toughening associated with crack-tip *shielding*. At several length-scales from the sub-micrometer to the tens of micrometers, the process of *microcracking* in bone provides the prevalent mechanism of microscale deformation [138–141]. Akin to many biological materials, structural features—such as those introduced by microcracks—provide a source of toughening which arises during crack growth, rather than crack initiation. To better understand this distinction, it is important to consider that many of the macroscopic fracture toughness properties of bone can be understood mechanistically in term of the behavior of the crack path. Certain features in the microstructure, such as interfaces, provide microstructurally 'weak' or preferred paths for cracking. As these features have a specific alignment in bone, this provides the basis for the marked anisotropy of the fracture properties of bone (*e.g.*, bone is easier to split than to break) [90, 142–144] and for the fact that the toughness is actually lower in shear than in tension [144–146]. In cortical bone, the path of least microstructural resistance is invariably along the cement lines [144–146], which are the hyper-mineralized interfaces between the sections of continuous bone matrix. These regions are therefore preferential sites for major microcracks to form, particularly as bone ages. These microcracks have a typical spacing in the tens to hundreds of micrometers and are primarily aligned along the long axis of the bone, an orientation which directly results in the strong anisotropy of toughness in bone.

The occurrence of microcracking is the source of extrinsic toughening in bone that is primarily associated with such mechanisms as crack deflection/twist and crack bridging [90, 142–144]. In fracture, crack trajectories result from the competition between the direction of maximum mechanical driving force (defined by the maximum strain energy release rate) and the path of "weakest" microstructure resistance [146, 147]. Where these two paths conflict, generally high toughness results. In contrast to the longitudinal orientations where these preferred mechanical and microstructural crack paths are nominally in the same direction, the opposite is true in the transverse (breaking) orientation, with the maximum driving force oriented parallel and directly ahead of the crack tip (which promotes coplanar cracking), and the "weakest" paths oriented perpendicular to the crack tip along the direction of the major microcracks. In human cortical bone, as the cement lines are oriented nominally along the longitudinal axis of the bone, preferred cracking paths will tend to be along this direction—this can lead to the significant (macroscopic) deflection of cracks which are attempting to propagate in the transverse direction. It is this phenomenon that makes the transverse orientation so much tougher and why bone is much more difficult to break than to split [144]. Recent fracture-mechanics measurements show that after only 500 μm of cracking, the fracture toughness, specifically the driving force for crack propagation, is more than five times higher in the transverse (breaking) direction than in the longitudinal (splitting) direction [144].

The major distinction in the toughness between these two orientations in bone can be directly attributed to a difference in crack paths, which in turn results in different toughening mechanisms, primarily involving crack deflection/twist in the transverse direction and crack bridging in the longitudinal directions. In the transverse direction, cement-line microcracks are roughly aligned perpendicular to the crack path, where they act as "delamination barriers" and blunt any growing cracks and generate highly tortuous crack paths, extremely rough fracture surfaces, and correspondingly high toughness. The latter follows because of the reduced local stress field due to crack blunting and the need to reinitiate the crack following local arrest at such delaminations [144, 145]. The crack path and consequent salient mechanisms of toughening in the longitudinal orientations are quite different. In human bone, the cement line microcracks in the longitudinal direction are coaxial to the growing crack, and thus form ahead of, and parallel to, the main crack tip. The nature of the coalescence of such microcracks to the growing crack leads to the formation of uncracked regions along the crack length which act to 'bridge' the crack and carry load that would otherwise be used to further crack propagation [115, 144, 145]. This mechanism of crack bridging, termed "uncracked-ligament bridging" also results in toughening but is a far less potent mechanism than that due to crack deflection/twist [90, 144].

It is apparent that each mechanism contributes to the toughness of bone, but yet common "tricks" are seen across scales. The "hidden length" due to unfolding of tropocollagen molecules is mimicked by fibrillar sliding and the unfolding of entire fibrils. Energy is dissipated at the tropocollagen/hydroxyapatite interface within fibrils, as well as the extracellular matrix across fibers. Even crack deflection and crack bridging—seemingly larger-scale mechanisms—has a molecular analogue in the nanoconfined hydroxyapatite mineral crystals. A possible concept to explain the observed length-scales at each hierarchical level is that they are a result of structural adaptation towards maximizing target material properties (*e.g.*, strength and dissipation) through geometric size effects (enhanced by the nanoconfinement of mineral phases).

In addition, dissipation mechanisms at all length scales have an added effect of spreading damage (or failure) across scales—deformation is not limited to molecules, fibrils, fibers, or bone, but shared synergistically amongst each hierarchy. To what benefit is this "multiscale dissipation"? One hypothesis is that it facilitates the *healing* capacity of bone. A remarkable property of bone is its well-known capacity of self-repair [148]. In bone, the combination of growth and remodeling (resorption and replacement of old material) occurs *via* specialized cells (osteoclasts) that are permanently removing material, while other cells (osteoblasts) are depositing new tissue. This continuous remodeling allows structural adaptation to changing external conditions, as well as the removal and replacement of damaged material [93, 149]. Damage shared across multiple scales may aid the healing process by minimizing the effort to repair any single "damage zone". Perhaps Nature has developed a way to subcontract the rehabilitation in a parallel (rather than serial) process, enabling function while remodeling occurs. Of course, this is only one potential hypothesis.

Even though individual mechanisms have been identified, there is still no integrated model of plastic deformation in bone. The overall hierarchical arrangement is important because it enables, through structures formed at characteristic geometric length-scales, superior molecular properties to be visible at larger biologically important intermediate mesoscales. By manipulating combinations of minerals and organic polymers into hierarchical structures spanning multiple length-scales, Nature has developed a wide range of bone-like hybrid materials (*e.g.*, shells, scales) with specific properties matched to function [91, 150, 151]. Indeed, these biological systems represent an inexhaustible source of inspiration to materials scientists in offering potential solutions for the development of new generations of structural materials [85, 152, 153].

### 8.3.4 General Design Principles

What lesson can we learn from bone, beyond the combination of collagen and hydroxyapatite? Like the simple spring model for silk, one possible route is to deconstruct the bone into simple(r) mechanical "building blocks" and learn from their contributions to overall behavior and performance (again, using abstraction to disconnect the functional system from the biological materials, a paradigm discussed in Chap. 4: Universality-Diversity Paradigm: Music, Materiomics, and Category Theory).

Jaeger and Fratzl [154] discussed the mineral platelet arrangement in collagen fibril and developed a simple mechanical model to estimate the stiffness of biocomposites, whereby the stiff mineral transfers tensile load and the soft, protein matrix transfers load between mineral crystals *via* shear (schematically represented by Fig. 8.21). The path of load transfer in the composite is thus simplified to a one-dimensional serial spring system consisting of mineral elements (tension) interspersed among protein elements (shear). A large aspect ratio of mineral platelets compensates for the low modulus of the protein phase. According to this simple model, the stiffness (Young's modulus) $E$ of the composite can be expressed as:

$$\frac{1}{E} = \frac{4(1-\Phi)}{G_p \Phi^2 \rho} + \frac{1}{\Phi E_m} \tag{8.6}$$

where $E_m$ is the Young's modulus of mineral, $G_p$ is the shear modulus of protein, $\Phi$ is the volume concentration of mineral, and $\rho$ is the aspect ratio of the mineral platelets [154]. The key insight of Eq. 8.6 is that the high stiffness of biocomposites is achieved by the large aspect ratio of mineral platelets. As the protein transfers stress between platelets *via* shear, it is more efficient to increase shear contact area rather than volume (*e.g.*, cross-sectional area) of the soft protein (see Fig. 8.21). This also indicates that the protein-mineral interface is critical (at least in terms of shear stress transfer), not necessarily the strength (or stiffness) of the protein phase. While we gain molecular insight by pulling tropocollagen molecules, for example, we must also scrutinize the collagen hydroxyapatite interface to fully understand a material such as bone. Indeed, the materiome is more than the sum of its building blocks.

Optimal "design" of such a composite (in terms of strength) would be achieved by ensuring simultaneous material failure—the mineral platelets must be able to sustain sufficient tensile stress whereas the protein layer and protein-mineral interface must sustain sufficient shear stress up to a concurrent failure state.[3] Assuming such optimization to have occurred naturally, the fracture toughness of the composite is a function of a single theoretical strength, $\sigma_{th}$ (*e.g.*, the mineral in tension, $\sigma_m$, or protein-mineral interface in shear, $\tau_p$). For robust design, one must consider the likelihood the composite contains crack-like flaws, and the theoretical strength

[3] This is similar to the principle of optimal design of engineered composite structures which states that all layers must fail simultaneously; no material strength is left unused.

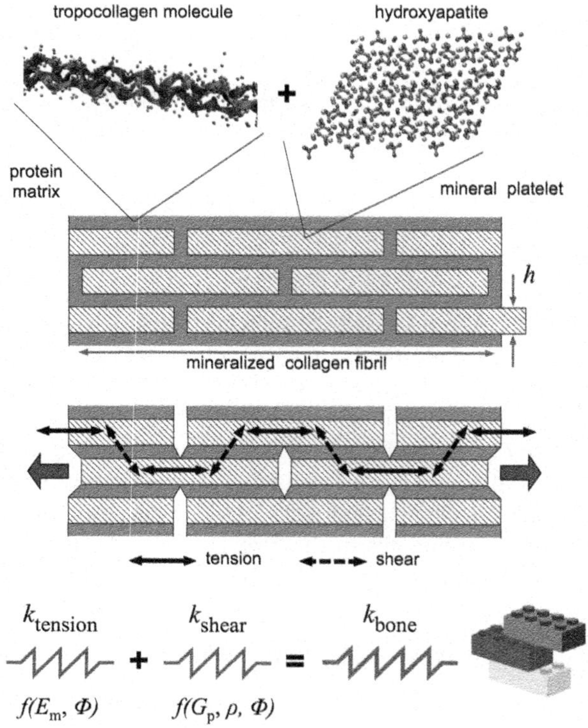


**Fig. 8.21** A simple mechanical model of a protein-mineral biocomposite (*e.g.*, bone). For bone, the molecular "building blocks" consist of tropocollagen molecules (soft) and hydroxyapatite crystals (stiff), which from a fibrillar "brick and mortar" structure, wherein staggered mineral crystals (of dimension $h$) are embedded in a protein matrix, forming a mineralized collagen fibril. Load transfer is achieved *via* tension in the mineral platelets, whereas the protein transfers load *via* the high shear zones between mineral layers (indicated by arrows). In general, the system can be represented as two serial springs, where one stiffness is a function of the mineral tensile stiffness ($E_m$) and the other is governed by the protein-mineral interfacial shear stiffness ($G_p$). Note that there is also dependence on the volume fraction ($\Phi$) and aspect ratio ($\rho$) of the mineral phase. Adapted from the works of Gao *et al.* [110] and Jaeger and Fratzl [154]

is never reached. Using an argument from linear elastic fracture mechanics,[4] Gao *et al.* investigated the interplay between the two phases of such bone-like composite materials [110]. The fracture strength of this "cracked" mineral platelet can be calculated from the Griffith criterion as:

$$\sigma_m^f = \alpha\sqrt{\frac{E_m\gamma}{h}} \tag{8.7}$$

[4]We recognize the absence of plasticity or other energy dissipative processes that could produce toughening in the vicinity of the crack front in this analysis.

where $\sigma_m^f$ is the fracture strength of the mineral phase, $E_m$ is again the Young's modulus of mineral, $\gamma$ the surface energy and $h$ is the *thickness* of mineral crystal. The parameter $\alpha$ is geometry dependent (*e.g.*, approximately equal to $\sqrt{\pi}$ for a half-cracked platelet, where the crack depth equals half of the platelet thickness). By setting $\sigma_{th} = \sigma_m^f$, it is apparent there exists a *critical length scale*:

$$h^* \approx \alpha^2 \frac{\gamma E_m}{\sigma_{th}^2} \tag{8.8}$$

below which the fracture strength of a *cracked* crystal is identical to that of a *perfect* crystal. Introducing numerical approximations for $E_m$ (100 GPa), $\gamma$ (1 J/m$^2$) and $\sigma_{th}$ (E/30) results in an estimate $h^*$ to be $\approx 30$ nm [110], the same magnitude of hydroxyapatite crystals found in bone. This length scale indicates that the nanometer size of mineral platelets in biocomposites may be the result of *fracture strength optimization*. When the mineral size exceeds this length scale, the fracture strength is sensitive to structural size and the material is sensitive to crack-like flaws and fails by stress concentration at crack tips. As the mineral size drops below this length scale, the strength of a perfect mineral platelet is maintained despite defects. The failure criterion is governed by theoretical strength, $\sigma_{th}$, rather than by the Griffith criterion, $\sigma_m^f$, and the material becomes insensitive to any pre-existing defects. In general, bone has become insensitive to flaws at the nanoscale [110]—the nanometer size of the mineral crystals in biocomposites is selected to ensure optimum fracture strength and maximum tolerance of flaws (for robustness). This finding is in drastic contrast to the classical engineering concept of stress concentration at macroscopic flaws.

The analysis can also be used to derive an optimal aspect ratio of the mineral platelets (assuming simultaneous failure, $\sigma_m = \rho\tau_p$):

$$\rho^* = \frac{\alpha}{\tau_p^f}\sqrt{\frac{E_m\gamma}{h}} \tag{8.9}$$

where $\rho^*$ is the optimal aspect ratio (note that this is independent of the critical length scale) and $\tau_p^f$ is the shear strength of protein matrix subject to fracture. It follows that a large mineral aspect ratio is required if the mineral strength is much higher than the shear strength of the protein or the protein mineral interface [110]. The mineral crystals in bone have thickness on the order of a few nanometers and aspect ratio 30 to 40 [110].

The composite material, through nanoconfinement of the mineral phase, expresses an intrinsic toughening mechanism which further enhances the resistance to fracture by forming larger local yield regions around crack-like defects, a mechanism that protects the integrity of the entire structure by allowing for localized failure (*e.g.*, single crystal fracture propagation). As a consequence, mineralized collagen fibrils are able to tolerate microcracks on the order of several hundred micrometers size without causing any macroscopic failure of the tissue. Specifically, the inclusion of the nanoscale mineral platelets on collagen fibrils not only increases their Young's modulus, but also their yield and fracture strengths.

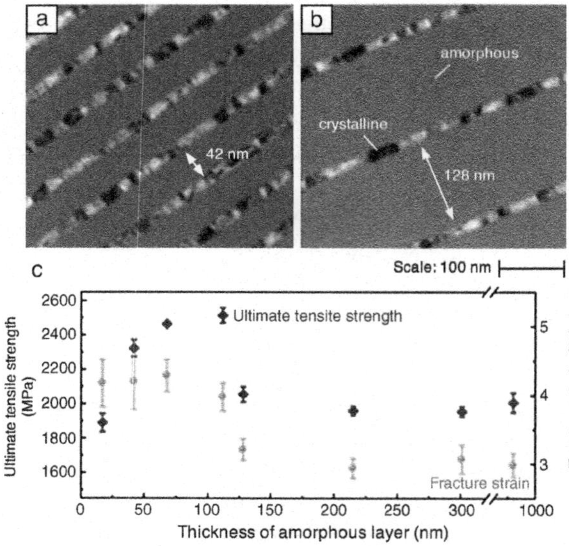


**Fig. 8.22** Enhancing mechanical properties *via* nanoconfinement: metallic glasses. (**a**) and (**b**) Nanolaminates with alternating layers of $Cu_{50}Zr_{50}$ amorphous metallic glass and nanocrystalline Cu, with amorphous layer thickness of 42 nm and 128 nm respectively. (**c**) Ultimate tensile strength and fracture strain of the nanolaminates as a function of the thickness of the amorphous layer. These plots demonstrate a clear change in mechanical behavior of the nanolaminates with metallic-glass thickness between 112 and 128 nm. Adapted from Greer *et al.* [164], used with permission, copyright © 2011 John Wiley & Sons, Inc.

We note that the emergence of a critical length scale to maximize strength is just one contribution within a more complex hierarchical system [155]. The energy dissipation mechanisms previously discussed span many scales, and improve the toughness and strength of bone in conjunction with the nanoscale structure (whether optimized or not), combined with other factors such as the stagger (or offset) between layers [156]. The fact that this principle was "learned from Nature" does not necessarily mean it was a driving factor for the adaptation and evolution of bone. Indeed, one may find counterexamples where nanoscale structural features are not a result of an innate resistance to very high stresses, and various strategies emerge to mitigate catastrophic failure completely independent of nanoscale structures [157–159]. However, such counterexamples do not preclude the use of this design principle for engineered materials, ranging from ceramic/polymer hybrid materials [160, 161] and metallic nanocomposites [162] and nanolaminate glasses [163, 164]. If the small size of the mineral particulates in bone is a key to strength, the same principle can be applied to the design of synthetic materials (see Fig. 8.22). Likewise, bone is not the only biological material to successfully combine "hard" and "soft" phases at the nanoscale—the shells of mollusks have developed similar means to enhance toughness.

### 8.3.5 Shells, Nacre, Biological Armor

Like bone, most hard biological materials incorporate minerals into soft matrices, mostly to achieve the stiffness required for structural support or armored protection. Mollusc shells are an excellent example of such high-performance natural materials [151]. Mollusks are composed of at least 95 % of minerals such as calcium carbonate (*e.g.*, calcite, aragonite), a brittle material, yet they exhibit a toughness approximately 3000 times higher than calcite, at the expense of a small reduction in stiffness. *How is such performance achieved? Can it be duplicated in artificial materials?*

Molluscs have a very soft body (*mollis* means soft in Latin) and most of them grow a hard shell for protection. The shell of molluscs is grown by the mantle, a soft tissue that covers the inside of the shell [165]. A great variety of shell structures has emerged from the evolution of molluscs. They include prismatic, foliated and cross lamellar structure, columnar and sheet nacre. All of these structures use either calcite or aragonite, with a small amount of organic material which never exceeds 5 % of the composition in weight. In order to provide an efficient protection, the shell must be both stiff and strong. Mechanical testing on various species of seashells revealed an elastic modulus ranging from 40 GPa to 70 GPa, and a strength in the range of 20 to 120 MPa [166]. By comparison, human femoral bone is softer ($E =$ 20 GPa) but stronger (150 to 200 MPa strength). Amongst all the structures found in shells, nacreous structures (known as mother-of-pearl) appear to be the strongest [167].

Like bone, *nacre* has a hierarchical structure, with specific structural features can be found at distinct length scales (see Fig. 8.23). At the millimeter scale the shell consists of a two-layer armor system, with a hard outer layer (large calcite crystals) and a softer but more ductile inner layer (nacre). Under external mechanical aggressions the hard calcite layer is difficult to penetrate, but is prone to brittle failure. Nacre, on the other hand, is relatively ductile and can maintain the integrity of the shell even if the outer layer is cracked, which is critical to protect the soft tissues of the animal. This design of hard ceramic used in conjunction with a softer backing plate is believed to be an optimal armor system [168]. Furthermore, within the nacreous layer itself there are a few sub-layers of weaker material, the so called 'growth lines' [169] which may act as crack deflectors [173], similar to the mechanism found in bone. The microscale architecture of nacre resembles a three dimensional "brick and mortar" wall, where the bricks are densely packed layers of microscopic aragonite polygonal tablets (about 5–8 μm in diameter for a thickness of about 0.5 μm) held together by 20 to 30 nm thick layers of organic materials. Remarkably, the arrangement and size of the tablets in nacre is highly uniform throughout the nacreous layer. Optical microscopy on a cleaved nacre surface reveals Voronoi-like contours, with no particular orientation within the plane of the layer [174, 175]. While the tablets are generally described and modeled as flat at the microscale [176, 177], they actually exhibit a significant waviness [175], and is highly conformal so that the tablets of adjacent layers fit perfectly together.

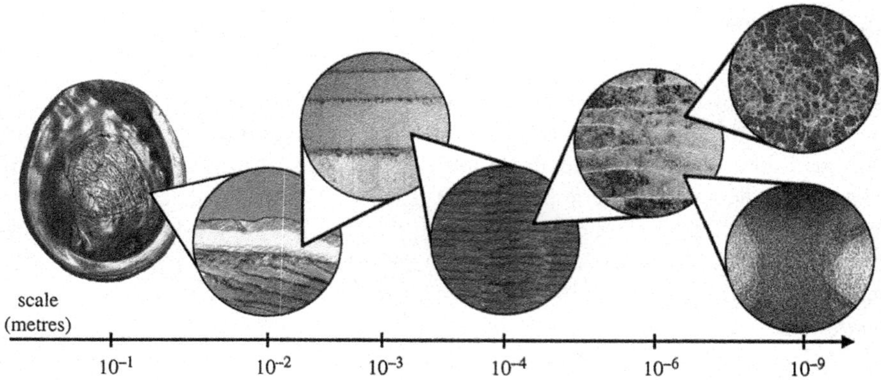


**Fig. 8.23** The hierarchical structure of nacre. The structure of a typical nacreous shell of red abalone, a marine gastropod, is depicted. At the largest scale, the adult shell is approximately 15 to 20 cm in diameter. It is composed of two layers: an outside layer made up of large crystals of calcite, which makes a hard, but brittle, material. The inner layer, nacre, is composed of 95 % volume of aragonite (one of the crystallographic forms of calcium carbonate) and 5 % volume of organic materials (proteins and polysaccharides). This two-layer arrangement is believed to be an ideal armour design [168]. Observations at higher magnification (millimeter scale) reveal a few lines that partition the nacreous layer. These so-called 'growth lines' mark pauses during the growth of the shell [169]. The bulk of nacre is made up of 0.5 mm thick layers, each layer being composed of a tiling of polygonal aragonite tablets approximately 5 to 8 mm in diameter. While they are often described as flat, the tablets actually show a rather convoluted surface, with a waviness that can reach up to half of the tablet thickness in amplitude. The interface between the tablets is a 30 nm thick complex system including several layers of organic materials, nanoasperities and direct mineral connections from one tablet to the next. The tablets themselves are composed of aragonite nanograins delimited by a fine three-dimensional network of organic material. The shell structure therefore contains six levels of hierarchy (growth line image from Menig *et al.* [170], nanograins from Rousseau *et al.* [171]). From Barthelat [172], used with permission, copyright © 2007

Nacre further exhibits structural features down to the nanoscale. While transmission electron microscopy suggest that the tablets are made of large aragonite grains with a few inclusions of nanograins [168, 178], recent scanning probe microscopy observations suggests, indirectly, that the tablet are nanostructured, with grains in the 30 nm range [171, 179]. These nanograins have all the same texture and they are delimited by a fine network of organic material [171]. At the 20 to 30 nm interfaces between the tablets, nanoscale features can also be found. The organic material that fills this space and bonds the tablets together is actually composed of several layers of various proteins [136]. These sheets of organic layers contain pores with a 20 to 100 nm spacing, leaving space for two types of aragonite structures: nanoasperities and direct aragonite connections across the interfaces (mineral "bridges" connecting tablets). These nanoscale features were observed using scanning probe microscopy [178, 180], scanning electron microscopy [167, 173] and transmission electron microscopy [173, 178, 181].

Good toughness is a necessary but not sufficient requirement for structural utility [167]. The more basic prerequisite is a deformation mechanism, and it is now widely recognized that tablet sliding is a key mechanism in the deformation of nacre [167, 175, 176, 182]. Because this mechanism is controlled by the interface between the tablets, many efforts have focused on investigating nanoscale mechanisms between the tablets [136, 167, 177, 181]. This remarkable behavior is achieved by the following microscopic mechanism: at a tensile stress of about 60 MPa the interfaces start to yield in shear and the tablets slide on one another, generating local deformation. This phenomenon spreads over large volumes throughout the specimen, which translates into relatively large strains at the macroscale. Once the potential sliding sites are exhausted, the specimen fails by pullout of the tablets, which occurs after local sliding distances of 100–200 nm. This type of micromechanism is unique to nacre, and it is the main source of its superior mechanical properties. For this reason numerous models were developed to capture this behavior [175, 176, 182].

In order to achieve such *structural* behavior, however, some *material* requirements must be met. First, the interface must be weaker than the tablets; otherwise, the tablets would fail in tension before any significant sliding could occur, which would lead to a brittle type of failure. Strong tablets are important in this regard, and it was shown that their small size confer them with increased tensile strength compared to bulk aragonite [110, 183]. It has also been suggested that the presence of nanograins provides some ductility to the tablets [184]. This would increase the tensile strength of the tablets, but would not significantly affect the deformation mode of nacre, which is dominated by tablet sliding. In addition, the aspect ratio of the tablets must be high enough to maximize sliding areas and produce strong cohesion within the material [185]. The platelet aspect ratio can be analyzed based on a classic composite shear lag load transfer model [182], where the critical length maximizes the load transferred to the platelet while averting platelet fracture and instead favors the observed failure mechanism of platelet pullout. Too thin tablets would lead to premature tablet failure and brittle behavior. Another fundamental requirement is that some hardening mechanism must take place at the tablet level in order to spread sliding throughout the material. As tablets start to slide, higher stresses are required to slide them further so that it is more favorable for the material to initiate new sliding sites, thus spreading deformation over large volumes. Since the tablets remain essentially elastic in this process, the hardening mechanism has to take place at the interfaces. Shear tests along the layers reveal a very strong hardening and failure at shear strains in excess of 15 % [175]. The full strain field, measured by image correlation technique, also captures a significant expansion across the layers [175]. This important observation suggested that the tablets have to climb obstacles in order to slide on one another. Either in tension or shear, strain hardening is the key to large deformation and is essential to the mechanical performance of nacre.

More recently, it was shown that these mechanisms, while necessary, are not sufficient to explain the behavior of nacre at the macroscale [175]. Another key mechanism is actually found at the microscale, where the waviness of tablets generates progressive locking, hardening and spreading of non-linear deformation around cracks and defects (see Fig. 8.24). The associated viscoplastic energy dissipation at the interfaces between tablets greatly enhances the toughness of nacre, arresting cracks before they become a serious threat to the shell and to the life of the animal.

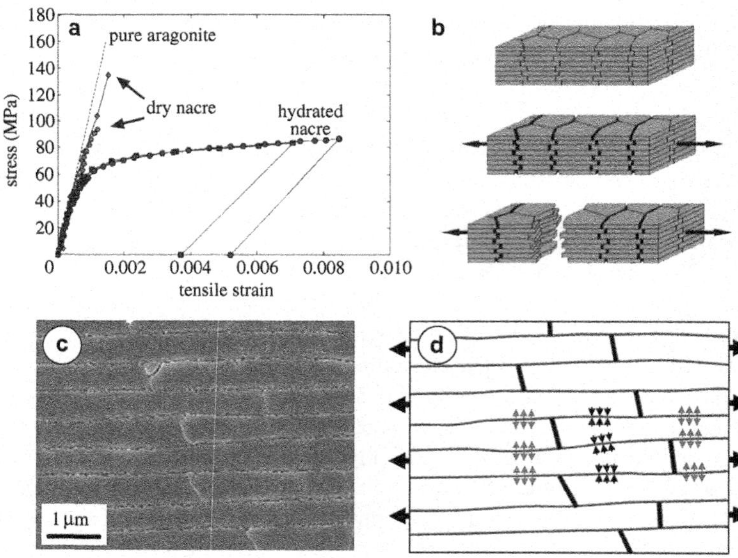


**Fig. 8.24** The deformation of nacre. (**a**) Stress–strain curve in tension along the tablets for pure aragonite, dry nacre and hydrated nacre. (**b**) Collective tablet sliding generates the relatively large deformations up to failure. (**c**) SEM image showing a dove-tail tablet ends. (**d**) Mechanism associated with the waviness: under tension, compressive stress builds up (*black arrows*), balanced by a tensile stress (*grey arrows*) outside of the sliding area. The result is progressive locking and local hardening. From Barthelat [172], used with permission, copyright © 2007

Nacre is a "perfect" example of a natural material, which developed a highly sophisticated microstructure for optimal performance, over millions of years of evolution. The attractive mechanical properties of nacre, bone and other materials have inspired the development of a large class of biomimetic materials and organic/inorganic composites [85, 151]. Nacre and other shell-like materials can exhibit remarkable combinations of stiffness, low weight, strength and toughness—the exact properties required for successful armor systems.

While nacre itself is a remarkable material, typically, such shell-like materials are integrated into a more complex system (*e.g.*, combining gradations of shell, soft tissue, and biological systems [186]) wherein the specific combination of different materials, microstructures, interfacial geometries, and layering are synergistically advantageous for penetration resistance, energy dissipation, mitigation of fracture and crack arrest, reduction of deflections, resistance to bending and tensile loads, and enable self-repair (healing). This is similar to how our own bones are integrated within our bodies, and function cooperatively with our muscles, tendons, and vasculature, for example. Again—to exploit the materiome—we must fully understand and take advantage of the inherent specificity of function associated with natural mechanical design principles [85].

## 8.4 Summary

In contrast to synthetic materials, evolutionary developments in biology have resulted in materials with remarkable structural properties, made out of relatively weak constituents, arranged in complex hierarchical patterns. We have outlined a series of investigations complementing and describing multi-scale phenomena from a material hierarchy perspective. Through analyzing bridging hierarchies within complex materials, material scientists will be able to narrow the knowledge gap existing in the description of materials on computational and experimental sides. We envision that such a scale-bridging and reciprocal *nano-to-macro-to-nano* approach to studying hierarchical materials, including materials other than silk, bone and nacre, will facilitate a better understanding at all levels of material composition and provide tools for effectively manipulating these properties to achieve superior material qualities for application in a variety of fields, from structural engineering to medicine. As the understanding gap narrows, we seek direct methods to be able to implement manipulation of material upon all length scales.

We see that silk, in its natural function as a structural element of a web, is not a "super-material" due to its strength, but rather due to its distinct nonlinear softening and stiffening behavior. Notably, while studies of other biomaterials (*e.g.*, of bone or nacre) have shown that their great mechanical robustness is due to the formation of rather large plastic regions (which facilitate the redistribution of mechanical energy over vast material volumes) the opposite is true at the scale of spider webs. Here, extreme localization of failure is the key to explain its overall mechanical performance. This result suggests a change in the paradigm of material design as a function of scale, where traditional perspectives of superior mechanical properties can be drastically enhanced by integrating the design of a material's nonlinear material behavior that is defined by a particular molecular makeup. Specifically, the material behavior itself can be more significant to a system than specific quantities such as ultimate strength upon failure. It is counterintuitive that the inherent softening of silk—often viewed as a material weakness—vastly enhances the structural robustness of a web. The unique strength of a cyclic *nano-to-macro-to-nano* analytic approach is, in addition to a more complete and accurate system description, the potential to link higher order scales to fundamental components at lowest hierarchical levels. The material behavior of silk is played out on two scales: the fibril and the web. At the fibril scale, nanoconfinement that stems from size-dependent homogeneity in load distribution dictates the fibril size. We make the distinction between nanoconfinement at this scale, which depends on the synergism of protein constituents, to the nanoconfinement of $\beta$-sheet crystals, a result of cooperative hydrogen bonding. At the web scale, we trace the ability of the web to effectively resist load at the observable macroscale behavior to molecular mechanisms.

Nanocomposites in Nature—integrating stiff phases with relatively soft protein phases—exhibit a generic mechanical structure in which the nanometer size of mineral particles is selected to ensure optimum strength and maximum tolerance of flaws (*e.g.*, robustness). This flaw tolerance explains a case of convergent evolution: the similarly small sizes of mineral particles in a wide range of biocomposites, from bone to nacre. These materials have in general relatively complex structures organized over several length scales (hierarchical structures) leading to mechanisms that operate over several length scales, down to the nanoscale.

As bone is such a complex biological material, we still possess only a limited quantitative understanding of how it deforms and fractures, together with the salient plasticity and toughening mechanisms active at the various structural length- and time-scales. The advent of multi-scale observations, high-resolution structural imaging using transmission electron microscopy and computed X-ray micro-tomography, and vastly more powerful theoretical/computational tools, however, now enables the analysis of bone as a "material" with rigorous physical science techniques. These types of studies that seek to uncover, model and quantify the fundamental nano to macroscale mechanisms that govern the properties of bone are quite distinct from those that have been traditionally followed in the bone and medical communities.

The concept of multi-scale hierarchical structures, where the microstructure at each level is adapted to local needs, allows the adaptation and optimization of the material form and structure at each level of hierarchy to meet specific functions, as demonstrated by both silk and bone. For synergistic performance, the behavior of each scale must seamlessly be integrated within a system. If the ultimate goal is the manipulation and design of such systems, holistic knowledge of the entire material system from nano to macro—the *materiome*—is necessary. The complexity and symbiosis of structural biological materials has generated enormous interest of late, primarily because these composite biological systems exhibit mechanical properties that are invariably far superior to those of their individual constituents. The advantage of "unlocking Nature" is not unveiled in copycat systems, but rather in the potential to create our own high-performance systems, be they intended for medicine, nanotechnology, energy, environmental or other "unnatural" applications.

## References

1. J.M. Gosline, M.E. Demont, M.W. Denny, The structure and properties of spider silk. Endeavour **10**(1), 37–43 (1986)
2. J.M. Gosline, P.A. Guerette, C.S. Ortlepp, K.N. Savage, The mechanical design of spider silks: from fibroin sequence to mechanical function. J. Exp. Biol. **202**(23), 3295–3303 (1999)
3. F. Vollrath, D.P. Knight, Liquid crystalline spinning of spider silk. Nature **410**(6828), 541–548 (2001)
4. N. Becker, E. Oroudjev, S. Mutz, J.P. Cleveland, P.K. Hansma, C.Y. Hayashi, D.E. Makarov, H.G. Hansma, Molecular nanosprings in spider capture-silk threads. Nat. Mater. **2**(4), 278–283 (2003)
5. F.K. Ko, J. Jovicic, Modeling of mechanical properties and structural design of spider web. Biomacromolecules **5**(3), 780–785 (2004)
6. Z.Z. Shao, F. Vollrath, Materials: surprising strength of silkworm silk. Nature **418**(6899), 741 (2002)

7. N. Du, X.Y. Liu, J. Narayanan, L.A. Li, M.L.M. Lim, D.Q. Li, Design of superior spider silk: from nanostructure to mechanical properties. Biophys. J. **91**(12), 4528–4535 (2006)
8. I. Agnarsson, M. Kuntner, T.A. Blackledge, Bioprospecting finds the toughest biological material: extraordinary silk from a giant riverine orb spider. PLoS ONE **5**(9) (2010)
9. F.G. Omenetto, D.L. Kaplan, New opportunities for an ancient material. Science **329**(5991), 528–531 (2010)
10. F. Vollrath, D. Porter, Spider silk as a model biomaterial. Appl. Phys. A, Mater. Sci. Process. **82**(2), 205–212 (2006)
11. C. Vepari, D.L. Kaplan, Silk as a biomaterial. Prog. Polym. Sci. **32**(8–9), 991–1007 (2007)
12. C.M. Mello, S. Arcidiacono, R. Beckwitt, J. Prince, K. Senecal, D.L. Kaplan, Nephila-clavipes dragline silk—approaches to a recombinantly produced silk protein. Biomol. Mater. Des. **330**, 37–42 (1994)
13. J. Perez-Rigueiro, C. Viney, J. Llorca, M. Elices, Silkworm silk as an engineering material. J. Appl. Polym. Sci. **70**(12), 2439–2447 (1998)
14. I. Agnarsson, A. Dhinojwala, V. Sahni, T.A. Blackledge, Spider silk as a novel high performance biomimetic muscle driven by humidity. J. Exp. Biol. **212**(13), 1989–1993 (2009)
15. B.O. Swanson, S.P. Anderson, C. DiGiovine, R.N. Ross, J.P. Dorsey, The evolution of complex biomaterial performance: the case of spider silk. Integr. Comp. Biol. **49**(1), 21–31 (2009)
16. F. Vollrath, D. Porter, Silks as ancient models for modern polymers. Polymer **50**(24), 5623–5632 (2009)
17. F. Vollrath, D. Porter, C. Holland, There are many more lessons still to be learned from spider silks. Soft Matter **7**(20), 9595–9600 (2011)
18. S. Cranford, A. Tarakanova, N. Pugno, M.J. Buehler, Nonlinear material behaviour of spider silk yields robust webs. Nature (2012)
19. J.P. O'Brien, S.R. Fahnestock, Y. Termonia, K.C.H. Gardner, Nylons from nature: synthetic analogs to spider silk. Adv. Mat. **10**(15), 1185 (1998)
20. A. Lazaris, S. Arcidiacono, Y. Huang, J.F. Zhou, F. Duguay, N. Chretien, E.A. Welsh, J.W. Soares, C.N. Karatzas, Spider silk fibers spun from soluble recombinant silk produced in mammalian cells. Science **295**(5554), 472–476 (2002)
21. C.Y. Hayashi, N.H. Shipley, R.V. Lewis, Hypotheses that correlate the sequence, structure, and mechanical properties of spider silk proteins. Int. J. Biol. Macromol. **24**(2-3), 271–275 (1999)
22. J.D. van Beek, S. Hess, F. Vollrath, B.H. Meier, The molecular structure of spider dragline silk: folding and orientation of the protein backbone. Proc. Natl. Acad. Sci. USA **99**(16), 10266–10271 (2002)
23. T. Lefevre, M.E. Rousseau, M. Pezolet, Protein secondary structure and orientation in silk as revealed by Raman spectromicroscopy. Biophys. J. **92**(8), 2885–2895 (2007)
24. S. Keten, M.J. Buehler, Nanostructure and molecular mechanics of spider dragline silk protein assemblies. J. R. Soc. Interface **7**(53), 1709–1721 (2010)
25. M.E. Rousseau, T. Lefevre, L. Beaulieu, T. Asakura, M. Pezolet, Study of protein conformation and orientation in silkworm and spider silk fibers using Raman microspectroscopy. Biomacromolecules **5**(6), 2247–2257 (2004)
26. M.J. Buehler, Y.C. Yung, Deformation and failure of protein materials in physiologically extreme conditions and disease. Nat. Mater. **8**(3), 175 (2009)
27. S. Keten, Z.P. Xu, B. Ihle, M.J. Buehler, Nanoconfinement controls stiffness, strength and mechanical toughness of beta-sheet crystals in silk. Nat. Mater. **9**(4), 359–367 (2010)
28. M. Sotomayor, K. Schulten, Single-molecule experiments in vitro and in silico. Science **316**(5828), 1144–1148 (2007)
29. D. Porter, F. Vollrath, The role of kinetics of water and amide bonding in protein stability. Soft Matter **4**(2), 328–336 (2008)
30. C.Y. Hayashi, R.V. Lewis, Evidence from flagelliform silk cDNA for the structural basis of elasticity and modular nature of spider silks. J. Mol. Biol. **275**(5), 773–784 (1998)

31. A.E. Brooks, H.B. Steinkraus, S.R. Nelson, R.V. Lewis, An investigation of the divergence of major ampullate silk fibers from nephila clavipes and argiope aurantia. Biomacromolecules **6**(6), 3095–3099 (2005)
32. G.P. Holland, M.S. Creager, J.E. Jenkins, R.V. Lewis, J.L. Yarger, Determining secondary structure in spider dragline silk by carbon-carbon correlation solid-state nmr spectroscopy. J. Am. Chem. Soc. **130**(30), 9871–9877 (2008)
33. B.L. Thiel, K.B. Guess, C. Viney, Non-periodic lattice crystals in the hierarchical microstructure of spider (major ampullate) silk. Biopolymers **41**(7), 703–719 (1997)
34. G. Bratzel, M.J. Buehler, Sequence-structure correlations and size effects in silk nanostructure: poly-ala repeat of n. clavipes masp1 is naturally optimized at a critical length scale. J. Mech. Behav. Biomed. Mat. **7**(3), 30–40 (2012)
35. G. Bratzel, M.J. Buehler, Molecular mechanics of silk nanostructures under varied mechanical loading. Biopolymers **97**(6), 408–417 (2012)
36. Y. Sugita, Y. Okamoto, Replica-exchange molecular dynamics method for protein folding. Chem. Phys. Lett. **314**(1–2), 141–151 (1999)
37. P. Bradley, K.M.S. Misura, D. Baker, Toward high-resolution de novo structure prediction for small proteins. Science **309**(5742), 1868–1871 (2005)
38. Y. Zhang, Progress and challenges in protein structure prediction. Curr. Opin. Struct. Biol. **18**(3), 342–348 (2008)
39. K.Y. Sanbonmatsu, A.E. Garcia, Structure of met-enkephalin in explicit aqueous solution using replica exchange molecular dynamics. Protein. Struct. Funct. Genet. **46**(2), 225–234 (2002)
40. F. Rao, A. Caflisch, Replica exchange molecular dynamics simulations of reversible folding. J. Chem. Phys. **119**(7), 4035–4042 (2003)
41. Y.M. Rhee, V.S. Pande, Multiplexed-replica exchange molecular dynamics method for protein folding simulation. Biophys. J. **84**(2), 775–786 (2003)
42. N. Miyashita, J.E. Straub, D. Thirumalai, Y. Sugita, Transmembrane structures of amyloid precursor protein dimer predicted by replica-exchange molecular dynamics simulations. J. Am. Chem. Soc. **131**(10), 3438 (2009)
43. S. Rammensee, U. Slotta, T. Scheibel, A.R. Bausch, Assembly mechanism of recombinant spider silk proteins. Proc. Natl. Acad. Sci. USA **105**(18), 6590–6595 (2008)
44. J. Gosline, M. Lillie, E. Carrington, P. Guerette, C. Ortlepp, K. Savage, Elastic proteins: biological roles and mechanical properties. Philos. Trans. R. Soc. Lond. B, Biol. Sci. **357**(1418), 121–132 (2002)
45. K.N. Savage, J.M. Gosline, The role of proline in the elastic mechanism of hydrated spider silks. J. Exp. Biol. **211**(12), 1948–1957 (2008)
46. D. Porter, F. Vollrath, Z. Shao, Predicting the mechanical properties of spider silk as a model nanostructured polymer. Eur. Phys. J. E **16**(2), 199–206 (2005)
47. C. Riekel, F. Vollrath, Spider silk fibre extrusion: combined wide- and small-angle x-ray microdiffraction experiments. Int. J. Biol. Macromol. **29**(3), 203–210 (2001)
48. J.E. Trancik, J.T. Czernuszka, F.I. Bell, C. Viney, Nanostructural features of a spider dragline silk as revealed by electron and x-ray diffraction studies. Polymer **47**(15), 5633–5642 (2006)
49. A.H. Simmons, C.A. Michal, L.W. Jelinski, Molecular orientation and two-component nature of the crystalline fraction of spider dragline silk. Science **271**(5245), 84–87 (1996)
50. J.D. van Beek, J. Kummerlen, F. Vollrath, B.H. Meier, Supercontracted spider dragline silk: a solid-state nmr study of the local structure. Int. J. Biol. Macromol. **24**(2-3), 173–178 (1999)
51. J.E. Jenkins, M.S. Creager, R.V. Lewis, G.P. Holland, J.L. Yarger, Quantitative correlation between the protein primary sequences and secondary structures in spider dragline silks. Biomacromolecules **11**(1), 192–200 (2010)
52. M.E. Rousseau, T. Lefevre, M. Pezolet, Conformation and orientation of proteins in various types of silk fibers produced by nephila clavipes spiders. Biomacromolecules **10**(10), 2945–2953 (2009)
53. M. Cetinkaya, S.B. Xiao, B. Markert, W. Stacklies, F. Grater, Silk fiber mechanics from multiscale force distribution analysis. Biophys. J. **100**(5), 1298–1305 (2011)

54. D. Porter, F. Vollrath, Nanoscale toughness of spider silk. Nano Today **2**(3), 6 (2007)
55. D. Porter, F. Vollrath, Silk as a biomimetic ideal for structural polymers. Adv. Mat. **21**(4), 487–492 (2009)
56. F. Vollrath, D. Porter, Spider silk as archetypal protein elastomer. Soft Matter **2**(5), 377–385 (2006)
57. A. Nova, S. Keten, N.M. Pugno, A. Redaelli, M.J. Buehler, Molecular and nanostructural mechanisms of deformation, strength and toughness of spider silk fibrils. Nano Lett. **10**(7), 2626–2634 (2010)
58. S. Shen, A. Henry, J. Tong, R.T. Zheng, G. Chen, Polyethylene nanofibres with very high thermal conductivities. Nat. Nanotechnol. **5**(4), 251–255 (2010)
59. T. Giesa, M. Arslan, N.M. Pugno, M.J. Buehler, Nanoconfinement of spider silk fibrils begets superior strength, extensibility, and toughness. Nano Lett. **11**(11), 5038–5046 (2011)
60. S. Frische, A.B. Maunsbach, F. Vollrath, Elongate cavities and skin-core structure in nephila spider silk observed by electron microscopy. J. Microsc. Oxford **189**, 64–70 (1998)
61. Z. Shao, X.W. Hu, S. Frische, F. Vollrath, Heterogeneous morphology of nephila edulis spider silk and its significance for mechanical properties. Polymer **40**(16), 4709–4711 (1999)
62. P. Poza, J. Perez-Rigueiro, M. Elices, J. Llorca, Fractographic analysis of silkworm and spider silk. Eng. Fract. Mech. **69**(9), 1035–1048 (2002)
63. F. Vollrath, T. Holtet, H.C. Thogersen, S. Frische, Structural organization of spider silk. Proc. R. Soc. Lond. B, Biol. Sci. **263**(1367), 147–151 (1996)
64. N. Du, Z. Yang, X.Y. Liu, Y. Li, H.Y. Xu, Structural origin of the strain-hardening of spider silk. Adv. Funct. Mater. **21**(4), 772–778 (2011)
65. A. Tarakanova, M.J. Buehler, A materiomics approach to spider silk. JOM (2012)
66. F. Vollrath, P. Selden, The role of behavior in the evolution of spiders, silks, and webs. Annu. Rev. Ecol. Evol. Syst. **38**, 819–846 (2007)
67. T. Kohler, F. Vollrath, Thread biomechanics in the 2 orb-weaving spiders araneus-diadematus (araneae, araneidae) and uloborus-walckenaerius (araneae, uloboridae). J. Exp. Zool. **271**(1), 1–17 (1995)
68. L.H. Lin, W. Sobek, Structural hierarchy in spider webs and spiderweb-type system. Struct. Eng. **76**(4), 59–64 (1998)
69. B.D. Opell, J.E. Bond, Changes in the mechanical properties of capture threads and the evolution of modern orb-weaving spiders. Evol. Ecol. Res. **3**(5), 567–581 (2001)
70. B.O. Swanson, T.A. Blackledge, J. Beltran, C.Y. Hayashi, Variation in the material properties of spider dragline silk across species. Appl. Phys. A, Mater. Sci. Process. **82**(2), 213–218 (2006)
71. M.S. Alam, M.A. Wahab, C.H. Jenkins, Mechanics in naturally compliant structures. Mech. Mater. **39**(2), 145–160 (2007)
72. Y. Aoyanagi, K. Okumura, Simple model for the mechanics of spider webs. Phys. Rev. Lett. **104**(3) (2010)
73. J.R. Rice, G.F. Rosengren, Plane strain deformation near a crack tip in a power-law hardening material. J. Mech. Phys. Solids **16**(1), 1 (1968)
74. A. Carpinteri, N. Pugno, Fracture instability and limit strength condition in structures with re-entrant corners. Eng. Fract. Mech. **72**(8), 1254–1267 (2005)
75. N. Pugno, A. Carpinteri, M. Ippolito, A. Mattoni, L. Colombo, Atomistic fracture: qfm vs. md. Eng. Fract. Mech. **75**(7), 1794–1803 (2008)
76. C. Dicko, F. Vollrath, J.M. Kenney, Spider silk protein refolding is controlled by changing ph. Biomacromolecules **5**(3), 704–710 (2004)
77. X. Wu, X.Y. Liu, N. Du, G.Q. Xu, B.W. Li, Unraveled mechanism in silk engineering: fast reeling induced silk toughening. Appl. Phys. Lett. **95**(9) (2009)
78. C.W.P. Foo, S.V. Patwardhan, D.J. Belton, B. Kitchel, D. Anastasiades, J. Huang, R.R. Naik, C.C. Perry, D.L. Kaplan, Novel nanocomposites from spider silk-silica fusion (chimeric) proteins. Proc. Natl. Acad. Sci. USA **103**(25), 9428–9433 (2006)
79. H.A. Currie, O. Deschaume, R.R. Naik, C.C. Perry, D.L. Kaplan, Genetically engineered chimeric silk-silver binding proteins. Adv. Funct. Mater. **21**(15), 2889–2895 (2011)

80. S. Gomes, K. Numata, I.B. Leonor, J.F. Mano, R.L. Reis, D.L. Kapan, Afm study of morphology and mechanical properties of a chimeric spider silk and bone sialoprotein protein for bone regeneration. Biomacromolecules **12**(5), 1675–1685 (2011)
81. X.X. Xia, Z.G. Qian, C.S. Ki, Y.H. Park, D.L. Kaplan, S.Y. Lee, Native-sized recombinant spider silk protein produced in metabolically engineered escherichia coli results in a strong fiber. Proc. Natl. Acad. Sci. USA **107**(32), 14059 (2010)
82. N.A. Ayoub, J.E. Garb, R.M. Tinghitella, M.A. Collin, C.Y. Hayashi, Blueprint for a high-performance biomaterial: full-length spider dragline silk genes. PLoS ONE **2**(6) (2007)
83. G.H. Altman, F. Diaz, C. Jakuba, T. Calabro, R.L. Horan, J.S. Chen, H. Lu, J. Richmond, D.L. Kaplan, Silk-based biomaterials. Biomaterials **24**(3), 401–416 (2003)
84. B. Blossman-Myer, W.W. Burggren, The silk cocoon of the silkworm, bombyx mori: macro structure and its influence on transmural diffusion of oxygen and water vapor. Comp. Biochem. Physiol., Part a Mol. Integr. Physiol. **155**(2), 259–263 (2010)
85. C. Ortiz, M.C. Boyce, Materials science—bioinspired structural materials. Science **319**(5866), 1053–1054 (2008)
86. J.P.R.O. Orgel, T.C. Irving, A. Miller, T.J. Wess, Microfibrillar structure of type i collagen in situ. Proc. Natl. Acad. Sci. USA **103**(24), 9001–9005 (1995)
87. D.J.S. Hulmes, T.J. Wess, D.J. Prockop, P. Fratzl, Radial packing, order, and disorder in collagen fibrils. Biophys. J. **68**(5), 1661–1670 (1995)
88. N. Sasaki, S. Odajima, Elongation mechanism of collagen fibrils and force-strain relations of tendon at each level of structural hierarchy. J. Biomech. **29**(9), 1131–1136 (1996)
89. S. Weiner, H.D. Wagner, The material bone: structure mechanical function relations. Annu. Rev. Mater. Sci. **28**, 271–298 (1998)
90. R.K. Nalla, J.H. Kinney, R.O. Ritchie, Mechanistic fracture criteria for the failure of human cortical bone. Nat. Mater. **2**(3), 164–168 (2003)
91. P. Fratzl, R. Weinkamer, Nature's hierarchical materials. Prog. Mater. Sci. **52**, 1263–1334 (2007)
92. R.K. Nalla, J.J. Kruzic, J.H. Kinney, M. Balooch, J.W. Ager, R.O. Ritchie, Role of microstructure in the aging-related deterioration of the toughness of human cortical bone. Mater. Sci. Eng. C **26**(8), 1251–1260 (2006)
93. J.D. Currey, *Bones* (Princeton University Press, Princeton, 2002)
94. M.A. Meyers, P.Y. Chen, A.Y.M. Lin, Y. Seki, Biological materials: structure and mechanical properties. Prog. Mater. Sci. **53**(1), 1–206 (2008)
95. J.D. Currey, The design of mineralised hard tissues for their mechanical functions. J. Exp. Biol. **202**(23), 3285–3294 (1999)
96. M.E. Launey, M.J. Buehler, R.O. Ritchie, On the mechanistic origins of toughness in bone. Annu. Rev. Mater. Sci. **40**, 25–53 (2010)
97. S. Weiner, W. Traub, H.D. Wagner, Lamellar bone: structure-function relations. J. Struct. Biol. **126**(3), 241–255 (1999)
98. M.M. Giraud-Guille, Twisted plywood architecture of collagen fibrils in human compact-bone osteons. Calcif. Tissue Int. **42**(3), 167–180 (1988)
99. J.Y. Rho, P. Zioupos, J.D. Currey, G.M. Pharr, Variations in the individual thick lamellar properties within osteons by nanoindentation. Bone **25**(3), 295–300 (1999)
100. W. Wagermaier, H.S. Gupta, A. Gourrier, M. Burghammer, P. Roschger, P. Fratzl, Spiral twisting of fiber orientation inside bone lamellae. Biointerphases **1**(1), 1–5 (2006)
101. W. Wagermaier, H.S. Gupta, A. Gourrier, O. Paris, P. Roschger, M. Burghammer, C. Riekel, P. Fratzl, Scanning texture analysis of lamellar bone using microbeam synchrotron x-ray radiation. J. Appl. Crystallogr. **40**, 115–120 (2007)
102. W.J. Landis, K.J. Hodgens, J. Arena, M.J. Song, B.F. McEwen, Structural relations between collagen and mineral in bone as determined by high voltage electron microscopic tomography. Microsc. Res. Tech. **33**(2), 192–202 (1996)
103. W.J. Landis, K.J. Hodgens, M.J. Song, J. Arena, S. Kiyonaga, M. Marko, C. Owen, B.F. McEwen, Mineralization of collagen may occur on fibril surfaces: evidence from conven-

tional and high-voltage electron microscopy and three-dimensional imaging. J. Struct. Biol. **117**(1), 24–35 (1996)
104. M.D. Grynpas, L.C. Bonar, M.J. Glimcher, X-ray diffraction radial distribution function studies on bone mineral and synthetic calcium phosphates. J. Mater. Sci. **19**(3), 723–736 (1984)
105. P. Fratzl, N. Fratzlzelman, K. Klaushofer, G. Vogl, K. Koller, Nucleation and growth of mineral crystals in bone studied by small angle x-ray scattering. Calcif. Tissue Int. **48**(6), 407–413 (1991)
106. P. Fratzl, H.S. Gupta, E.P. Paschalis, P. Roschger, Structure and mechanical quality of the collagen-mineral nano-composite in bone. J. Mater. Chem. **14**(14), 2115–2123 (2004)
107. G.N. Ramachandran, G. Kartha, Structure of collagen. Nature **176**(4482), 593–595 (1955)
108. K.E. Kadler, D.F. Holmes, J.A. Trotter, J.A. Chapman, Collagen fibril formation. Biochem. J. **316**, 1–11 (1996)
109. D.J. Prockop, K.I. Kivirikko, Collagens: molecular biology, diseases, and potentials for therapy. Annu. Rev. Biochem. **64**, 403–434 (1995)
110. H. Gao, B. Ji, I.L. Jäger, E. Arzt, P. Fratzl, Materials become insensitive to flaws at nanoscale: lessons from nature. Proc. Natl. Acad. Sci. USA **100**(10), 5597–5600 (2003)
111. S.L. Hui, C.W. Slemenda, C.C. Johnston, Age and bone mass as predictors of fracture in a prospective study. J. Clin. Invest. **81**(6), 1804–1809 (1988)
112. G.M. Kiebzak, Age-related bone changes. Exp. Gerontol. **26**(2–3), 171–187 (1991)
113. R.P. Heaney, Is the paradigm shifting? Bone **33**(4), 457–465 (2003)
114. E.A. Zimmermann, E. Schaible, H. Bale, H.D. Barth, S.Y. Tang, P. Reichert, B. Busse, T. Alliston, J.W. Ager, R.O. Ritchie, Age-related changes in the plasticity and toughness of human cortical bone at multiple length scales. Proc. Natl. Acad. Sci. USA **108**(35), 14416–14421 (2011)
115. R.K. Nalla, J.J. Kruzic, R.O. Ritchie, On the origin of the toughness of mineralized tissue: microcracking or crack bridging? Bone **34**(5), 790–798 (2004)
116. A.J. Bailey, Molecular mechanisms of ageing in connective tissues. Mech. Ageing Dev. **122**(7), 735–755 (2001)
117. D.R. Eyre, I.R. Dickson, K. Vanness, Collagen cross-linking in human-bone and articular-cartilage—age-related-changes in the content of mature hydroxypyridinium residues. Biochem. J. **252**(2), 495–500 (1988)
118. D.R. Sell, V.M. Monnier, Structure elucidation of a senescence cross-link from human extracellular-matrix—implication of pentoses in the aging process. J. Biol. Chem. **264**(36), 21597–21602 (1989)
119. P. Odetti, S. Rossi, F. Monacelli, A. Poggi, M. Cirnigliaro, M. Federici, A. Federici, Advanced glycation end products and bone loss during aging, Maillard Reaction: Chemistry at the Interface of Nutrition, Aging, and Disease. Ann. N.Y. Acad. Sci. **1043**, 710–717 (2005)
120. D. Vashishth, G.J. Gibson, J.I. Khoury, M.B. Schaffler, J. Kimura, D.P. Fyhrie, Influence of nonenzymatic glycation on biomechanical properties of cortical bone. Bone **28**(2), 195–201 (2001)
121. X. Wang, X. Shen, X. Li, C.M. Agrawal, Age-related changes in the collagen network and toughness of bone. Bone **31**(1), 1–7 (2002)
122. J.W. Ager, G. Balooch, R.O. Ritchie, Fracture, aging, and disease in bone. J. Mater. Res. **21**(8), 1878–1892 (2006)
123. M.J. Buehler, Nanomechanics of collagen fibrils under varying cross-link densities: atomistic and continuum studies. J. Mech. Behav. Biomed. Mater. **1**(1), 59–67 (2008)
124. Y. Tang, R. Ballarini, M.J. Buehler, S.J. Eppell, Deformation micromechanisms of collagen fibrils under uniaxial tension. J. R. Soc. Interface **7**(46), 839–850 (2010)
125. M.J. Buehler, Atomistic and continuum modeling of mechanical properties of collagen: elasticity, fracture, and self-assembly. J. Mater. Res. **21**(8), 1947–1961 (2006)
126. M.J. Buehler, Nature designs tough collagen: explaining the nanostructure of collagen fibrils. Proc. Natl. Acad. Sci. USA **103**(33), 12285–12290 (2006)
127. S.G.M. Uzel, M.J. Buehler, Molecular structure, mechanical behavior and failure mechanism of the c-terminal cross-link domain in type i collagen. J. Mech. Behav. Biomed. Mater. **4**(2),

153–161 (2011)
128. T. Siegmund, M.R. Allen, D.B. Burr, Failure of mineralized collagen fibrils: modeling the role of collagen cross-linking. J. Biomech. **41**(7), 1427–1435 (2008)
129. M.J. Buehler, Molecular nanomechanics of nascent bone: fibrillar toughening by mineralization. Nanotechnology **18**(29) (2007)
130. H.S. Gupta, W. Wagermaier, G.A. Zickler, D.R.B. Aroush, S.S. Funari, P. Roschger, H.D. Wagner, P. Fratzl, Nanoscale deformation mechanisms in bone. Nano Lett. **5**(10), 2108–2111 (2005)
131. H.S. Gupta, J. Seto, W. Wagermaier, P. Zaslansky, P. Boesecke, P. Fratzl, Cooperative deformation of mineral and collagen in bone at the nanoscale. Proc. Natl. Acad. Sci. USA **103**(47), 17741–17746 (2006)
132. Y.L. Sun, Z.P. Luo, A. Fertala, K.N. An, Stretching type ii collagen with optical tweezers. J. Biomech. **37**(11), 1665–1669 (2004)
133. A. Gautieri, M.J. Buehler, A. Redaelli, Deformation rate controls elasticity and unfolding pathway of single tropocollagen molecules. J. Mech. Behav. Biomed. Mater. **2**(2), 130–137 (2009)
134. A.S. Craig, M.J. Birtles, J.F. Conway, D.A.D. Parry, An estimate of the mean length of collagen fibrils in rat tail tenson as a function of age. Connect. Tissue Res. **19**(1), 51–62 (1989)
135. G.E. Fantner, T. Hassenkam, J.H. Kindt, J.C. Weaver, H. Birkedal, L. Pechenik, J.A. Cutroni, G.A.G. Cidade, G.D. Stucky, D.E. Morse, P.K. Hansma, Sacrificial bonds and hidden length dissipate energy as mineralized fibrils separate during bone fracture. Nat. Mater. **4**(8), 612–616 (2005)
136. B.L. Smith, T.E. Schaffer, M. Viani, J.B. Thompson, N.A. Frederick, J. Kindt, A. Belcher, G.D. Stucky, D.E. Morse, P.K. Hansma, Molecular mechanistic origin of the toughness of natural adhesives, fibres and composites. Nature **399**(6738), 761–763 (1999)
137. K. Tai, F.J. Ulm, C. Ortiz, Nanogranular origins of the strength of bone. Nano Lett. **6**(11), 2520–2525 (2006)
138. P. Zioupos, J.D. Currey, The extent of microcracking and the morphology of microcracks in damaged bone. J. Mater. Sci. **29**(4), 978–986 (1994)
139. D. Vashishth, J.C. Behiri, W. Bonfield, Crack growth resistance in cortical bone: concept of microcrack toughening. J. Biomech. **30**(8), 763–769 (1997)
140. D. Vashishth, K.E. Tanner, W. Bonfield, Contribution, development and morphology of microcracking in cortical bone during crack propagation. J. Biomech. **33**(9), 1169–1174 (2000)
141. D. Vashishth, K.E. Tanner, W. Bonfield, Experimental validation of a microcracking-based toughening mechanism for cortical bone. J. Biomech. **36**(1), 121–124 (2003)
142. Y.N. Yeni, T.L. Norman, Calculation of porosity and osteonal cement line effects on the effective fracture toughness of cortical bone in longitudinal crack growth. J. Biomed. Mater. Res. **51**(3), 504–509 (2000)
143. Y.N. Yeni, D.P. Fyhrie, A rate-dependent microcrack-bridging model that can explain the strain rate dependency of cortical bone apparent yield strength. J. Biomech. **36**(9), 1343–1353 (2003)
144. K.J. Koester, J.W. Ager, R.O. Ritchie, The true toughness of human cortical bone measured with realistically short cracks. Nat. Mater. **7**(8), 672–677 (2008)
145. R.K. Nalla, J.S. Stolken, J.H. Kinney, R.O. Ritchie, Fracture in human cortical bone: local fracture criteria and toughening mechanisms. J. Biomech. **38**(7), 1517–1525 (2005)
146. E.A. Zimmermann, M.E. Launey, H.B. Barth, R.O. Ritchie, Mixed-mode fracture of human cortical bone. Biomaterials **30**(29), 5877–5884 (2009)
147. R.O. Ritchie, R.M. Cannon, B.J. Dalgleish, R.H. Dauskardt, J.M. McNaney, Mechanics and mechanisms of crack-growth at or near ceramic-metal interfaces: interface engineering strategies for promoting toughness. Mater. Sci. Eng. A, Struct. Mater.: Prop. Microstruct. Process. **166**(1–2), 221–235 (1993)
148. R.B. Martin, D.B. Burr, *Structure, Function, and Adaptation of Compact Bone* (Raven Press, New York, 1989)

149. D. Taylor, J.G. Hazenberg, T.C. Lee, Living with cracks: damage and repair in human bone. Nat. Mater. **6**(4), 263–268 (2007)
150. J. Aizenberg, J.C. Weaver, M.S. Thanawala, V.C. Sundar, D.E. Morse, P. Fratzl, Skeleton of euplectella sp.: structural hierarchy from the nanoscale to the macroscale. Science **309**(5732), 275–278 (2005)
151. H.D. Espinosa, J.E. Rim, F. Barthelat, M.J. Buehler, Merger of structure and material in nacre and bone—perspectives on de novo biomimetic materials. Prog. Mater. Sci. **54**(8), 1059–1100 (2009)
152. G. Mayer, Rigid biological systems as models for synthetic composites. Science **310**(5751), 1144–1147 (2005)
153. J. Aizenberg, P. Fratzl, Biological and biomimetic materials. Adv. Mat. **21**(4), 387–388 (2009)
154. I. Jager, P. Fratzl, Mineralized collagen fibrils: a mechanical model with a staggered arrangement of mineral particles. Biophys. J. **79**(4), 1737–1746 (2000)
155. Z.Q. Zhang, Y.W. Zhang, H.J. Gao, On optimal hierarchy of load-bearing biological materials. Proc. R. Soc. Lond. B, Biol. Sci. **278**(1705), 519–525 (2011)
156. Z.Q. Zhang, B. Liu, Y. Huang, K.C. Hwang, H. Gao, Mechanical properties of unidirectional nanocomposites with non-uniformly or randomly staggered platelet distribution. J. Mech. Phys. Solids **58**(10), 1646–1660 (2010)
157. S. Kamat, X. Su, R. Ballarini, A.H. Heuer, Structural basis for the fracture toughness of the shell of the conch strombus gigas. Nature **405**(6790), 1036–1040 (2000)
158. S. Kamat, H. Kessler, R. Ballarini, M. Nassirou, A.H. Heuer, Fracture mechanisms of the *Strombus gigas* conch shell: II—micromechanics analyses of multiple cracking and large-scale crack bridging. Acta Mater. **52**(8), 2395–2406 (2004)
159. R. Ballarini, R. Kayacan, F.J. Ulm, T. Belytschko, A.H. Heuer, Biological structures mitigate catastrophic fracture through various strategies. Int. J. Fract. **135**(1-4), 187–197 (2005)
160. L.J. Bonderer, A.R. Studart, L.J. Gauckler, Bioinspired design and assembly of platelet reinforced polymer films. Science **319**(5866), 1069–1073 (2008)
161. E. Munch, M.E. Launey, D.H. Alsem, E. Saiz, A.P. Tomsia, R.O. Ritchie, Tough, bio-inspired hybrid materials. Science **322**(5907), 1516–1520 (2008)
162. N.C. Broedling, A. Hartmaier, M.J. Buehler, H.J. Gao, The strength limit in a bio-inspired metallic nanocomposite. J. Mech. Phys. Solids **56**(3), 1086–1104 (2008)
163. D.C. Jang, J.R. Greer, Transition from a strong-yet-brittle to a stronger-and-ductile state by size reduction of metallic glasses. Nat. Mater. **9**(3), 215–219 (2010)
164. J.-Y. Kim, D. Jang, J.R. Greer, Nanolaminates utilizing size-dependent homogeneous plasticity of metallic glasses. Adv. Funct. Mater. **21**(23), 4550–4554 (2011)
165. A.Y.M. Lin, P.Y. Chen, M.A. Meyers, The growth of nacre in the abalone shell. Acta Biomater. **4**(1), 131–138 (2008)
166. J.D. Currey, J.D. Taylor, Mechanical behavior of some molluscan hard tissues. J. Zool. **173**(Jul), 395–406 (1974)
167. R.Z. Wang, Z. Suo, A.G. Evans, N. Yao, I.A. Aksay, Deformation mechanisms in nacre. J. Mater. Res. **16**(9), 2485–2493 (2001)
168. M. Sarikaya, I.A. Aksay, *Biomimetics: Design and Processing of Materials*. AIP Series in Polymers and Complex Materials (AIP Press, Woodbury, 1995)
169. X.W. Su, A.M. Belcher, C.M. Zaremba, D.E. Morse, G.D. Stucky, A.H. Heuer, Structural and microstructural characterization of the growth lines and prismatic microarchitecture in red abalone shell and the microstructures of abalone "flat pearls". Chem. Mater. **14**(7), 3106–3117 (2002)
170. R. Menig, M.H. Meyers, M.A. Meyers, K.S. Vecchio, Quasi-static and dynamic mechanical response of haliotis rufescens (abalone) shells. Acta Mater. **48**(9), 2383–2398 (2000)
171. M. Rousseau, E. Lopez, P. Stempfle, M. Brendle, L. Franke, A. Guette, R. Naslain, X. Bourrat, Multiscale structure of sheet nacre. Biomaterials **26**(31), 6254–6262 (2005)
172. F. Barthelat, Biomimetics for next generation materials. Philos. Trans. R. Soc., Math. Phys. Eng. Sci. **365**(1861), 2907–2919 (2007)

173. A. Lin, M.A. Meyers, Growth and structure in abalone shell. Mater. Sci. Eng. A, Struct. Mater.: Prop. Microstruct. Process. **390**(1-2), 27–41 (2005)
174. H. Tang, F. Barthelat, H.D. Espinosa, An elasto-viscoplastic interface model for investigating the constitutive behavior of nacre. J. Mech. Phys. Solids **55**(7), 1410–1438 (2007)
175. F. Barthelat, H. Tang, P.D. Zavattieri, C.M. Li, H.D. Espinosa, On the mechanics of mother-of-pearl: a key feature in the material hierarchical structure. J. Mech. Phys. Solids **55**(2), 306–337 (2007)
176. S.P. Kotha, Y. Li, N. Guzelsu, Micromechanical model of nacre tested in tension. J. Mater. Sci. **36**(8), 2001–2007 (2001)
177. A.G. Evans, Z. Suo, R.Z. Wang, I.A. Aksay, M.Y. He, J.W. Hutchinson, Model for the robust mechanical behavior of nacre. J. Mater. Res. **16**(9), 2475–2484 (2001)
178. F. Barthelat, C.M. Li, C. Comi, H.D. Espinosa, Mechanical properties of nacre constituents and their impact on mechanical performance. J. Mater. Res. **21**(8), 1977–1986 (2006)
179. X.D. Li, W.C. Chang, Y.J. Chao, R.Z. Wang, M. Chang, Nanoscale structural and mechanical characterization of a natural nanocomposite material: the shell of red abalone. Nano Lett. **4**(4), 613–617 (2004)
180. B.J.F. Bruet, H.J. Qi, M.C. Boyce, R. Panas, K. Tai, L. Frick, C. Ortiz, Nanoscale morphology and indentation of individual nacre tablets from the gastropod mollusc trochus niloticus. J. Mater. Res. **20**(9), 2400–2419 (2005)
181. F. Song, Y.L. Bai, Effects of nanostructures on the fracture strength of the interfaces in nacre. J. Mater. Res. **18**(8), 1741–1744 (2003)
182. A.P. Jackson, J.F.V. Vincent, R.M. Turner, The mechanical design of nacre. Proc. R. Soc. Lond. B, Biol. Sci. **234**(1277), 415 (1988)
183. J.D. Currey, Mechanical-properties of mother of pearl in tension. Proc. R. Soc. Lond. B, Biol. Sci. **196**(1125), 443 (1977)
184. X.D. Li, Z.H. Xu, R.Z. Wang, In situ observation of nanograin rotation and deformation in nacre. Nano Lett. **6**(10), 2301–2304 (2006)
185. B.H. Ji, H.J. Gao, Mechanical properties of nanostructure of biological materials. J. Mech. Phys. Solids **52**(9), 1963–1990 (2004)
186. H.M. Yao, M. Dao, T. Imholt, J.M. Huang, K. Wheeler, A. Bonilla, S. Suresh, C. Ortiz, Protection mechanisms of the iron-plated armor of a deep-sea hydrothermal vent gastropod. Proc. Natl. Acad. Sci. USA **107**(3), 987–992 (2010)

# Chapter 9
# Pathological Materiomics

**Abstract** The performance of a biological material or system—be it mechanical, optical, thermal or otherwise physiological—is associated with normal, or healthy, functioning. Variations in material or structure (such as molecular defects and genetic mutations) can propagate across scales and be expressed unpredictably, resulting in disorders, diseases, and afflictions (such as Alzheimer's disease, Progeria syndrome, sickle cell disease, or *osteogenesis imperfecta*). Here we review relevant materiomic investigations of such afflictions, termed pathological materiomics, encompassing the characterization of genetic disease, viral infections, or injuries/trauma that have a pathological basis in mechanical behavior. Consideration of how material properties change—the mechanics of disease—can lead to a new paradigm that may expand beyond the focus on biochemical readings alone, and contribute to disease diagnosis and treatment. Moreover, abnormal material variation can shed insight into healthy interactions and function across the materiome—we can potentially reverse engineer from the smallest atomistic scale, tracking the progression of disease, and learn from "failure".

*When things go right it's hard to figure out why, but when things go wrong it's really easy.*

*Steven Soderbergh, American Film director, producer, screenwriter*

## 9.1 Introduction

The above quotation by a film director has nothing to do with biology, materials science, or science in general. Yet, it encompasses the popular adage that when things go wrong, we can "*learn from mistakes*". Undoubtedly, it can be easier to see how things work (or fail) in hindsight—such as the forensic process of piecing together a crime or the explanation of a system failure in retrospect—rather than anticipate potential problems before they occur. Regardless, when systems fail, or "things go wrong", reverse engineering of the problem can give rise to a fuller understanding of the ideal, normal functioning of the system. We gain a unique opportunity to learn from mistakes, and sometimes we can only recognize critical components *after* they fail.

S.W. Cranford, M.J. Buehler, *Biomateriomics*, Springer Series in Materials Science 165,
DOI 10.1007/978-94-007-1611-7_9, 

In industry, such "mistakes" are often identified and weeded out in trials, prototypes, and test cases—the colloquial research and development (R&D) stage. Commercial products are developed by pushing a system to its limits, removal and redesign of components, and/or substitution of various materials (or vendors). This is one motivation for the scrutiny of failure mechanisms in biological materials. We can learn a lot of information from mistakes or when things go wrong, *i.e. failure*. Nature, however, throws other "mistakes" our way. Missing components, substituted materials and malfunctioning parts are rampant in biological systems. They are more commonly referred to as *disease*.

Materiomics can help crack the secrets of how Nature designs materials with unique properties such as extreme toughness, hardness, or compliance, but can also shed light on how diseases—specifically those that induce minute changes of molecular structure of proteins—can impact behavior at larger physiological scales. Here, we focus on a selection of case studies to examine material phenomena in the progression of disease. From this holistic perspective, a *de novo* basis for understanding biological processes can be used to develop new approaches for treating medical disorders. It may change the way biologists and medical doctors think of biological material systems—from a multiscale mechanistic approach—to help to treat and cure diseases.

## 9.2 Mechanics of Disease

Throughout the past century, the focus on addressing diseases has derived primarily from a biochemical approach. However, advancements and increased understanding has yielded an enhanced appreciation for the role of materials science of protein materials in variegated medical disorders. The development of links between failure of protein materials in the context of disease is currently a little explored aspect with great potential to greatly enhance our understanding of the role of materials phenomena in biological systems. The holistic characterization of properties and structure-function relationships for biological materials across multiple scales can play a crucial role in developing a better understanding of diseases, an application we term *pathological materiomics*.

Injuries and genetic diseases are often caused by structural changes in protein materials (*e.g.*, defects, flaws, changes to the molecular structure), resulting in failure of the material's intended function. The disease can be thought of as defects or flaws in a complex structural system or machine. Commonly, we may isolate the root cause of a problem (a known point mutation, for example), but how that manifests into symptoms is unpredictable (and uncontrollable). A materiomic perspective can simplify the problem by introducing a platform independent of biological and physiochemical interactions, with a focus on material properties and mechanical functionality, uncovering similarities and connections previously unseen. While minor details differ, a design flaw in a building made of steel members is similar to a design flaw in a wooden structure, and removal or failure of similar components will make them both come crashing down. Yet biological materials are tackled piece-wise—a different mechanism for collagen, a different therapy for neurological disorders. These systems are more inherently complex, but that does not necessarily infer complete separation of behavior and function. For biological materials, we currently have no "universal building code" to follow, yet universal aspects of these materials arise over and over again (as discussed in Chap. 3: The Challenges of Biological Materials and Chap. 4: Universality-Diversity Paradigm: Music, Materiomics, and Category Theory). The linking of such observations may eventually provide common explanations of the molecular origin of certain diseases, which exhibit changes in material properties. Undoubtedly, material properties play an essential role in biological systems, and that the current paradigm of focusing on biochemistry alone as the cause of diseases is insufficient.

**Pathological Materiomics:** The characterization of material properties as manifested by genetic disease (*e.g.*, point mutations, cellular defects), viral infections (*e.g.*, malaria), or injuries/trauma that have a pathological basis in mechanical behavior, resulting in failure of the material system's intended function, linking fundamental molecular effects to macroscopic physiological response.

It is envisioned that the long-term potential impact of biomateriomics can be used to predict diseases in the context of diagnostic tools by measuring material properties, rather than focusing on symptomatic chemical readings alone. The scale-bridging paradigm of materiomics can potentially emerge as a critical niche in the development of links between failure of biological materials in the context of genetic and infectious disease. Such diseases can be characterized by single point mutations, genetic deficiencies or alterations, or chromosomal aberrations, inherently molecular triggers that lead to dramatic, catastrophic effects at significantly larger scales. A materiomics approach can be advantageous when pathological conditions ultimately lead to a change in mechanical behavior, providing potential to greatly enhance out understanding of the role of materials phenomenon in biological systems through both experimental and computational investigations. Consideration of how mechanical behavior and material properties change in diseases could lead to new pathological insights that expand beyond biochemical signals and interactions. Additionally, conventional models of failure and disease that only consider one level of the material's structure do not capture the full range of relevant hierarchies and mechanisms and as such remain limited in their ability to describe material breakdown processes associated with disease.

The following sections present and discuss case studies where the change of material properties, structure, or function is a crucial element in the expressed disease. Translation of this knowledge could enable detection of diseases by measuring material properties rather than by focusing on symptomatic biochemical readings alone. Altogether, understanding the role of different hierarchical levels of protein materials in diseases could potentially bring about a new paradigm of approaches to address medical disorders; however, further research is needed to elucidate the underlying multi-scale failure mechanisms.

## 9.3 Sickle Cell Disease: The "First" Molecular Disease

Hemoglobin, the most abundant molecule found in red blood cells (RBCs), is one of the most important proteins as it transports oxygen from the respiratory organs (lungs in humans) to where it is needed, releasing the oxygen to burn nutrients to provide energy to power the functions of the organism, and assists in the transportation of the resultant carbon dioxide to be dispensed (*e.g.*, exhaled). The oxygen binding capacity of hemoglobin increases the total blood oxygen capacity seventy-fold compared to dissolved oxygen in blood [1]. It is undoubtedly the most critical protein in your circulatory system. It comes with no surprise that a minuscule change in the sequence of hemoglobin (*e.g.*, a mutation) can have dire consequences.

Sickle-cell disease (SCD), or sickle-cell anemia[1] is an autosomal recessive genetic blood disorder, characterized by red blood cells that assume an abnormal, rigid, or "sickle" shape (see Fig. 9.1), and has been recognized since the early 20th century. Sickling decreases the cells' flexibility—a characteristic mechanical expression—and results in a risk of various complications. For instance, because of increased stiffness and cell adherence to the endothelium, the circulation of sickle cells through the body's narrow blood vessels (arterioles, capillaries, venules) is often obstructed resulting in tissue and organ damage [2].

The oxygen dependence of sickling suggested that hemoglobin was probably involved in the process, causing the cells to acquire their distorted shape. In 1949, America's leading physical chemist, Linus Pauling along with his colleagues were the first to demonstrate that sickle-cell disease occurs as a result of an abnormality in the hemoglobin molecule—a single gene mutation in the hemoglobin protein sequence [3] (see Fig. 9.1). This was the first time a genetic disease was linked to a mutation of a specific protein, a milestone in the history of molecular biology, and it was published in their paper "*Sickle Cell Anemia, a Molecular Disease*" [3].

Pauling's paper established SCD as a genetic disease in which affected individuals have a different form of the metalloprotein hemoglobin in their blood. The paper reports a difference in electrophoretic mobility between hemoglobin from healthy individuals and those with sickle-cell anemia, with those with sickle cell trait having a mixture of the two types. The paper suggests that the difference in electrophoretic mobility is due to a different number of ionizable amino acid residues in the protein portion of hemoglobin (which was later confirmed [4, 5]), and that this change in molecular structure is responsible for the sickling process. The paper introduced the

[1]There are now known to be a cluster of sickle cell diseases, but we will focus here on the classic sickle cell anemia, a specific form of sickle-cell disease in which there is homozygosity for the mutation that causes the Hemoglobin variant HbS.

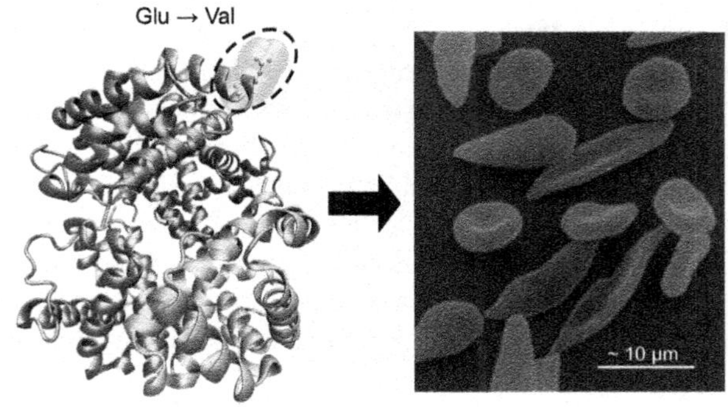


**Fig. 9.1** Molecular structure of hemoglobin protein (*left*), depicting the location of the single point mutation ($\beta6$ Glu → Val) that results in the "sickling" of red blood cells upon deoxygenization, compared to the regular, discoid shape of healthy red blood cells (indicated on *right*). We note the change in scale, from the mutation at the atomistic (Ångstrom) scale to ultimate expression at the cellular (micrometer) scale—a difference of four orders of magnitude!

concept of a "molecular disease", and is considered a major impetus to the development of molecular medicine. There was a causal link—not a mere correlation-between the existence of "defective" hemoglobin molecules and the pathological consequences of sickle cell disease. The paper helped establish that genes control not just the presence or absence of enzymes (as genetics had shown in the early 1940s) but also the specific structure of protein molecules.

**Molecular Medicine**: A broad medical field, where physical, chemical, biological and medical techniques are used to describe molecular structures and mechanisms, identify fundamental molecular and genetic errors of disease, and to develop molecular interventions to correct them. The molecular medicine perspective emphasizes cellular and molecular phenomena and interventions rather than the previous conceptual and observational focus on patients and their organs.

The study of SCD is commonly considered groundbreaking in two ways [6]:

1. It showed that the cause of a disease could be traced to an alteration in the molecular structure of a protein, raising the possibility that many diseases could be explained in this way.
2. As this disease was known to be inherited (*e.g.*, an autosomal recessive gene), SCD illustrated that genes precisely determine the structure of proteins (a debatable concept at the time).

While Pauling's seminal paper indicated the molecular cause of SCD, another challenge was to follow the expression of the mutation from the amino acid substitution, to the 'sickled' shape of a red blood cell, and how that ultimately affects the circulatory system and is expressed as symptoms. This inherently multiscale problem has been widely investigated. The gene defect is a known mutation of a single nucleotide of the $\beta$-globin gene, which results in glutamic acid (Glu) being substituted by valine (Val) at position six (*i.e*, $\beta 6$ position). Glutamic acid is a polar, hydrophilic molecule, but valine is nonpolar and strongly hydrophobic. Hemoglobin with this mutation are referred to as HbS, as opposed to the normal adult HbA. The genetic disorder is due to the mutation of a single nucleotide, from a GAG to GTG mutation, becoming a GUG by transcription. Oddly, this is normally a benign mutation, causing no apparent effects on the secondary, tertiary, or quaternary protein structure of hemoglobin in conditions of normal oxygen concentration (*e.g.*, no protein "misfolding").

What does occur, under conditions of low oxygen concentration, is the polymerization of the HbS molecules. The HbS form of hemoglobin exposes a hydrophobic patch on the protein between adjacent helices (termed here a "receptor"). The hydrophobic residues of the valine at the $\beta 6$ position in hemoglobin acts as a "donor" and are able to associate with the hydrophobic patch, causing HbS molecules to aggregate and form fibrous precipitates. Each molecule thus interacts with one lateral neighbor as a donor, and the other lateral neighbor as a receptor (see Fig. 9.2). Moreover, these "coupled fibers" aggregate further into bundles consisting of characteristic 14-stranded fibers (seven double strands) assembled and twisted in a rope-like fashion [8]. The structural basis for fibril nucleation is yet another complex problem not completely understood [9]. Simply put, the defective hemoglobin tetramers adhere together along a common axis, forming stiff fibers which subsequently gel within red blood cells (RBCs). Instead of a loose collection of hemoglobin molecules, the RBCs are afflicted with a network of fibers, which distort the cell and change its biomechanical properties (decrease deformability and increase viscosity). The entire process is also reversed upon oxygenization, wherein HbS fibers and molecules disassociate and the normal shape returns to the RBC. Because the HbS fibers play a vital role in the formation of the sickle-shaped RBC, the material properties and biomechanical behaviors of polymerized HbS fibers is a subject of intense research interest.

The elasticity of red blood cells is critical to their function, as they deform to allow passage in capillaries of diameters smaller than their own. Thus a mechanical understanding (and more critically, an understanding of the differences between normal and mutated hemoglobin) can shed insight in the pathophysiology of the disease. This includes the characterization of the hemoglobin molecules, resulting fibers, fibrous aggregates, afflicted RBC, gel rheology, and cell-tissue interactions.

The basis of gel rheology (the study of flow and deformation), potentially measured at the macroscale of blood flow, is inherently linked to the micromechanics of individual fibers. Since the distortion and increased rigidity of abnormal RBC are caused by the formation of HbS fibers and fiber domains, the properties of the HbS fiber have been extensively studied experimentally. For instance, the bending moduli and persistence lengths of individual fibers were obtained by experimental studies isolating HbS fibers [10], with persistence lengths on the order of millimeters—much larger than the diameter of a typical red blood cell (one the order of 6 to 8 μm). A direct consequence is that induced bending of such fibers (*e.g.*, attempting

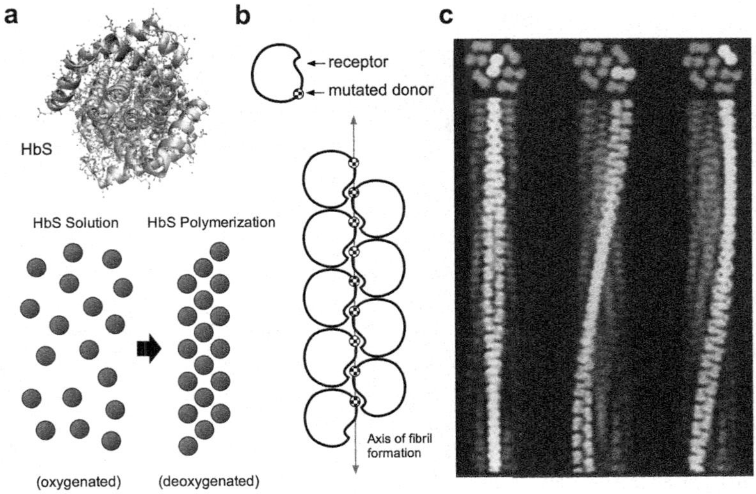


**Fig. 9.2** Formation of hemoglobin fibrils. (**a**) Molecular representation of HbS protein structure (*top*). In mutated form, the substituted amino acid causes no apparent effects on the secondary, tertiary, or quaternary structure of hemoglobin in conditions of normal oxygen concentration (HbS solution). Under low oxygen conditions, polymerization of the molecules occurs (HbS polymerization). (**b**) Schematic of the polymerization mechanism. Under conditions of low oxygen concentration the HbS form of hemoglobin exposes a hydrophobic patch on the protein between helices (*receptor*). The hydrophobic residues of the valine at the $\beta 6$ position of the hemoglobin (*mutated donor*) are able to associate with the hydrophobic patch, causing HbS molecules to aggregate and form fibrous precipitates. These fibrils further aggregate into 14-stranded fibers assembled and twisted in a rope-like fashion. (**c**) Experimental imaging of hemoglobin fibrils ($\approx 25$ nm in diameter), highlighting the double strands. The three unique double strand trajectories are highlighted in the model fiber electron density projections. The trajectories of the remaining three double strands are related by symmetry to those depicted above. The rest of the fiber is shown at lower intensity for reference. Experimental images from Lewis *et al.* [7], used with permission, copyright © 1994 John Wiley and Sons, Inc.

to squeeze through a capillary) would impart significant forces on the host cell and microvasculature. Indeed, the external pressure required to buckle these HbS fibers is more than four times the magnitude of typical blood pressure [10]. Studies that compare the stiffness and strength of single HbS fibers with the strength of aggregated HbS ropes suggest that only the hierarchical rope structure can overcome the tension of the RBC membrane [11], whereas single HbS fiber as long as the diameter of a RBC would buckle and not be able to deform the membrane. Thus, the aggregation (hierarchical assembly) is a critical mechanism of the disease.

Other characteristics associated with HbS fiber aggregation or self-assembly include fiber cross-links and associated binding energies [12] (and effected by the helical nature of the fibers). Moreover, sickle cells formed homotypic aggregations (self-adhesion) and heterotypic adhesion to endothelial cells and white blood cells, creating a host of complex interactions [13, 14]. These events contribute to the obstruction of blood flow and vaso-occlusion that are the hallmarks of SCD. Fully understanding such behaviors across multiple scales can lead to novel therapeutic treatments. For example, the specific geometry of fiber aggregation can be exploited in attempts to develop drugs with designed molecular affinities, *i.e.* finding the "weak links" in the fiber chain (if they exist).

The results on domain formation in HbS gels in a recent study [15] provide strong support for the conclusion that the rate of homogeneous nucleation determines the density of fiber domains, and therefore the shape of the cell (as proposed by Ferrone *et al.* on the basis of theoretical arguments [16]). Indeed, kinetics plays an extremely important role in the pathophysiology of SCD as the hemoglobin polymerization (fibril formation) and resulting cell "sickling" occurs upon deoxygenation of the molecule. Unfortunately, this deoxygenation occurs in the capillaries to facilitate oxygen transport to tissues and organs, and, necessarily, the smallest of the blood vessels involved in microcirculation. When polymerization is sufficiently rapid to reach equilibrium (saturation), almost all RBC would contain fibers, be sickled, and thus much less deformable. If polymerization is delayed, the cells escape the narrow vessels of the microcirculation before any significant polymerization has begun [17]. At intermediate rates, the cells undergo varying degrees of distortion, dependent on the intracellular protein concentration [18]. The effect of polymerization rate is illustrated by Fig. 9.3.

For SCD, fiber structure, polymerization kinetics and equilibria are well characterized and intimately related to pathogenesis. The problem is inherently multiscale, encompassing the genetic mutation (atomistic scale, Ångstroms), HbS polymerization (molecular scale; nanometers), fiber bundle and gel formation (cellular scale; nanometers to micrometers), the interaction between fiber bundles and the RBC membrane (cellular scale; micrometers), to the interaction of the sickled RBCs with blood vessels (tissue level; micrometers to millimeters). As such, many approaches have been used in the investigation of SCD. Computationally, this include early full atomistic molecular dynamics (MD) simulations comparing the crystal (or quaternary) structures of normal hemoglobin (HbA) to the mutated (HbS) variant [19, 20], molecular level investigations of the hemoglobin polymerization [21, 22] (relating molecular contact points, interactions, and geometry), to more recent, scale-bridging coarse-grain molecular dynamic approaches of the fibers [23] (similar to coarse-grain approaches as discussed in Chap. 6: Computational Approaches and Simulation). Experimental investigators have applied a multitude of diverse biomechanical techniques across scales to study the complex and abnormal rheological behavior of sickle RBCs and sickle blood (see reviews by Ballas and Mohandas [24], and Stuart and Nash [25]). Viscometry, filtration, ektacytometry, flow chambers, and microfluidic devices are techniques used for sickle blood and blood flow at the macroscale, whereas cell-based assays utilize micropipette aspiration, optical tweezers, and/or microfluidics at the microscale (see Chap. 5: Experimental Approaches). Each of these techniques differs in sensitivity and specificity and provides insights related to diverse aspects of the biomechanical behavior of cells throughout scales.

We note that the above discussion of sickle-cell disease is not intended to be comprehensive (for extensive reviews see Barabino *et al.* [26], Stuart and Nagel [27] or Bunn [28]). As one of the first characterized "*molecular*" diseases, there has been over 50 years of research pertaining to many facets of the simple mutation of the $\beta 6$ position within the hemoglobin molecule, ranging from the atomistic

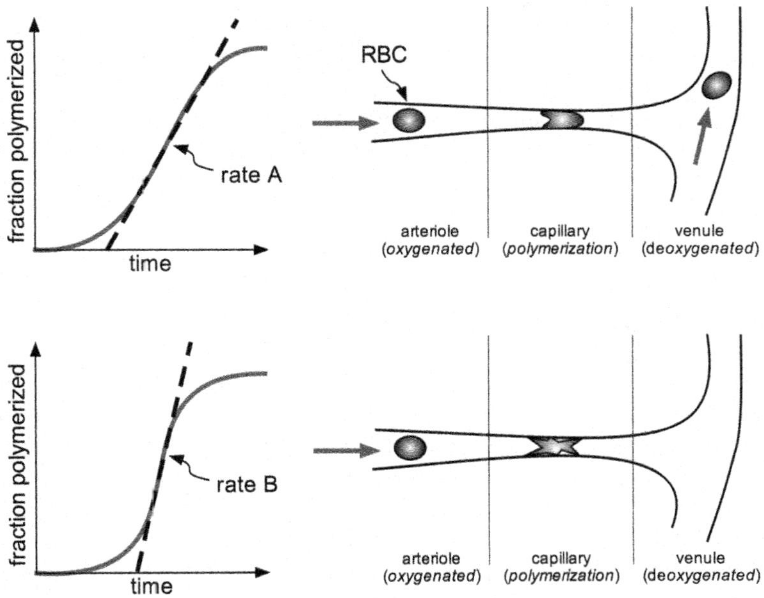


**Fig. 9.3** Importance of kinetics in expression of SCD. The delay time of HbS polymerization involves a critical balance between the rate and extent of deoxygenization, transit of red blood cells though the microcirculation, and HbS concentration. As the rate of polymerization is increased from A to B, there is an increase in the likelihood of obstruction within the microvasculature (see plots, *left*). Simplified schematics of microcirculation depict a cell passing from the arteriole to the venous circulation. *Top*: A red cell deforms as it passes through a capillary and escapes into the large vessels on the venous side without intracellular polymerization because the transit time is shorter than the delay time (rate A). *Bottom*: The delay time is shorter than the capillary transit time (rate B > rate A), and polymerization occurs while the cell is in the capillary, with the possibility of blocking the circulation. Figure derived from Eaton *et al.* [15]

(protein) to macroscale (blood circulatory) level. The importance of material and mechanical properties, rooted in the single point mutation, is befitting a materiomic perspective. The characterization of the biomechanical properties of sickle RBCs and sickle blood contribute to a better understanding of the pathophysiology of the disease. While SCD provides a platform to investigate the effects of a single known mutation, the knowledge is gained in retrospect, learning from "mistakes" in biology. Such a condition highlights the need for predictive understanding—while SCD arises from a mutation in the $\beta 6$ position, can we control and predict the effects of mutations in other conditions? Can we reverse, prevent, or treat such genetic disorders through more complete understanding? This is the ultimate goal shared by the exemplary researchers in the field.

An intriguing fact of the epidemiology of SCD is that the highest frequency of sickle-cell disease is found in tropical regions, particularly sub-Saharan Africa, India and the Middle-East—specifically in regions where the infectious disease malaria is (or was) common. The malaria parasite has a complex life cycle and spends part of it in red blood cells. Oddly enough, there is a benefit in carrying a single sickle-cell gene (sickle-cell trait, or SCT)—while not totally resistant to malaria, those with the sickle-cell trait are more tolerant to infection and show less severe symptoms when infected. What characteristics of the SCT are inherently beneficial in suppressing the infection and symptoms of malaria? What links a genetic mutation with a mosquito-borne infectious disease? Fortunately, there are numerous studies of the biomechanics of malaria infected red blood cells that may shed light on the issue.

## 9.4 Malaria and Red Blood Cells

There are routes other than genetic mutations that affect the material functionality and mechanistic response of biological tissues. Similar to SCD, we can consider the material behavior of human red blood cells in the context of an infectious disease—specifically *Plasmodium falciparum* malaria. The cause of malaria, a disease that affects nearly 8 % of the work population and causing nearly 2 to 3 million deaths annually [29], is the protoctistan parasite *Plasmodium.*

The life cycle of Plasmodium begins in the liver [30, 31] prior to infecting the blood [30, 32], wherein symptoms become apparent, associated with overt clinical signs including prostration, respiratory distress, pulmonary edema, convulsions, circulatory collapse, abnormal bleeding, jaundice, hemoglobinuria, severe anemia, and/or impaired consciousness [33, 34]. Within the blood, the Plasmodium parasites penetrate red blood cells and continually change the structure of the cell's spectrin network [35]. Pioneering experimental investigations by Suresh, Dao, Lim *et al.* [36, 37] have indicated that during the course of a 48 hour period after invasion of the red blood cell by the parasite, the effective stiffness of the cell increases by more than a factor of ten [37].

The combination of the reduction of deformability and a marked increase in the adhesion of the red blood cells results in obstructed flow of cells through the microvasculature [39]. A recent study [38] illustrated this behavior through the ingenious use of elastomeric microchannels—which can be thought of as synthetic arteries, capillaries, and other blood vessels—to characterize complex behaviors of single red blood cells, under flow, in multicellular blockages. Various channel widths (from 2 to 8 μm—the diameter of a typical RBC) were shown to be easily traversed by the highly elastic, uninfected healthy red blood cells. As the cells became infected with Plasmodium, the successful passing of the channel was dependent on the stage of parasitic development (known as ring, trophozoite, and schizont stages). Early development stage (ring stage) cells could traverse the channels at all widths, similar to uninfected cells. In the trophozoite stage, most cells could not traverse the smaller, 2 and 4 μm channels. Some that passed through the 4 μm channels emerged from constricted space with deformations whose shape-recovery could be observed in real time. The final development schizont forms failed to traverse even 6 μm channels and rapidly formed a capillary blockage. Interestingly, individual uninfected red blood cells readily squeezed through the blockages formed by immobile schizonts in a 6 μm capillary—the channels were not completed blockaded

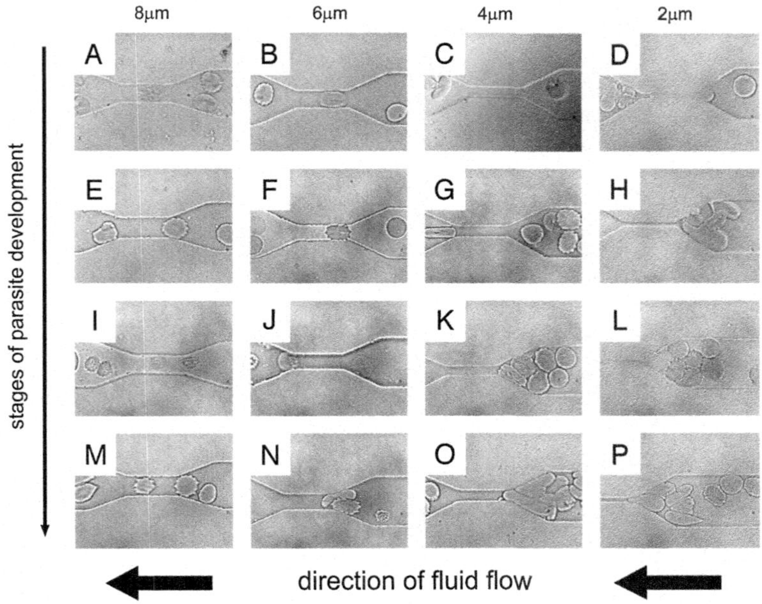


**Fig. 9.4** Blocking the flow: four sequences of images of four stages of malaria-infected red blood cells (early ring stage, early trophozoite, late trophozoite, and schizont) passing through microchannel constrictions. (**A–D**) Ring-stage infected erythrocytes (RBCs) retained much of the structural characteristics of normal erythrocytes and were able to pass through all constricted channels. Early trophozoite (**E–H**) and late trophozoite (**I–L**) infected cells passed through the larger 8 μm and 6 μm channels (**E** and **F**; **I** and **J**) but eventually blocked the smaller 4 μm and 2 μm channels (**G** and **H**; **K** and **L**). (**M–P**) Schizont stage infected erythrocytes blocked all but the 8 μm channel (**M**). From Rathod, Chiu *et al.* [38], used with permission, copyright © 2003 National Academy of Sciences

by the infected cells (see Fig. 9.4). These mechanical factors associated with cell deformability and cytoadherence are considered to be key mechanistic pathways in the pathogenic basis of the disease [33].

Appropriating many of the techniques used to examine blood and red blood cells for SCD, current investigative techniques to probe malaria-infected red blood cells have been a combination of experimental and computational methods across a multitude of scales, exemplifying an integrated materiomics framework. The mechanical deformation characteristics of red blood cells have been experimentally determined through such techniques as micropipette aspiration [40], optical tweezers [36] (see Fig. 9.5), and the discussed flow studies through microfluidic channels [38].

Such biomechanical assays are required to provide critical information regarding disease progression and treatment by monitoring the systematic alterations of cell structure and response. Further, the obtained empirical data may provide the necessary information to parameterize computational models (either independently or in correlation with bottom-up atomistic methods) and facilitate the development of deterministic models [41]. In the case of malaria, infection leads to marked changes in

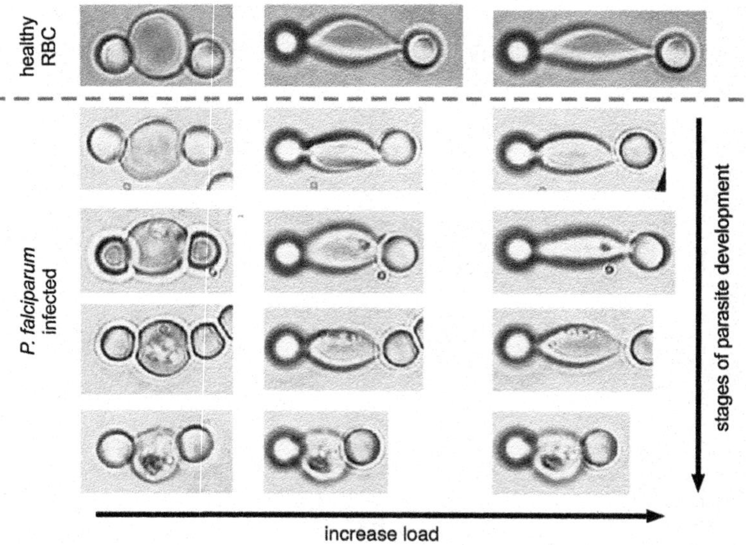


**Fig. 9.5** Optical images of healthy red blood cell and malaria (*Plasmodium falciparum*) infected cells, subject to tensile stretching *via* optical tweezers. *Left*: Prior to tensile stretching (zero force). *Middle*: A constant force of 68 ± 12 pN. *Right*: A constant force of 151 ± 20 pN. Note the presence of the *P. falciparum* parasite inside the infected RBCs. Clearly, the parasite-invaded cell loses its ability to stretch. Adapted from Suresh *et al.*, [37], used with permission, copyright © 2005 Elsevier

the molecular structure of the red blood cell. The experimental investigations cited were not focused on the molecular interactions and triggers initiated by the parasite at the nanoscale, but rather the manifested mechanistic effects including increased rigidity and cytoadherence, compromised motility and sequestion in microvasculature. Such information can lead to the development of continuum models for mechanics of red blood cells *via* multiscale analysis [36, 42], which can then be applied in numerous computational assays and linked to the pathology of *P. falciparum* (see Fig. 9.6). In the derivation of such models, the differences between infected and noninfected blood cells would necessarily be translated to specific variations in constitutive responses—the unique material relationship between stress and strain. This could quantify, for example, the precise increase in stiffness and rigidity (or decrease in deformability) that would induce blockage of capillaries. As shown in Fig. 9.4, the early development ring stage of infected cells can still traverse relatively small channels of blood flow (a mechanism that facilitates the spread of the parasite throughout the body). Exploiting the precise changes in rigidity and cytoadherence could potential lead to novel, filtration/removal systems to segregate infected from healthy cells [43, 44].

Aforementioned in the previous section, sickle human hemoglobin (HbS) confers a survival advantage to individuals living in endemic areas of malaria—in areas where malaria is a problem, people's chances of survival actually increase if they carry sickle-cell trait. While both afflictions lead to a more rigid cell, the mech-

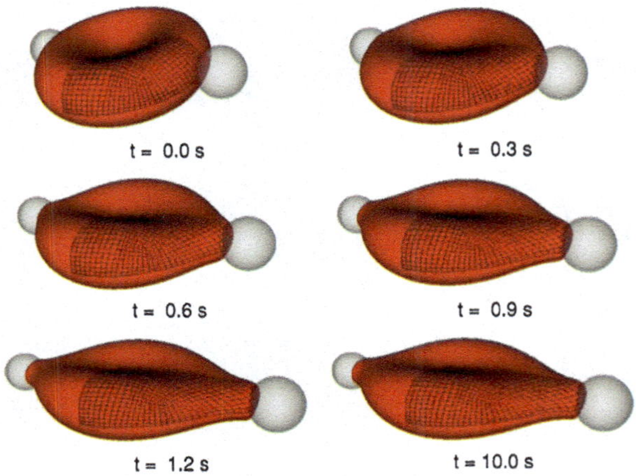


**Fig. 9.6** Finite element modeling (FEM) of red blood cell. Figure depicts the time evolution of a RBC subject to a constant force applied *via* optical tweezers (similar to conditions depicted in Fig. 9.5). Such models are implemented to replicate the empirical relation between applied forces and axial as well as transversal diameters obtained from several experiments. Accurate models can then be used in deterministic computational simulations (such as blood flow). Reverse engineering the constitutive laws can indicated mechanistic differences between healthy and infected red blood cells. Used with permission, from Hartmann [42], copyright © 2010 Springer

anisms by which erythrocytes (RBCs) containing abnormal HbS hemoglobins are partially protected against *P. falciparum* infections are not fully understood. It is known that during the peripheral blood stage of replication malaria parasites have a high rate of oxygen consumption and ingest large amounts of hemoglobin [45, 46]. Mutated HbS is characteristically deoxygenated and potentially more difficult to digest upon polymerization.

A recent study on mice [47] successfully linked the sickle-cell trait to suppression of infection. Sickle hemoglobin induces the expression of heme in blood cells—the iron carrying molecule within RBCs. Once released from the hemoglobin molecule, free heme becomes cytotoxic [48, 49]. However, this effect is countered by the expression of heme oxygenase-1 (HO-1) [48–50], an enzyme that catabolizes free heme into biliverdin, iron and carbon monoxide (CO). During the blood stage of Plasmodium infection cell-free hemoglobins are unnaturally released from infected red blood cells. The CO byproduct binds these cell-free hemoglobin and inhibits oxidation, preventing further accumulation of circulating free heme after Plasmodium infection, suppressing the pathogenesis of malaria [50]. Such findings provide insight into molecular mechanisms *via* which sickle hemoglobin confers host tolerance to severe forms of malaria, potentially inspiring related treatments and therapies with similar *modus operandi*.

With a focus on the entire cell structure, the multi-scale bridging previously stressed as critical to a biomateriomics approach appears to be abandoned. However, the full comprehension of red blood cell mechanics is far from complete, and provides an example of a *top-down* approach to the investigation of biological mate-

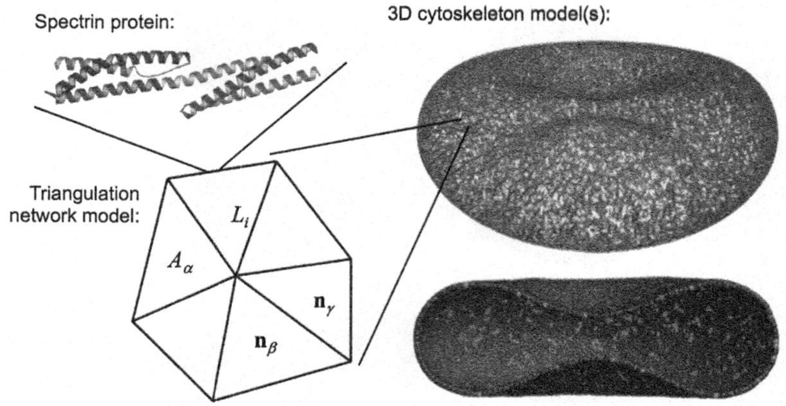


**Fig. 9.7** A "molecular-structure-informed" RBC model, using a coarse-grain molecular dynamics approach (CGMD) to explicitly represent the spectrin network composing the cytoskeleton of a RBC. The triangular elements are parameterized to reflect the bending rigidity of the underlying lipid membrane, and the entropic worm-like chain (WLC) behavior of the spectrin proteins. A fully 3D representation of the cytoskeleton can then be modeled, informed through AFM studies, and mechanically loaded, mimicking experimental assays. Adapted from Li *et al.* [53], used with permission, copyright © 2005 Elsevier Inc.

rials. We note that top-down approaches are complementary to a molecular bottom-up perspective, and both, ideally, should agree for a description of the *materiome*. The cellular level expression of malaria is befitting of such a top-down methodology, as it is not a "molecular disease" such as SCD (*e.g.*, the result of a mosquito carrying the *P. falciparum* parasite, and not a protein mutation).

Reducing the structural scale, the basic building block of the RBC cytoskeleton is the *spectrin* heterodimer, consisting of intertwined $\alpha$ and $\beta$ polypeptide chains running antiparallel to one other [51]. Early mechanical investigations explicitly represented the topological organization of spectrin molecules *via* random Delaunay triangulation, where the elasticity of an individual spectrin molecule is represented by a single spring constant for each linear spring element [52] (effectively an elastic network model, ENM, as discussed in Chap. 6: Computational Approaches and Simulation). These "molecular-structure-informed models" can account for protein-scale structural features, providing an intermediate link between atomistically informed mechanical behavior and continuum level models. More recently, Li *et al.* undertook a three-dimensional computational study of whole-cell equilibrium shape and deformation of human red blood cell (RBC), using spectrin-level energetics directly incorporating the architecture of the network [53] (Fig. 9.7). These models were informed by cytoskeletal networks as imaged by AFM [54]. Assuming the mechanical deformation of the RBC was dominated by the cytoskeleton, coarse-grain potentials were implemented to account for the bending rigidity of the lipid membrane and the entropic worm-like chain (WLC) behavior of the spectrin proteins.

The next step is to parameterize such coarse-grain potentials from full atomistic molecular investigations [55–57], to realize a fully-informed RBC cytoskeleton model, and track potential changes induced by malaria infection, for example. Model refinements, such as introducing breakable protein associations (*e.g.*, actin-spectrin interactions), have successfully elucidated the roles of shear stress, specific chemical agents, and thermal fluctuations in cytoskeleton remodeling [58]. Such models demonstrate that specific biochemical activation is a possible means for RBC deformability to be actively "tuned" during its circulation in microvasculature.

Red blood cells afflicted with malaria parasites demonstrate the need for the apt consideration of biochemical processes, structure-property-process relationships, and ultimate mechanical response of biological materials within a comprehensive framework [59]. Lykotrafitis and co-workers recently analyzed force-displacement curves of SCD RBCs and quantified material stiffness *via* AFM [60] using two theoretical models derived from the Hertz model [60, 61]. It was concluded that deoxygenation and therapeutic treatment have a significant impact on the stiffness significantly increasing the rigidity of SCD erythrocytes [61]. In contrast to normal RBCs, SCD erythrocytes substantially deviate from the characteristic biconcave shape and possess highly irregular morphologies characterized by protrusions, elongation, bumps, and enlargement associated with aggregation of HbS fibres—deviation from idealized spectrin networks. Indeed, the concurrent investigations of SCD and malaria, both expressed and manifested in the mechanical properties of red blood cells, benefit from integrated materiomic methods, wherein similarities and differences in the root molecular causes can be discovered.

## 9.5 Progeria: A Precocious Mutation

Turning away from blood cells, we focus on another disease with a root cause in a single point mutation that manifests in mechanical changes at the cellular level: Hutchinson-Gilford progeria syndrome (also known simply as *progeria*). Progeria is a rare disorder that causes extremely rapid aging (precocious symptoms resembling aspects of aging are manifested at an early age), typically resulting in cardiovascular disease and premature death before age sixteen. The etiology of progeria is marked by a structural defect in the lamin nuclear membrane [62–65]: the deletion of 50 amino acids near the end of the lamin A protein. This particular section of the lamin A tail helps provide structural support for the cell's nuclear membrane (for hierarchical depiction of the cell and membrane, see Fig. 9.8). Unlike SCD, for example, the genetic condition occurs as a new mutation, and is rarely inherited. Similar to SCD, however, progeria is caused by a known and particular point mutation: a replacement of cytosine with thymine in position 1824 of the LMNA gene (responsible for producing lamin proteins), creating a form of the structural protein which cannot be processed properly and accumulates in the nuclear lamina. Seemingly counterintuitive, the flawed material is stronger, not weaker, than its healthy counterpart. Studies have shown that in these mutated proteins, the nuclear skeleton becomes brittle and rigid, which leads to cellular dysfunction, and rapid

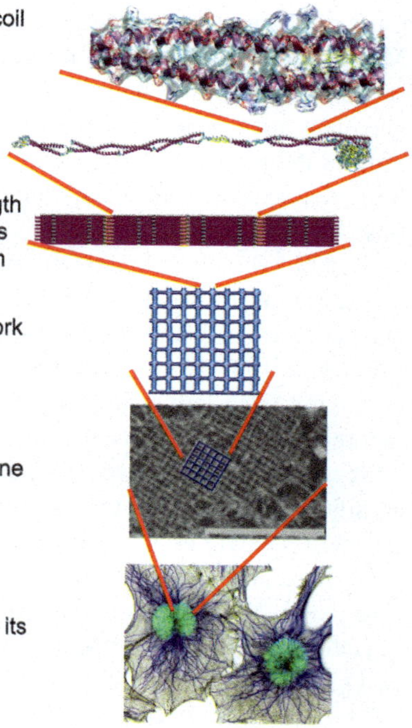


**Fig. 9.8** Hierarchical structure of an intermediate filament meshwork as found in the inner layer of nuclear lamina of *Xenopus* oocytes, depicting organizational structure ranging from nano to macro. Typical molecular structural features of intermediate filament meshwork include hydrogen bonds, coiled alpha-helices constituting a lamin A dimer composed of coiled-coils and linkers, bundles of dimers that fuse laterally to form full length filaments with a diameter of ∼ 10 nm, which form an orthogonal meshwork attached to the inner nuclear membrane with a lattice constant of ∼ 50 nm [66]. The scale bar is 1 μm. Adapted from [67, 68]

breakdown—a process similar to "aging". Simply put, in a structural system, more rigid components take more load. Their rapid deterioration can be compared to uneven brake pads in a car—areas of a brake pad called upon to do more work wear away faster. It is difficult to bridge the gap from the molecular level—where the mutation impacts the protein's configuration and function—to the cellular level where the impact of the misfolding is evident.

The nuclear lamina consists of a two-dimensional matrix of proteins dominated by lamin proteins which provide strength and stability. Many other genetic diseases resulting from structural flaws in the lamin protein network have been identified, generally referred to as "laminopathies" [62–64]. Besides providing mechanical support, the nuclear lamina regulates important cellular events such as DNA replication and cell division. Thus, it can be thought of as both a structural shell (providing mechanical integrity to the cell nucleus) and a gateway for materials/supplies necessary for cellular functioning. Both structural integrity and function are affected by the mutation causing progeria, complicating the understanding of disease expression and mechanisms.

In progeria, the single point mutation in the LMNA gene is indeed a precocious overachiever—it activates a cryptic splice site causing 50 amino acids to be deleted! The resulting mutant protein is called $\Delta 50$ lamin A (or $\Delta 50$ LA) [69]. The deletion in $\Delta 50$ LA includes the effective cleavage site, which is suggested to be responsible for the accumulation of $\Delta 50$ LA at the inner nuclear membrane. However, lack of cleavage cannot explain all of the molecular changes in progeria. Recently, *in vitro* binding assays have shown differential binding of $\Delta 50$ LA to nuclear proteins and

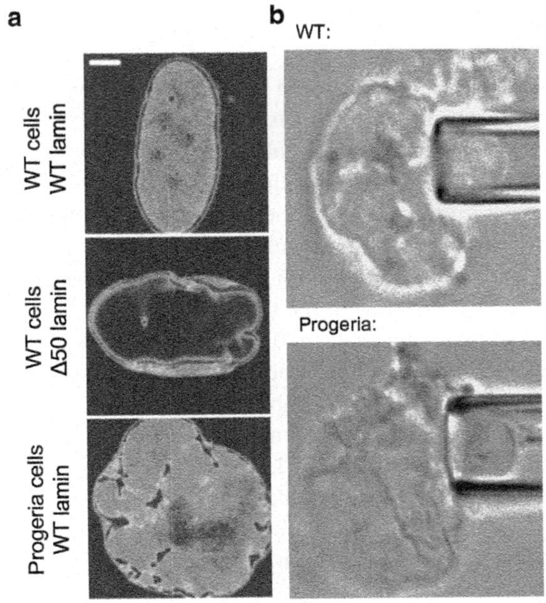


**Fig. 9.9** Effect of Hutchinson-Gilford progeria syndrome on nuclear lamina. (**a**) Wild type (WT) and progeria afflicted cells expressing either WT or $\Delta 50$ LA. (**b**) Micropipetted aspiration of WT (control) and progeria inflected nuclear lamina. The control nuclei buckled relatively smoothly outside the pipette as aspiration increased, indicating that mechanical stresses were distributed across the entire lamina. In striking contrast, the lamina of progeria nuclei collapsed along long prominent fault lines, appearing crumpled or folded. Adapted from Dahl *et al.* [65], used with permission, copyright © 2006 National Academy of Sciences

chromatin [70]. The loss of 50 amino acids from the lamin A tail may alter the protein more than simply effecting cleavage [71]. Complicating the matter, the mutated tail domain of lamin A is mostly disordered and demonstrates the characteristic qualities of intrinsically disordered proteins [72] including a promiscuity in protein binding [73, 74] and a tendency to aggregate [75]. It is difficult to predict how the removal of 50 amino acids in a region lacking secondary structure will affect the overall structure of the protein domain.

The progression of this disease has been associated with mechanical failure of the cell's nuclear membrane in tissues subject to mechanical loading (see Fig. 9.9). A recent study based on live-cell imaging and micropipette aspiration has shown that progeria nuclear membranes display a reduced deformability and feature the formation of fractures upon application of mechanical load [65]. The cause of these fractures is attributed to changes in the lamin microstructure, where filaments form more ordered domains that prevent the dissipation of mechanical stress. This structural alteration leads to a change in the deformation mechanism, from a dissipative mode ("ductile") in healthy cells to a catastrophic, localized failure mode ("brittle") in diseased cells. These mechanisms appear predominantly in cells that are subjected to mechanical deformation, particularly in endothelial and smooth muscle cells of the vascular system. Perhaps a loss of mechanical integrity in the cell's nuclear membrane can influence gene regulation by triggering a wide range of biochemical processes that lead to the rapid aging phenomenon. Linking the behavior of the nuclear membrane to the 50 "missing" amino acids is nontrivial. The exact molecular failure mechanisms remain unknown and its investigation represents an opportunity where a materiomic approach could make important contribution.

Recently, using a combination of theoretical and experimental methods, the atomistic conformations and nanomechanical stabilities of the tail domain of healthy lamin tail and a mutant associated with progeria (the $\Delta 50$ LA tail) was investigated [76]. The study demonstrated that the $\Delta 50$ LA tail is more compact than the healthy lamin tail domain (see Fig. 9.10). More importantly the unfolding of the $\Delta 50$ LA tail domain requires a higher energy barrier to overcome unfolding, likely due to transitioning through a pseudo-stable intermediate state. This phenomenon is in agreement with experimental measurements.

Such findings highlight the need to keep the protein geometry in mind when trying to understand the molecular mechanisms of the disease. Previously, experimental works of healthy and progeria mutated cells have shown a stiffening of cell nuclei from progeria patients and resistance to mechanical pressure, and alteration in the mechanical properties is due to the presence and overaccumulation of the $\Delta 50$ LA protein and associated filaments at the inner nuclear membrane [77, 78]. The etiology of progeria has long been thought to be a function of affected cleavage [79], but molecular results suggest that changes in protein stability may be contributing to the disease pathology.

Furthermore, experimental and computational results quantitatively and consistently show that the $\Delta 50$ LA tail is a more stable structure, as well as a more compact molecule. This suggests an explanation for the increased compactness of *in vitro* reconstituted filaments of full length $\Delta 50$ LA than healthy lamin [80]. This compact filament structure, likely caused by a smaller and more stable $\Delta 50$ LA tail, may also partly be responsible for reduced turnover of proteins from the nucleoskeleton [77, 78]. Filaments that include the more stable $\Delta 50$ LA may more readily aggregate, show reduced turnover, and due to their stronger binding appear mechanically more rigid and therefore "brittle", thereby directly explaining earlier experimental findings [65, 77, 81] based on the geometric and mechanistic insight developed in molecular level studies.

The "structure" of intrinsically disordered proteins is described as an ensemble of conformations which may represent conformations allowing for protein-protein binding to different binding partners and have been shown to be necessary for specific mechanical functions of biological materials including spider silk and talin rods [82, 83]. Given lamin A's proposed role in mechanotransduction, and the multiple binding partners of the tail domain, further study of the conformations of the tail domain relating to mechanical function should be explored in future work [84]. It is not (yet) possible to determine the precise mechanisms responsible for the premature aging process in progeria patients because this requires a much more extensive understanding of the structure and function of nuclear lamins at all levels in the material's makeup. Progeria offers an expedient example of how material failure due to

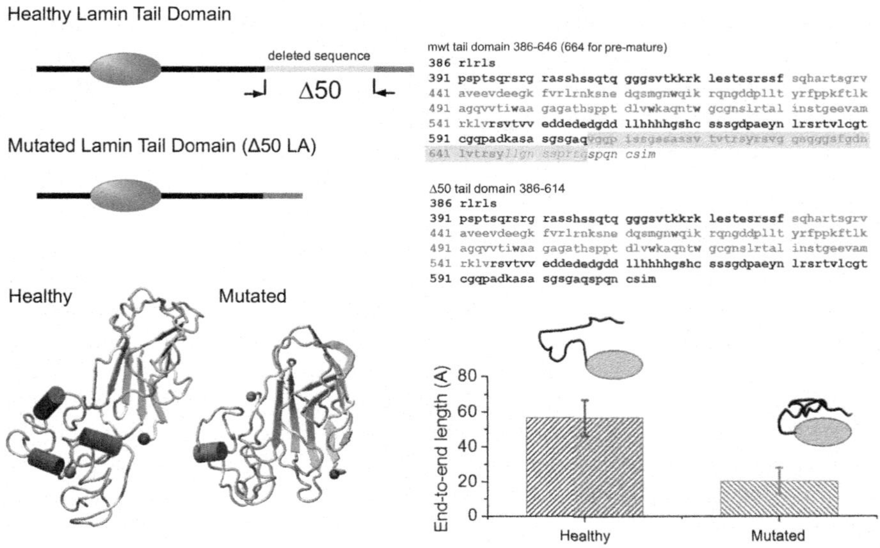


**Fig. 9.10** Molecular effects of progeria mutation. *Top*: Schematic of the amino acid sequences of lamin tail domains for healthy and progeria mutated proteins, indicating the segment of amino acids deleted in the $\Delta 50$ LA tail. The $\Delta 50$ LA, also called progerin, has lost 50 amino acids (shaded). *Bottom left*: Selected conformations and structures of tail domains for normal and disease cases with high significance factors. As the structural conformation of the lamin tails are dynamic and stochastic, these molecular configurations are statistical representations determined by replica exchange molecular dynamics methods. *Bottom right*: The end-to-end length of all the conformations of the healthy and $\Delta 50$ LA tail domains, respectively. The $\Delta 50$ LA tail domain is significantly more compact, even when considering the loss of amino acids in the truncation. Adapted from [76]

structural flaws (missing amino acids) within a protein material (lamin) can lead to the breakdown of critical biological components (nuclear lamina) through complex structure-property relationships. Thus far, we have limited our discussion to changes in cellular function. We shall see this is not always the case.

## 9.6 Brittle Bones

In cases where a disease or condition arises from a genetic miscue—such as sickle-cell disease or progeria—the aim of materiomics is to link the mutations at the nanoscale with the behavior at the macroscale with a bottom-up approach. The expression of SCD (as well as malaria) and progeria are ultimately at the cellular level (red blood cells and nuclear lamin, respectively), the "mechanistic" response of the mutation was uncovered by research efforts. For other mutations, the expression and change in material behavior is much more clear and apparent—such is the case of *osteogenesis imperfecta*, also known as brittle bone disease.

Genetic disorders in collagenous tissues have been linked to the alteration of the material structure due to mutations in the genes that encode the tropocollagen molecule. Diseases associated with collagen abnormalities have severe consequences, because collagens are major structural proteins within most human body tissues (*e.g.*, tendon, skin, bone) that are crucial for the mechanical integrity of organisms. *Osteogenesis imperfecta* (abbreviated OI) is a genetic disorder in collagen characterized by mechanically weakened tendon, fragile bones, skeletal deformities, and in severe cases, prenatal death [85]. Patients affected by OI exhibit an array of symptoms, including short stature, loose joints, blue sclerae, dentinogenesis imperfecta, hearing loss, and neurological and pulmonary complications [85]. In addition to the effects of OI mutations on bone tissue, there are also severe changes to mechanical properties of other connective tissues such as tendon.

As discussed in previous chapters, bone itself is a complicated composite combining very hard and brittle material, the mineral, with much softer material, the collagen fibrils [86]. Yet a single mutation in the softer material—not the stronger mineral—fundamentally changes the properties of bone and enhances the susceptibility for catastrophic brittle failure. Some abnormalities in the composite hierarchical structure from the nanometer scale to the macroscopic scale have been suggested as possible sources for the brittleness of OI bone. These include the presence of both hypo- or hypermineralized regions [87] and changes in collagen cross-links [88], mineral crystallinity [88], and collagen fibril diameters [89].

The origin of this disease resides in changes to the structure of tropocollagen molecules due to the substitution of a single glycine amino acid [90, 91] (see Fig. 9.11). The classification of OI is based commonly on clinical features that leads to four major groups, from mild (OI type I) to severe (OI type III and IV) to perinatal lethal (OI type II) [92]. The genetic basis for approximately 90 % of all known forms of this disease lies in mutations of type I collagen genes [93], as tabulated in the database of human collagen mutations [94]. Missense mutations that alter a glycine (Gly) codon in the genes encoding the characteristic collagen triple helix are the most common causes of OI [93]. The replacement of either guanine (G) residue in the glycine codon GGC can theoretically result in the replacement of eight different amino acids: serine (Ser), cysteine (Cys), alanine (Ala), valine (Val), aspartic acid (Asp), glutamic acid (Glu), arginine (Arg), and tryptophan (Trp). All possibilities have been described in association with OI, although the frequency by which the different mutations occur varies considerably, with tryptophan replacements being exceedingly rare. While each mutation has been linked to OI, the mechanism underlying the expression is not understood, and only general correspondence between the specific mutations and the severity of OI has been reported.

Some collagen mutations prevent the formation of triple helical molecules (procollagen suicide), while other mutations cause structural changes to tropocollagen molecules, leading to bending (*e.g.*, due to kinks induced by amino acid substitutions), reduced mechanical stiffness (*e.g.*, due to changes of the volume and hydrophobicity), or changes in the intermolecular adhesion (*e.g.*, due to changes in surface charges) [90, 91, 95]. At mesoscopic length-scales, these molecular-level changes lead to poor fibril packing [96, 97] and a decrease in cross-link density

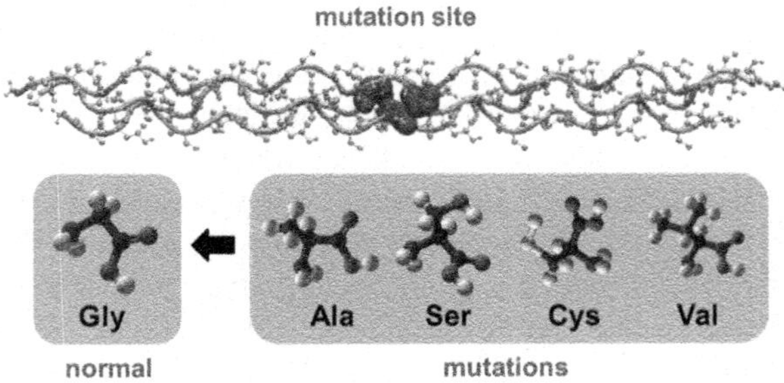


**Fig. 9.11** Molecular geometry of the tropocollagen molecule, indicating the residue that is replaced in the *osteogenesis imperfecta* mutation (thick visualization). Tropocollagen-like peptides share a common structure, consisting of three chains made of Gly-Pro-X triplets (where X may be any of various other amino acid residues). The *osteogenesis imperfecta* mutation replaces a single Gly position of each chain (highlighted by thick visualization) with one of eight replacing residues related to *osteogenesis imperfecta* (four possible mutations shown)

[97, 98]. Changes in the size and shape of mineral crystals in bone (*e.g.*, less organized, more round-shaped crystals) have also been reported [99–101], which might be related to a change in the ability of tropocollagen to bind to the mineral phase of bone [85, 97]. At larger length-scales, the effects of *osteogenesis imperfecta* mutations lead to inferior mechanical properties of tendon and bone [102].

It remains unclear at what level in the tissue structure and how a single point mutation influences the behavior. Previous studies have provided a general correspondence between the specific mutation types with phenotypic severity, however, the molecular and mesoscale mechanics by which a single point mutation influences the mechanical behavior of collagenous tissues at multiple length scales remain unknown. For instance, a mechanically inferior collagen matrix in addition to an increased and less organized mineral content, and an overall reduced bone volume [103], might explain the phenomenon of brittle bones [97]. Other studies have shown that—assuming prestrained collagen molecules carry most of the tensile load in normal bone while the mineral confers rigidity and compression stability—the inherent brittleness could be linked to tensile behavior of the collagen (*via* a dramatic reduction of ultimate tensile strain) [102]. To address these points, investigations at all relevant hierarchical levels must be carried out, beginning at the molecular level.

Using current multiscale approaches a series of systematic computational experiments were recently reported, focused on pure collagenous tissue and collagen fibrils to investigate the effect of *osteogenesis imperfecta* mutations on single molecule properties, changes to intermolecular interactions, and changes to the mechanical properties of collagen fibrils [95, 104]. The fundamental question addressed is how it is possible that a single point mutation at the level of a single tropocollagen molecule can lead to the brittle behavior and failure of macroscopic tissue? The preliminary step is to identify the mechanisms and changes in behavior of the tropocollagen molecules (atomistic scale) and collagen fibrils (nanoscale). It was demonstrated that the mutations cause a fundamental change in stress distribution within the mutated collagen fibrils, due to the formation of nano-cracks that cause local stress concentrations at the mutations that induce material failure *via* intermolecular slip. Mutations that lead to the most severe OI phenotype correlate with the strongest effects, leading to weakened intermolecular adhesion, increased intermolecular spacing, reduced stiffness, as well as a reduced failure strength of collagen fibrils. Ultimately, the molecular-scale models predict a softening that the emergence of these stress concentrations could play a role in the physiological effects of the disease as they lead to macroscopically weaker connective tissue, including tendon and in particular bone.

Three major effects of OI mutations on the mechanical properties of collagenous tissues were identified. First, at the single molecule level, where molecular softening occurs as the disease severity increases. Second, at the intermolecular level, where mutations lead to a weakening of intermolecular adhesion and increase of intermolecular equilibrium spacing as disease severity increases, leading to a reduction in likelihood of cross-link formation. Third, at the collagen fibril level, where OI mutations lead to reduction in strength through a change in the stress distribution within fibrils, and specifically the formation of localized stress concentrations that lead to early material failure. Note that each effect can be tied to a specific structural hierarchy—from single molecular behavior, to the interaction between two molecules, to a collection of molecules, wherein the mutations induce different effects accordingly (see Fig. 9.12).

The observed changes in the intermolecular interaction and in the mechanical properties of collagen fibrils could be a major cause of structural and property changes of collagenous tissues, leading to a mechanical deterioration of fibrils and mineralized fibrils. For collagen fibril properties, the location of the mutations is a crucial factor in determining their impact on material properties, with mutations close to the ends of molecules featuring the strongest effects. Similar effects of the glycine replacement location along the triple helix on the severity of the resulting type of OI have been previously reported in experiment, showing that mutations located closer to the terminals (in particular the C-terminus) of the tropocollagen molecule lead to more severe phenotypes [105, 106], in agreement with simulated cases [104]. Such effects are important to better understand the molecular-scale based foundations of this disease. Namely, elucidation of how nanoscopic events involving only tens to hundreds of atoms can trigger tissue failure with potentially system wide catastrophic consequences. Several key features of mutated tropocollagen molecules provide potential (yet preliminary) explanations:

- *Poor fibril packing*. The alternation of collagen strands with normal interactions and strands with increased repulsion due to the presence of glycine replacements may hinder normal fibril packing, leading to a partial loss of the lateral crystallinity, a feature often observed in OI tissues [96, 97].
- *Decrease of cross-link density*. At a larger length scale, the molecular effects of OI mutations observed can help to explain the poor mechanical properties of tendons: the poorer fibril organization and the larger distance between tropocollagen molecules (that is, less resistant material per unit area) in conjunction with the lower number of cross-links may account for the reduced mechanical strength

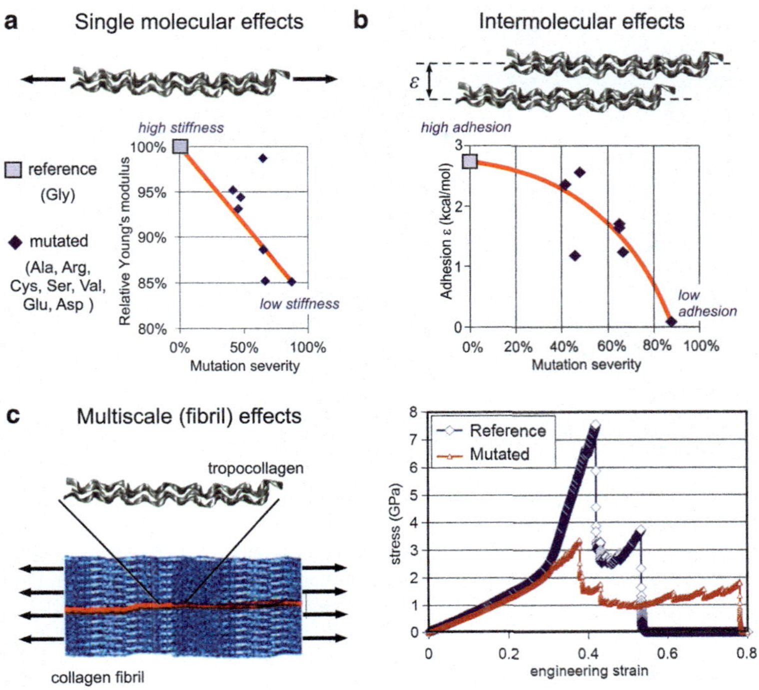


**Fig. 9.12** Hierarchical multi-scale effects of *osteogenesis imperfecta* (OI) mutation. (**a**) Single tropocollagen molecule: reduction of Young's modulus as a function of the glycine replacement and OI severity. The elastic properties of a tropocollagen molecule are depicted as decreasing as amino acid residues are substituted, which are ordered based on the resulting disease severity (lowest to highest severity: glycine (Gly; reference value), alanine (Ala), arginine (Arg), cysteine (Cys), serine (Ser), valine (Val), aspartic acid (Asp), glutamic acid (Glu) [95]). More severe mutations tend to lead to a greater reduction of the elastic modulus, with greatest reduction of 15 %. (**b**) Intermolecular effects: change of tropocollagen adhesion profile. Plot of adhesion energy $\varepsilon$ as a function of OI severity, including the reference case (Gly). As the mutations become more severe, the adhesion energy decreases. The decrease correlates directly with the clinically measured severity of various types of mutations [102]. (**c**) Collagen fibril: influence of OI mutations on the mechanical properties of a collagen fibril, leading to a significant reduction of mechanical strength and yield strain. The stress-strain curve for the reference (healthy) case and the mutated (OI) case, showing that the presence of mutations can severely influence the overall mechanical signature (highly cross-linked fibril). Intermolecular sliding sets in at 30 % vs. 42 % in the mutated fibril, and the maximum stress is significantly reduced (to ~50 % of its reference value). The small-deformation elastic modulus is reduced by ~15 % under the presence of the OI mutation. Adapted from [104]

of tendons typical of OI [102]. Eventually, a mechanically inferior collagen matrix plus an increased and less organized mineral content might lead to the phenomenon of brittle bones, an important and physiologically evident feature of OI [97]. Further studies will be required to provide more mechanistic insight into these specific mechanisms beyond the pure collagen phase. Such studies should

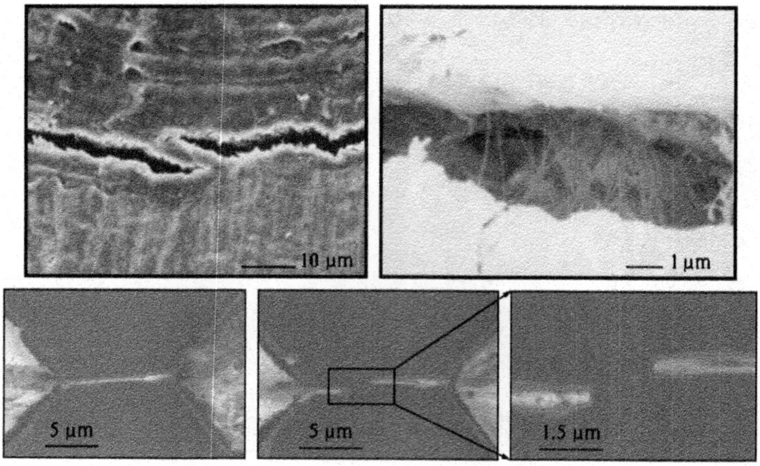


**Fig. 9.13** Experimental investigations exploring the fracture in bone. *Top*: Evidence of extrinsic toughening mechanisms in bone fracture, depicting crack bridging (*left*), and collagen fibril based bridging (*right*). *Bottom*: SEM images of a strained and broken collagen fibril. Note that these examples were performed on healthy, non-mutated collagen and bone. Adapted from Eppell *et al.* [107], used with permission, copyright © 2008 Elsevier Inc.

entail an explicit consideration of the interface between collagen and hydroxyapatite mineral crystals and it changes under the presence of mutations.

- *Changes in mineral crystals size and shape.* A poorer lateral packing of collagen molecules may affect the growth of hydroxyapatite crystals. In an environment characterized by poor fibril packing, mineral crystals fail to grow highly organized in the fibril axis direction. Rather, they might be less organized and round shaped instead of platelet-like. This feature would be consistent with the observed mineral crystals size and shape found in OI [99–101].

Such bottom-up nanomechanical mechanistic analysis of the effects of OI are critical in understanding behavior and properties at larger scales from a materials failure perspective (compared to the failure of healthy bone/collagen depicted, for example, in Fig. 9.13). A better understanding of the molecular mechanisms underlying OI (and other such "molecular diseases") could eventually enable a direct link from the scale of genetic mutations to the macroscale phenomenon of brittle bones. It should be emphasized that, while the mutation occurs in the collagen sequence, additional work is necessary to study the specific role of the mineral phase in this disease in particular for bone tissues. Undoubtedly, the severity of the OI mutation types and their effect on intermolecular interaction are correlated. The development of understanding of how structural modifications lead to the changes in the mechanical properties of higher hierarchical levels is critically important on the path of forming a more holistic picture of OI. The results of such computational studies of collagen at the fibril level might be directly accessible to experimental studies using MEMS devices [107], for example, allowing validation of proposed mechanisms (Fig. 9.13).

Such types of investigations might help to shed further light on mechanical effects of OI mutations, especially those expressed at larger scales. Other future studies could focus on the effect the position of the mutated protein, and what the effect of other triples surrounding the mutated glycine (that is, instead of the stabilizing Gly-Pro-Hyp sequence) would be. OI is an autosomal dominant disease, and thus mutated structures arise even if only one of the two copies of the collagen gene encoding is mutated. Depending on the assembly of the chains into tropocollagen trimers, we may find molecules with zero, one, two, or three mutations at the same position. It remains an open question what the effect of different numbers of mutations is. Future studies could be focused on the question whether or not mutation effects are additive (that is, two mutations lead to more severe effects than one mutation alone), or if it is an on/off process (that is, the effects do not change even if there is more than one mutation in the tropocollagen molecule).

We stress that referenced studies are the culmination of both multiscale hierarchical constitutive material models for collagenous tissues [108, 109] as well as known clinical and experimental data for normal collagen and the fundamental point mutations of *osteogenesis imperfecta* [92, 94, 105]. The citations referenced are again neither intended to be exhaustive nor canonical, and merely scratch the surface of the research that has pieced together the puzzle of brittle bone disease. It is because of these advancements in both simulation and experimental approaches, combined with the growing knowledge base of the behavior of biological protein materials that a fundamental multiscale understanding of OI is even possible. The integration and reciprocation of computational methods with empirical data across scales is a keystone to biomateriomics.

The benefits of a deterministic mechanistic model from nanoscale point mutations to macroscale effects (such as brittle bones) are self-evident, and similar approaches could be used for many other diseases and processes in which materials failure due to a drastic change of the materiome behavior play a crucial role in disease initiation and progression. Indeed, similar to OI and discussed in Chap. 8: Unlocking Nature: Case Studies, a deterioration in the bone toughness has been correlated with aging [110, 111], as a result of a variety of nano- and/or microstructural changes including progressively larger osteonal dimensions and densities [112], increased cross-link densities [113], and increased microcracking. The recent study by Ritchie and co-workers [114] helped elucidate that age-related structural changes can significantly degrade the fracture resistance (bone *quality*) over multiple length scales, using X-ray scattering, diffraction and tomography [114] (this particular investigation was also discussed in the previous chapter). Knowledge of the mechanical deterioration of healthy aging bone, combined with the mechanisms of brittle bone disease, are inherently combined in the materiomic description of the material—we must fully understand the idiosyncrasies of functioning tissues and biological systems to comprehend (treat and/or prevent) diseases (failure). The key insight put forth here is that for a comprehensive understanding of disease states such as brittle bone disease, an integrated view of material and structure at multiple scales is critical to link physiological mechanisms, clinical evidence, and to develop potential treatment options.

While we have discussed the afflictions that fundamentally change the behaviors and normal functioning of existing biological components (red blood cells, cellular lamina, and tendon or bone), we now turn to the case in which Nature's mistakes introduce an entirely foreign material into the mix. Such is the case for misfolded and aggregated protein materials and cancer—the accumulation and growth of extraneous materials that have no redeeming or useful natural functions in a healthy system.

## 9.7 Protein Aggregation and Alzheimer's Disease

What can occur when proteins stop functioning properly? Alzheimer's, Parkinson's, type II diabetes and prion diseases [120–123] have been linked to the formation of foreign material deposits in tissues—previously benign and functional proteins that "misfold" and aggregate. A wide range of natural and artificial peptides and proteins possess an intrinsic propensity to self-assemble into fibrillar nanostructures. These fibrils are generally organized in a similar manner at the molecular level—they are characterized by $\beta$-strands that are oriented perpendicularly to the fibril axis, and connected through a dense hydrogen-bonding network, which results in supramolecular $\beta$-sheets that often extend continuously over thousands of molecular units [124–126]. Functional amyloid-like materials were discovered in varying roles throughout Nature [116–119, 127, 128]. Amyloid materials, which are present in cells and in the extracellular space, represent a class of nanoscale structures that have various functional and pathological roles (see Fig. 9.14). As such, amyloid nanostructures are increasingly viewed as a general alternative form of protein structure that is different from, but in many cases no less organized than, the native states of proteins. Moreover, this type of structure does not depend primarily on highly evolved side-chain interactions (*diversity*)—associated disease states are not dependent on specific mutations. Rather, *universal* physical and chemical characteristics that are inherent in the nature of all polypeptide molecules (*e.g.*, a propensity for hydrogen bonding in the backbone) result in the structure formation [124]. Figure 9.15 shows the hierarchical structure of amyloid materials, and also presents an analogy to the structure of language (see also the discussion in Chap. 4: Universality-Diversity Paradigm: Music, Materiomics, and Category Theory) [115].

Such fibrillar structures first received attention through their association with diseases related to protein misfolding, including neurodegenerative disorders such as Alzheimer's and Parkinson's diseases, where normally soluble proteins are deposited pathologically into obdurate aggregates known as amyloid fibrils [129]. This link is a relatively recent discovery—in 1985 it was first suggested that the neurodegeneration in Alzheimer's disease may be caused by deposition of amyloid $\beta$-peptide (A$\beta$) in plaques in brain tissue [130], where accumulation of A$\beta$ in the brain is the primary influence driving the disease pathogenesis. The rest of the disease process, such as the formation of tau protein neurofibrillary tangles, is proposed to result from the aggregation of amyloid $\beta$-peptides [121]. This marked the begin-

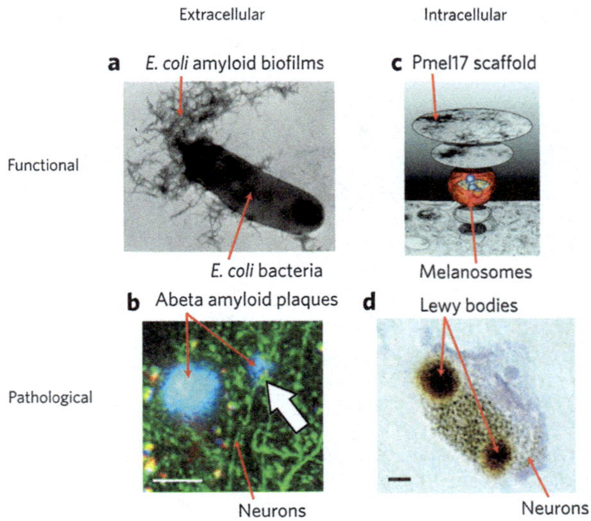


**Fig. 9.14** Classification of amyloid materials. Amyloid materials can be extracellular (**a**, **b**) or intracellular (**c**, **d**), and functional (**a**, **c**) or pathological (**b**, **d**). (**a**) Functional amyloid in biofilms produced by bacterial species such as *E. coli* and certain *Salmonella spp.* (**b**) Amyloid plaques as seen in a mouse model of Alzheimer's disease (scale bar, 20 μm). The large white arrow shows a newly formed plaque. (**c**) Transmission electron microscope image of Pmel17 scaffolds in melanosomes involved in the biosynthesis of melanin. (**d**) Lewy bodies, pathological protein aggregates that develop in neurons in Parkinson's disease (scale bar, 8 μm) (From [115], *insets* reproduced with permission: (**a**) [116], © 2007 National Academy of Sciences; (**b**) [117], © 2008 Nature Publishing Group; (**c**) [118], © 2009 ASBMB; (**d**) [119], © 1997 Nature Publishing Group)

ning of the modern era of research on this common, devastating neurodegenerative disease, and the first steps in understanding the *molecular etiology* of dementia.

These accidental material deposits, referred to as amyloid plaques, are highly ordered hierarchical assemblies of $\beta$-sheet protein domains that form spontaneously [133–137]. Once formation is initiated, amyloid plaques grow uncontrollably to length-scales of micrometers and persist under a wide range of pH conditions. The self-driven assembly of such materials has motivated the use of amyloid fibrils in synthetic systems, such as biocompatible films. Triggered at the molecular level, the growth and self-assembly of amyloidogenic materials and similar protein aggregates is a promising technique of material synthesis [115], requiring little control or energetic expense (such fabrication techniques are discussed further in Chap. 10: Synthesis and Design). However, it is this self-driven assembly that creates a problem within biological systems: once these delinquent proteins begin to misfold and aggregate, the process is very difficult to halt or reverse. Amyloid fibers and their various manifestations illustrate the importance of being able to predict the folding and unfolding of natural protein-materials, predicated by the amino acid sequence. In the words of Chris Dobson (University of Cambridge) [124]:

> The manner in which a newly synthesized chain of amino acids transforms itself into a perfectly folded protein depends both on the intrinsic properties of the amino-acid sequence

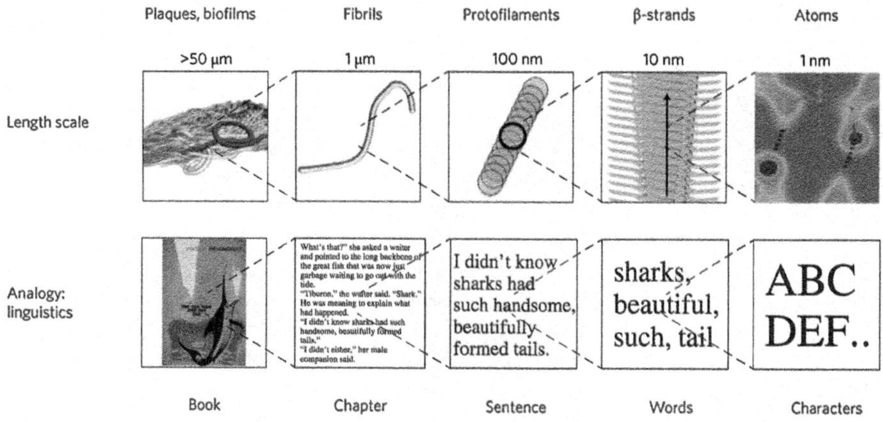


**Fig. 9.15** The hierarchical structure of amyloid materials. *Upper panel*: Five different levels of hierarchy in the structure of amyloid materials from atoms to plaques. *Bottom panel*: Hierarchical structure in linguistics as an analogy to demonstrate how functional properties emerge owing to the hierarchical assembly of simple building blocks. Adapted from [115]

> and on multiple contributing influences from the crowded cellular milieu... Aggregation of misfolded proteins that escape the cellular quality-control mechanisms is a common feature of a wide range of highly debilitating and increasingly prevalent diseases.

Only recent progress in the application of solid-state NMR (ssNMR) [138–140] and in growing peptide elongated microcrystals has provided detailed structural and biochemical information on amyloids [141]. In particular, these studies revealed that the molecules composing the fibrils posses a degree of order that has previously only been associated with *crystalline* materials (such as metals or ceramics), and not with proteins. In fact, X-ray diffraction patterns of aligned amyloid fibrils allowed to define the structural core as a set of $\beta$-sheets oriented in parallel to the fibril axis, with their strands perpendicular to this axis [141]. Since this regularity can be detected through the resonance lines of ssNMR, it is expected that complete atomistic-level structures will be available for many other amyloidic systems (whose detailed description is currently still missing). Despite progress in structure identification, many of the fundamental properties are not fully understood. This is partly due to the fact that larger scale structural models of amyloid fibrils remain elusive, preventing bottom-up theoretical and computational studies to describe their nanomechanical properties. Most known structural models of amyloid fibrils are limited to lengths of several nanometers (or a few layers of protein). However, relevant functional and biologically active properties (and properties for potential nanotechnology applications) emerge at scales of hundreds of nanometers and micrometers, encompassing interactions of amyloid plaques with neuron cells and other tissues (see Fig 9.16).

In the last few years, significant progress has been made in theoretical and computational studies providing the first models of the peptide-level structure of amyloids, as well as for the formation mechanisms of the amyloid fibrils [142] (see Fig 9.17). These studies have been focused on a range of different systems that in-

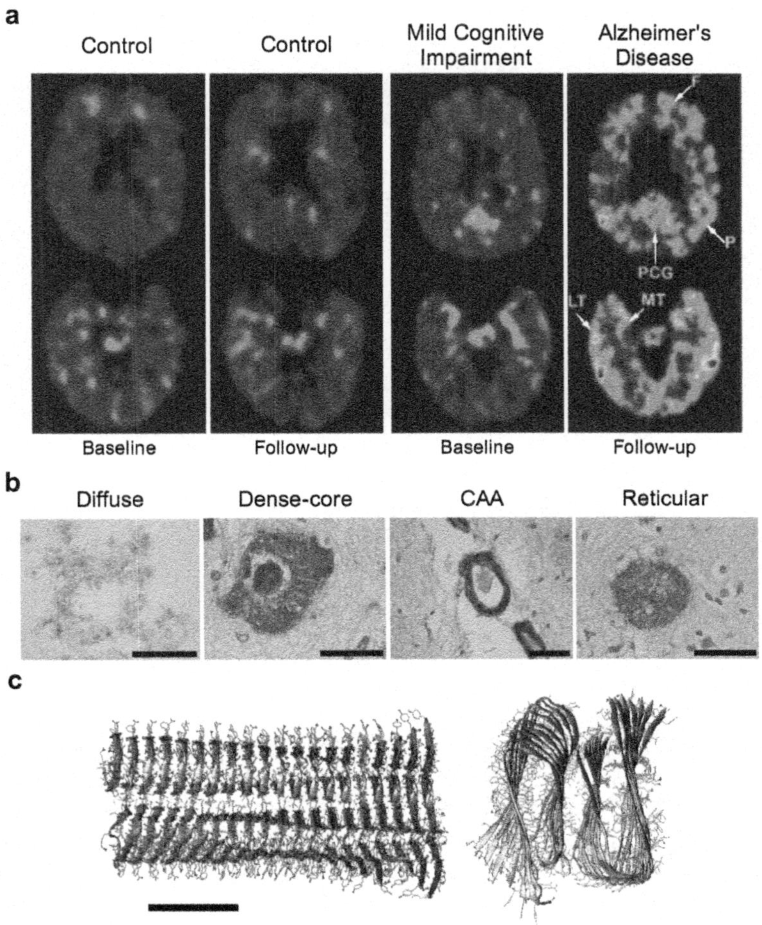


**Fig. 9.16** Multiscale depictions of amyloidogenesis and Alzheimer's Disease (AD). (**a**) PET scans in the parietal region (*top*) and the temporal region (*bottom*) in one control subject and one subject with mild cognitive impairment who was reclassified on follow-up as having Alzheimer's Disease. Scans of the subject with mild cognitive impairment, who was reclassified as having Alzheimer's disease, showed increased amyloid and tau aggregation (*shaded*) in the frontal, parietal, and lateral temporal regions. From [131], used with permission, copyright © 2006 The New England Journal of Medicine. (**b**) Types of amyloid deposits frequently observed in AD patients. Scale bars represent 25 μm. From [132], used with permission, copyright © 2011 Elsevier Inc. (**c**) Molecular model of amyloid fibril, depicting characteristic cross-$\beta$ structure. Scale bar represents $\sim 2.5$ nm

clude primarily smaller sequences, such as polyalanine peptides [143], A$\beta$-peptides [144, 145], calcitonine [146] and $\beta$2-microglobuline [147–149]. Some studies focused on larger structures like prion proteins [150]. While diverse sequences, such structures show universal behaviors in terms of fibril formation and aggregation, contributing to the understanding of the disease-specific peptides and the distinct functionality of the amyloid $\beta$-peptide. In 2002, a computational study [151] and

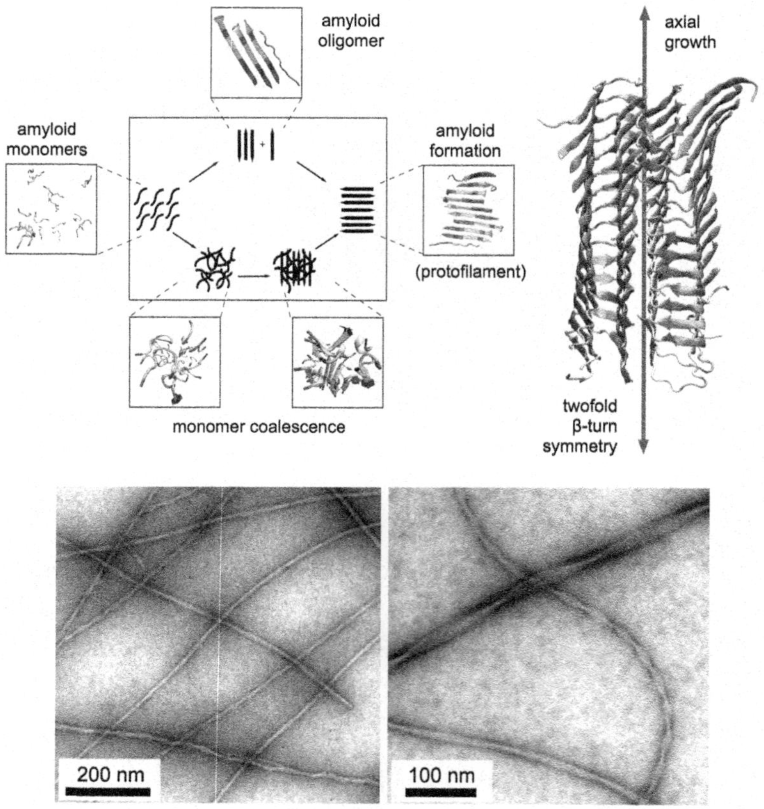


**Fig. 9.17** Amyloid fibril formation. *Top*: Schematic diagram of two mechanistic pathways to protofilament/fibril formation (formation of $\beta$-stacked oligomers). *Upper path*: An effective one-step process involving the assembly of monomers directly into $\beta$-sheet rich oligomers, followed by sequential (piecewise) growth. *Lower path*: A more general two-step process, where the monomers coalesce to form oligomer aggregates before undergoing a process of conformational conversion. The boxes show representative structures from simulations. Schematic adapted from Cheon *et al.* [137]. *Right*: Molecular model of one possible fibril structure with twofold $\beta$-unit symmetry, depicting characteristic cross-$\beta$ axial growth. *Bottom*: Transmission electron microscope images of amyloid fibrils formed by the 40-residue $\beta$-amyloid peptide associated with Alzheimer's disease A$\beta$(1-40)

an experimental investigation [138] independently proposed a rather similar model for the A$\beta$(1-42) peptide amyloid. Theory and experiment agreed rather well, and both found that the most basic peptide building block is a U-turn bent $\beta$-sheet (also referred to as "$\beta$-unit"; Fig. 9.17).

This A$\beta$(1-42) peptide is directly related to Alzheimer's disease, formed after sequential cleavage of the Amyloid Precursor Protein (APP) that is expressed in many tissues and concentrated in the synapses of neurons. The successive actions of the $\alpha$- and $\gamma$-secretases generates the A$\beta$ proteins that can be present in many different isoforms. Among them, the most common are the A$\beta$(1-40) and A$\beta$(1-42) structures. These peptides, and in particular their associated amyloidic fibrils, are thus being extensively studied especially by ssNMR, and a wide range of information on their atomistic structures and configurations are now available [138, 152]. The basic structure is a single $\beta$-unit stabilized by a salt bridge and the overall fiber consists of the repetition of units of these along the fiber axis. These formed layers stack in register, and the stability of the resulting fibrous structure depends on a dense and highly organized network of hydrogen bonds (similar to the $\beta$-crystalline regions found in spider silk). In a recent article, the presence of two distinct morphologies for the A$\beta$(1-40) protein has been suggested [140], a twofold and threefold morphology. Specifically, Tycko and coworkers observed two types of amyloid fibers that are periodically twisted around the growth axis, confirming the existence of polymorphism in amyloid fibrils [140]. Both morphologies contain in register parallel $\beta$-sheets, but differ in the overall symmetry (that is, twofold versus threefold), the conformation of non-$\beta$-strand traits, as well as certain quaternary contacts. According to experimental imaging, the twofold fibril morphology is characterized by a striated ribbon, while most of the threefold fibrils display a periodically twisted morphology. In the last case, a molecular-level model of short segments of amyloid fibrils composed of 6 layers (of 3 nm length) has recently been described in the literature based on ssNMR data [138, 140]. However, this study did not result in larger-scale structural models of amyloid fibrils, preventing a direct comparison with TEM based imaging of amyloid fibrils. More recently, the twist angle, periodicity and atomic structure at scales of hundreds of nanometers has been determined by even larger scale atomistic simulations [153–155], made possible by the advancement in computational power and the molecular details determined by previous computational and experimental research. Such investigations provide the data necessary to parameterize larger scale models [156], providing a bridge between atomistic- and molecular-level simulations and fibril aggregations, plaques, and plaque-tissue deterministic modeling.

The conversion of biologically functional proteins from the native soluble state to the amyloid configuration causes the catastrophic loss of their solubility, and most importantly, their functionality. It is understood that the formation of amyloid plaque deposits leads to neurotoxicity, which interferes with the biological function of the native tissue [120, 121]. The mechanical robustness of amyloid plaques has been attributed to the large number of hydrogen-bonds as well as to steric and hydrophobic interactions between different parts of the $\beta$-sheet structures [124, 136, 157]. However, the specific functional role of amyloids in medical disorders such as Alzheimer's disease, Parkinson's disease, and type II diabetes is not well understood, resulting in a need to further identify their structure and properties to clarify formation mechanisms and structure-property links across all scales. This requires a combined experimental, theoretical, and computational perspective, delineating the similarities between such misfolded protein structures and the resulting effects that propagate across scales (affecting, in the case of Alzheimer's disease, the functionality of neurological tissues, for example).

The most "guilty" A$\beta$(1-40) and A$\beta$(1-42) amyloid structures provide a starting block to study the links between amyloids and disease, and the properties of fibrils have been begun to be probed *via* AFM techniques and molecular dynamics simulations [135, 158, 159]. However, this is still merely an initial step in the understanding of the pathology of a complex affliction such as Alzheimer's which *cannot* be explained by the *sole* characterization of amyloid $\beta$-peptides. For example, even at the molecular scale one must fully consider the interaction with tau protein neurofibrillary tangles, or the affects on $\alpha$-, $\beta$-, and $\gamma$-secretase cleavage [121] (which produce the $\beta$-peptides), let alone the larger-scale interactions between aggrieved synapses and neurites of brain neurons. This problem is far from fully solved, but the necessary toolbox is beginning to take shape.

Being said, the mechanical robustness of amyloid plaques and the body's failure to eliminate these material deposits remains as a primary reason for our inability to reverse the progression of this disease. An improved understanding of how the hierarchical structure of amyloid plaques contribute to their extreme mechanical stability could lead to new strategies for treatment through targeting selective breakdown of these material deposits *in vivo*. Through this understanding, and further trials of anti-amyloid therapies, we can learn the fundamental basis of Alzheimer's Disease, the most devastating and widespread disease affecting the human intellect.

## 9.8 Outlook and Summary

The previous discussions were focused of a very small selection of specific diseases in which there has already been significant progress in linking molecular changes to disease expression and pathogenesis. Since the advent of "molecular medicine" through the investigation of sickle-cell disease, there have been continuous gains in understanding of the genetic and molecular etiology of many diseases. Recent advancements in experimental and computational abilities have further opened the floodgates to such investigations, for the first time having the potential to link atomistic changes (such as an interchange of amino acids) to system-level behaviors (such as brittle bones). Knowing how such minute deviations progress and intensify across scales is essentially to developing new therapies and treatments that can themselves target the smallest and least evasive scales. For example, scientists are currently developing nanoparticles to transport time-release anticancer drugs directly to cancerous cells, developing drugs that fight diseases without damaging healthy tissues and cells, and improving new predictive models of disease [160].

Approaches as discussed in this chapter have also been employed for the more complex condition of cancer. Due to the variegated and heterogeneous classes of cancer, it is difficult to choose a representative case for discussion. Suffice to say, there are many previous and ongoing studies exploring cancer across a multitude of scale. For instance, the change of cell stiffness in cancer cells has been shown by using a microfluidic optical cell stretcher [161]. Studies of AFM indentation of cells confirmed that cancer cells display a reduced stiffness compared with that of healthy cells [162]. This mechanical signature of cancer cells could perhaps be used to define novel diagnostic approaches in cancer detection. The approach based on microfluidic stretching [161] may provide an effective high-throughput platform for cancer diagnosis. Complementary, the simulation of cancerous tumors and cells across multiple biological scales in space and time is increasingly being recognized as a powerful tool to refine hypotheses, focus experiments, and enable more accurate predictions and theories through the combination of biophysical, biochemical, and biomechanical factors (for a recent review of multiscale cancer modeling, see Deisboeck, Wang, *et al.* [163]). More theoretical approaches—such as the formulation of growth laws for tumors [164, 165] or cancer invasiveness [166, 167]—can be validated through experimental observations and directly applied for diagnostic purposes.

Consideration of how material properties change in diseases could lead to a new paradigm that may expand beyond the focus on biochemical readings alone. This will eventually be important for both disease diagnosis and treatment. The development of constitutive material models for protein materials in the context of genetic disease provides an initial understanding of the disease state, but it is not the entire story. Further work must be done to validate such models through quantitative comparison with experiments, to link the predictions to larger tissue levels, and to incorporate additional structural features into the model. Similar approaches to those presented could be used for many other diseases in which materials failure due to a drastic change of the constitutive behavior of a material component plays a crucial role in disease initiation and progression.

While sickle-cell disease, progeria, and *osteogenesis imperfecta* have well-defined biological "targets" (red blood cells, nuclear lamina, and collagen, respectively), the future of pathological materiomics lies in more complex disease states (such as cancer), wherein the symptoms cannot be explained by a multiscale "snapshot" of a stricken cell or tissue. As we have seen in the case of Alzheimer's Disease, for example, extensive knowledge of the formation of amyloid fibrils and plaques is only a primary step in understanding the condition. Similarities and differences must be discovered that link the amyloidogenic routes of Alzheimer's Disease, Parkinson's Disease, type II diabetes, and other afflictions associated with fibrous protein aggregates. Through the abstraction and categorization theories discussed in Chap. 4: Universality-Diversity Paradigm: Music, Materiomics, and Category Theory, isolation of the fundamental (and universal) building blocks and functionalities characterizing amyloidosis may uncover a common connection, and lead to a collective cure, unattainable by attacking such affliction in a piecemeal or disjoint manner. This is not to say that disease-specific investigations are gratuitous—they are necessary to provide a complete understanding, and, as the saying goes, "the devil is in the details". It is only recently that the technology exists to uncover the data required to understand such complex biological failures, to reverse engineer from the smallest atomistic scale and learn from mistakes. Immediate progress lies in the amalgamation of the details, the convergence of disciplines for a shared objective, and a holistic perspective promoted by materiomics. Pathological materiomics provides a suitable framework to reconcile the multi-scale mechanisms of disease and tissue failure that could prove beneficial for diagnostics and treatment, complementary to current physiological approaches.

## 9.9 Suggested Readings

For more information on the specific diseases discussed:

- B.S. Pace, *Renaissance of Sickle Cell Disease Research in the Genome Era* (World Scientific Publishing, 2007)
- G.R. Serjeant, B.E. Serjeant, *Sickle Cell Disease*. Oxford Medical Publications (Oxford University Press, 2001)
- P.D. O'Malley (ed.), *New Developments in Sickle Cell Disease Research* (Nova Science, 2006)
- D.L. Doolan (ed.), *Malaria Methods and Protocols*. Methods in Molecular Medicine (Humana Press (Springer), 2002)
- I.W. Sherman (ed.), *Molecular Approaches to Malaria* (ASM Press, 2005)
- M. Berrios (ed.), *Nuclear Structure and Function* (Academic Press, 1997)
- E.S. Roach, V.S. Miller (eds.), *Neurocutaneous Disorders* (Cambridge University Press, 2004)
- M.B. Fossel, *Cells, Aging, and Human Disease* (Oxford University Press, 2004)
- C.-P. Adler, *Bone Diseases* (Springer, 2000)
- P. Fratzl (ed.), *Collagen: Structure and Mechanics* (Springer, 2008)
- M.M. Verbeek, R.M. de Waal, H.V. Vinters (eds.), *Cerebral Amyloid Angiopathy in Alzheimer's Disease and Related Disorders* (Springer, 2010)
- R. Harris, F. Fahrenholz (eds.), *Alzheimer's Disease: Cellular and Molecular Aspects of Amyloid beta*. Subcellular Biochemistry (Springer, 2005)

On molecular medicine and protein-based disease:

- R.J. Trent, *Molecular Medicine: An Introductory Text*, 3rd edn. (Elsevier, 2005)
- M.S. Runge, C. Patterson (eds.), *Principles of Molecular Medicine* (Humana Press (Springer), 2006)
- W.B. Coleman, G.J. Tsongalis (eds.), *Molecular Pathology: The Molecular Basis of Human Disease* (Elsevier, 2009)
- B. Bergeron, *Case Studies in Genes and Disease: A Primer for Clinicians* (American College of Physicians, 2004)

## References

1. E.D. Devillota, M.T.G. Carmona, J.J. Rubio, S.R. Deandres, Equality of the in vivo and in vitro oxygen-binding capacity of hemoglobin in patients with severe respiratory-disease. Br. J. Anaesth. **53**(12), 1325–1328 (1981)
2. F.A. Ferrone, A. Aprelev, M.A. Rotter, Z. Etzion, R.M. Bookchin, R.W. Briehl, The effects of erythrocyte membranes on the nucleation of sickle hemoglobin. Biophys. J. **88**(4), 2815–2822 (2005)
3. L. Pauling, H.A. Itano, S.J. Singer, I.C. Wells, Sickle cell anemia, a molecular disease. Science **110**(2865), 543–548 (1949)
4. V.M. Ingram, Specific chemical difference between the globins of normal human and sickle-cell anaemia haemoglobin. Nature **178**(4537), 792–794 (1956)

5. V.M. Ingram, Gene mutations in human haemoglobin—chemical difference between normal and sickle cell haemoglobin. Nature **180**(4581), 326–328 (1957)
6. B.J. Strasser, Molecular medicine—"sickle cell anemia, a molecular disease". Science **286**(5444), 1488–1490 (1999)
7. M.R. Lewis, L.J. Gross, R. Josephs, Cryoelectron microscopy of deoxy-sickle hemoglobin fibers. Microsc. Res. Tech. **27**(5), 459–467 (1994)
8. G. Dykes, R.H. Crepeau, S.J. Edelstein, 3-Dimensional reconstruction of fibers of sickle-cell hemoglobin. Nature **272**(5653), 506–510 (1978)
9. R. Mirchev, F.A. Ferrone, The structural link between polymerization and sickle cell disease. J. Mol. Biol. **265**(5), 475–479 (1997)
10. R.W. Briehl, J.C. Wang, M.S. Turner, G. Agarwal, S. Kwong, R. Josephs, F.A. Ferrone, Micromechanics of isolated sickle cell hemoglobin fibers: bending moduli and persistence lengths. J. Mol. Biol. **315**(4), 601–612 (2002)
11. L. Mahadevan, A.E. Cohen, Kinks, rings, and rackets in filamentous structures. Proc. Natl. Acad. Sci. USA **100**(21), 12141–12146 (2003)
12. C.W. Jones, J.C. Wang, F.A. Ferrone, R.W. Briehl, M.S. Turner, Interactions between sickle hemoglobin fibers. Faraday Discuss. **123**, 221–236 (2003)
13. P.S. Frenette, Sickle cell vaso-occlusion: multistep and multicellular paradigm. Curr. Opin. Hematol. **9**(2), 101–106 (2002)
14. D.K. Kaul, M.E. Fabry, R.L. Nagel, Microvascular sites and characteristics of sickle-cell adhesion to vascular endothelium in shear-flow conditions—pathophysiological implications. Proc. Natl. Acad. Sci. USA **86**(9), 3356–3360 (1989)
15. G.W. Christoph, J. Hofrichter, W.A. Eaton, Understanding the shape of sickled red cells. Biophys. J. **88**(2), 1371–1376 (2005)
16. F.A. Ferrone, J. Hofrichter, W.A. Eaton, Kinetics of sickle hemoglobin polymerization .2. a double nucleation mechanism. J. Mol. Biol. **183**(4), 611–631 (1985)
17. A. Mozzarelli, J. Hofrichter, W.A. Eaton, Delay time of hemoglobin-s polymerization prevents most cells from sickling in vivo. Science **237**(4814), 500–506 (1987)
18. J.D. Corbett, W.E. Mickols, M.F. Maestre, Effect of hemoglobin concentration on nucleation and polymer formation in sickle red-blood-cells. J. Biol. Chem. **270**(6), 2708–2715 (1995)
19. K. Kuczera, J. Gao, B. Tidor, M. Karplus, Free-energy of sickling—a simulation analysis. Proc. Natl. Acad. Sci. USA **87**(21), 8481–8485 (1990)
20. M. Prabhakaran, M.E. Johnson, Molecular-dynamics of sickle and normal hemoglobins. Biopolymers **33**(5), 735–742 (1993)
21. A. Roufberg, F.A. Ferrone, A model for the sickle hemoglobin fiber using both mutation sites. Protein Sci. **9**(5), 1031–1034 (2000)
22. F.A. Ferrone, Polymerization and sickle cell disease: a molecular view. Microcirculation **11**(2), 115–128 (2004)
23. H. Li, H. Vi, G. Lykotrafitis, Modeling sickle hemoglobin fibers as one chain of coarse-grained particles mechanics of biological systems and materials, vol. 2. Conf. Proc. Soc. Exper. Mech. Ser. **9999**, 219–224 (2011)
24. S.K. Ballas, N. Mohandas, Sickle red cell microrheology and sickle blood rheology. Microcirculation **11**(2), 209–225 (2004)
25. J. Stuart, G.B. Nash, Red-cell deformability and hematological disorders. Blood Rev. **4**(3), 141–147 (1990)
26. G.A. Barabino, M.O. Platt, D.K. Kaul, Sickle cell biomechanics. Annu. Rev. Biomed. Eng. **12**, 345–367 (2010)
27. M.J. Stuart, R.L. Nagel, Sickle-cell disease. Lancet **364**(9442), 1343–1360 (2004)
28. H.F. Bunn, Mechanisms of disease—pathogenesis and treatment of sickle cell disease. N. Engl. J. Med. **337**(11), 762–769 (1997)
29. B.M. Cooke, N. Mohandas, R.L. Coppel, Malaria and the red blood cell membrane. Sem. Hematol. **41**(2), 173–188 (2004)
30. L. Schofield, G.E. Grau, Immunological processes in malaria pathogenesis. Nat. Rev. Immunol. **5**(9), 722–735 (2005)

31. M. Prudencio, A. Rodriguez, M.M. Mota, The silent path to thousands of merozoites: the plasmodium liver stage. Nat. Rev., Microbiol. **4**(11), 849–856 (2006)
32. K. Haldar, S.C. Murphy, D.A. Milner, T.E. Taylor, Malaria: mechanisms of erythrocytic infection and pathological correlates of severe disease. Ann. Rev. Pathol.-Mech. Dis. **2**, 217–249 (2007)
33. L.H. Miller, D.I. Baruch, K. Marsh, O.K. Doumbo, The pathogenic basis of malaria. Nature **415**(6872), 673–679 (2002)
34. R. Idro, N.E. Jenkins, C.R.J.C. Newton, Pathogenesis, clinical features, and neurological outcome of cerebral malaria. Lancet Neurol. **4**(12), 827–840 (2005)
35. B.M. Cooke, N. Mohandas, R.L. Coppell, The malaria-infected red blood cell: structural and functional changes. Adv. Parasitol. **50**, 1–86 (2001)
36. S. Suresh, M. Dao, C.T. Lim, Mechanics of the human red blood cell deformed by optical tweezers. J. Mech. Phys. Solids **51**(11–12), 2259–2280 (2003)
37. S. Suresh, J. Spatz, J.P. Mills, A. Micoulet, M. Dao, C.T. Lim, M. Beil, T. Seufferlein, Connections between single-cell biomechanics and human disease states: gastrointestinal cancer and malaria. Acta Biomater. **1**(1), 15–30 (2005)
38. P.K. Rathod, J.P. Shelby, J. White, K. Ganesan, D.T. Chiu, A microfluidic model for single-cell capillary obstruction by plasmodium falciparum infected erythrocytes. Proc. Natl. Acad. Sci. USA **100**(25), 14618–14622 (2003)
39. R. Udomsangpetch, R. Suwanarusk, B.M. Cooke, A.M. Dondorp, K. Silamut, J. Sattabongkot, N.J. White, The deformability of red blood cells parasitized by plasmodium falciparum and p-vivax. J. Infect. Dis. **189**(2), 190–194 (2004)
40. E.A. Evans, Bending elastic-modulus of red-blood-cell membrane derived from buckling instability in micropipet aspiration tests. Biophys. J. **43**(1), 27–30 (1983)
41. C.T. Lim, E.H. Zhou, S.T. Quek, Mechanical models for living cells—a review. J. Biomech. **39**(2), 195–216 (2006)
42. D. Hartmann, A multiscale model for red blood cell mechanics. Biomech. Model. Mechanobiol. **9**(1), 1–17 (2010)
43. L.J. Cardo, J. Salata, D. Wilder, Removal of plasmodium falciparum-infected red blood cells from whole blood by leukoreduction filters. Transfusion **49**(2), 337–346 (2009)
44. J.Y. Han, H.W. Hou, A.A.S. Bhagat, A.G.L. Chong, P. Mao, K.S.W. Tan, C.T. Lim, Deformability based cell margination—a simple microfluidic design for malaria-infected erythrocyte separation. Lab Chip **10**(19), 2605–2613 (2010)
45. D.A. Elliott, M.T. McIntosh, H.D. Hosgood, S. Chen, G. Zhang, P. Baevova, K.A. Joiner, Four distinct pathways of hemoglobin uptake in the malaria parasite plasmodium falciparum. Proc. Natl. Acad. Sci. USA **105**(7), 2463–2468 (2008)
46. A.B. Vaidya, M.W. Mather, Mitochondrial evolution and functions in malaria parasites. Annu. Rev. Microbiol. **63**, 249–267 (2009)
47. M.P. Soares, A. Ferreira, I. Marguti, I. Bechmann, V. Jeney, A. Chora, N.R. Palha, S. Rebelo, A. Henri, Y. Beuzard, Sickle hemoglobin confers tolerance to plasmodium infection. Cell **145**(3), 398–409 (2011)
48. G. Balla, H.S. Jacob, J. Balla, M. Rosenberg, K. Nath, F. Apple, J.W. Eaton, G.M. Vercellotti, Ferritin—a cytoprotective antioxidant strategem of endothelium. J. Biol. Chem. **267**(25), 18148–18153 (1992)
49. R. Gozzelino, V. Jeney, M.P. Soares, Mechanisms of cell protection by heme oxygenase-1. Annu. Rev. Pharmacol. Toxicol. **50**, 323–354 (2010)
50. M.P. Soares, A. Ferreira, J. Balla, V. Jeney, G. Balla, A central role for free heme in the pathogenesis of severe malaria: the missing link? J. Mol. Med. **86**(10), 1097–1111 (2008)
51. J.C. Winkelmann, B.G. Forget, Erythroid and nonerythroid spectrins. Blood **81**(12), 3173–3185 (1993)
52. J.C. Hansen, R. Skalak, S. Chien, A. Hoger, An elastic network model based on the structure of the red blood cell membrane skeleton. Biophys. J. **70**(1), 146–166 (1996)
53. J. Li, M. Dao, C.T. Lim, S. Suresh, Spectrin-level modeling of the cytoskeleton and optical tweezers stretching of the erythrocyte. Biophys. J. **88**(5), 3707–3719 (2005)

54. F. Liu, J. Burgess, H. Mizukami, A. Ostafin, Sample preparation and imaging of erythrocyte cytoskeleton with the atomic force microscopy. Cell Biochem. Biophys. **38**(3), 251–270 (2003)
55. X. Periole, M. Vendruscolo, A.E. Mark, Molecular dynamics simulations from putative transition states of alpha-spectrin sh3 domain. Protein. Struct. Funct. Bioinform. **69**(3), 536–550 (2007)
56. D.T. Mirijanian, J.W. Chu, G.S. Ayton, G.A. Voth, Atomistic and coarse-grained analysis of double spectrin repeat units: the molecular origins of flexibility. J. Mol. Biol. **365**(2), 523–534 (2007)
57. D.T. Mirijanian, G.A. Voth, Unique elastic properties of the spectrin tetramer as revealed by multiscale coarse-grained modeling. Proc. Natl. Acad. Sci. USA **105**(4), 1204–1208 (2008)
58. J. Li, G. Lykotrafitis, M. Dao, S. Suresh, Cytoskeletal dynamics of human erythrocyte. Proc. Natl. Acad. Sci. USA **104**(12), 4937–4942 (2007)
59. S. Suresh, Mechanical response of human red blood cells in health and disease: some structure-property-function relationships. J. Mater. Res. **21**(8), 1871–1877 (2006)
60. J.L. Maciaszek, G. Lykotrafitis, Sickle cell trait human erythrocytes are significantly stiffer than normal. J. Biomech. **44**(4), 657–661 (2011)
61. J.L. Maciaszek, B. Andemariam, G. Lykotrafitis, Microelasticity of red blood cells in sickle cell disease. J. Strain Anal. Eng. Des. **46**(5), 368–379 (2011)
62. B. Burke, C.L. Stewart, Life at the edge: the nuclear envelope and human disease. Nat. Rev. Mol. Cell Biol. **3**(8), 575–585 (2002)
63. J. Lammerding, P.C. Schulze, T. Takahashi, S. Kozlov, T. Sullivan, R.D. Kamm, C.L. Stewart, R.T. Lee, Lamin a/c deficiency causes defective nuclear mechanics and mechanotransduction. J. Clin. Invest. **113**(3), 370–378 (2004)
64. Y. Gruenbaum, A. Margalit, R.D. Goldman, D.K. Shumaker, K.L. Wilson, The nuclear lamina comes of age. Nat. Rev. Mol. Cell Biol. **6**(1), 21–31 (2005)
65. K.N. Dahl, P. Scaffidi, M.F. Islam, A.G. Yodh, K.L. Wilson, T. Misteli, Distinct structural and mechanical properties of the nuclear lamina in Hutchinson-Gilford progeria syndrome. Proc. Natl. Acad. Sci. USA **103**(27), 10271–10276 (2006)
66. U. Aebi, J. Cohn, L. Buhle, L. Gerace, The nuclear lamina is a meshwork of intermediate-type filaments. Nature **323**(6088), 560–564 (1986)
67. Z. Qin, M.J. Buehler, Dynamic failure of a lamina meshwork in cell nuclei under extreme mechanical deformation. BioNanoScience **1**(1–2), 14–23 (2011)
68. Z. Qin, M.J. Buehler, Flaw tolerance of nuclear intermediate filament lamina under extreme mechanical deformation. ACS Nano **5**(4), 3034–3042 (2011)
69. N. Levy, A. De Sandre-Giovannoli, R. Bernard, P. Cau, C. Navarro, J. Amiel, I. Boccaccio, S. Lyonnet, C.L. Stewart, A. Munnich, M. Le Merrer, Lamin a truncation in Hutchinson-Gilford progeria. Science **300**(5628), 2055 (2003)
70. I. Duband-Goulet, F. Bruston, E. Delbarre, C. Ostlund, H.J. Worman, B. Buendia, Loss of a DNA binding site within the tail of prelamin a contributes to altered heterochromatin anchorage by progerin. FEBS Lett. **584**(14), 2999–3004 (2010)
71. S.G. Young, M. Meta, S.H. Yang, L.G. Fong, Prelamin a farnesylation and progeroid syndromes. J. Biol. Chem. **281**(52), 39741–39745 (2006)
72. R. Pomes, S. Rauscher, Molecular simulations of protein disorder. Biochem. Cell Biol. Biochim. Biol. Cellul. **88**(2), 269–290 (2010)
73. R. Foisner, E.C. Schirmer, Proteins that associate with lamins: many faces, many functions. Exp. Cell Res. **313**(10), 2167–2179 (2007)
74. K.L. Wilson, M.S. Zastrow, S. Vlcek, Proteins that bind a-type lamins: integrating isolated clues. J. Cell Sci. **117**(7), 979–987 (2004)
75. L. Serrano, R. Linding, J. Schymkowitz, F. Rousseau, F. Diella, A comparative study of the relationship between protein structure and beta-aggregation in globular and intrinsically disordered proteins. J. Mol. Biol. **342**(1), 345–353 (2004)
76. Z. Qin, A. Kalinowski, K. Dahl, M.J. Buehler, Structure and stability of the lamin a tail domain and HGPS mutant. J. Struct. Biol. **175**, 425–433 (2011)

77. D.E. Discher, K.N. Dahl, S.M. Kahn, K.L. Wilson, The nuclear envelope lamina network has elasticity and a compressibility limit suggestive of a molecular shock absorber. J. Cell Sci. **117**(20), 4779–4786 (2004)
78. R.D. Goldman, D.K. Shumaker, M.R. Erdos, M. Eriksson, A.E. Goldman, L.B. Gordon, Y. Gruenbaum, S. Khuon, M. Mendez, R. Varga, F.S. Collins, Accumulation of mutant lamin a causes progressive changes in nuclear architecture in Hutchinson-Gilford progeria syndrome. Proc. Natl. Acad. Sci. USA **101**(24), 8963–8968 (2004)
79. F.S. Collins, B.C. Capell, M. Olive, M.R. Erdos, K. Cao, D.A. Faddah, U.L. Tavarez, K.N. Conneely, X. Qu, H. San, S.K. Ganesh, X. Chen, H. Avallone, F.D. Kolodgie, R. Virmani, E.G. Nabel, A farnesyltransferase inhibitor prevents both the onset and late progression of cardiovascular disease in a progeria mouse model. Proc. Natl. Acad. Sci. USA **105**(41), 15902–15907 (2008)
80. R.D. Goldman, P. Taimen, K. Pfleghaar, T. Shimi, D. Moller, K. Ben-Harush, M.R. Erdos, S.A. Adam, H. Herrmann, O. Medalia, F.S. Collins, A.E. Goldman, A progeria mutation reveals functions for lamin a in nuclear assembly, architecture, and chromosome organization. Proc. Natl. Acad. Sci. USA **106**(49), 20788–20793 (2009)
81. B. Buendia, E. Delbarre, M. Tramier, M. Coppey-Moisan, C. Gaillard, J.C. Courvalin, The truncated prelamin a in Hutchinson-Gilford progeria syndrome alters segregation of a-type and b-type lamin homopolymers. Hum. Mol. Genet. **15**(7), 1113–1122 (2006)
82. M.P. Sheetz, A. del Rio, R. Perez-Jimenez, R.C. Liu, P. Roca-Cusachs, J.M. Fernandez, Stretching single talin rod molecules activates vinculin binding. Science **323**(5914), 638–641 (2009)
83. M.J. Buehler, S. Keten, Z.P. Xu, B. Ihle, Nanoconfinement controls stiffness, strength and mechanical toughness of beta-sheet crystals in silk. Nat. Mater. **9**(4), 359–367 (2010)
84. D.E. Ingber, N. Wang, J.D. Tytell, Mechanotransduction at a distance: mechanically coupling the extracellular matrix with the nucleus. Nat. Rev. Mol. Cell Biol. **10**(1), 75–82 (2009)
85. F.H. Glorieux, F. Rauch, Osteogenesis imperfecta. Lancet **363**(9418), 1377–1385 (2004)
86. P. Fratzel, *Collagen: Structure and Mechanics* (Springer, New York, 2008)
87. W. Traub, T. Arad, U. Vetter, S. Weiner, Ultrastructural studies of bones from patients with osteogenesis imperfecta. Matrix Biol. **14**(4), 337–345 (1994)
88. U. Vetter, M.A. Weis, M. Morike, E.D. Eanes, D.R. Eyre, Collagen cross-links and mineral crystallinity in bone of patients with osteogenesis imperfecta. J. Bone Miner. Res. **8**(2), 133–137 (1993)
89. J.P. Cassella, S.Y. Ali, Abnormal collagen and mineral formation in osteogenesis imperfecta. Bone Miner. **17**(2), 123–128 (1992)
90. P.H. Byers, G.A. Wallis, M.C. Willing, Osteogenesis imperfecta: translation of mutation to phenotype. J. Med. Genet. **28**(7), 433–442 (1991)
91. D.J. Prockop, K.I. Kivirikko, Collagens: molecular biology, diseases, and potentials for therapy. Annu. Rev. Biochem. **64**, 403–434 (1995)
92. D.O. Sillence, A. Senn, D.M. Danks, Genetic-heterogeneity in osteogenesis imperfecta. J. Med. Genet. **16**(2), 101–116 (1979)
93. S.A. Goldstein, K.M. Kozloff, A. Carden, C. Bergwitz, A. Forlino, T.E. Uveges, M.D. Morris, J.C. Marini, Brittle iv mouse model for osteogenesis imperfecta iv demonstrates postpubertal adaptations to improve whole bone strength. J. Bone Miner. Res. **19**(4), 614–622 (2004)
94. R. Dalgleish, The human type i collagen mutation database. Nucleic Acids Res. **25**(1), 181–187 (1997)
95. M.J. Buehler, A. Gautieri, S. Vesentini, A. Redaelli, Single molecule effects of osteogenesis imperfecta mutations in tropocollagen protein domains. Protein Sci. **18**(1), 161–168 (2009)
96. D.J. McBride, V. Choe, J.R. Shapiro, B. Brodsky, Altered collagen structure in mouse tail tendon lacking the alpha2(i) chain. J. Mol. Biol. **270**, 275–284 (1997)
97. E. Miller, D. Delos, T. Baldini, T.M. Wright, N.P. Camacho, Abnormal mineral-matrix interactions are a significant contributor to fragility in oim/oim bone. Calcif. Tissue Int. **81**(3), 206–214 (2007)

98. T.J. Sims, C.A. Miles, A.J. Bailey, N.P. Camacho, Properties of collagen in oim mouse tissues. Connect. Tiss. Res. **44**(Suppl 1), 202–205 (2003)
99. P. Fratzl, O. Paris, K. Klaushofer, W.J. Landis, Bone mineralization in an osteogenesis imperfecta mouse model studied by small-angle x-ray scattering. J. Clin. Invest. **97**, 396–402 (1996)
100. N.P. Camacho, L. Hou, T.R. Toledano, W.A. Ilg, C.F. Brayton, C.L. Raggio, L. Root, A.L. Boskey, The material basis for reduced mechanical properties in oim bones. J. Bone Miner. Res. **14**(2), 264–272 (1999)
101. B. Grabner, W.J. Landis, P. Roschger, S. Rinnerthaler, H. Peterlik, K. Klaushofer, P. Fratzl, Age- and genotype-dependence of bone material properties in the osteogenesis imperfecta murine model (oim). Bone **29**(453–457) (2001)
102. K. Misof, W.J. Landis, K. Klaushofer, P. Fratzl, Collagen from the osteogenesis imperfecta mouse model (oim) shows reduced resistance against tensile stress. J. Clin. Invest. **100**(1), 40–45 (1997)
103. P. Chavassieux, E. Seeman, P.D. Delmas, Insights into material and structural basis of bone fragility from diseases associated with fractures: how determinants of the biomechanical properties of bone are compromised by disease. Endocrine Rev. **28**(2), 151–164 (2007)
104. M.J. Buehler, A. Gautieri, S. Uzel, S. Vesentini, A. Redaelli, Molecular and mesoscale mechanisms of osteogenesis imperfecta disease in collagen fibrils. Biophys. J. **97**(3), 857–865 (2009)
105. B. Brodsky, K. Beck, V.C. Chan, N. Shenoy, A. Kirkpatrick, J.A.M. Ramshaw, Destabilization of osteogenesis imperfecta collagen-like model peptides correlates with the identity of the residue replacing glycine. Proc. Natl. Acad. Sci. USA **97**(8), 4273–4278 (2000)
106. J. Baum, X.Y. Liu, S. Kim, Q.H. Dai, B. Brodsky, Nuclear magnetic resonance shows asymmetric loss of triple helix in peptides modeling a collagen mutation in brittle bone disease. Biochemistry **37**(44), 15528–15533 (1998)
107. S.J. Eppell, Z.L. Shen, M.R. Dodge, H. Kahn, R. Ballarini, Stress-strain experiments on individual collagen fibrils. Biophys. J. **95**(8), 3956–3963 (2008)
108. M.J. Buehler, Atomistic and continuum modeling of mechanical properties of collagen: elasticity, fracture, and self-assembly. J. Mater. Res. **21**(8), 1947–1961 (2006)
109. M.J. Buehler, Nature designs tough collagen: explaining the nanostructure of collagen fibrils. Proc. Natl. Acad. Sci. USA **103**(33), 12285–12290 (2006)
110. P. Zioupos, J.D. Currey, Changes in the stiffness, strength, and toughness of human cortical bone with age. Bone **22**(1), 57–66 (1998)
111. R.K. Nalla, J.J. Kruzic, J.H. Kinney, M. Balooch, J.W. Ager, R.O. Ritchie, Role of microstructure in the aging-related deterioration of the toughness of human cortical bone. Mat. Sci. Eng. C Biomim. Supramol. Syst. **26**(8), 1251–1260 (2006)
112. R.W. Mccalden, J.A. Mcgeough, M.B. Barker, C.M. Courtbrown, Age-related-changes in the tensile properties of cortical bone—the relative importance of changes in porosity, mineralization, and microstructure. J. Bone Jt. Surg., Am. Vol. **75A**(8), 1193–1205 (1993)
113. D.R. Sell, V.M. Monnier, Structure elucidation of a senescence cross-link from human extracellular-matrix—implication of pentoses in the aging process. J. Biol. Chem. **264**(36), 21597–21602 (1989)
114. E.A. Zimmermann, E. Schaible, H. Bale, H.D. Barth, S.Y. Tang, P. Reichert, B. Busse, T. Alliston, J.W. Ager, R.O. Ritchie, Age-related changes in the plasticity and toughness of human cortical bone at multiple length scales. Proc. Natl. Acad. Sci. USA **108**(35), 14416–14421 (2011)
115. T.P. Knowles, M.J. Buehler, Nanomechanics of functional and pathological amyloid material. Nat. Nanotechnol. **6**, 469–479 (2011)
116. N.D. Hammer, J.C. Schmidt, M.R. Chapman, The curli nucleator protein, csgb, contains an amyloidogenic domain that directs csga polymerization. Proc. Natl. Acad. Sci. USA **104**(30), 12494–12499 (2007)
117. M. Meyer-Luehmann, T.L. Spires-Jones, C. Prada, M. Garcia-Alloza, A. de Calignon, A. Rozkalne, J. Koenigsknecht-Talboo, D.M. Holtzman, B.J. Bacskai, B.T. Hyman, Rapid ap-

pearance and local toxicity of amyloid-beta plaques in a mouse model of Alzheimer's disease. Nature **451**(7179), 720–725 (2008)

118. B. Watt, G. van Niel, D.M. Fowler, I. Hurbain, K.C. Luk, S.E. Stayrook, M.A. Lemmon, G. Raposo, J. Shorter, J.W. Kelly, M.S. Marks, N-terminal domains elicit formation of functional pmel17 amyloid fibrils. J. Biol. Chem. **284**(51), 35543–35555 (2009)
119. M.G. Spillantini, M.L. Schmidt, V.M.Y. Lee, J.Q. Trojanowski, R. Jakes, M. Goedert, alpha-synuclein in Lewy bodies. Nature **388**(6645), 839–840 (1997)
120. D.J. Selkoe, Alzheimer's disease: genes, proteins, and therapy. Physiol. Rev. **81**(2), 741–766 (2001)
121. D.J. Selkoe, J. Hardy, Medicine—the amyloid hypothesis of Alzheimer's disease: progress and problems on the road to therapeutics. Science **297**(5580), 353–356 (2002)
122. P. Mesquida, C.K. Riener, C.E. MacPhee, R.A. McKendry, Morphology and mechanical stability of amyloid-like peptide fibrils. J. Mater. Sci., Mater. Med. **18**(7), 1325–1331 (2007)
123. V.A. Iconomidou, S.J. Hamodrakas, Natural protective amyloids. Curr. Protein Pept. Sci. **9**(3), 291–309 (2008)
124. C.M. Dobson, Protein folding and misfolding. Nature **426**(6968), 884–890 (2003)
125. T. Luhrs, C. Ritter, M. Adrian, D. Riek-Loher, B. Bohrmann, H. Doeli, D. Schubert, R. Riek, 3d structure of Alzheimer's amyloid-beta(1-42) fibrils. Proc. Natl. Acad. Sci. USA **102**(48), 17342–17347 (2005)
126. M.R. Sawaya, S. Sambashivan, R. Nelson, M.I. Ivanova, S.A. Sievers, M.I. Apostol, M.J. Thompson, M. Balbirnie, J.J.W. Wiltzius, H.T. McFarlane, A.O. Madsen, C. Riekel, D. Eisenberg, Atomic structures of amyloid cross-beta spines reveal varied steric zippers. Nature **447**(7143), 453–457 (2007)
127. J.W. Kelly, W.E. Balch, Amyloid as a natural product. J. Cell Biol. **161**(3), 461–462 (2003)
128. D.M. Fowler, A.V. Koulov, C. Alory-Jost, M.S. Marks, W.E. Balch, J.W. Kelly, Functional amyloid formation within mammalian tissue. PLoS Biol. **4**(1), 100–107 (2006)
129. M. Bucciantini, E. Giannoni, F. Chiti, F. Baroni, L. Formigli, J.S. Zurdo, N. Taddei, G. Ramponi, C.M. Dobson, M. Stefani, Inherent toxicity of aggregates implies a common mechanism for protein misfolding diseases. Nature **416**(6880), 507–511 (2002)
130. C.L. Masters, G. Simms, N.A. Weinman, G. Multhaup, B.L. Mcdonald, K. Beyreuther, Amyloid plaque core protein in Alzheimer-disease and down syndrome. Proc. Natl. Acad. Sci. USA **82**(12), 4245–4249 (1985)
131. G.W. Small, V. Kepe, L.M. Ercoli, P. Siddarth, S.Y. Bookheimer, K.J. Miller, H. Lavretsky, A.C. Burggren, G.M. Cole, H.V. Vinters, P.M. Thompson, S.C. Huang, N. Satyamurthy, M.E. Phelps, J.R. Barrio, Pet of brain amyloid and tau in mild cognitive impairment. N. Engl. J. Med. **355**(25), 2652–2663 (2006)
132. S. Kumar-Singh, D. Pirici, C. Van Cauwenberghe, C. Van Broeckhoven, Fractal analysis of amyloid plaques in Alzheimer's disease patients and mouse models. Neurobiol. Aging **32**(9), 1579–1587 (2011)
133. T.S. Burkoth, T.L.S. Benzinger, V. Urban, D.M. Morgan, D.M. Gregory, P. Thiyagarajan, R.E. Botto, S.C. Meredith, D.G. Lynn, Structure of the beta-amyloid((10-35)) fibril. J. Am. Chem. Soc. **122**(33), 7883–7889 (2000)
134. A. Dutt, M.G.B. Drew, A. Pramanik, Beta-sheet mediated self-assembly of dipeptides of omega-amino acids and remarkable fibrillation in the solid state. Org. & Biomol. Chem. **3**(12), 2250 (2005)
135. J.F. Smith, T.P.J. Knowles, C.M. Dobson, C.E. MacPhee, M.E. Welland, Characterization of the nanoscale properties of individual amyloid fibrils. Proc. Natl. Acad. Sci. USA **103**(43), 15806–15811 (2006)
136. T.P. Knowles, A.W. Fitzpatrick, S. Meehan, H.R. Mott, M. Vendruscolo, C.M. Dobson, M.E. Welland, Role of intermolecular forces in defining material properties of protein nanofibrils. Science **318**(5858), 1900–1903 (2007)
137. G. Favrin, M. Cheon, I. Chang, S. Mohanty, L.M. Luheshi, C.M. Dobson, M. Vendruscolo, Structural reorganisation and potential toxicity of oligomeric species formed during the assembly of amyloid fibrils. PLoS Comput. Biol. **3**(9), 1727–1738 (2007)

138. R. Tycko, A.T. Petkova, Y. Ishii, J.J. Balbach, O.N. Antzutkin, R.D. Leapman, F. Delaglio, A structural model for Alzheimer's beta-amyloid fibrils based on experimental constraints from solid state nmr. Proc. Natl. Acad. Sci. USA **99**(26), 16742–16747 (2002)
139. C.E. MacPhee, C.P. Jaroniec, V.S. Bajaj, M.T. McMahon, C.M. Dobson, R.G. Griffin, High-resolution molecular structure of a peptide in an amyloid fibril determined by magic angle spinning nmr spectroscopy. Proc. Natl. Acad. Sci. USA **101**(3), 711–716 (2004)
140. R. Tycko, A.K. Paravastu, R.D. Leapman, W.M. Yau, Molecular structural basis for polymorphism in Alzheimer's beta-amyloid fibrils. Proc. Natl. Acad. Sci. USA **105**(47), 18349–18354 (2008)
141. D. Eisenberg, R. Nelson, M.R. Sawaya, M. Balbirnie, A.O. Madsen, C. Riekel, R. Grothe, Structure of the cross-beta spine of amyloid-like fibrils. Nature **435**(7043), 773–778 (2005)
142. R. Nussinov, B.Y. Ma, Simulations as analytical tools to understand protein aggregation and predict amyloid conformation. Curr. Opin. Chem. Biol. **10**(5), 445–452 (2006)
143. C.K. Hall, H.D. Nguyen, Spontaneous fibril formation by polyalanines; discontinuous molecular dynamics simulations. J. Am. Chem. Soc. **128**(6), 1890–1901 (2006)
144. G. Hummer, N.V. Buchete, R. Tycko, Molecular dynamics simulations of Alzheimer's beta-amyloid protofilaments. J. Mol. Biol. **353**(4), 804–821 (2005)
145. Y.D. Wu, W. Han, A strand-loop-strand structure is a possible intermediate in fibril elongation: long time simulations of amylold-beta peptide (10-35). J. Am. Chem. Soc. **127**(44), 15408–15416 (2005)
146. R. Nussinov, N. Haspel, D. Zanuy, B.Y. Ma, H. Wolfson, A comparative study of amyloid fibril formation by residues 15-19 of the human calcitonin hormone: a single beta-sheet model with a small hydrophobic core. J. Mol. Biol. **345**(5), 1213–1227 (2005)
147. N.J. Deng, L. Yan, D. Singh, P. Cieplak, Molecular basis for the cu2+ binding-induced destabilization of beta 2-microglobulin revealed by molecular dynamics simulation. Biophys. J. **90**(11), 3865–3879 (2006)
148. J.G. Saven, S. Park, Simulation of ph-dependent edge strand rearrangement in human beta-2 microglobulin. Protein Sci. **15**(1), 200–207 (2006)
149. V. Daggett, R.S. Armen, Characterization of two distinct beta(2)-microglobulin unfolding intermediates that may lead to amyloid fibrils of different morphology. Biochemistry **44**(49), 16098–16107 (2005)
150. S. Santini, P. Derreumaux, Helix h1 of the prion protein is rather stable against environmental perturbations: molecular dynamics of mutation and deletion variants of prp(90-231). Cell. Mol. Life Sci. **61**(7–8), 951–960 (2004)
151. R. Nussinov, B.Y. Ma, Stabilities and conformations of Alzheimer's beta-amyloid peptide oligomers (a beta(16-22 ') a beta(16-35 ') and a beta(10-35)): sequence effects. Proc. Natl. Acad. Sci. USA **99**(22), 14126–14131 (2002)
152. R. Tycko, Insights into the amyloid folding problem from solid-state nmr. Biochemistry **42**(11), 3151–3159 (2003)
153. M.J. Buehler, R. Paparcone, J. Sanchez, Comparative study of polymorphous Alzheimer's a beta(1-40) amyloid nanofibrils and microfibers. J. Comput. Theor. Nanosci. **7**(7), 1279–1286 (2010)
154. M.J. Buehler, R. Paparcone, S. Keten, Atomistic simulation of nanomechanical properties of Alzheimer's a beta(1-40) amyloid fibrils under compressive and tensile loading. J. Biomech. **43**(6), 1196–1201 (2010)
155. M.J. Buehler, R. Paparcone, Failure of A-beta(1-40) amyloid fibrils under tensile loading. Biomaterials **32**(13), 3367–3374 (2011)
156. M.J. Buehler, R. Paparcone, S.W. Cranford, Self-folding and aggregation of amyloid nanofibrils. Nanoscale **3**(4), 1748–1755 (2011)
157. F. Chiti, C.M. Dobson, Protein misfolding, functional amyloid, and human disease. Annu. Rev. Biochem. **75**, 333–366 (2006)
158. W. Hwang, S.G. Zhang, R.D. Kamm, M. Karplus, Kinetic control of dimer structure formation in amyloid fibrillogenesis. Proc. Natl. Acad. Sci. USA **101**(35), 12916–12921 (2004)

159. H.D. Nguyen, C.K. Hall, Molecular dynamics simulations of spontaneous fibril formation by random-coil peptides. Proc. Natl. Acad. Sci. USA **101**(46), 16180 (2004)
160. K. Cho, X. Wang, S. Nie, Z.G. Chen, D.M. Shin, Therapeutic nanoparticles for drug delivery in cancer. Clin. Cancer Res. **14**(5), 1310–1316 (2008)
161. J. Guck, S. Schinkinger, B. Lincoln, F. Wottawah, S. Ebert, M. Romeyke, D. Lenz, H.M. Erickson, R. Ananthakrishnan, D. Mitchell, J. Kas, S. Ulvick, C. Bilby, Optical deformability as an inherent cell marker for testing malignant transformation and metastatic competence. Biophys. J. **88**(5), 3689–3698 (2005)
162. S.E. Cross, Y.-S. Jin, J. Rao, J.K. Gimzewski, Nanomechanical analysis of cells from cancer patients. Nat. Nanotechnol. **2**, 780–783 (2007)
163. T.S. Deisboeck, Z. Wang, P. Macklin, V. Cristini, Multiscale cancer modeling. Annu. Rev. Biomed. Eng. **13**, 127–155 (2011)
164. C. Guiot, P.P. Delsanto, A. Carpinteri, N. Pugno, Y. Mansury, T.S. Deisboeck, The dynamic evolution of the power exponent in a universal growth model of tumors. J. Theor. Biol. **240**(3), 459–463 (2006)
165. P.P. Delsanto, C.A. Condat, N. Pugno, A.S. Gliozzi, M. Griffa, A multilevel approach to cancer growth modeling. J. Theor. Biol. **250**(1), 16–24 (2008)
166. C. Guiot, N. Pugno, P.P. Delsanto, Elastomechanical model of tumor invasion. Appl. Phys. Lett. **89**(23) (2006)
167. C. Guiot, N. Pugno, P.P. Delsanto, T.S. Deisboeck, Physical aspects of cancer invasion. Phys. Biol. **4**(4), 1–6 (2007)

# Chapter 10
# Synthesis and Design

**Abstract** Understanding the "materiome" could provide a blueprint for bio-inspired, high performance functional materials. Successful synthesis and design—materiomic engineering—must extend beyond biomimitics and bio-inspired systems, exploit fundamental mechanisms of self-assembly and hierarchies, and integrate widely available building blocks such as amyloid proteins and DNA. The ultimate goal is not to produce one-off nanodevices or precision made materials, but rather materials that can self-assemble and adapt independently, befitting a range of functions from a common set of building blocks and molecular components.

*Until man duplicates a blade of grass, Nature can laugh at his so-called scientific knowledge.*

*Thomas Edison (1847–1931)*

## 10.1 Introduction

As previous chapters have shown, materiomics provides a powerful means to investigate complex biological materials, whether to more completely understand natural materials in their intended functional roles (such as silk) or mechanistic explanations of molecular missteps (such as the point mutation source of progeria). Revisiting the music analogy described in Chap. 4: Universality-Diversity Paradigm: Music, Materiomics, and Category Theory, this is unfortunately akin to being a world renowned musical critic without being able to write a single note or play an instrument. One may understand and appreciate the complexity, but we want to be able to compose our own symphonies! There lies vast opportunity in the creation and invention of new, possibly exotic and surprising materials—materials that have never existed and (potentially) would never be stumbled upon by conventional materials science and engineering approaches or through natural processes in evolution.

While we may not be able to completely understand the intricacies of Nature's design (yet), we can learn from Nature's successes and develop our own methods as a route for material design and synthesis. Recently, there is a great deal of interest in

S.W. Cranford, M.J. Buehler, *Biomateriomics*, Springer Series in Materials Science 165,
DOI 10.1007/978-94-007-1611-7_10, © Springer Science+Business Media Dordrecht 2012

the possibility that complex nanoscale material systems and devices can be designed and engineered. Such material systems required an integrated multiscale approach central to materiomics, and knowledge of interactions and relationships throughout the materiome. Exploiting, manipulating, and designing a complex materiome for intended functionality is a concept we term *materiomic engineering*.

The development of such material systems and devices will lead to the development of new materials, electronics and smart drugs. It is hoped that such nanocomponents are to have wide applicability in various devices and have enough flexibility to integrate into a large number of potentially very different environments (features of *universality*). Such robust performance is essential for the success of biological materials, which lack controlled laboratory conditions and must perform under a wide array of conditions. As such, the mere existence of biological "nanomachines" in Nature in the form of proteins (*e.g.*, enzymes) suggests that such tailored nanocomponents will be possible to produce. As the material from which Nature's nanomachines are made, proteins seem ideal to form the basis of engineered components of such nanodevices. Protein molecules have the ability to form a rich variety of natural and structures and materials and thus provide a flexible platform for development of novel synthetic materials.

**Materiomic Engineering**: Materiomic approaches to material system synthesis, by utilization hierarchical structures, self-assembly and/or self-organization processes, and knowledge of the entire materiome of the designed system, to explicitly tune mechanistic parameters and behavior, controlling nanoscale components and attain desired macroscopic responses.

## 10.2 Materials Inspired by Complex Biology

The ultimate ability to design synthetic materials at the same level of complexity of Nature has been a fundamental challenge for science for many years (referred to as *biomimetics*). As stressed throughout this text, Nature is highly skilled at producing tough, resilient materials out of simple components, and then actively improve on them over time, weeding out the systems that fail and optimizing the systems that succeed, both in terms of organism survival and adaptation. Engineers have not yet figured out how to use such weak components for high-tech applications, let alone replicate the robust properties and complexity of biological systems. However, Nature has a very big advantages over any scientific research: plenty of time. Nature has been researching over the entire course of natural history—successful product development and release in one continuous evolving process. Only recently has technology evolved to unlock the secrecy of (some) biological self-assembly processes while allowing direct (limited) manipulation of material components at the molecular scale.

**Biomimetics**: Design and engineering of synthetic processes, substances, devices, materials, or systems that imitate Nature, inspired by a biological method or analog. Biomimetic research can be focused on the investigation of natural materials, or on the processes that optimize the structure of materials in a manner similar to that occurring in Nature.

As discussed, control of a material system's structure and function (the *materiome*) ultimately requires control and design at the molecular scale. In this regard, current technologies such as scanning probe techniques have essentially answered this challenge: both atomic force microscopy (AFM) [1] and scanning tunneling microscopy (STM) [2, 3] allow the manipulation individual atoms. Yet, the piece-wise placement of individual atoms is not a practical route to synthesize bulk quantities of material—even with a horde of graduate students working overtime in the lab! There are simply too many atoms to work with—imagine painting a house one milliliter (or one nanoliter) of paint at a time! While beautiful works of art can be created with such care—Michelangelo's *The Last Judgement* on the altar wall of the Sistine Chapel took four painstaking years to complete—so to can astounding molecular structures (a good example is the 'nanodragster', assembled molecular piece by molecular piece, one reaction at a time [4], see Fig. 10.1).

While these "one-off" creations are quite exquisite, practical implementation and manufacture of "bulk material" (*e.g.*, systems on the order of micrometers) requires a more efficient means of assembly. As a result, we again turn to Nature, which has perfected the art of self-assembly, which offers inexpensive, parallel synthesis of nanostructures: works of art with one carefully planned stroke of a brush. Thus, materiomic engineering appears to be a two-pronged problem:

1. To understand the complexity of the desired system across scales (*i.e.*, designing the materiome). Or, in other words, the blueprints and purpose of the system (*e.g.*, *design*).
2. To develop the means for practical synthesis (*i.e.*, efficient, low-cost assembly and manufacturing), akin to a construction schedule, or how the system will be assembled (*e.g.*, *build*).

The two issues are intimately linked, and one cannot be considered without the other. This is similar to a design-build procurement route of structural engineering, wherein a single party is responsible for design and construction of a project. Complex systems necessitate such a "single contract" approach, where full understanding of the system architecture and assembly protocol is necessary for success. The researchers presented in the following sections have developed ingenious methods to exploit molecular mechanisms to synthesize new materials from the bottom-up: atoms and molecules direct their self-organization, without the need for external direction. While these materials are not "grown" in a biological sense, they are imitating Nature. Approaches that incorporate these concepts into novel products open new doors for innovative manufacturing, characterized by a higher degree of flexibility, less dependence on resources, or the ability to produce locally.

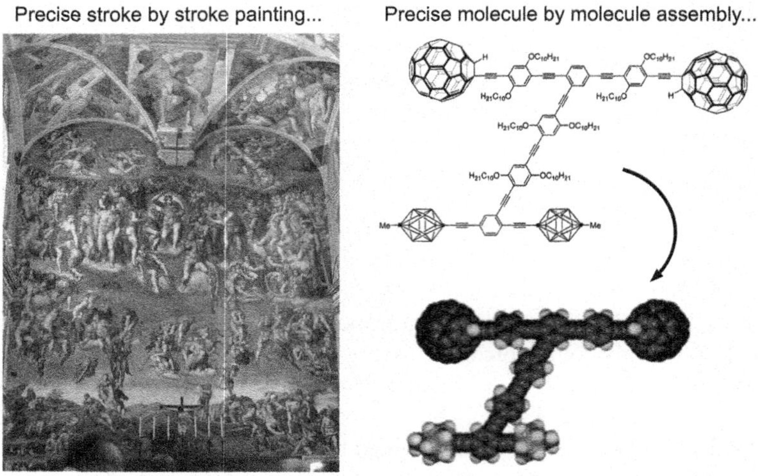


**Fig. 10.1** Complexity attained *via* precise control. *Left*: Michelangelo's *The Last Judgement* on the altar wall of the Sistine Chapel in Vatican City. The fresco took four years to complete and was executed from 1537 to 1541. *Right*: Precise manipulation of molecules and chemical reactions resulted in the successful synthesis of a mixed wheeled nanovehicle composed of a *p*-carborane small-wheeled short front axle and a $C_{60}$ large-wheeled long rear axle that has been termed a nanodragster due to the structural relation and resemblance to a dragster. Initial STM imaging has indicated that the two type of wheels establish linear rolling motion (Adapted with permission from Kelly *et al.*, Organic Letters 11(24), 5602 (2009) [4]. Copyright © 2009 American Chemical Society)

All biomolecules (including peptides and proteins) interact and self-organize to form well-defined structures that are associated with functionality [5]. While these molecules exist in biological organisms and may seem "alive" under observation, they follow the laws of physics and chemistry just like synthetic polymers, thus, in theory, they can be mechanistically understood. The problem is indeed complex, but that does not infer unsolvability. By observing the processes by which supramolecular architectures are assembled in Nature [5], we can begin attempt to exploit self-assembly for the synthesis of entirely novel synthetic materials by taking advantage of the "basic" processes.[1] Peptides and proteins provide versatile *building blocks* for fabricating materials. While there are a relatively limited number of such components (limited by the finite set of 20 amino acids, a few nucleotides, lipids, and available sugars, for instance), their specific combinatorial functions have yet to be clearly defined. Indeed, biological molecules have long been recognized as potentially useful components of nanodevices because of their structural and sequence flexibility and because of their ability to self-assemble (some examples given in Fig. 10.2). Interestingly, molecular self-assembly is ubiquitous in the natural world

[1] Again, we note the rudimentary airplane of the Wright brothers in 1902 does not compare to the complicated design of a Boeing 787, but it flew nevertheless. In similar fashion, we can currently use the self-assembly processes of Nature in simple, single component or single function systems, with refinement and introduction of complexity in the future.

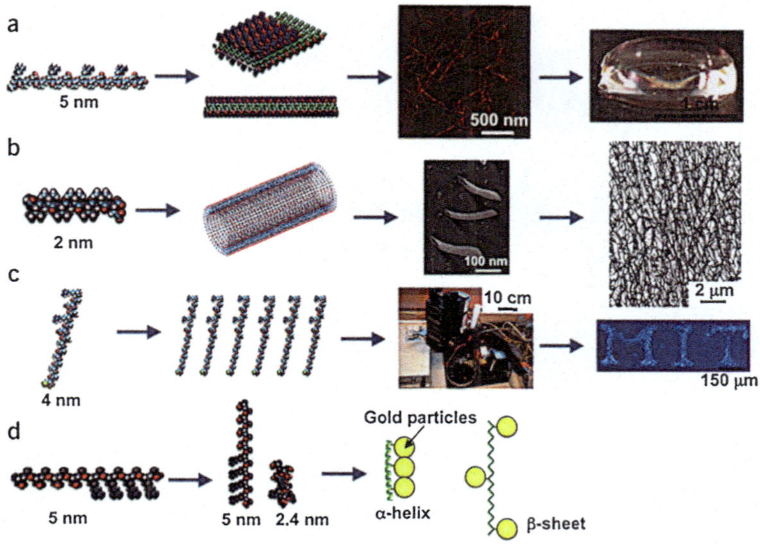


**Fig. 10.2** Examples of self-assembled material systems inspired by Nature. (**a**) Ionic self-complementary peptides with 16 amino acids, 5 nm in size, with an alternating polar and nonpolar pattern. They undergo self-assembly to form nanofibers with the nonpolar residues inside, and + and − charged residues form complementary ionic interactions, like a checkerboard. These nanofibers form interwoven matrices that further form a scaffold hydrogel with very high water content. (**b**) A type of surfactant-like peptide, 2 nm in size, that has a distinct head charged group, either positively charged or negatively charged, and a nonpolar tail consisting of six hydrophobic amino acids. The peptides can self-assemble into nanotubes and nanovehicles with a diameter of 30–50 nm. These nanotubes go on to form an inter-connected network similar to what has been observed in carbon nanotubes. (**c**) Surface nanocoating peptide. This type of peptide has three distinct segments: a functional segment, which interacts with other proteins and cells; a linker segment that not only can be either flexible or stiff, but also sets the distance from the surface; and an anchor for covalent attachment to the surface. These peptides can be used as ink for an inkjet printer to print directly on a surface, instantly creating any arbitrary pattern, as shown here. (**d**) Molecular switch peptide, a type of peptide with strong dipoles that can undergo drastic conformation changes, between $\alpha$-helix and $\beta$-strand or $\beta$-sheet, under external stimuli. It is conceivable that metal nanocrystals could be attached to these dipolar peptides to fabricate them into tiny switches (from S. Zhang [6], used with permission, copyright © 2003 Nature Publishing Group)

yet is mediated by weak, noncovalent bonds—hydrogen bonds, ionic bonds (electrostatic interactions), hydrophobic interactions, van der Waals interactions, and water-mediated hydrogen bonds. Yet Nature has made such weak bonding reliable and repeatable. Although these bonds are relatively insignificant in isolation, when combined together as a whole (cooperativity and hierarchical arrangements), they govern the structural conformation of all biological macromolecules and influence their interaction with other molecules.

Synthesis techniques have advances to the point where the complexity of nanostructures that can be fabricated rivals that of naturally occurring structures [6–8]. However, the scaling of hierarchical materials from nanoscale building blocks to macroscale functionality is a nontrivial progression. Nature, of course, takes advan-

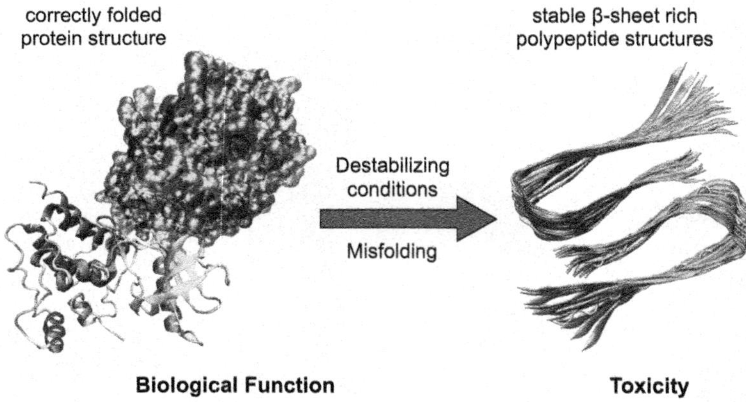


**Fig. 10.3** Nature's mistake results in stable material building block. The polypeptide chains of proteins display the intrinsic tendency to interact to each other with basic features dominated by hydrogen bonds between main chain atoms, giving rise to polymers rich in $\beta$ structure. Evolution has selected amino acid sequences able to organize into monomeric structures (secondary and tertiary globular structures) where the main chain is folded in a unique way and the closely packed side chains prevent its inherent tendency to generate $\beta$-crystallization. As the main chain is preserved as a common feature of all natural proteins, these can revert to this tendency under conditions loosening the closely packed tertiary structure. Hence, natural proteins can be thought of as a special group of evolved polymers where the specific interactions among side-chains dictate the globular structure (dictating functionality), whereas amyloid fibrils can be seen as the products of the intrinsic tendency of the polypeptide chains to organize into the polymer-like repeating structures, toxic in their natural environment (living organism). Figure inspired by [13]

tage of natural self-assembly over a relatively large timeframe to "guide the production" of material systems. A possible route to the development of bio-inspired and biomimetic systems is the understanding and exploitation of this self-assembly. One such example is the production of nanostructured biofilms from amyloidogenic proteins [9].

## 10.3 Benefitting from Nature's Misfolds

Amyloids are hierarchical protein materials, formed by insoluble fibrous protein aggregates (the result of uncontrolled protein misfolding), and observed in connection with severe degenerative disorders [10, 11]. Amyloids can form from diverse protein sequences [12] through a process called amyloidogenesis, where proteins lose their native functional configuration (*e.g.*, as enzyme or hormone) and form fibers with a characteristic hydrogen bonded cross-beta-sheet structure (Fig. 10.3). Hence, peptides and proteins possess a common hidden structure, resulting from the intrinsic tendency of the peptide backbone to polymerize into hydrogen bonded $\beta$-sheet rich assemblies displaying the characteristic cross-beta structure of amyloid fibrils, toxic to their natural function [13]. This inherent tendency becomes evident when proteins are destabilized so as their native folded structures are loosened and partially folded unstable molecules are significantly populated. This occurs when the conditions are right (or not right, depending on your perspective), and once formed are difficult to revert back to the intended protein states.

Compared to the native state of most proteins—where the structure is strongly dependent on the specific amino acid sequence and biological environment—the amyloidogenic configuration is a generic, widely accessible, stable structure of peptides and proteins [14]. Moreover, due to the remarkable mechanical and thermal stability of amyloids, combined with chemical properties that can be tuned *via* the introduction of additional elements (including enzymes, metal ions, fluorophores, biotin or cytochromes) they offer a promising platform for material assembly.

The beta-rich secondary structure of the fibrils correspond to a first level of hierarchical organization that many other materials in Nature feature, including silk, bone and hair. This conversion implies the dramatic loss of both structure and function of the proteins. Essentially they transition from active biological components to passive agglomerates. It has been shown that a great variety of amino acid sequences can lead to the formation of amyloid fibers—an unintentional and "rogue guise" [10] of proteins—provided they share some characteristic features such as an elongated, unbranched morphology, as well as a core structure that consists of a set of sheets oriented in parallel to the fibril axis, with their strands perpendicular to this axis [15]. These characteristic features are universal to the broader family of all amyloids, despite different protein sequences, which has been linked to their high stability, stiffness and capacity to provide structural templates across length-scales [16, 17], that render them good candidates as new novel biomaterials (see Fig. 10.4). Diverse features, such as variations in the protein sequence, give rise to biochemical properties specific to particular amyloids. As discussed previously (Chap. 9: Pathological Materiomics) while a promising candidate for biomaterials, the aberrant folding of proteins and peptides and subsequent aggregation of amyloid fibrils is associated with a range of pathological conditions from the systemic amyloidoses to neurodegenerative disorders—such as Alzheimer's and Parkinson's disease—thereby providing the initial motivation for the characterization of structural and mechanical behavior. *But how can we exploit this natural assembly and aggregation*?

Recent progress in the application of solid-state NMR [21, 22] and in growing elongated amyloid micro-crystals has provided detailed structural and biochemical information on the molecular-level structure amyloids [15]. In particular, these studies revealed that the molecules composing the fibrils posses a degree of uniformity that has previously only been associated with crystalline materials. However, many of their fundamental physical properties, specifically their great strength, sturdiness and elasticity, are not fully understood. This is partly due to the fact that larger-scale structural models of amyloid fibrils remain elusive, preventing bottom-up studies to describe the link between their hierarchical structure and physical properties. Recently, the structure of Alzheimer A$\beta$(1-42) fibrils was predicted by atomistic-based multi-scale analysis [23], providing a structural model of amyloid fibers with lengths of hundreds of nanometers at atomistic resolution. Note that the structure of the fibrils was only made possible by the *convergence* of experimental and computational

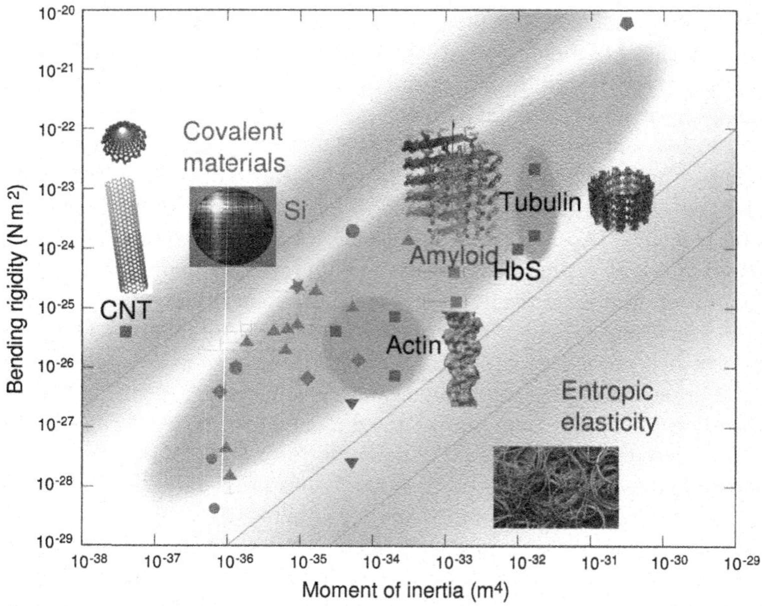


**Fig. 10.4** Mechanical properties of amyloid fibrils in comparison to biological and inorganic or non-biological materials. Plot depicts bending rigidity, $EI$, versus moment of inertia, $I$, for covalent materials, strong non-covalent interactions (such as H-bonds) and weak non-covalent interactions. Stars for Alzheimer's amyloid $\beta$-peptide residues 1–40 simulation results [18], squares show data for experimental results [19]. From [20], reprinted with permission, copyright © 2011 Nature Publishing Group

advances—the combination of solid-state NMR data [21, 22] and full atomistic MD simulations [23].

One or more fibrils arrange to form a protofilament [24, 25] and the assembly of multiple protofilaments results in morphologies ranging from twisted rope-like structures to flat-tapes with nanometer-scale diameters [26, 27]. There are examples of amyloids used as bionanomaterials in the form of nanowires [28, 29], scaffolds and (bio)templates [30, 31], liquid crystals [32], adhesives [33] and films [9]. These diverse hierarchical organizations suggests that the observed properties at the macro-scale might directly depend on the interactions and structural details at the atomistic level. In particular, the transfer of the mechanical properties of amyloids through different levels of organization remains unknown and critical to the understanding of the tunable aspects of amyloids, as well as possible candidates for application in nanotechnology (Fig. 10.5).

Approaches have recently emerged that exploit the stability and accurate self-assembly [6, 9] of artificial amyloid-like structures for technological applications [28]. As an example of using amyloidogenesis in the context of material assembly, Welland *et al.* elegantly exploit the unique properties of amyloids to form multifunctional materials by controlling the interplay of universality and diversity at different levels of hierarchy [9, 34]. The assembly procedure results in an intentional hier-

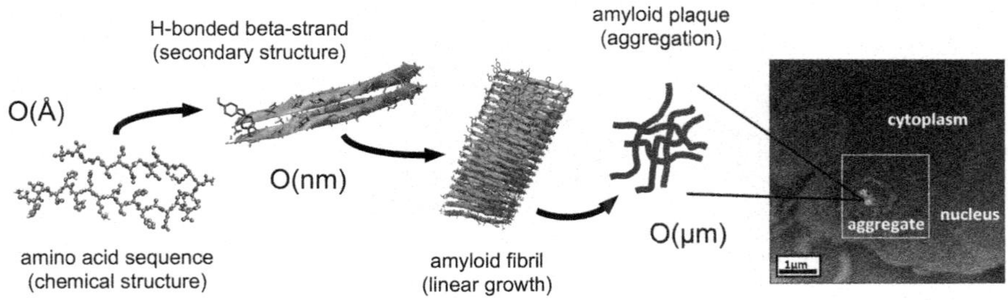


**Fig. 10.5** Hierarchical representation of amyloidogenic structures, from the constitutive polypeptide chains (amino acid sequence on the order of Ångstrom), to the formation of H-bonded beta-strands (secondary structures on the order of nanometers), which assemble to form stable amyloid fibrils (protofilaments on the order of hundreds of nanometers). Specifically, the molecular structures depicted here are amyloid fibrils formed by the amyloid peptide A$\beta$(1-40) associated with Alzheimer's Disease. The amyloid fibrils self-assemble and aggregate to form disordered plaques, characteristic of severe degenerative disorders such as Alzheimer's, type II diabetes, and Parkinson's disease. *Inset* depicts HAADF-STEM image of a amloidogenic aggregate in the cell cytoplasm (from McGuire *et al.*, copyright © 2010 IOP Publishing)

archy of length scales: nanometer ordering within the fibrils and micrometer scale ordering in the stacking of the fibrils (see Fig. 10.6). At a fundamental level, the protein sequence can be altered to design the biochemical properties of amyloids. Similarly, chemical functionalization can be added to realize hierarchical structures of molecules that would not naturally form. At larger hierarchical levels, the arrangement of amyloid fibrils in the material developed by Welland and coworkers can be controlled by introducing plasticizing molecules, enabling one to precisely tune the material's mechanical properties—films produced through protein self-assembly are highly rigid with a Young's modulus up to 5–7 GPa, comparable to the highest values for proteinaceous materials found in Nature (see Figs. 10.4 and 10.9). These structural alterations make it possible to create a broad range of functional properties based on a limited set of elements through manipulation of assembly and control of structural hierarchies, *i.e.*, the explicit control of the materiome.

Computational approaches provide a novel way forward to link the amino acid sequence to structural properties with measurable geometric effects at much larger hierarchical levels in the material. For example, sequence variations could be studied *in vitro* and *in silico* and then their properties could be compared in experiment [17, 36]. Such studies may help to understand fundamental issues related to amyloid growth and structure, and elucidate the molecular processes that underlie the normal and aberrant effects of protein self-assembly and mis-assembly in living systems.

Moreover, the properties of predicted amyloid fiber structures can be studied using molecular, coarse-grained, or continuum-based simulation approaches [18, 23, 37–39] as discussed in Chap. 6: Computational Approaches and Simulation, which could result in identifying the stiffness, elasticity and strength properties across scales. Similarly, electronic structure calculations could be utilized to elucidate electronic, magnetic, and optical properties of amyloid fibers.

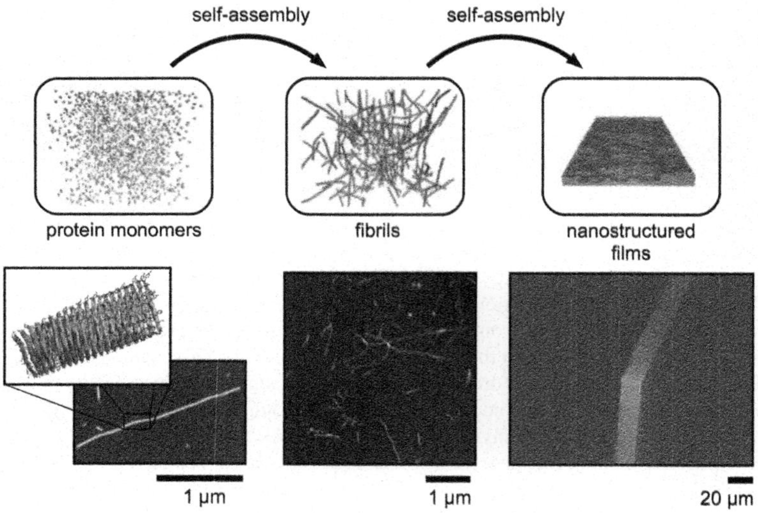


**Fig. 10.6** Directing the self-assembly of amyloids, amyloid fibrils, and aggregation into films. *Top*: Self-assembly process. A simple nanoscale motif consisting of protein molecules are first assembled into amyloid fibrils under controlled external conditions where the formation of intermolecular interactions is favored over intramolecular ones. The fibrils are then cast into free-standing films. During the casting process, the fibrils align in the film plane and can stack with nematic order, resulting in materials that have a hierarchy of length scales: nanometer ordering within the fibrils and micrometer-scale ordering in the stacking of the fibrils. *Bottom*: Images of hierarchical structures. Atomic force micrographs (AFM) of single and multiple fibrils (scale bars 1 μm); *inset* of atomistic model; scanning electron micrograph (SEM) of the resulting free-standing protein film. Leftmost AFM image from [35], used with permission, copyright © 2006 by the National Academy of Sciences; all other images from Knowles *et al.* [9], copyright © 2010 Nature Publishing Group

It is again noted that such amyloid structures were initially investigated due to the pathological association with Alzheimer's and Parkinson's diseases [11, 40]. Not only are there inherent benefits in understanding amyloid-type materials from a diagnostic perspective, the ability to manipulate and engineer the properties such a protein material prompts the natural utilization as a potential tissue scaffold. The presence of amyloid structures, albeit in a disease state, substantiates suitable biochemical and physico-chemical factors, which perhaps can be utilized to improve or replace biological functions through tissue engineering.

Tissue engineering is associated with applications that repair or replace portions of or whole tissues (*i.e.*, bone, cartilage, blood vessels, etc.). Often, the tissues involved require certain mechanical and structural properties for proper functioning, which can be realized though a materiomic framework. The development of suitable scaffolds for tissue engineering involves an implicit balance between architecture and specific tissue type by consideration of both microstructure and microenvironment [43]. For example, a critical parameter of tissue scaffolds is the relative pore size to facilitate nutrient transport [44]. Commonly, there is an inverse relation between pore size and mechanical stability which can be reconciled though a complete understanding of the materials structure-property-process relations—the

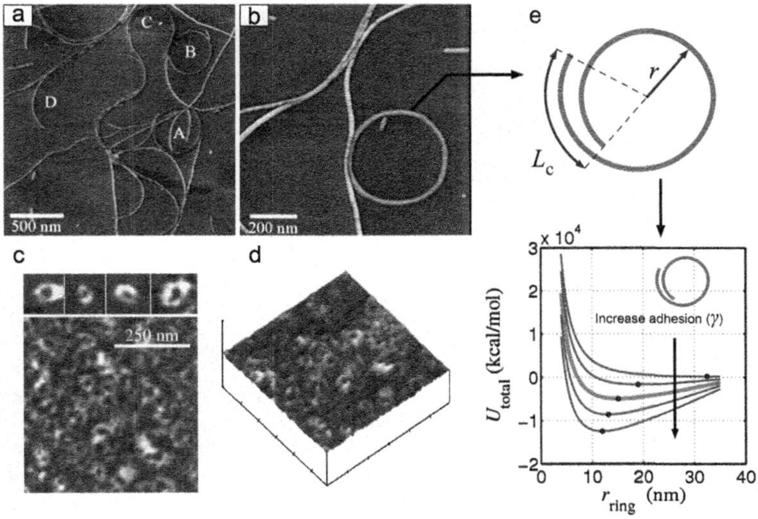


**Fig. 10.7** Directing self-assembly to attain variations in structure. (**a**) and (**b**) Typical morphological features of insulin amyloid grown at 1500 bar. (**a**) Closed (**A** and **B**), open (**C**), and partly bent (**D**) fibres. (**b**) The most abundant circular form has a diameter of 340–420 nm and is accompanied by bent stretches of fibrils that are typically 20–100 nm long (from [41], used with permission, copyright © 2004 Elsevier). (**c**) and (**d**) AFM images of the ring-shaped amyloid structures of equine lysozyme. (**c**) Height image of the ring-shaped structures. Insertions above show a selection of individual rings. (**d**) Three-dimensional height image of the same structures. The z-scale shown above background is 50 nm (from [42], used with permission, copyright © 2003 Elsevier). (**e**) Simple mechanical model of circular morphology, based on a balance of mechanical bending energy and adhesion energy, in which equilibrium radius can be predicted from adhesion strength (*plot*). Adapted from [39]

same relations that materiomics aims to elucidate. The structure-property relations of the constituent protein components are fundamental in the development of tissue scaffolds, to accurately mimic the natural composition of biomaterials, and to understand the role of material stiffness changes in disease states, such as those associated with amyloidogenesis. We can also exploit the self-assembly of amyloid films to induce topological changes in the microstructure, and tune relative pore size and other topological features. For example, the formation of annular or ring-like protofilament structures has been observed as a function of pH and metal ion concentration [42], as well as a function of pressure [41]. Simple mechanical models of such circular morphologies can be used to predict the stiffness and adhesion energy of such fibrils under varying environmental conditions [39]. Thus, we may be able to control the inevitability of amyloidogenesis to create natural scaffolds for tissue growth rather than the toxic environment that currently propagates (Fig. 10.7).

An intriguing aspect of amyloid assembly is the manner by which weak non-covalent inter-molecular chemical bonds generate mechanically strong materials. Functionality is generated by compiling few simple-and often lower-performance-elements into assemblies where structures are created at multiple length-scales

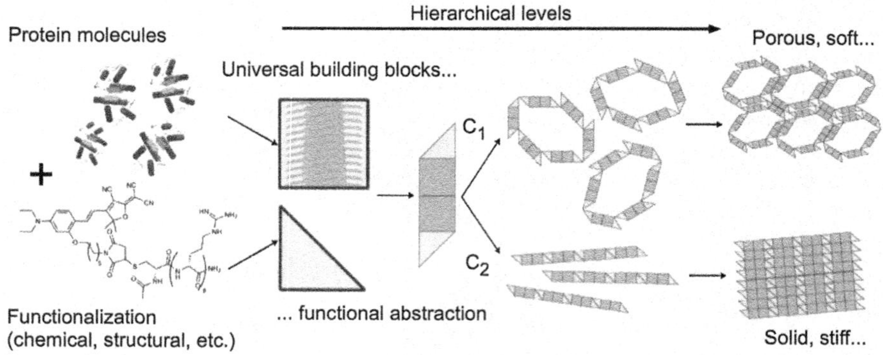


**Fig. 10.8** Universality-diversity paradigm applied to amyloid materials. Amyloid formation from a multiple native proteins results in universal building blocks that can be assembled and functionalized (for example, with fluorophores or metal particles) into larger and more diverse structures. Due to similar characteristics (*e.g.*, beta-rich secondary structure) these larger components can be classified as stable "building blocks" based on common functionality, irrespective of particular chemical make-up. Different structures can be achieved through changes in pH, external pressure or other processing conditions (*e.g.*, $C_1$ or $C_2$). The end material system can achieve disparate mechanical properties (stiff versus soft, solid versus porous, yet is composed of the identical constituent materials). Adapted from [20]

(Fig. 10.8). Here, we explicitly realize the universality-diversity concept as defined in Chap. 4: Universality-Diversity Paradigm: Music, Materiomics, and Category Theory. Natural proteinaceous materials are typically generated through the biological design paradigm where multifunctionality is created not through the use of high quality, or use of a large number of distinct building blocks as is commonly done in traditional engineering. *In silico* studies show that the key to create strength out of weak hydrogen-bonding is their grouping into clusters, arranged in beta-strands with a relatively short length. This universal design feature allows hydrogen-bonds to work cooperatively and reach maximum strength [45, 46].

An advantage of biological self-assembly is the ability to generate and control ordering on the nanoscale, as shown schematically in Fig. 10.8. This natural propensity for nanoscale organisation into fibrils [17, 47] can be used to template other materials that do not on their own possess a propensity to form ordered structures at that scale, such as metal particles [48]. This strategy is a realization of the concept of using universal building blocks, arranged in hierarchical structures, to create diverse functional materials. This approach opens up opportunities towards hierarchical *de novo* material design to mimic and exceed functional properties realized in natural materials. The power of this principle has been demonstrated in the fabrication of conductive nanowires, where the self-assembling peptide fibrils act as templates for the deposition of metals on the outside of structures to yield electrically conducting wires [8, 28, 49]. The propensity of proteins to undergo multi-level hierarchical assembly opens up the possibility of such larger scale structures to be generated through self-assembly, while maintaining the accurate control of nanoscale organization.

The multi-scale investigations and integration of such protein-based materials is naturally within the realm of materiomics, with recent studies being undertaken out for collagen [50, 51], elastin [52, 53], and fibrin [54], for instance. The integration of the result of each investigation ultimately contributes to the materiomic information, required for successful tissue engineering and biomaterial development. The mechanical properties of materials reflect intimately the nature of the fundamental inter-molecular interactions that bind their constituents into larger-scale hierarchical systems. In amyloid materials this connection offers general insights into the characteristics of interactions in natural or artificial proteinaceous materials. Amyloid fibrils can be seen to form a reference class of structures where the core of fibrils consists of arrays of beta-strands that involve essentially every residue in the core domain, resulting in continuous beta-sheets that can extend over thousands of molecular units, a unique situation in biological materials. Many amyloid fibrils, especially those formed from proteins that do not undergo the assembly process towards the amyloidogenic state in Nature, have not been influenced directly by evolutionary pressures. As such, these proteins give access to the inherent mechanical properties of beta-sheets on supra-molecular length-scales without the influence of evolutionary adaptation that typically pervades the properties of natural protein materials. In a comparison with other biological materials (Fig. 10.9) amyloid fibrils are remarkably strong and possess Young's moduli comparable to the most rigid proteinaceous materials in Nature.

## 10.4 Multipurpose Folding DNA

Nature typically employs 'bottom-up fabrication', which exploits the intrinsic properties of atoms and molecules to direct their self-organization (*i.e.*, growth). In practice, it can be used to make relatively simple nanostructures, but the difficulty arises in the bottom-up fabrication of complex hierarchies—the self-assembly of self-assembled structures, so-to-speak. The challenge lies in the difficulty of engineering diverse yet specific binding interactions that most self-assembled structures contain and result in just a few unique positions (folded or unfolded, bonded or unbonded, for example). Biological materials require multiple or intermediate configurations, allowing more complex structures across many scales. Precise and predictable self-assembly is required to create nanostructures of high complexity.

The stability of amyloids can be considered a by-product of the misfolding of proteins. More useful would be the adoption of predictable and repeatable folding as assembly units. Luckily, a more robust structure and a potential means for self-assembly is found in one of the more common building blocks in Nature: deoxyribonucleic acid, or, as it is more commonly known, DNA. While the primary job of DNA in cells is to carry genetic information from one generation to the next, the highly stable and programmable molecule is an promising building material for nanoscale structures that could be used to deliver drugs, act as biosensors, perform artificial photosynthesis, and more. Modified by functional groups, DNA nanostructures can serve as scaffolds to control the positioning of other molecular species,

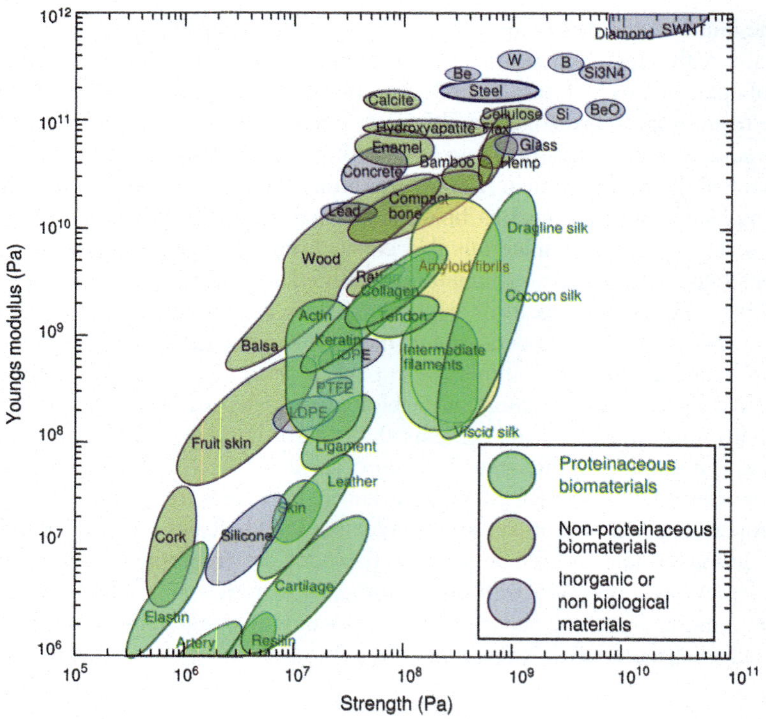


**Fig. 10.9** Mechanical properties of amyloid fibrils in comparison to biological and inorganic or non-biological materials. Plot depicts Young's modulus, $E$, versus strength for proteinaceous and non-proteinaceous biomaterials, as well as inorganic (non-biological) materials. Covalent and metallic bonding results in the stiffest and strongest materials, with diamond and single-wall carbon nanotubes (SWNTs) being the best performers. Silks are the strongest and stiffest protein materials, followed by amyloid and collagen; and significantly more-rigid materials (for example, bone) contain minerals. From [20], used with permission, copyright © 2011 Nature Publishing Group

which opens opportunities to study intermolecular synergies, such as protein-protein interactions, as well as to build artificial multicomponent nanomachines [55]. The self-assembly of DNA strands has been used to create linear structures and as a method for creating patterns [56, 57]. Nucleic acids provide a robust platform for self-assembly: the exquisite specificity of Watson-Crick base pairing allows a combinatorially large set of nucleotide sequences to be used when designing binding interactions. The emerging field of "DNA nanotechnology" [58, 59] has exploited this property to create a number of more complex nanostructures, including two-dimensional arrays [60], as well as three-dimensional shapes such as a cube [61] and truncated octahedron [62].

For early successes in the manipulation of DNA, it was assumed that precise control over the molecules was necessary, rather than letting the molecules perform the work. Sequences were highly optimized to avoid secondary structure or undesired binding interactions, strands were highly purified, and strand concentra-

**Fig. 10.10** DNA origami shapes. (Left to right): square; rectangle; star; disk with three holes (smiley face); triangle with rectangular domains; sharp triangle with trapezoidal domains and bridges between them. Top row, diagrams showing the bend of helices at crossovers (where helices touch) and away from crossovers (where helices bend apart). Bottom two rows, AFM images. All images and panels without scale bars are the same size, 165 nm × 165 nm. Scale bars for lower AFM images: rectangle, 1 mm; stars/disks/triangles, 100 nm (from Rothemund [57], used with permission, copyright © 2006 Nature Publishing Group)

tion/stoichiometry was precisely tuned and constrained [63]. While such criteria was important for the formation of many DNA nanostructures, it was counter to the principles of Nature—a robust self-assembly process with little control requirements.

Rothemund and colleagues [57] developed a versatile and simple method for using numerous short single strands of DNA to direct the folding of a long, single strand of DNA into desired shapes, roughly 100 nm in diameter and with a spatial resolution of about 6 nm (see Fig. 10.10). Through a methodology termed "scaffolded DNA origami", it is possible to assemble arbitrary shapes, such as squares, triangles and five-pointed stars. Moreover, assembly of groups of these basic shapes enables the creation of structures with arbitrarily shaped holes or surface patterns (*i.e.*, controlled and self-assembled levels of hierarchy). Rothemund achieved patterns of DNA shapes with a complexity tenfold higher than that of any previously self-assembled arbitrary pattern, and comparable to that achieved using AFM and STM manipulation of single atoms and molecules. In addition, the method is easy to implement (relative to previous methods), high yield and relatively inexpensive (in terms of time and energy).

The design of a DNA origami is performed *via* the astute combination of geometric modeling, constituent oligonucleotides, and balance between DNA folding, strand spacing, and cross-overs. Scaffolded DNA origami can be adapted to create more complex or larger structures such as the design of three dimensional structures. The design for a desired shape is made by filling the shape with a 7-kilobase single-stranded DNA scaffold and by choosing over 200 short oligonu-

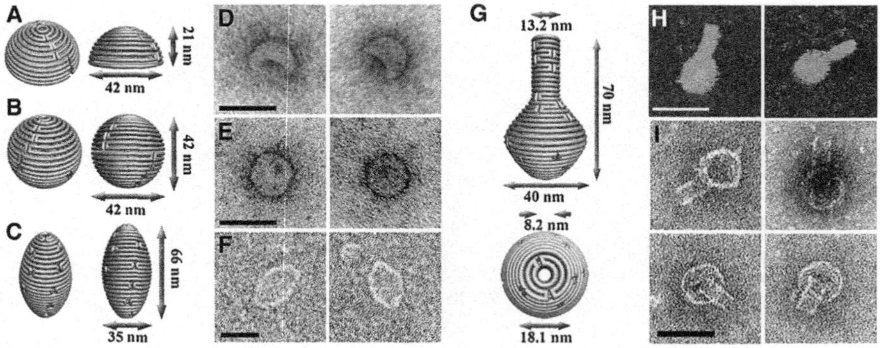


**Fig. 10.11** DNA nanostructures with complex 3D curvatures. (**A**) Schematic representation of the hemisphere. (**B**) Schematic representation of the sphere. (**C**) Schematic representation of the ellipsoid. (**D**) TEM images of the hemisphere, randomly deposited on TEM grids. The concave surface is visible as a dark area. (**E**) TEM images of the sphere, randomly deposited on TEM grids. Due to the spherical symmetry, the orientation can not be determined. (**F**) TEM images of the ellipsoid. The outline of the ellipsoid is visible. Scale bar for the TEM images in (**D**), (**E**) and (**F**) is 50 nm. (**G**) Schematic representation of the nanoflask. (**H**) AFM images of the nanoflask. Scale bar is 75 nm. (**I**) TEM images of the nanoflask, randomly deposited on TEM grids. The cylindrical neck and rounded bottom of the flask are clearly visible in the images. Scale bar is 50 nm (from Han *et al.* [64], used with permission, copyright © 2011 American Association for the Advancement of Science)

cleotide 'staple strands' to hold the scaffold in place. Once synthesized and mixed, the staple and scaffold strands self-assemble in a single step. Because each oligonucleotide can serve as a kind of 6-nm "pixel", the structures can be programmed to bear complex patterns such as words and images on their surfaces. Extending the technique, William Shih's lab at Harvard Medical School translated this concept to three dimensions, allowing design of complex curved and bent structures that opened new avenues for synthetic biological design at the nanoscale. More recently, self-assembling DNA nanostructures that define intricate curved surfaces in three-dimensional (3D) space have been constructed using the DNA origami folding technique [64] as depicted in Fig. 10.11. Enhanced availability of diverse single-stranded scaffolds combined with longer scaffold alternatives could result in even larger attainable objects with more complicated features.

There is, however, a caveat in the current approach: the requirement for unique sequence information means that the method cannot be scaled up arbitrarily; whenever structures above a critical size or level of complexity are desired, it will therefore be necessary to combine scaffolded DNA origami with hierarchical self-assembly. The extent of scaffolded DNA origami is thus limited to a few hierarchies (DNA strand, DNA structure, basic assembled shape, patterned shapes), bridging a few scales (on the order of 100s of nanometers), but demonstrates the approach necessary for successful self-assembly.

From another perspective the self-assembly of DNA results in the sufficient production of structural building blocks, which can then be exploited in larger-scale assemblies (such as top-down fabrication techniques). Indeed, practical components

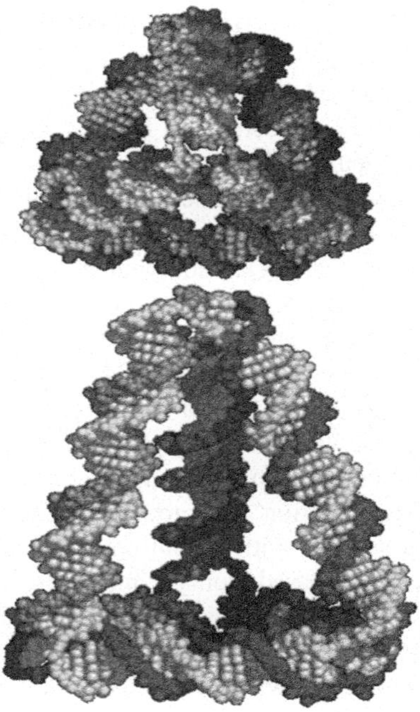

**Fig. 10.12** DNA tetrahedra assembled as macromolecular "building blocks". Design of a DNA tetrahedron formed by annealing four oligonucleotides in complementary hybridized subsequences (corresponding to each face of the tetrahedra forming interwoven helices). The backbone of each oligonucleotide is highlight shading (from Goodman *et al.* [66], used with permission, copyright © 2005 American Association for the Advancement of Science)

for three-dimensional molecular nanofabrication must be simple to produce, rigid, and adaptable, leading to the widespread investigation of potential "nanocomponents" such as carbon nanotubes [65]. DNA provides an ideal candidate for nanofabrication of rigid structures because it is inherently biocompatible, and, as mentioned, assembly can be controlled by base-pairing.

Turberfield *et al.* have successfully synthesized and mechanically characterized a family of DNA tetrahedra that have been designed to self-assemble in a single step in only a few seconds [66]. The DNA tetrahedron is designed to be mechanically robust; it consists of rigid triangles of DNA helices covalently joined at the vertices (Fig. 10.12). The four component oligonucleotides each run around one face and hybridize to form the double-helical edges. In contrast to the challenging syntheses of DNA cubes [61] and octahedral [62], the synthesis of tetrahedra is extremely simple. In addition, atomic force microscopy (AFM) was used to both image the tertiary structure of individual tetrahedra and to demonstrate their mechanical rigidity. Such tetrahedra are intended as versatile building blocks for 3D nanofabrication, *via* programmable DNA linkers with controllable functionality. In essence, the atomistic scale building blocks of amino acids have been "upscaled" to molecular structural components. One could imagine producing a vast supply of these "molecular bricks", similar to the standardized structural steel components.

## 10.5 Piecewise Controlled Self-assembly

Knowledge of molecular interactions and structures will enable nanoscale devices to be engineered and incorporate desired functionalities but, as shown, due to the difficulties of controlling matter precisely at the nanoscale with current technology, the nanodevice components must self-assemble. It is clear that biological molecules offer exciting possibilities for construction of new complex, self assembled nanodevice components whose structures can be finely controlled, but rather than "bulk" assembly (such as the relatively homogeneous assembly of amyloids or DNA origami), can we direct the assembly piece by piece and still have a self-driven process? With clever building blocks, it seems we can.

As mentioned, a unifying trait of many of these amazing materials is the relatively weak building blocks in which they are composed—soft and flexible strands of proteins bonded and attracted by electrostatic and hydrogen bonds. We now turn our attention again to DNA and the exploitation of H-bonding for supramolecular assembly, with the introduction of non-protein building components (specifically, carbon-based aromatic molecules).

In the context of supramolecular self-assembly, a major challenge is to control the size and shape of the resulting molecular structures. Unlike proteins, macromolecules without specified amino acid or nucleobase pairing have a propensity to assemble in disordered, random arrangements (such as those derived from layer-by-layer disposition of polyelectrolytes). Fine tuning the molecular architecture in such self-driven processes is both an intellectual and practical problem. Many bottom-up approaches have been undertaken to direct the self-assembly of macromolecules—such as templates [67, 68], electric-field-assisted alignment [69], and base modification [70]—but the majority of the superstructures (*e.g.*, 1D nanowires, fibrils and ribbons, 2D crystals, membranes and tapes) lack fine-positioning of functional units and defined architectures. One potential approach is the use of H-bonding functional units—mimicking the assembled alpha-helica, beta-sheet, and DNA structures of Nature. Through design of specific side-chain sequences, the acceptor-donor mechanism of H-bonding allows "targeted" groups to locate each other and bind in desired positions and geometries.

Recent work by Surin and Schenning encompasses a two tiered approach—full atomistic simulation combined with experimental measures [71], as shown in Fig. 10.13, again, illustrating the theme of convergence between empirical and computational approaches. The study involves the use of water-soluble naphthalene (NT) and an oligo($p$-phenylene vinylene) (OPV) guest derivatives end-capped with one diaminotriazine (DT) H-bonding unit, which binds to a single-strand oligothymine template (single-strand DNA, or ssDNA) in aqueous solution. Essentially, one can consider a the fundamental "building blocks", in which the ssDNA provides a structural backbone, whereas assembly is facilitated by with $\pi$-stacking of the NT or OPV guests and the H-bonding of the DT groups (see Fig. 10.13). Such an abstract perspective is powerful, as this approach, in principle, can be applied to any functional molecule equipped with an appropriate hydrogen-bonding moiety to create uniform well-organized nanoscale objects.

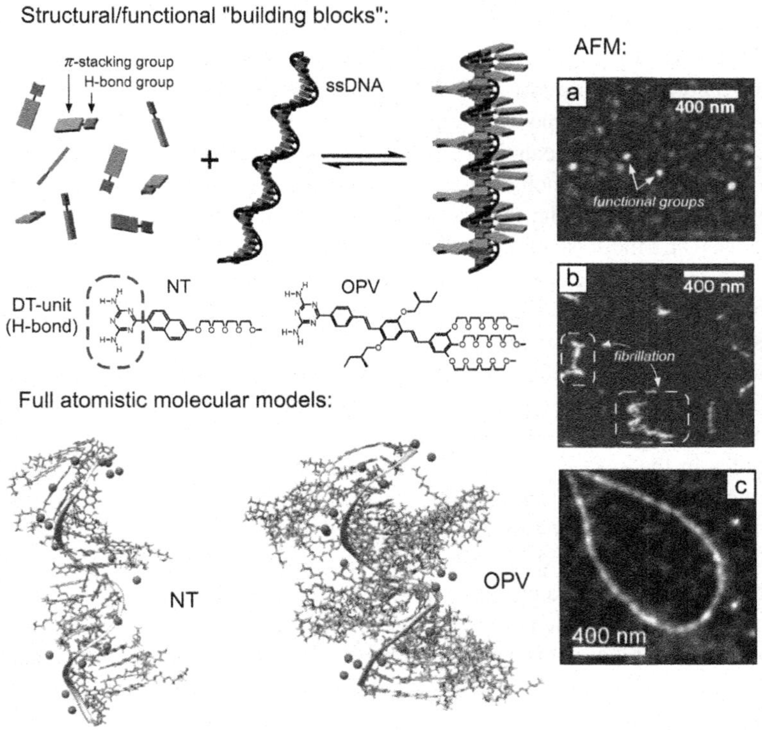


**Fig. 10.13** Two tiered analysis of supramolecular self-assembly. *Top*: Schematic representation of DNA templated self-assembly, where the DNA can be considered the structural backbone and the NT or OPV represent the binding blocks (consisting H-bonding attachment sections, and $\pi$-stacking regions). *Bottom*: Full atomistic model snapshots at the end of molecular dynamics for DNA–NT and DNA–OPV supramolecules. Simulations allow the analysis of system properties (such as free energy and atomic distances) inaccessible by experimental methods. *Right*: Liquid-cell TM-AFM height images of aqueous mixtures deposited surfaces of (**a**) and (**b**) DNA-NT and (**c**) DNA-OPV, depicting the longer lengths attainable by OPV functional groups

Self-assembly of this specific system has been shown to lead to helical DNA-like hybrid structures, as indicated by circular-dichroism (CD) spectroscopy [72]. However, the spectroscopic data fails to indicate detailed information on the influence of the molecular structure of the guest on the structural order and intermolecular interactions of the system. Thus, additional molecular modeling simulations were undertaken to confirm structural models that can be related to CD spectroscopy and AFM measurements (see Fig. 10.13). Full atomistic MD simulations of the organization of conjugated oligomers and single-strand DNA H-bonded complexes provide detailed structural information unattainable by experimental means, which is a key issue for the understanding of dominant interactions in such structures. For example, the binding energy between adjacent NT molecules (−4 kcal/mol) and OPV molecules (−29 kcal/mol) can be directly compared to the binding energy for the thymine-diaminotriazine H-bonded dimer (−13.5 kcal/mol). Structural information such as helical pitch and interlayer spacing can also be extracted from a geometric analysis of the MD simulations. For NT, the relatively weak interactions do not alter the "natural" structure of the ssDNA backbone, while stronger interactions and steric effects of OPV impose a significantly distorted DNA structure. Complimentary CD spectra analysis can then be used to confirm such structures experimentally. Surin and co-workers conclude that, for larger interactions between the guests in the complex (NT or OPV), the H-bond interaction between the guests and the template diminishes, and the $\pi$-stacking of the functional guests will have stronger influence on the structure of the template. In the structure with OPV, the stacking interaction prevails—some H-bonds are broken in favor of better $\pi$-stacking—while the opposite is true for NT-based structures.

Additional AFM studies revealed that the assembly on surfaces and the persistence length of these hierarchical complexes is also determined by the interaction between the functional guest molecules. For large interactions between conjugated molecules in the stacks, the complex can further aggregate along the strand axis, inducing longer lengths. These insights are important for the future design of templated assemblies, and the understanding of their spectroscopic properties, for instance, electron-donor and -acceptor conjugated units carrying different H-bonding moieties can be aligned and fine-positioned with a sequence determined by the sequence of oligonucleotide complementary bases.

The caveat here, of course, is that such assembly mechanisms are frequently only applicable on a single scale. While the supramolecules can be reliably assembled at the nanoscale, they are not arranged in larger hierarchical structures. Thus, another driving mechanism must be developed to transcend higher orders of magnitude to form hierarchical structures and control geometries. A better understanding of the nature and amplitude of the intermolecular interactions is essential for exploiting these functional supramolecules. Controlled and "preprogrammed" self-assembled with precise molecular architectures is an important goal for the attainment of functional materials with predictable physico-chemical properties. Exploitation of H-bond propensity has been used successfully to form both mono- and multicomponent [73, 74] $\pi$-conjugated architectures as discussed here.

It is interesting to note that this self-assembly methodology exploits three distinct behaviors at the molecular level:

1. Implementing DNA as a structural backbone. The specific sequencing and helicity of DNA results in a promising template to organize molecular architectures in a "bottom-up" approach for construction of functional nanosized objects.
2. Weak interactions through H-bonding (atom-atom interactions).
3. Weak interactions through $\pi$-stacking (molecule-molecule interactions).

While the first two behaviors are rampant throughout Nature, exploitation of the intrinsic $\pi$-stacking of large aromatic structures is an ingenious addition to the self-assembly "toolbox". Engineers do have at least one important advantage over Nature: we can choose our materials. Nature, by contrast, often has to make do with whatever is readily available locally, and whatever structures have been created through the lengthy trial-and-error of evolution. A spider or a cell doesn't have great resources. It can't import materials and uses what is immediately available. With the current influx of research concerning carbon materials such as graphene and graphene oxide, understanding the interactions with biological interfaces will play an important role in combination of biological materials (such as protein strands and DNA) with synthetic materials (such as carbon nanotubes, metals, and inorganic polymers) that will further extend the design space in terms of achievable mechanical, optical, electrical, thermal, and chemical properties and functionalities.

## 10.6 Summary

The preceding discussion was limited to a few choice biological material systems—namely amyloids and DNA—to elucidate the promise of self-assembly and potential of weak bonding as a means for robust hierarchical structures. It is with little surprise that these biological materials have had ample research in recent years, with the widespread occurrence of amyloidogenesis throughout a spectrum of native proteins, and the universality of DNA within animals and organisms. This is not to say, however, that such materials are unique. Indeed, we could have focused on the synthesis of collagenous materials, silk-inspired fibers, engineered tissues, or bio-inspired polymeric platforms as illustrative examples, to name a few. The key feature lies in the hierarchical self-assembly requiring the insights provided by materiomics—taking advantage of the complexity of Nature's design. How can we smartly choose a suitable system for an intended functionality? One approach to the successful discovery, selection and development of suitable "building blocks" lies in a brute force screening of combinatorial peptide and existing protein libraries [75, 76]. While successful in the discovery of compounds with desired interactions and functional proclivities [75, 77], materiomics enables a more systematic approach, wherein the desired molecular characteristics are known *a priori* and can thereby be *constructed* atom by atom or amino acid by amino acid.

Materiomics provides an exciting opportunity for the engineering of complex biological, bio-inspired, and synthetic material systems, based on quantitative insight into their fundamental physical and chemical features. A rigorous understanding may enable us eventually to integrate concepts from living systems into engineering materials design, seamlessly. Optical, mechanical and electrical properties at ultra-small material scales, their control, synthesis and analysis as well as their theoretical description represent major scientific and engineering opportunities. However, as in the case of conventional "engineered" materials, these breakthroughs may only be possible provided that their fundamental concepts are well understood and appropriate models developed. Characterization of materials found in biology—in particular in the context of living systems—enabled through a rigorous materiomics approach is aimed towards the elucidation of these fundamental principles of assembly, deformation and possible failure.

The long-term goal is to develop a new engineering paradigm that encompasses the analysis and design of structures and materials, starting from the molecular level, in order to create new materials that mimic and exceed the properties of biological ones by utilizing material concepts discovered in biological materials. Such work can lead to the development of a new set of tools that can be applied, together with synthetic biological and self-assembly methods, to select, design, and produce a new class of materials, similar to the approaches used today in computer aided design of buildings, cars and machines. The availability of multifunctional and changeable materials reduces the necessity for the use of different materials to achieve different properties, and as such, may provide significant savings in weight and cost. The utilization of abundant natural building blocks such as organic (*e.g.*, peptides, proteins, sugars) or inorganic (*e.g.*, minerals, inorganic molecules) constituents, combined with novel synthesis techniques based on self-assembly, could lead to new lightweight materials for structural applications in cars, airplanes, and buildings that could reduce the overall energy consumption and ecological footprint of materials.

Such an integrated and holistic approach will enable more sophisticated and complex new designs, unattainable by traditional materials science and engineering approaches. Just as biology has done, humans could engineer materials with desired properties such as strength or flexibility by using abundant and cheap materials. Using cleverly designed structures, materials with almost any kind of desired properties could be produced, even using a very limited, and almost arbitrary set of components. Though the field is still in its infancy, the self-assembly of biological molecules ultimately promises materials, devices and technologies beyond our current imaginations.

## 10.7 Suggested Readings

On amyloids, amyloidogenesis, and related proteins:


- E.M. Sigurdsson (ed.), *Amyloid Proteins*, Methods in Molecular Biology (Humana Press, 2010)
- R. Wetzel, I. Kheterpal (eds.), *Amyloid, Prions, and Other Protein Aggregates, Part B*, Methods in Enzymology, vol. 412 (Elsevier, 2006)
- J.M. Squire, D.A.D. Parry, A. Kajava, *Fibrous Proteins: Amyloids, Prions and Beta Proteins*, Advances in Protein Chemistry, vol. 73 (Elsevier, 2006)
- W. Xia, H. Xu (eds.), *Amyloid Precursor Protein: A Practical Approach* (CRC Press, 2005)


On DNA and DNA-based materials:


- J. Jin, J. Grote (eds.), *Materials Science of DNA* (CRC Press, 2011)
- G. Zuccheri, B. Samorí (eds.), *DNA Nanotechnology: Methods and Protocols*, Methods in Molecular Biology (Humana Press, 2011)
- A. Bhambhani, C.V. Kumar, *Novel Enzyme/DNA/Inorganic Materials: A New Generation of Biomaterials* (Nova Science Publishers, 2008)


On the synthesis, processing and properties of biological materials:

- D.L. Kaplan (ed.), *Biopolymers from Renewable Resources* (Springer, 1998)
- K. MecGrath, D.L. Kaplan (eds.), *Protein-Based Materials* (Birkhäuser, 1997)

On biomimicry and bio-inspired material systems:

- J.M. Benyus, *Biomimicry: Innovation Inspired by Nature* (Harper Collins, 1997)

On nanodevices, nanomachines, and molecular assembly:

- B.W. Ninham, P. Lo Nostro, *Molecular Forces and Self Assembly: In Colloid, Nano Sciences and Biology*, Cambridge Molecular Science (Cambridge University Press, 2010)
- J.A. Pelesko, *Self Assembly: The Science of Things That Put Themselves Together* (Taylor & Francis, 2007)
- M. Gross, *Travels to the Nanoworld: Miniature Machinery in Nature and Technology* (Harper Collins, 1999)

## References

1. T. Junno, K. Deppert, L. Montelius, L. Samuelson, Controlled manipulation of nanoparticles with an atomic-force microscope. Appl. Phys. Lett. **66**(26), 3627–3629 (1995)
2. D.M. Eigler, E.K. Schweizer, Positioning single atoms with a scanning tunneling microscope. Nature **344**(6266), 524–526 (1990)
3. A.J. Heinrich, C.P. Lutz, J.A. Gupta, D.M. Eigler, Molecule cascades. Science **298**(5597), 1381–1387 (2002)
4. K.F. Kelly, G. Vives, J.H. Kang, J.M. Tour, Molecular machinery: synthesis of a "nanodragster". Org. Lett. **11**(24), 5602–5605 (2009)
5. C.-I. Brndn, J. Tooze, *Introduction to Protein Structure*, 2nd edn. (Garland, New York, 2009)
6. S.G. Zhang, Fabrication of novel biomaterials through molecular self-assembly. Nat. Biotechnol. **21**(10), 1171–1178 (2003)
7. N. Kol, L. Adler-Abramovich, D. Barlam, R.Z. Shneck, E. Gazit, I. Rousso, Self-assembled peptide nanotubes are uniquely rigid bioinspired supramolecular structures. Nano Lett. **5**(7), 1343–1346 (2005)
8. W. Lu, C.M. Lieber, Nanoelectronics from the bottom up. Nat. Mater. **6**(11), 841–850 (2007)
9. T.P.J. Knowles, T.W. Oppenheim, A.K. Buell, D.Y. Chirgadze, M.E. Welland, Nanostructured films from hierarchical self-assembly of amyloidogenic proteins. Nat. Nanotechnol. **5**(3), 204–207 (2010)
10. M. Fandrich, M.A. Fletcher, C.M. Dobson, Amyloid fibrils from muscle myoglobin—even an ordinary globular protein can assume a rogue guise if conditions are right. Nature **410**(6825), 165–166 (2001)
11. C.M. Dobson, Protein folding and misfolding. Nature **426**(6968), 884–890 (2003)
12. R. Kayed, E. Head, J.L. Thompson, T.M. McIntire, S.C. Milton, C.W. Cotman, C.G. Glabe, Common structure of soluble amyloid oligomers implies common mechanism of pathogenesis. Science **300**(5618), 486–489 (2003)
13. M. Stefani, Protein misfolding and aggregation: new examples in medicine and biology of the dark side of the protein world. Biochim. Biophys. Acta, Mol. Basis Dis. **1739**(1), 5–25 (2004)
14. F. Chiti, C.M. Dobson, Protein misfolding, functional amyloid, and human disease. Annu. Rev. Biochem. **75**, 333–366 (2006)
15. R. Nelson, M.R. Sawaya, M. Balbirnie, A.O. Madsen, C. Riekel, R. Grothe, D. Eisenberg, Structure of the cross-beta spine of amyloid-like fibrils. Nature **435**(7043), 773–778 (2005)
16. I.W. Hamley, Peptide fibrillization. Angew. Chem., Int. Ed. **46**(43), 8128–8147 (2007)

17. T.P. Knowles, A.W. Fitzpatrick, S. Meehan, H.R. Mott, M. Vendruscolo, C.M. Dobson, M.E. Welland, Role of intermolecular forces in defining material properties of protein nanofibrils. Science **318**(5858), 1900–1903 (2007)
18. R. Paparcone, S. Keten, M.J. Buehler, Atomistic simulation of nanomechanical properties of Alzheimer's a beta(1-40) amyloid fibrils under compressive and tensile loading. J. Biomech. **43**(6), 1196–1201 (2010)
19. C. Sachse, N. Grigorieff, M. Fandrich, Nanoscale flexibility parameters of Alzheimer amyloid fibrils determined by electron cryo-microscopy. Angew. Chem., Int. Ed. **49**(7), 1321–1323 (2010)
20. T.P.J. Knowles, M.J. Buehler, Nanomechanics of functional and pathological amyloid materials. Nat. Nanotechnol. **6**(7), 469–479 (2011)
21. A.T. Petkova, Y. Ishii, J.J. Balbach, O.N. Antzutkin, R.D. Leapman, F. Delaglio, R. Tycko, A structural model for Alzheimer's beta-amyloid fibrils based on experimental constraints from solid state nmr. Proc. Natl. Acad. Sci. USA **99**(26), 16742–16747 (2002)
22. A.K. Paravastu, R.D. Leapman, W.M. Yau, R. Tycko, Molecular structural basis for polymorphism in Alzheimer's beta-amyloid fibrils. Proc. Natl. Acad. Sci. USA **105**(47), 18349–18354 (2008)
23. R. Paparcone, M.J. Buehler, Microscale structural model of Alzheimer a beta(1-40) amyloid fibril. Appl. Phys. Lett. **94**(24), (2009)
24. L.C. Serpell, M. Sunde, M.D. Benson, G.A. Tennent, M.B. Pepys, P.E. Fraser, The protofilament substructure of amyloid fibrils. J. Mol. Biol. **300**(5), 1033–1039 (2000)
25. J.L. Jimenez, E.J. Nettleton, M. Bouchard, C.V. Robinson, C.M. Dobson, H.R. Saibil, The protofilament structure of insulin amyloid fibrils. Proc. Natl. Acad. Sci. USA **99**(14), 9196–9201 (2002)
26. A. Aggeli, I.A. Nyrkova, M. Bell, R. Harding, L. Carrick, T.C.B. McLeish, A.N. Semenov, N. Boden, Hierarchical self-assembly of chiral rod-like molecules as a model for peptide beta-sheet tapes, ribbons, fibrils, and fibers. Proc. Natl. Acad. Sci. USA **98**(21), 11857–11862 (2001)
27. T.P.J. Knowles, J.F. Smith, A. Craig, C.M. Dobson, M.E. Welland, Spatial persistence of angular correlations in amyloid fibrils. Phys. Rev. Lett. **96**(23), (2006)
28. T. Scheibel, R. Parthasarathy, G. Sawicki, X.M. Lin, H. Jaeger, S.L. Lindquist, Conducting nanowires built by controlled self-assembly of amyloid fibers and selective metal deposition. Proc. Natl. Acad. Sci. USA **100**(8), 4527–4532 (2003)
29. M.F. Perutz, J.T. Finch, J. Berriman, A. Lesk, Amyloid fibers are water filled nanotubes. Proc. Natl. Acad. Sci. USA **99**, 5591–5595 (2002)
30. A. Baldwin, R. Bader, J. Christodoulou, C. MacPhee, C. Dobson, P. Barker, Cytochrome display on amyloid fibrils. J. Am. Chem. Soc. **128**(7), 2162–2163 (2006)
31. M. Yemini, M. Reches, J. Risphon, E. Gazit, Novel electrochemical biosensing platform using self-assembled peptide nanotubes. Nano Lett. **5**, 183–186 (2005)
32. A. Corrigan, C. Mueller, M. Krebs, The formation of nematic liquid crystal phases by hen lysozyme amyloid fibrils. J. Am. Chem. Soc. **128**(46), 14740–14741 (2006)
33. A. Mostaert, M. Higgins, T. Fukuma, F. Rindi, S. Jarvis, Nanoscale mechanical characterisation of amyloid fibrils discovered in a natural adhesive. J. Biol. Phys. **32**, 393–401 (2006)
34. M.J. Buehler, Nanomaterials: strength in numbers. Nat. Nanotechnol. **5**(3), 172–174 (2010)
35. C.M. Dobson, J.F. Smith, T.P.J. Knowles, C.E. MacPhee, M.E. Welland, Characterization of the nanoscale properties of individual amyloid fibrils. Proc. Natl. Acad. Sci. USA **103**(43), 15806–15811 (2006)
36. R. Paparcone, M.A. Pires, M.J. Buehler, Mutations alter the geometry and mechanical properties of Alzheimer's a beta(1-40) amyloid fibrils. Biochemistry **49**(41), 8967–8977 (2010)
37. R. Paparcone, S. Cranford, M.J. Buehler, Compressive deformation of ultralong amyloid fibrils. Acta Mech. Sin. **26**(6), 977–986 (2010)
38. R. Paparcone, M.J. Buehler, Failure of a beta(1-40) amyloid fibrils under tensile loading. Biomaterials **32**(13), 3367–3374 (2011)

39. R. Paparcone, S.W. Cranford, M.J. Buehler, Self-folding and aggregation of amyloid nanofibrils. Nanoscale **3**(4), 1748–1755 (2011)
40. J.W. Kelly, Towards an understanding of amyloidogenesis. Nat. Struct. Biol. **9**(5), 323–325 (2002)
41. R. Winter, R. Jansen, S. Grudzielanek, W. Dzwolak, High pressure promotes circularly shaped insulin amyloid. J. Mol. Biol. **338**(2), 203–206 (2004)
42. L.A. Morozova-Roche, M. Malisauskas, V. Zamotin, J. Jass, W. Noppe, C.M. Dobson, Amyloid protofilaments from the calcium-binding protein equine lysozyme: formation of ring and linear structures depends on ph and metal ion concentration. J. Mol. Biol. **330**(4), 879–890 (2003)
43. B.J. Papenburg, J. Liu, G.A. Higuera, A.M.C. Barradas, J. de Boer, C.A. van Blitterswijk, M. Wessling, D. Stamatialis, Development and analysis of multi-layer scaffolds for tissue engineering. Biomaterials **30**(31), 6228–6239 (2009)
44. T.S. Karande, J.L. Ong, C.M. Agrawal, Diffusion in musculoskeletal tissue engineering scaffolds: design issues related to porosity, permeability, architecture, and nutrient mixing. Ann. Biomed. Eng. **32**(12), 1728–1743 (2004)
45. S. Keten, M.J. Buehler, Geometric confinement governs the rupture strength of h-bond assemblies at a critical length scale. Nano Lett. **8**(2), 743–748 (2008)
46. S. Keten, Z. Xu, B. Ihle, M.J. Buehler, Nanoconfinement controls stiffness, strength and mechanical toughness of beta-sheet crystals in silk. Nat. Mater. **9**(4), 359–367 (2010)
47. C.E. MacPhee, C.M. Dobson, Formation of mixed fibrils demonstrates the generic nature and potential utility of amyloid nanostructures. J. Am. Chem. Soc. **122**(51), 12707–12713 (2000)
48. M. Reches, E. Gazit, Casting metal nanowires within discrete self-assembled peptide nanotubes. Science **300**(5619), 625–627 (2003)
49. O. Carny, D.E. Shalev, E. Gazit, Fabrication of coaxial metal nanocables using a self-assembled peptide nanotube scaffold. Nano Lett. **6**(8), 1594–1597 (2006)
50. Y.L. Sun, Z.P. Luo, A. Fertala, K.N. An, Stretching type ii collagen with optical tweezers. J. Biomech. **37**(11), 1665–1669 (2004)
51. M.J. Buehler, Atomistic and continuum modeling of mechanical properties of collagen: elasticity, fracture, and self-assembly. J. Mater. Res. **21**(8), 1947–1961 (2006)
52. A. Valiaev, D.W. Lim, S. Schmidler, R.L. Clark, A. Chilkoti, S. Zauscher, Hydration and conformational mechanics of single, end-tethered elastin-like polypeptides. J. Am. Chem. Soc. **130**(33), 10939–10946 (2008)
53. J. Djajamuliadi, T.F. Kagawa, K. Ohgo, K.K. Kumashiro, Insights into a putative hinge region in elastin using molecular dynamics simulations. Matrix Biol. **28**(2), 92–100 (2009)
54. A.E.X. Brown, R.I. Litvinov, D.E. Discher, P.K. Purohit, J.W. Weisel, Multiscale mechanics of fibrin polymer: gel stretching with protein unfolding and loss of water. Science **325**(5941), 741–744 (2009)
55. H. Yan, C. Lin, Y. Liu, Designer DNA nanoarchitectures. Biochemistry **48**(8), 1663–1674 (2009)
56. H. Yan, T.H. LaBean, L.P. Feng, J.H. Reif, Directed nucleation assembly of DNA tile complexes for barcode-patterned lattices. Proc. Natl. Acad. Sci. USA **100**(14), 8103–8108 (2003)
57. P.W.K. Rothemund, Folding DNA to create nanoscale shapes and patterns. Nature **440**(7082), 297–302 (2006)
58. A. Chworos, I. Severcan, A.Y. Koyfman, P. Weinkam, E. Oroudjev, H.G. Hansma, L. Jaeger, Building programmable jigsaw puzzles with RNA. Science **306**(5704), 2068–2072 (2004)
59. N.C. Seeman, P.S. Lukeman, Nucleic acid nanostructures: bottom-up control of geometry on the nanoscale. Rep. Prog. Phys. **68**(1), 237–270 (2005)
60. S.H. Park, C. Pistol, S.J. Ahn, J.H. Reif, A.R. Lebeck, C. Dwyer, T.H. LaBean, Finite-size, fully addressable DNA tile lattices formed by hierarchical assembly procedures. Angew. Chem., Int. Ed. **45**(5), 735–739 (2006)
61. J.H. Chen, N.C. Seeman, Synthesis from DNA of a molecule with the connectivity of a cube. Nature **350**(6319), 631–633 (1991)

62. Y.W. Zhang, N.C. Seeman, Construction of a DNA-truncated octahedron. J. Am. Chem. Soc. **116**(5), 1661–1669 (1994)
63. N.C. Seeman, De novo design of sequences for nucleic-acid structural-engineering. J. Biomol. Struct. Dyn. **8**(3), 573–581 (1990)
64. D.R. Han, S. Pal, J. Nangreave, Z.T. Deng, Y. Liu, H. Yan, DNA origami with complex curvatures in three-dimensional space. Science **332**(6027), 342–346 (2011)
65. R.H. Baughman, A.A. Zakhidov, W.A. de Heer, Carbon nanotubes—the route toward applications. Science **297**(5582), 787–792 (2002)
66. R.P. Goodman, I.A.T. Schaap, C.F. Tardin, C.M. Erben, R.M. Berry, C.F. Schmidt, A.J. Turberfield, Rapid chiral assembly of rigid DNA building blocks for molecular nanofabrication. Science **310**(5754), 1661–1665 (2005)
67. S.R. Bull, L.C. Palmer, N.J. Fry, M.A. Greenfield, B.W. Messmore, T.J. Meade, S.I. Stupp, A templating approach for monodisperse self-assembled organic nanostructures. J. Am. Chem. Soc. **130**(9), 2742–2743 (2008)
68. O. Ikkala, G. ten Brinke, Hierarchical self-assembly in polymeric complexes: towards functional materials. Chem. Commun., 2131–2137 (2004)
69. L. Sardone, V. Palermo, E. Devaux, D. Credgington, M. de Loos, G. Marletta, F. Cacialli, J. van Esch, P. Samor, Electric-field-assisted alignment of supramolecular fibers. Adv. Mater. **18**(10), 1276–1280 (2006)
70. K. Tanaka, A. Tengeiji, T. Kato, N. Toyama, M. Shionoya, A discrete self-assembled metal array in artificial DNA. Science **299**(5610), 1212–1213 (2003)
71. M. Surin, P.G.A. Janssen, R. Lazzaroni, P. Leclere, E.W. Meijer, A.P.H.J. Schenning, Supramolecular organization of ssDNA-templated $\pi$-conjugated oligomers via hydrogen bonding. Adv. Mater. **21**(10–11), 1126–1130 (2009)
72. P.G.A. Janssen, J. Vandenbergh, J.L.J. van Dongen, E.W. Meijer, A.P.H.J. Schenning, ssDNA templated self-assembly of chromophores. J. Am. Chem. Soc. **129**(19), 6078–6079 (2007)
73. P. Jonkheijm, A. Miura, M. Zdanowska, F.J.M. Hoeben, S. De Feyter, A.P.H.J. Schenning, F.C. De Schryver, E.W. Meijer, p-conjugated oligo-(p-phenylenevinylene) rosettes and their tubular self-assembly. Angew. Chem., Int. Ed. **43**(1), 74–78 (2004)
74. F. Wurthner, Z. Chen, F.J.M. Hoeben, P. Osswald, C.-C. You, P. Jonkheijm, J.v. Herrikhuyzen, A.P.H.J. Schenning, P.P.A.M. van der Schoot, E.W. Meijer, E.H.A. Beckers, S.C.J. Meskers, R.A.J. Janssen, Supramolecular p-n-heterojunctions by co-self-organization of oligo(p-phenylene vinylene) and perylene bisimide dyes. J. Am. Chem. Soc. **126**(34), 10611–10618 (2004)
75. A. Furka, Combinatorial chemistry: 20 years on .... Drug Discov. Today **7**(1), 1–4 (2002)
76. M.H. Hecht, A. Das, A. Go, L.H. Bradley, Y.N. Wei, De novo proteins from designed combinatorial libraries. Protein Sci. **13**(7), 1711–1723 (2004)
77. S.E. Blondelle, E. Takahashi, R.A. Houghten, E. PerezPaya, Rapid identification of compounds with enhanced antimicrobial activity by using conformationally defined combinatorial libraries. Biochem. J. **313**, 141–147 (1996)

# Chapter 11
# The Future of Biomateriomics

**Abstract** Necessitated by the complexity of biological materials, understanding the relationships between material structure and function, and Nature's persistent use of universality, cooperativity, and multiscale hierarchies, biomateriomics unifies efforts to understand and ultimately exploit "Nature's secrets". Indeed, with a keen eye and a new holistic perspective, we recognize glimpses of materiomics in current research, unlocking novel functional materials, both biological and synthetic.

*Any sufficiently advanced technology is indistinguishable from magic.*

*Sir Arthur C. Clarke, British science fiction author, inventor, and futurist (1917–2008)*

## 11.1 The Inevitability of Complexity and Convergence

Many science fiction fans will recognize Arthur C. Clarke's dictum above. Science has progressed in many fields, increasing our understanding ranging from quantum level neutrinos to galaxy spanning quasars. Yet, there is hesitation in pursuing mechanistic interpretations of biological systems, regressing to the adage that "life is complex" and suggesting it is beyond deterministic science. Complexity, while challenging, does not infer "magic", just a larger problem to solve.

The differences between biology and technology (and between organisms) are obvious, particularly at the molecular and device level. Nevertheless, convergent evolution, a well-established concept in both engineering and evolutionary biology, yields remarkable similarities at higher levels of organization. Recently, engineering systems have begun to achieve almost biological levels of complexity. For example, a popular smartphone seamlessly combines applications as portable media players, compact digital cameras, pocket video cameras, and Global Positioning System (GPS) navigation units (*e.g.*, functions) and typically include high-resolution touchscreens, keyboards, and web browsers (*e.g.*, interfaces), including high-speed data access *via* Wi-Fi and mobile broadband (*e.g.*, communication). In terms of cost and complexity, a smartphone is essentially a computer that just happens to make phone

S.W. Cranford, M.J. Buehler, *Biomateriomics*, Springer Series in Materials Science 165, DOI 10.1007/978-94-007-1611-7_11, © Springer Science+Business Media Dordrecht 2012 

calls. All for approximately \$200.[1] The consequence of good design is that its complexity is hidden from users (except when the phone is dropped). Being said, the internal activity level is staggering: at a data rate of ~50 Mbit/s, a typical smartphone can record the equivalent of the data of the human genome in about two minutes (assuming 2 bits for each base pair). At the same time, your children will most likely mock the cell phones that were once "high tech" in the late twentieth century. Cellular phones are not the only systems undergoing such explosions in complexity as a result of advanced controls and embedded networking; virtually all technologies are evolving similarly. It has been claimed that this technological evolution of complexity is convergent with that of biology [1].

At the same time, we want to be able to *engineer* such complexity. Understanding the mechanics of complex materials—the ability to withstand pulling, twisting and heavy loads—at the atomistic level can be likened to analysis of a building structure. Instead of beams, there are bonds, instead of columns, there are molecules. Both can be investigated and understood *via* mechanics. A grand challenge in researching these materials is to characterize their structure and mechanical properties in a practical manner. Apart from the issue that the structure is inherently complicated, structures on the nanoscale cannot be solved directly through tried and trusted techniques at the macroscale. We don't have the robust tools (yet) for the complete understanding the complex nanomaterials that we want to engineer.

Regardless of the widespread acceptance and adoption of the term "materiomics", the directions and trends described throughout this text will nevertheless continue to be developed, extended, advanced and applied to complex biological and synthetic material systems alike. Indeed, Feynman's rhetorical query about atom-by-atom manipulation in the 1960's [2] resulted in the predicting of the "new age of nanotechnology" by Drexler in 1986, heralding thorough and inexpensive control of the structure of matter [3, 4]. More recently, advances in molecular biology and chemistry advocate the coming "Kuhnian revolution" [5] and the potential of "reverse engineering of biology" [1], stressing the importance of system-level and holistic understanding of complex systems. Aizenberg and Fratzl [6], linking materials science and biology, defined "biomimetic materials research" as a progression of three steps:

1. *Simplification*: learning from the original; elucidation of the structure–function relationships in biological materials.
2. *Adaption*: Extract the general underlying principles (physical, chemical, optical, electronic, thermal, mechanical, etc.) of this structure–function relationship (using both experiments and theory), in order to make them available as concepts useful in materials science and engineering.
3. *Extension*: Develop pathways for the synthesis and the manufacturing of *de novo* materials based on these principles, taking into account the existing capabilities and constraints imposed by engineering and by the economy.

---

[1] Or free, with contract.

In a similar vein, the challenge in biomateriomics is to understand the underlying design principles and mechanisms that determine the optimized structural organization in biological material systems across scales, from molecule to organism, and the associated relationship to function. More immediate, in 2011 the government of the United States, through the White House, unveiled an ambitious Materials Genome Initiative,[2] an effort to "double the speed with which we discover, develop, and manufacture new materials", a direct nod to the Human Genome Project, which accelerated a range of biological sciences by identifying and deciphering the basic building blocks of the human genetic code.

Undoubtedly, as discussed in the opening chapter, there is an ongoing *convergence*[3] in the scientific community that cannot be ignored [7]. While the benefits of integrating biology with engineering and materials science is clear, more abstract constructs can potentially be "borrowed" from the arts (such as music [8–10]), or perhaps even economic or sociological models [11]. As demonstrated in Chap. 4: Universality-Diversity Paradigm: Music, Materiomics, and Category Theory, universal patterns arise throughout Nature [12], and there is a commonality underlying such complex systems—be they instruments playing a riff or amino acids forming a protein. The implication is clear—most hierarchically structured systems follow similar design rules. *What does this mean for the future of biomateriomics*?

## 11.2 The Future of Materiomics

The advent of materiomics is not an attempt to introduce a *new* field of science. Rather, it is a *unifying* proposition motivated by the convergence of many fields towards a fundamental integrated description of materials and their functional roles. Materiomics takes a materials science perspective towards complex biological systems, explicitly accounting for feedback loops that link functional requirements (and changes thereof) to altered material components and structure, at different scales in both time and length. The incentive for materiomics is twofold:

- Primarily, to amalgamate the advancing knowledge base of chemistry, biology, materials science, and mechanics to a single, holistic description of a material system from nano to macro. This complete description, the materiome, contains the information required to manipulate mechanical function and properties and development of new, predictive materiomic theories and models, which specifically include the hallmark of living systems-feedback loops that facilitate an autonomous sensing, structural change, and as a result, adaptation to altered environmental conditions and functional requirements. Investigative methods developed from a multi-disciplinary perspective for multi-scale analysis can be applied to a vast amount of material systems, both current and future, offering new

[2]Materials Genome Initiative for Global Competitiveness, http://www.whitehouse.gov/blog/2011/06/24/materials-genome-initiative-renaissance-american-manufacturing.

[3]The third revolution: the convergence of the life sciences, physical sciences, and engineering, http://web.mit.edu/dc/Policy/MIT%20White%20Paper%20on%20Convergence.pdf.

insights and theoretical formulations unavailable to past biochemists, medical engineers, and materials scientists.

- Secondly, the application of materiomics can unlock fundamental design principles of Nature's materials based on high-throughput computational, theoretical and experimental methods, and utilize this insight in the development of advanced materials, biological and synthetic, micro- and macroscopic. This harnessing of Nature's "trade secrets" could usher in new technologies that are currently unattainable without the integrative approaches a materiomics perspective provides. Indeed, such integrative approaches are already being implemented as discussed, albeit by different research groups with disparate goals. Materiomics highlights the similarities and promotes a mutually beneficial relation between all researchers and scientists concerned with biomimetic materials, advanced composite design, nanotechnology, medical engineering, tissue engineering, mechanisms of disease, genetic defects, and any field requiring the complete description and understanding of a specific materiome.

This text has endorsed a scale-bridging approach to characterizing the materiome—from experimental techniques such as AFM and nanoindentation, to large-scale computational means that coarse-grain full atomistic behaviors—but this need not be the case. Whether the materiome is uncovered by a multi-scale approach using experiment and computation, or more conventional assays that determine pertinent interactions (such as high-throughput screening of materials [13–15], recombinant DNA sequencing [16, 17], or tissue engineering applications [18–20]) is inconsequential—the extension of understanding and connection to the system function is the key requirement.

As illustrated throughout this text, materiomics is a powerful tool to enhance the understanding of materials in biology, at multiple scales and in a variety of functional contexts, contributing to the development of a holistic understanding of biological systems. It overcomes the barrier that currently separates the understanding at different length and time-scales, through the development of new experimental synthesis and characterization methods, novel model systems and an enhanced appreciation for a multi-scale view of materials in general, in order to fully understand multi-scale or cross-scale interactions of the materiome—be it silk, bone, or carbon nanotubes. Indeed, a materiomics perspective can be taken for other materials and material systems (such as a reinforced concrete member or a steel truss), but it may then be considered a rather trivial application. More complex systems, such as stimuli-responsive polymer composites, can be (and have been) understood through multi-scale holistic perspectives that materiomics promotes. Beyond structure-property relations, materiomics further wishes to encompass cross-scale interactions, multi-functionality, feedback, and adaptation that is common to biological systems, and currently being developed in advanced material systems and emerging nanoscale devices. Such a unifying approach is only made possible by the concurrent advances in multi-scale analysis, chemical characterization, computational methods and experimental techniques developed across a spectrum of disciplines.

The field of biomateriomics provides a powerful integrated theoretical framework for complex hierarchical biological materials, which enables to define future scientific hypotheses in the field of biology, synthetic materials and nanotechnology in a systematic way. Such hypothesis must be proved through a unified approach, leading to a detailed understanding of how Nature successfully links materials and structure, properties and function, simultaneously, from nano-to-macro. While Nature has proven adept at turning weakness into strength, we can still improve on Nature's design. The ultimate goal of materiomics is to reduce the complexity of such systems and to enable effective engineering solutions to shape tomorrow's technology and innovation.

## References


1. M.E. Csete, J.C. Doyle, Reverse engineering of biological complexity. Science **295**(5560), 1664–1669 (2002)
2. R.P. Feynman, There's plenty of room at the bottom. Caltech Eng. Sci. **23**, 22–36 (1960)
3. K.E. Drexler, *Engines of Creation: The Coming Era of Nanotechnology*, 1st edn. (Anchor Press/Doubleday, Garden City, 1986)
4. K.E. Drexler, *Nanosystems: Molecular Machinery, Manufacturing, and Computation* (Wiley, New York, 1992)
5. R.C. Strohman, The coming Kuhnian revolution in biology—commentary. Nat. Biotechnol. **15**(3), 194 (1997)
6. J. Aizenberg, P. Fratzl, Biological and biomimetic materials. Adv. Mater. **21**(4), 387–388 (2009)
7. P.A. Sharp, C.L. Cooney, M.A. Kastner, J. Lees, R. Sasisekharan, M.B. Yaffe, S.N. Bhatia, T.E. Jacks, D.A. Lauffenburger, R. Langer, P.T. Hammond, M. Sur, The third revolution: the convergence of the life sciences, physical sciences, and engineering. Massachusetts Institute of Technology (2011)
8. M.J. Buehler, Tu(r)ning weakness to strength. Nano Today **5**(5), 379–383 (2010)
9. S. Cranford, M.J. Buehler, Materiomics: biological protein materials, from nano to macro. Nanotechnol. Sci. Appl. **3**, 127–148 (2010)
10. T. Giesa, D.I. Spivak, M.J. Buehler, Reoccurring patterns in hierarchical protein materials and music: the power of analogies. BioNanoScience **1**(4), 153–161 (2011)
11. D.I. Spivak, T. Giesa, E. Wood, M.J. Buehler, Category theoretic analysis of hierarchical protein materials and social networks. PLoS ONE **6**(9), e23911 (2011)
12. T. Ackbarow, M.J. Buehler, Hierarchical coexistence of universality and diversity controls robustness and multi-functionality in protein materials. J. Comput. Theor. Nanosci. **5**(7), 1193–1204 (2008)
13. R. Potyrailo, K. Rajan, K. Stoewe, I. Takeuchi, B. Chisholm, H. Lam, Combinatorial and high-throughput screening of materials libraries: review of state of the art. ACS Comb. Sci. **13**(6), 579–633 (2011)
14. C.G. Simon, S. Lin-Gibson, Combinatorial and high-throughput screening of biomaterials. Adv. Mater. **23**(3), 369–387 (2011)
15. H.V. Unadkat, M. Hulsman, K. Cornelissen, B.J. Papenburg, R.K. Truckenmuller, G.F. Post, M. Uetz, M.J.T. Reinders, D. Stamatialis, C.A. van Blitterswijk, J. de Boer, An algorithm-based topographical biomaterials library to instruct cell fate. Proc. Natl. Acad. Sci. USA **108**(40), 16565–16570 (2011)
16. T.L. Blundell, B.L. Sibanda, M.J.E. Sternberg, J.M. Thornton, Knowledge-based prediction of protein structures and the design of novel molecules. Nature **326**(6111), 347–352 (1987)

17. C.W.P. Foo, D.L. Kaplan, Genetic engineering of fibrous proteins: spider dragline silk and collagen. Adv. Drug Deliv. Rev. **54**(8), 1131–1143 (2002)
18. M.M. Stevens, J.H. George, Exploring and engineering the cell surface interface. Science **310**(5751), 1135–1138 (2005)
19. D.E. Discher, P. Janmey, Y.L. Wang, Tissue cells feel and respond to the stiffness of their substrate. Science **310**(5751), 1139–1143 (2005)
20. H. Fernandes, L. Moroni, C. van Blitterswijk, J. de Boer, Extracellular matrix and tissue engineering applications. J. Mater. Chem. **19**(31), 5474–5484 (2009)

# Index



S.W. Cranford, M.J. Buehler, *Biomateriomics*, Springer Series in Materials Science 165, DOI 10.1007/978-94-007-1611-7, © Springer Science+Business Media Dordrecht 2012

The manufacturer's authorised representative in the EU is Springer Nature Customer Service Centre GmbH, Europaplatz 3, 69115 Heidelberg, Germany. If you have any concerns regarding our products, please contact ProductSafety@springernature.com

Printed and bound by CPI Group (UK) Ltd, Croydon, CR0 4YY
22/09/2026
02220980-0002